Genomic Solutions for Breeding Climate-resilient Crops: Unlocking Nature's Blueprint for Sustainable Agriculture

CABI Biotechnology Series

Biotechnology, in particular the use of transgenic organisms, has a wide range of applications including agriculture, forestry, food and health. There is evidence that it could make a major impact in producing plants and animals that are able to resist stresses and diseases, thereby increasing food security. There is also potential to produce pharmaceuticals in plants through biotechnology, and provide foods that are nutritionally enhanced. Genetically modified organisms can also be used in cleaning up pollution and contamination. However, the application of biotechnology has raised concerns about biosafety, and it is vital to ensure that genetically modified organisms do not pose new risks to the environment or health. To understand the full potential of biotechnology and the issues that relate to it, scientists need access to information that not only provides an overview of and background to the field, but also keeps them up to date with the latest research findings.

This series, which extends the scope of CABI's successful "Biotechnology in Agriculture" series, addresses all topics relating to biotechnology including transgenic organisms, molecular analysis techniques, molecular pharming, in vitro culture, public opinion, economics, development and biosafety. Aimed at researchers, upper-level students and policy makers, titles in the series provide international coverage of topics related to biotechnology, including both a synthesis of facts and discussions of future research perspectives and possible solutions.

Titles Available

1. Animal Nutrition with Transgenic Plants
Edited by G. Flachowsky

2. Plant-derived Pharmaceuticals: Principles and Applications for Developing Countries
Edited by K.L. Hefferon

3. Transgenic Insects: Techniques and Applications
Edited by M.Q. Benedict

4. Bt Resistance: Characterization and Strategies for GM Crops Producing Bacillus thuringiensis Toxins
Edited by Mario Soberón, Yulin Gao and Alejandra Bravo

5. Plant Gene Silencing: Mechanisms and Applications
Edited by Tamas Dalmay

6. Ethical Tensions from New Technology: The Case of Agricultural Biotechnology
Edited by Harvey James

7. GM Food Systems and their Economic ImpactTatjana Brankov and Koviljko Lovre

8. Endophyte Biotechnology: Potential for Agriculture and Pharmacology
Edited by Alexander Schouten

9. Forest Genomics and Biotechnology
Richard Meilan and Matias Kirst

10. Transgenic Insects: Techniques and Applications, 2nd Edition
Edited by M.Q. Benedict and Maxwell J. Scott

11. Plant Omics: Advances in Big Data Biology
Edited by Hajime Ohyanagi, Eiji Yamamoto, Ai Kitazumi and Kentaro Yano

12. Next-generation Sequencing and Agriculture
Edited by Phillip Bayer and Dave Edwards

13. Aquaculture and Fisheries Biotechnology: Genetic Approaches
Rex A Dunham

14. Algal Biotechnology
Edited by Qiang Wang

15. Genetic Engineering of Vegetable Crops
Edited by Pritam Kalia

16. Plant Breeding Technology: Future Trends and Challenges
Edited by Zeba Khan

17. Stress-Resilient Crops: Coordinated Omics-CRISPR-Nanotechnology Strategies
Edited by Jen-Tsung Chen

18. High Throughput Plant Metabolomics
Edited by Jen-Tsung Chen

19. Plant Biotechnology and Sustainable Agriculture: Bridging the Gap for Global Food Security
Edited by Tariq Aftab

20. Biotechnology of Orchids and Other Ornamentals
Edited by Pablo Bolaños-Villegas, Jian-Zhi Huang

21. Deep Learning in Plant Science and Agricultural Biotechnology
Edited by Jen-Tsung Chen

22. AI-Enhanced Plant-omics
Edited by Jen-Tsung Chen

23. Plant Spatial Biology
Edited by Jen-Tsung Chen

24. Genomic Solutions for Breeding Climate-resilient Crops: Unlocking Nature's Blueprint for Sustainable Agriculture
Edited by Rajeev K. Varshney, Rutwik Barmukh and Richard Harper

Genomic Solutions for Breeding Climate-resilient Crops: Unlocking Nature's Blueprint for Sustainable Agriculture

---

Edited by: Rajeev K. Varshney, Rutwik Barmukh and Richard J. Harper

CABI

CABI is a trading name of CAB International

CABI
Nosworthy Way
Wallingford
Oxfordshire OX10 8DE
UK

Tel: +44 (0)1491 832111
E-mail: info@cabi.org
Website: www.cabi.org

CABI
200 Portland Street
Boston
MA 02114
USA

Tel: +1 (617)682-9015
E-mail: cabi-nao@cabi.org

A catalogue record for this book is available from the British Library, London, UK.

ISBN-13: 9781800627161 (hardback)
9781800627178 (ePDF)
9781800627185 (ePub)

DOI: 10.1079/9781800627185.0000

Commissioning Editor: Rebecca Stubbs
Editorial Assistant: Emma McCann
Production Editor: James Bishop

Typeset by Straive, Pondicherry, India
Printed and bound in the UK by CPI Group (UK) Ltd, Croydon, CR0 4YY

Contents

Preface

Agriculture today faces unprecedented challenges due to both its greenhouse gas emissions and the increasing impact of climatic change. Rising temperatures, erratic rainfall, drought, and the emergence of new pests and pathogens are threatening crop productivity and stability globally. These stressors not only endanger food and nutritional security but also risk the livelihoods of millions of smallholder farmers in vulnerable regions. Nitrous oxide from agriculture is a major contributor to total global emissions. Ensuring sustainable agricultural production under changing climatic conditions has, therefore, become one of the top priorities of the global research and development community.

Crop breeding for climate resilience has traditionally relied on conventional selection approaches, which, although successful in improving yields and adaptability, often require long breeding cycles and offer limited precision. The advent of genomics and allied omics technologies has revolutionized our understanding of plant biology, enabling researchers to uncover the complex genetic and molecular networks underlying stress tolerance, productivity, and adaptation. These advances have provided powerful tools to accelerate the development of climate-smart crops capable of maintaining productivity under multiple climate-induced stresses. Similarly, these tools can also boost crop nitrogen use efficiency, potentially reducing the contributions of nitrous oxide to the atmosphere.

Over the past two decades, enormous progress has been made in sequencing and annotating the genomes of major food and industrial crops. The availability of high-quality reference genomes, together with advances in re-sequencing, transcriptomics, proteomics, metabolomics, and phenomics, has transformed the landscape of modern breeding. Integration of these data-driven technologies has facilitated the identification of key genes, allelic variants, and regulatory pathways associated with stress resilience and resource-use efficiency. Moreover, genome editing and genomic selection approaches are now being used to accelerate the breeding process, leading to the development of cultivars with improved tolerance to drought, heat, and biotic stresses.

In this context, the book *Genomic Solutions for Breeding Climate-Resilient Crops: Unlocking Nature's Blueprint for Sustainable Agriculture* provides a comprehensive overview of the genomic innovations that are transforming modern agriculture and moreover their role in tackling the impacts of climate change. This book brings together leading scientists, breeders, and policy experts from around the world to present cutting-edge research, technological developments, and case studies that highlight the role of genomics in developing climate-resilient crop varieties. The book covers diverse aspects, from climate change projections and impacts on crop production, to genomic interventions, multi-omics integration, genome editing, and microbiome research. It also explores emerging technologies such as nanotechnology and the evolving policy and governance frameworks that support the development and deployment of resilient agricultural systems.

The book aims to serve as a valuable reference for researchers, breeders, students, and policy-makers interested in harnessing genomic innovations for sustainable crop improvement. It provides a consolidated understanding of how genomics and allied technologies can be leveraged to mitigate climate risks, improve resource efficiency, and enhance agricultural productivity in an environmentally sustainable manner.

The editors are grateful to all the contributing authors for their scholarly efforts, insights, and dedication in preparing high-quality chapters within a tight timeframe. We extend our sincere appreciation to the reviewers (Drs. Ayomide Fadiji, Daniel Edge-Garza, Jingjuan Zhang, Kar-Chun Tan, Mehanathan Muthamilarasan, Sunil Gangurde, Vanika Garg, and Yogendra Kalenahalli) for their constructive comments and suggestions that have significantly improved the scientific quality of the book chapters. Our gratitude also goes to the CABI editorial and production teams for their continuous support, professionalism, and commitment in bringing this book to fruition.

We gratefully acknowledge the encouragement and institutional support provided by Murdoch University, Australia, which greatly facilitated collaboration and coordination among contributors across continents. Rajeev Varshney sincerely acknowledges the unwavering support of his wife (Monika), son (Prakhar), and daughter (Preksha), whose patience and understanding enabled him to fulfill his editorial responsibilities alongside research and administrative duties at Murdoch University. Rutwik Barmukh extends his heartfelt gratitude to his parents (Mr. Ketan Barmukh and Mrs. Krutika Barmukh), for their constant encouragement and moral support during the preparation of this book, in addition to his ongoing research commitments at Murdoch University. Richard Harper deeply appreciates the support and patience of his family throughout the development of this work. The editors also express their thanks to friends and collaborators from the wider research community for their encouragement, cooperation, and valuable support in numerous ways.

We hope that this book will serve as a timely and authoritative resource, promoting further research collaborations and inspiring innovative approaches toward breeding climate-resilient crops. By harnessing the power of genomics and related technologies, we believe agriculture can be made more productive, resilient, and sustainable – ensuring food security and environmental stability for future generations.

Rajeev K. Varshney
Rutwik Barmukh
Richard J. Harper
Murdoch University, Australia

Contributors

Oluwaseun Deborah Adedeji, Department of Economics; Agricultural and Food Economics, Swedish University of Agricultural Sciences, Uppsala, Sweden oluwaseun.adedeji@slu.se

Oluwaseun J. Akinlade, WA State Agricultural Biotechnology Centre, Centre for Crop and Food Innovation, Food Futures Institute, Murdoch University, Murdoch 6150, Western Australia, Australia

Gaurav Augustine, School of Agricultural Biotechnology, Punjab Agricultural University, Ludhiana-141004, India

M J Baig, ICAR-National Rice Research Institute, Cuttack, India, 753006

Rutwik Barmukh, WA State Agricultural Biotechnology Centre, Centre for Crop and Food Innovation, Food Futures Institute, Murdoch University, Murdoch, WA 6150, Australia

Sarita Jane Bennett, The School of Molecular and Life Sciences, Curtin University Sarita.Bennett@curtin.edu.au

S. Brundavanam, Murdoch Applied Innovation Nanotechnology Research Group, School of Mathematics and Statistics, Chemistry and Physics, Murdoch University, 90 South Street, Perth WA 6150, Western Australia, Australia

Wisut Chamsa-ard, School of Renewable Energy and Smart Grid Technology (SGtech), Naresuan University, Phitsanulok, Thailand, 65000

Annapurna Chitikineni, WA State Agricultural Biotechnology Centre, Centre for Crop and Food Innovation, Murdoch University, Murdoch, WA 6150, Australia

D Fawcett, Murdoch Applied Innovation Nanotechnology Research Group, School of Mathematics and Statistics, Chemistry and Physics, Murdoch University, 90 South Street, Perth WA 6150, Western Australia, Australia

Vasundhara Gaddam, Global Research Program-Accelerated Crop Improvement, International Crops Research Institute for the Semi-Arid Tropics (ICRISAT), Patancheru 502324, India

Nicholas A. George, The School of Molecular and Life Sciences, Curtin University

Chandana Ghosh, ICAR-National Rice Research Institute, Cuttack, India, 753006

Muskan Gupta, School of Agricultural Biotechnology, Punjab Agricultural University, Ludhiana-141004, India

Tanmoy Halder, ICAR-National Rice Research Institute, Cuttack, India, 753006

AFM Halim, Murdoch Applied Innovation Nanotechnology Research Group, School of Mathematics and Statistics, Chemistry and Physics, Murdoch University, 90 South Street, Perth WA 6150, Western Australia, Australia

Richard Harper, Centre for Crop and Food Innovation, Food Futures Institute, Murdoch University, Murdoch, WA 6150, Australia

Kerryn Hawke, School of Environmental and Conservation Sciences, Murdoch University, 6150, Western Australia Kerryn.Hawke@murdoch.edu.au

Jagdish Jaba, Global Research Program-Accelerated Crop Improvement, International Crops Research Institute for the Semi-Arid Tropics (ICRISAT), Patancheru 502324, India

Murukarthick Jayakodi, Texas A&M AgriLife Research Center, Dallas, TX, USA; Department of Soil & Crop Sciences, Texas A&M University, College Station, TX, USA muru.jayakodi@ag.tamu.edu

Ramesha H Jayaramaiah, SoilsWest, Centre for Sustainable Farming Systems, Food Futures Institute, Murdoch University, Perth, WA 6150, Australia Ramesha.HJayaramaiah@murdoch.edu.au

Abdulqader Jighly, AgriSapiens PTY LTD, Melbourne, 3108 VIC, Australia a.jighly@agrisapiens.com.au

Uday Jha, Indian Council for Agricultural Research-Indian Institute of Pulses Research, Kanpur, Uttar Pradesh, India

Reem Joukhadar, AgriSapiens PTY LTD, Melbourne, 3108 VIC, Australia r.joukhadar@agrisapiens.com.au

Jatin Kala, School of Environmental and Conservation Sciences, Murdoch University, 6150, Western Australia J.Kala@murdoch.edu.au

Anil Namdev Kale, Department of Agricultural Biotechnology and Molecular Biology, Dr. Rajendra Prasad Central Agricultural University, Pusa, Bihar, India

Sivasakthi Kaliamoorthy, Global Research Program-Accelerated Crop Improvement, International Crops Research Institute for the Semi-Arid Tropics (ICRISAT), Patancheru 502324, India

Yogendra Kalenahalli, Global Research Program-Accelerated Crop Improvement, International Crops Research Institute for the Semi-Arid Tropics (ICRISAT), Patancheru 502324, India

Ajaya K. Karkee, WA State Agricultural Biotechnology Centre, Centre for Crop and Food Innovation, Food Futures Institute, Murdoch University, Murdoch 6150, Western Australia, Australia

Jaismeen Kaur, School of Agricultural Biotechnology, Punjab Agricultural University, Ludhiana-141004, India

Simranjit Kaur, Austrian Institute of Technology, Tulln, 3430, Austria

Ekta Kharche, School of Agricultural Biotechnology, Punjab Agricultural University, Ludhiana-141004, India

Bhargava Kotte, WA State Agricultural Biotechnology Centre, Centre for Crop and Food Innovation, Food Futures Institute, Murdoch University, Murdoch 6150, Western Australia, Australia

Himabindu Kudapa, Global Research Program-Accelerated Crop Improvement, International Crops Research Institute for the Semi-Arid Tropics (ICRISAT), Patancheru 502324, India Himabindu.Kudapa@icrisat.org

Aman Kumar, School of Agricultural Biotechnology, Punjab Agricultural University, Ludhiana-141004, India

Ashish Kumar, Jawaharlal Nehru Agricultural University, Jabalpur, Madhya Pradesh, India

Vinay Kumar, Sharma Department of Agricultural Biotechnology and Molecular Biology, Dr. Rajendra Prasad Central Agricultural University, Pusa, Bihar, India

Anusha Kutala, Global Research Program-Accelerated Crop Improvement, International Crops Research Institute for the Semi-Arid Tropics (ICRISAT), Patancheru 502324, India

Mahesh Madineni, Global Research Program-Accelerated Crop Improvement, International Crops Research Institute for the Semi-Arid Tropics (ICRISAT), Patancheru 502324, India

Premchand Maisnam, Kellog's Biological Station, Michigan State University, Michigan, USA

Martin Mascher, Leibniz Institute of Plant Genetics and Crop Plant Research (IPK) Gatersleben, Seeland, Germany; German Centre for Integrative Biodiversity Research (iDiv) Halle-Jena-Leipzig, Leipzig, Germany

Venugopal Mendu, Department of Agriculture, Agribusiness and Environmental Sciences, Texas A&M University-Kingsville, Texas, USA

Reyazul Rouf Mir, WA State Agricultural Biotechnology Centre, Centre for Crop and Food Innovation, Murdoch University, Murdoch, WA 6150, Australia

Kutubuddin A. Molla, ICAR-National Rice Research Institute, Cuttack, India, 753006 kutubuddin.molla@icar.gov.in

Caitlin E. Moore, The School of Agriculture and Environment, The University of Western Australia; Centre for Water and Spatial Science, The University of Western Australia

Arielís Morales-Salvá, Department of Plant Sciences and Landscape Architecture, University of Maryland, Maryland, USA

Yogesh Dashrath Naik, Department of Agricultural Biotechnology and Molecular Biology, Dr. Rajendra Prasad Central Agricultural University, Pusa, Bihar, India; New Mexico State University, Agricultural Science Center, Clovis, 88101, USA

Richardson Uchenna Okechukwu, International Institute of Tropical Agriculture, Onne, Nigeria

Akinbode Okunola, Institute of Agriculture and Natural Resources, University of Nebraska-Lincoln, NE, USA

Omotola Dorcas Olaoye, West Africa Centre for Crop Improvement (WACCI), College of Basic and Applied Sciences, University of Ghana, Legon, Ghana; International Institute of Tropical Agriculture, Ibadan, Nigeria

Babatunde Olasege, The Institute for Molecular Bioscience, The University of Queensland, Brisbane, Queensland, Australia

Sonali Panda, ICAR-National Rice Research Institute, Cuttack, India, 753006; Ravenshaw University, Cuttack, India, 753006

Manish Pandey, International Crops Research Institute for the Semi-Arid Tropics (ICRISAT), Patancheru, Hyderabad, Telangana, India

Gérrard Eddy Jai Poinern, Murdoch Applied Innovation Nanotechnology Research Group, School of Mathematics and Statistics, Chemistry and Physics, Murdoch University, 90 South Street, Perth WA 6150, Western Australia, Australia G.Poinern@murdoch.edu.au

Somashekhar Punnuri, College of Agriculture, Family Sciences and Technology, 1005 State University Dr, Fort Valley State University, Fort Valley, GA, USA punnuris@fvsu.edu

Naveen Puppala, New Mexico State University, Agricultural Science Center, Clovis, 88101, USA

Senthil Ramachandran, Global Research Program-Accelerated Crop Improvement, International Crops Research Institute for the Semi-Arid Tropics (ICRISAT), Patancheru 502324, India

Sagar Krushnaji Rangari, Department of Agricultural Biotechnology and Molecular Biology, Dr. Rajendra Prasad Central Agricultural University, Pusa, Bihar, India; New Mexico State University, Agricultural Science Center, Clovis, 88101, USA; International Crops Research Institute for the Semi-Arid Tropics (ICRISAT), Patancheru, Hyderabad, Telangana, India

Nidhi Rawat, Department of Plant Sciences and Landscape Architecture, University of Maryland, Maryland, USA nidhirwt@umd.edu

Ali Raza, Guangdong Key Laboratory of Plant Epigenetics, College of Life Sciences and Oceanography, Shenzhen University, Shenzhen, China alirazamughal143@gmail.com

Someswar Rao Sagurthi, Department of Genetics and Biotechnology, Osmania University, Hyderabad 500007, India

Bijayalaxmi Sahoo, ICAR-National Rice Research Institute, Cuttack, India, 753006

Nitika Sandhu, School of Agricultural Biotechnology, Punjab Agricultural University, Ludhiana-141004, India nitikasandhu@pau.edu

Mamta Sharma, Global Research Program-Accelerated Crop Improvement, International Crops Research Institute for the Semi-Arid Tropics (ICRISAT), Patancheru 502324, India

Stuti Sharma, Jawaharlal Nehru Agricultural University, Jabalpur, Madhya Pradesh, India

Hyeonah Shim, Leibniz Institute of Plant Genetics and Crop Plant Research (IPK) Gatersleben, Seeland, Germany

Harsh Shukla, School of Agricultural Biotechnology, Punjab Agricultural University, Ludhiana-141004, India

Brajesh K Singh, Hawkesbury Institute for the Environment, Western Sydney University, Penrith, NSW 2753, Australia

Jasneet Singh, School of Agricultural Biotechnology, Punjab Agricultural University, Ludhiana-141004, India

Kuldeep Singh, Global Research Program-Accelerated Crop Improvement, International Crops Research Institute for the Semi-Arid Tropics (ICRISAT), Patancheru 502324, India

Pankaj K Singh, The UWA Institute of Agriculture and School of Agriculture & Environment, The University of Western Australia, Perth WA 6001, Australia

Rakesh Srivastava, Global Research Program-Accelerated Crop Improvement, International Crops Research Institute for the Semi-Arid Tropics (ICRISAT), Patancheru 502324, India

Cassandria G. Tay Fernandez, WA State Agricultural Biotechnology Centre, Centre for Crop and Food Innovation, Food Futures Institute, Murdoch University, Murdoch, WA, Australia

Mahendar Thudi, College of Agriculture, Family Sciences and Technology, 1005 State University Dr, Fort Valley State University, Fort Valley, GA, USA; Centre for Crop Health and School of Agriculture and Environmental Science, University of Southern Queensland, Toowoomba, Australia mahendar.thudi@fvsu.edu

Vijay Tiwari, Department of Plant Sciences and Landscape Architecture, University of Maryland, Maryland, USA

Rajeev K Varshney, WA State Agricultural Biotechnology Centre, Centre for Crop and Food Innovation, Food Futures Institute, Murdoch University, Murdoch, WA 6150, Australia Rajeev.Varshney@murdoch.edu.au

Xinyue Zhang, WA State Agricultural Biotechnology Centre, Centre for Crop and Food Innovation, Food Futures Institute, Murdoch University, Murdoch, WA, Australia

Reviewers

Daniel Edge-Garza, Department of Primary Industries & Regional Development, Perth, Australia

Ayomide Fadiji, Hawkesbury Institute for the Environment, Western Sydney University, Richmond, NSW, Australia

Sunil Gangurde, CIMMYT-China Shandong Wheat and Maize Research Center, College of Agronomy, Shandong Agricultural University, Taian, Shandong, China

Vanika Garg, WA State Agricultural Biotechnology Centre, Centre for Crop and Food Innovation, Food Futures Institute, Murdoch University, Murdoch, WA 6150, Australia

Yogendra Kalenahalli, International Crops Research Institute for the Semi-Arid Tropics (ICRISAT), Hyderabad, 502324, Telangana, India

Mehanathan Muthamilarasan, Department of Plant Sciences, School of Life Sciences, University of Hyderabad, Hyderabad, Telangana, 500046, India

Kar-Chun Tan, Centre for Crop and Disease Management,School of Molecular and Life Sciences, Curtin University, Bentley, Australia

Jingjuan Zhang, WA State Agricultural Biotechnology Centre, Centre for Crop and Food Innovation, Food Futures Institute, Murdoch University, Murdoch, WA 6150, Australia

Editor's Biography

Rajeev K. Varshney

Rajeev Kumar Varshney FRS FAA is an agricultural scientist specialising in genomics, genetics, molecular breeding and capacity building in developing countries. As the Director of the Centre for Crop and Food Innovation, the Director of the WA State Agricultural Biotechnology Centre, and the International Chair - Agriculture and Food Security at the Food Futures Institute (FFI) at Murdoch University, Australia, Prof Varshney is at the forefront of pioneering agricultural research.

His research focuses on delivering translational scientific outcomes that help solve the most significant challenges facing global food production, and he has played an instrumental role in determining the genetic traits and underlying mechanisms responsible for climate-resilient crops. His work has led to the decoding of 15 crop genomes, the development of over 20 high-yielding legume crop varieties and the establishment of superior seed systems. Throughout his career, Prof Varshney has established and led international programmes that use genomic variation to breed new crop varieties, that address global malnutrition and food insecurity, improve sustainability in agriculture, and enhance farmer profitability.

He holds Adjunct/Honorary/Visiting Professor positions at 10 academic institutions in Australia, China, Ghana, Hong Kong and India. His current research focus is on improving wheat, legume, and horticultural crops for a range of agronomic and abiotic stress tolerance traits by developing and deploying novel genomics and pre-breeding approaches, including pangenomics, haplotype cataloguing and functional genomics approaches.

Rutwik Barmukh

Rutwik Barmukh is a plant geneticist and genomics researcher and currently serves as a Postdoctoral Research Fellow at the Centre for Crop & Food Innovation, Murdoch University, Australia. His research focuses on developing genomics-assisted and AI-guided breeding strategies to accelerate genetic gain for yield, enhance abiotic stress tolerance, and improve symbiotic nitrogen fixation in major legume crops. His work integrates large-scale genome sequencing, haplotype cataloguing, genomic prediction, and speed breeding to deliver improved pre-breeding resources for crop improvement programs. He completed his PhD in Genetics at the International Crops Research Institute for the Semi-Arid Tropics (ICRISAT) in association with Osmania University, India, where his research addressed the genetic and molecular basis of drought adaptation and insect resistance in chickpea. His doctoral work contributed to the identification and deployment of key drought-responsive genomic regions and candidate genes in breeding pipelines.

Dr. Barmukh has held visiting scientist positions at the University of Leeds, UK, and Università Politecnica delle Marche, Italy, gaining experience in functional and population genomics. He has delivered several invited presentations at key international conferences and has also contributed to organizing scientific workshops at conferences such as the International Plant & Animal Genome (PAG) Conference, Australia. He is a recipient of several prestigious awards and fellowships, including the Crawford Fund Scholar Program Award (Australia), NASA's Spaceflight Technology, Applications, and Research (STAR) Program Award (USA), the Erasmus+ AdaptNET Scholarship (European Commission), and multiple Best Poster Awards at international scientific conferences. He has published more than 40 peer-reviewed articles in high-impact journals, with his research work widely cited in the fields of crop genomics and molecular breeding. His research aims to support the development of climate-resilient, high-yielding crop varieties to strengthen agricultural sustainability and food security.

Richard Harper

Richard Harper is a Professor in the Food Futures Institute within Murdoch University. His research focuses on developing innovative strategies for managing soil, water, and biodiversity—fields that face numerous persistent and complex challenges globally. His primary emphasis is on leveraging carbon markets to facilitate large-scale changes in landscapes; this approach not only addresses these complex issues but also tackles global greenhouse gas emissions, enabling both governments and industries to fulfill their ambitious commitments to reduce emissions. His research has explored the significance of carbon sequestration and bioenergy in various contexts, including agricultural lands, plantations, natural forests, rangelands, and coastal environments. Concurrently, a related program is examining solutions for soil water repellency, which poses significant challenges in agricultural and forested lands. He became a part of Murdoch University in 2009, after accumulating two decades of scientific and policy experience with the Western Australian Government. He was a lead author on the Intergovernmental Panel on Climate Change's Fifth Assessment Report and has served on a range of Australian Government committees.

1 From Climate Challenges to Genomic Solutions: A Framework for Resilient Agriculture

Rutwik Barmukh[1], Rajeev K. Varshney[1,*] and Richard J. Harper[2]
[1]*WA State Agricultural Biotechnology Centre, Centre for Crop and Food Innovation, Food Futures Institute, Murdoch University, Murdoch, WA 6150, Australia;* [2]*Centre for Crop and Food Innovation, Food Futures Institute, Murdoch University, Murdoch, WA 6150, Australia*

Abstract

Climate change poses a critical challenge to global food security by threatening agricultural productivity and sustainability. Moreover, agricultural emissions contribute to climate change. This chapter provides an integrative overview of modern strategies for developing climate-resilient crops. It highlights the role of genetic interventions, genebank genomics, predictive breeding and multi-omics approaches in unlocking genetic diversity and understanding complex stress-response mechanisms. Advances in genome editing, microbiome genomics and nanotechnology are discussed as transformative tools for enhancing crop adaptation, nutrient use efficiency and productivity. The chapter further underscores the importance of enabling policy frameworks, governance mechanisms and digital innovations that facilitate the translation of scientific breakthroughs into sustainable agricultural practices. Taken together, these interdisciplinary approaches provide a roadmap for accelerating genetic gain, improving adaptation to climate change, reducing agricultural emissions and ensuring global food and nutrition security through sustainable, science-driven crop improvement.

1.1 Climate Change and Agricultural Challenges

Climate change, driven primarily by greenhouse gas emissions and widespread land-use change, has emerged as one of the most significant global challenges in recent decades. It has led to a steady rise in mean global temperatures, altered rainfall patterns, increased evapotranspiration and water demand, shifts in crop–pest dynamics and more frequent extreme weather events (Figure 1.1; Schiermeier, 2018; Armstrong-McKay *et al.*, 2022). Notably, meteorological variables, particularly temperature and rainfall, account for up to 50% of the observed yield variability in global crop production (Bernacchi *et al.*, 2025). In addition, agriculture makes a considerable contribution to the greenhouse gas emissions which cause climate change, via land-use change, methane emissions from animal agriculture, and nitrous oxide associated with fertilizer use for crop production. The latter is estimated to emit 1.8±1.1 Gt CO_2-eq/yr or 3.2% of total emissions (Nabuurs *et al.*, 2022), and is thus the focus of policy debates.

*Corresponding author. Email: rajeev.varshney@murdoch.edu.au

 Genomic Solutions for Breeding Climate-resilient Crops (R.K. Varshney, R. Barmukh and R.J. Harper)
DOI: 10.1079/9781800627185.0001

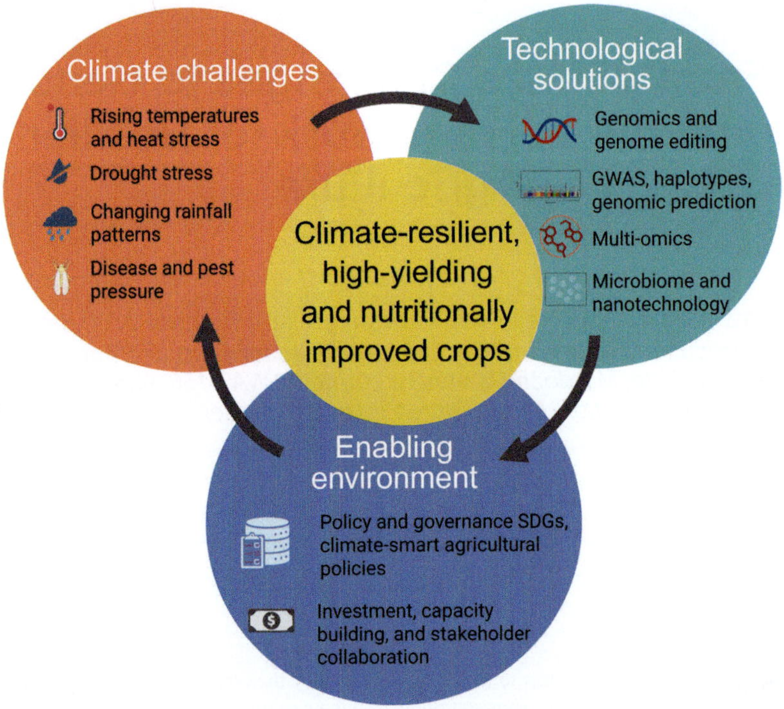

Figure 1.1. Integrative framework for climate-resilient crop improvement. This conceptual framework illustrates the interconnected strategies for developing climate-resilient, high-yielding, and nutritionally improved crops. Climate challenges, such as heat, drought, changing rainfall patterns, and pest pressure, drive the need for advanced genomic and multi-omics interventions, including genome-wide association studies (GWAS), genomic prediction, genome editing (CRISPR-Cas9, base/prime editing), and integrated multi-omics approaches. Complementary innovations such as microbiome engineering and nanotechnology (nano-fertilizers, nano-pesticides, nanosensors) enhance crop performance and sustainability. The enabling environment, encompassing supportive policy frameworks, governance mechanisms, capacity building, and digital innovations, facilitates the translation and adoption of these technologies. Arrows indicate the flow from challenges to solutions, leading to climate-resilient crop outcomes, while feedback loops emphasize the continuous interaction between policy support and technological innovation for sustainable agriculture. SDGs: Sustainable Development Goals. (Authors' own image.)

Alongside these climatic challenges, global food demand is projected to double over the next three decades, driven by population growth and changing dietary preferences (Ray *et al.*, 2013). Without effective mitigation and adaptation strategies, fluctuations in environmental conditions are expected to further reduce agricultural productivity by intensifying abiotic and biotic stresses, disrupting growing season lengths and timing, and increasing the frequency of extreme events that complicate regular farm management (Anderson *et al.*, 2020).

The development of reliable climate change projections relies heavily on climate models, which in turn depend on assumptions about the trajectory of future greenhouse gas emissions. These assumptions are influenced by multiple socio-economic and technological factors, including population growth, economic development, and uptake of innovation. Chapter 2, "*Climate Change Projections: Regional Variations and Emission Scenario Dependencies*" by Kala and Hawke, provides a comprehensive overview of the evolution of climate change emission scenarios and their underlying assumptions. The chapter discusses recent advances in modeling crop impacts under various Representative Concentration Pathways (RCPs) and Shared Socioeconomic Pathways (SSPs), identifies key uncertainties in predicting agricultural impacts, and examines how different future climate scenarios may affect water availability and irrigation across major agricultural regions.

Understanding both the current and projected impacts of climate change on global food systems is crucial for developing strategies that enhance agricultural resilience. Chapter 3, "*Impacts of Climate Change on Crop Production*" by George and colleagues, explores how rising temperatures, shifting rainfall patterns, the co-occurrence of heat and drought stress and changing pest–disease dynamics collectively influence crop productivity. It also highlights broader implications, including threats to crop wild relatives and genetic diversity, as well as socio-economic vulnerabilities. Importantly, the chapter discusses potential positive outcomes, such as the expansion of double cropping opportunities in warmer regions or the northward extension of subtropical and tropical crops into areas that were previously unsuitable (Stokes and Howden, 2010). Similarly, higher temperatures and increased precipitation in high-latitude regions, such as northern Europe, may enhance crop yields and expand production zones, as already evidenced in recent decades (Redhead *et al.*, 2025).

Given the scale and complexity of climate change impacts, addressing agricultural challenges requires systemic and transformational responses. Building climate-resilient agriculture will depend on an integrated suite of strategies encompassing advanced breeding technologies, improved soil and water management, diversified cropping systems and sustainable land-use practices. Within this broader framework, crop genomics remains a cornerstone of adaptation. The development and deployment of improved crop varieties continue to offer a practical, scalable, and farmer-driven pathway to adaptation – one that complements other mitigation and management strategies while enhancing the long-term sustainability of agricultural systems.

1.2 Genomic Solutions for Climate Resilience

Agriculture is now transitioning into a post-Green Revolution era, where genomic innovation has become an important driver of yield gains and sustainability. The pre-genome sequencing era served as a crucial bridge between classical genetics and modern genomics, laying the foundation for today's high-throughput sequencing platforms (Giani *et al.*, 2020). The subsequent genome sequencing revolution, fueled by advances in genotyping and sequencing technologies, has profoundly transformed our understanding of complex traits, accelerating progress in trait discovery, gene mapping and crop improvement across both cereals and legumes.

Contemporary genomic approaches such as genome-wide association studies (GWAS), genomic selection (GS), and pangenomics have enabled rapid identification of loci associated with seed yield, stress tolerance, and adaptation to changing environments (Figure 1.1; Varshney *et al.*, 2021a). In parallel, genome editing, protein engineering, artificial intelligence (AI), nanotechnology and synthetic biology are empowering researchers to precisely tailor crop traits and optimize biological pathways for enhanced resilience and efficiency (Li *et al.*, 2025). Chapter 4, "*Genomic Interventions for Future-Proofing Crops*" by Thudi and colleagues, explores the integration of these advanced genomic tools with multidisciplinary approaches to strengthen crop resilience in the face of mounting food security and climate challenges.

Among the most pressing consequences of climate change are multiple abiotic stresses, including drought, waterlogging, and extreme temperatures combined with other stressors such as soil salinity, metal toxicities and nutrient deficiencies, among others that threaten global crop yields and stability (Figure 1.1). These stresses can cause yield losses of up to 50–70%, jeopardizing both food security and economic sustainability (Benitez-Alfonso *et al.*, 2023). Deciphering the genomic basis of abiotic stress tolerance offers unprecedented opportunities to identify key regulatory genes, networks and adaptive traits that underpin resilience (Barmukh *et al.*, 2022). Recent advances in genomics-assisted breeding (GAB) have greatly accelerated the discovery of molecular markers, genes, and pathways associated with stress tolerance. The integration of GAB with modern breeding and AI-driven models enhances both the speed and precision of developing "stress-smart" crops. Chapter 5, "*Genomics-assisted Breeding for Abiotic Stress Tolerance in Crops*" by Raza and colleagues, presents a detailed discussion of the economic and biological dimensions of abiotic stress, and outlines innovative genomic strategies to develop resilient crop varieties capable of thriving under multiple, concurrent stresses.

In parallel, genomic interventions for pest and disease management have opened new avenues for safeguarding agricultural productivity. Genomic analyses have unraveled the genetic architecture and regulatory pathways underlying resistance mechanisms (Barmukh *et al.*, 2022; Zhang *et al.*, 2022), while approaches such as mutagenesis and mutant analysis expand the pool of genetic variation available for resistance breeding. Chapter 6, "*Genomic Solutions for Pest and Disease Management in Wheat*" by Morales-Salvá and colleagues, details the genomic advances in pest and pathogen management in wheat. The chapter highlights applications ranging from genetic mapping, marker-assisted and genomic selection, positional cloning and association mapping, to TILLING/MutRenSeq, gene editing, and RNA interference (RNAi). By presenting specific examples of these approaches, the authors emphasize how genomics is revolutionizing sustainable pest and disease management and contributing to global food security.

Beyond stress tolerance and pest resistance, enhancing nutrient use efficiency is fundamental to sustainable crop production. Nitrogen (N), a key macronutrient for plant growth, remains one of the most limiting and environmentally challenging inputs in agriculture. Its inefficient utilization not only constrains yields but also contributes to environmental pollution and elevated greenhouse gas emissions (Gu *et al.*, 2023). Improving Nitrogen Use Efficiency (NUE) through genomic approaches represents a critical step toward sustainable food production under climate stress. Chapter 7, "*Optimizing Nitrogen Use Efficiency: Genomic Insights for Reducing Carbon Footprint in a Changing Climate*" by Sandhu and colleagues, provides a comprehensive overview of the genomic and molecular mechanisms underlying nitrogen uptake, assimilation, and remobilization. The authors discuss advances in marker-assisted and gene-editing technologies, highlight key quantitative trait loci (QTLs) and regulatory networks associated with enhanced NUE, and emphasize the integration of these innovations with sustainable agronomic practices and supportive policies to balance productivity with environmental stewardship.

Taken together, the innovations discussed across these chapters underscore the transformative potential of genomics in building climate-resilient, resource-efficient and sustainable agricultural systems. By integrating cutting-edge genomic tools with breeding, systems biology and data-driven approaches, researchers and breeders can develop crop varieties with enhanced abiotic and biotic stress tolerance, improved nutrient use efficiency and reduced environmental footprints, ensuring global food security in an increasingly resource-constrained and climate-unstable world.

1.3 Harnessing Genebank Genomics and Predictive Breeding for Crop Improvement

As global challenges such as climate change intensify, agricultural productivity faces growing threats from a range of abiotic and biotic stresses. The vast plant genetic diversity conserved in genebank collections holds immense potential to address these challenges by providing genetic solutions for improved stress resilience, adaptability and nutritional enhancement in crops (Varshney *et al.*, 2021b). Recent advances in next-generation sequencing technologies have revolutionized the way genetic resources are characterized and utilized, transforming traditional genebanks into digital repositories of genomic diversity (Figure 1.1). These digital platforms enable not only efficient conservation and curation but also systematic characterization and targeted deployment of valuable alleles into modern breeding programs (Milner *et al.*, 2019; Varshney *et al.*, 2021b).

Chapter 8, "*Genebank Genomics for Adapting Crops to Climate Change*" by Shim and colleagues, presents a detailed overview of how genomics-driven approaches are redefining the use of genebank collections. The chapter highlights how integrating genomic data with historical documentation, phenotypic records and environmental metadata can facilitate the discovery of novel adaptive traits and identification of resilient genotypes that can be mobilized into breeding pipelines. Such approaches bridge the gap between long-term conservation and real-world utilization of genetic resources to enhance agricultural resilience under changing climatic conditions.

Despite these advances, phenotyping remains a major bottleneck in the effective utilization of plant genetic resources. Traditional phenotypic evaluations are time-consuming, resource-intensive and often limited by environmental variability. To overcome this, genomic prediction offers a transformative framework that enables early and accurate selection of superior genotypes using genome-wide markers in combination with machine learning (ML) algorithms (Crossa *et al.*, 2025). Chapter 9, "*Genomic Prediction for Climate Resilience Breeding*" by Akinlade and colleagues, delves into the application of genomic prediction in breeding stress-tolerant crops. The chapter emphasizes the value of multi-environment trials (METs) and the integration of environmental data to enhance predictive accuracy. It further discusses current challenges and future directions, highlighting how the convergence of multi-omics, pangenomics and gene-editing technologies can further strengthen the predictive power of genomic prediction models. Together, these innovations position genomic prediction as a cornerstone of climate-resilient breeding and future food security.

While genomic prediction has demonstrated remarkable potential, its predictive accuracy often declines in novel or highly variable environments, largely due to the incomplete representation of genotype × environment (G×E) interactions (Burgueño *et al.*, 2012). Addressing this limitation requires a paradigm shift from genotype-centric approaches toward integrated predictive breeding, which captures the complex interplay between genetics, environment and management. Incorporating environmental characterization (envirotyping) into genomic prediction models enables breeders to better understand and exploit G×E interactions, predict performance under diverse conditions, and design environment-specific or broadly adapted crop varieties (Heslot *et al.*, 2014; Prado *et al.*, 2024).

Chapter 10, "*Predictive Breeding for Crop Genetic Improvement*" by Jighly and Joukhadar, provides a comprehensive synthesis of emerging methodologies that integrate genomic and environmental data within predictive frameworks. The chapter discusses the application of linear, non-linear, and hybrid modeling approaches, and explores future directions centered on the incorporation of multi-omics and phenomics data, advanced AI and deep learning methods, and the development of user-friendly predictive breeding tools. These innovations aim to bridge the gap between research and breeding practice, accelerating the development of crops that can thrive under variable and uncertain climatic conditions.

By harnessing the vast untapped diversity stored in genebanks and integrating it with powerful predictive models and environmental intelligence, researchers and breeders can design the next generation of climate-resilient, high-yielding, and sustainable crop varieties – a critical step toward ensuring global food and nutritional security in the face of a changing climate.

1.4 Expanding the Resilience Toolbox: Multi-omics, Genome Editing and Beyond

Multi-omics approaches, encompassing genomics, transcriptomics, proteomics, metabolomics, epigenomics, and phenomics, provide a comprehensive framework for understanding the complex biological networks that underpin plant responses to abiotic and biotic stresses (Figure 1.1). By integrating datasets from these complementary omics layers, researchers can uncover key genes, proteins, metabolites and regulatory pathways involved in stress tolerance and adaptation (Kudapa *et al.*, 2024; Wang *et al.*, 2024). Chapter 11, "*Integrated Multi-Omics Approaches for Climate-Resilient Crop Improvement*" by Gaddam and colleagues, provides a detailed overview of how multi-omics integration accelerates the discovery of adaptive traits and informs breeding strategies. This chapter highlights how multi-omics data help bridge the gap between genotype and phenotype, enabling the identification of molecular markers and biomarkers that complement traditional genetic markers in breeding programs. The integration of multi-omics with bioinformatics, computational modeling, and systems biology thus holds enormous potential to design nutrient-dense, climate-resilient and high-yielding crops capable of sustaining productivity under dynamic environmental conditions.

The genomic regions and candidate genes identified through multi-omics analyses can be directly utilized in precise genome editing to validate their functions and enhance desirable agronomic traits. Modern genome editing tools, such as CRISPR-Cas9, TnpB, base editing and prime editing, offer unprecedented precision in introducing targeted genetic modifications, from single-nucleotide substitutions to large-scale chromosomal rearrangements (Li *et al.*, 2024; Sun *et al.*, 2025). Chapter 12, "*Genome Editing for Crop Improvement and Sustainable Agriculture*" by Halder and colleagues, reviews the rapid progress and broad applications of genome editing technologies in modern agriculture. It emphasizes their potential to improve crop resilience, productivity and sustainability while also identifying emerging research areas where genome editing could drive transformative innovations in crop design. The chapter underscores how integrating genome editing with biotechnology, systems biology and advanced breeding frameworks can provide powerful and cost-effective solutions for global food security.

Beyond the plant genome itself, microbial communities play a pivotal role in determining plant health, stress tolerance, and ecosystem sustainability. Understanding the plant–microbiome–environment interactions is therefore essential for developing microbiome-based strategies to enhance crop resilience and productivity under climate stress. Advances in metagenomics, metatranscriptomics and synthetic biology have significantly deepened our understanding of how microbial diversity and function contribute to nutrient cycling, disease suppression and abiotic stress mitigation (Fadiji *et al.*, 2023; Vannier *et al.*, 2023; Liu *et al.*, 2024). Chapter 13, "*Microbiome Genomics and Climate Resilience*" by Jayaramaiah and colleagues, examines the functional roles of plant and soil microbiomes in conferring drought tolerance, heat resistance and salinity adaptation. The chapter further outlines strategies to optimize beneficial microbial functions for sustainable agriculture and discusses the challenges of translating genomic insights into scalable field applications. By emphasizing scalability, accessibility and environmental stewardship, this chapter highlights how microbiome genomics can be harnessed to build resilient and sustainable agricultural systems in the face of accelerating climate change.

Complementing these biological approaches, nanotechnology offers innovative tools to enhance agricultural efficiency, precision and sustainability. Nanomaterials can be applied via foliage, stems, or soil to reach target plant tissues and improve resource utilization (Figure 1.1). Nanotechnology enables the development of nanosensors that monitor soil health, detect pathogens and optimize irrigation, thereby supporting the implementation of precision agriculture (Dutta *et al.*, 2022; El-Saadony *et al.*, 2022). Furthermore, nano-fertilizers and nano-pesticides have shown potential to increase nutrient use efficiency, reduce chemical inputs and mitigate environmental contamination (Javaid *et al.*, 2024). Nanoparticles have also demonstrated the ability to enhance plant tolerance to drought, salinity and other stresses by improving water retention and nutrient uptake (Giri *et al.*, 2023). Chapter 14, "*Nanotechnology in Agriculture: Climate-Responsive Applications*" by Poinern and colleagues, explores these innovations, highlighting how nanotechnology can serve as a transformative tool for climate-smart agriculture. The chapter also discusses the convergence of nanotechnology with biotechnology and genome editing platforms, such as CRISPR-Cas9, as a frontier for developing next-generation climate-resilient crops. At the same time, it emphasizes the importance of addressing biosafety, regulatory, economic and public acceptance challenges to ensure the responsible and sustainable deployment of nanotechnologies in agriculture.

Together, these chapters illustrate how multi-omics integration, genome editing, microbiome genomics and nanotechnology collectively represent a powerful suite of innovations driving the next frontier of climate-resilient agriculture. The synergistic application of these technologies will enable precise trait engineering, enhanced stress resilience, efficient resource use and sustainable productivity – paving the way for a new era of data-driven, environmentally responsible and climate-smart crop improvement.

1.5 Enabling Environments: Policy, Governance and Sustainability

Genetic interventions and agroecological practices have demonstrated considerable potential in enhancing soil health, suppressing pests and stabilizing yields under variable climatic conditions (Altieri and Nicholls, 2017; Roorkiwal *et al.*, 2020). However, while these sustainable approaches have helped buffer the impacts of climate stress, they remain insufficient in isolation. Their transformative potential can only be fully harnessed when supported by coherent, forward-looking policy frameworks that promote integration and scalability.

The successful adoption of climate-smart agricultural innovations, including genomics-enabled breeding and precision technologies (such as genome editing), depends on targeted investments, institutional capacity-building and coordinated action among public, private and community stakeholders. Proactive and inclusive policy interventions are therefore essential to catalyse the systemic transformation required for agriculture to remain productive, sustainable, and adaptive in the face of rapid climate change (Zahnow *et al.*, 2025). Embedding climate resilience within agricultural policy frameworks and governance mechanisms allows for risk mitigation, resource optimization and the building of long-term adaptive capacity (Figure 1.1).

Chapter 15, "*Policy, Governance, and Future Pathways for Climate-resilient Crop Production*" by Adedeji and colleagues, delves into the critical intersections of policy, governance and innovation that underpin climate-resilient agriculture. It examines global and national policy initiatives designed to support agricultural adaptation and explores emerging frameworks that leverage science, technology and innovation to enhance resilience. Aligning with global sustainability commitments, particularly Sustainable Development Goal (SDG) 2 (Zero Hunger) and SDG 13 (Climate Action), the chapter highlights forward-looking strategies such as AI-driven precision farming and digital decision-support tools for climate adaptation.

Ultimately, achieving climate resilience in agriculture requires more than technological advancement, it demands a systemic shift built upon long-term vision, multi-stakeholder collaboration and adaptive governance. Only through the integration of sound science, inclusive policies, and sustainable practices can global food systems be safeguarded against the uncertainties of a changing climate.

1.6 Concluding Remarks

Ensuring global food and nutritional security under a changing climate demands a holistic, systems-level approach that integrates cutting-edge science, sustainable practices and enabling policies. From foundational advances in genomics, multi-omics and genome editing to the exploration of microbiome functions and nanotechnology applications, the book highlights how modern biotechnological and digital tools are reshaping the future of crop improvement. These innovations, when combined with precision breeding, bioinformatics and AI-driven decision-making, offer unprecedented opportunities to accelerate the development of climate-resilient, high-yielding, and nutritionally enhanced crop varieties.

Equally important is the recognition that scientific innovation alone is not sufficient. The transformative potential of these technologies can only be realized within supportive policy and governance frameworks that foster equitable access, sustainability and resilience. Strengthening institutional capacities, promoting knowledge exchange and aligning with global sustainability goals will be critical to translating scientific advances into tangible benefits for farmers and societies. Together, the insights presented across the chapters illustrate a pathway toward sustainable agriculture – one that leverages genomics, technology and policy coherence to build resilient food systems capable of meeting the challenges of tomorrow.

References

Altieri, M.A. and Nicholls, C.I. (2017) The adaptation and mitigation potential of traditional agriculture in a changing climate. *Climatic Change* 140, 33–45. DOI: 10.1007/s10584-013-0909-y.

Anderson, R., Bayer, P.E. and Edwards, D. (2020) Climate change and the need for agricultural adaptation. *Current Opinion in Plant Biology* 56, 197–202. DOI: 10.1016/j.pbi.2019.12.006.

Armstrong-McKay, D.I., Staal, A., Abrams, J.F., Winkelmann, R., Sakschewski, B. *et al.* (2022) Exceeding 1.5 C global warming could trigger multiple climate tipping points. *Science* 377, eabn7950. DOI: 10.1126/science.abn7950.

Barmukh, R., Roorkiwal, M., Garg, V., Khan, A.W., German, L. *et al.* (2022) Genetic variation in CaTIFY4b contributes to drought adaptation in chickpea. *Plant Biotechnology Journal* 20, 1701–1715. DOI: 10.1111/pbi.13840.

Benitez-Alfonso, Y., Soanes, B.K., Zimba, S., Sinanaj, B., German, L. *et al.* (2023) Enhancing climate change resilience in agricultural crops. *Current Biology* 33, R1246–R1261. DOI: 10.1016/j.cub.2023.10.028.

Bernacchi, C.J., Long, S.P. and Ort, D.R. (2025) Safeguarding crop photosynthesis in a rapidly warming world. *Science* 388, 1153–1160. DOI: 10.1126/science.adv5413.

Burgueño, J., Campos, G. de los, Weigel, K. and Crossa, J. (2012) Genomic prediction of breeding values when modeling genotype × environment interaction using pedigree and dense molecular markers. *Crop Science* 52, 707–719. DOI: 10.2135/cropsci2011.06.0299.

Crossa, J., Martini, J.W., Vitale, P., Pérez-Rodríguez, P., Costa-Neto, G. *et al.* (2025) Expanding genomic prediction in plant breeding: Harnessing big data, machine learning, and advanced software. *Trends in Plant Science* 30, 756–774. DOI: 10.1016/j.tplants.2024.12.009.

Dutta, S., Pal, S., Panwar, P., Sharma, R.K. and Bhutia, P.L. (2022) Biopolymeric nanocarriers for nutrient delivery and crop biofortification. *ACS Omega* 7, 25909–25920. DOI: 10.1021/acsomega.2c02494.

El-Saadony, M.T., Saad, A.M., Soliman, S.M., Salem, H.M., Desoky, E.S.M. *et al.* (2022) Role of nanoparticles in enhancing crop tolerance to abiotic stress: A comprehensive review. *Frontiers in Plant Science* 13, 946717. DOI: 10.3389/fpls.2022.946717.

Fadiji, A.E., Barmukh, R., Varshney, R.K. and Singh, B.K. (2023) Exploring the connectivity between rhizosphere microbiomes and the plant genes: A way forward for sustainable increase in primary productivity. *Journal of Sustainable Agriculture and Environment* 2, 424–443. DOI: 10.1002/sae2.12081.

Giani, A.M., Gallo, G.R., Gianfranceschi, L. and Formenti, G. (2020) Long walk to genomics: History and current approaches to genome sequencing and assembly. *Computational and Structural Biotechnology Journal* 18, 9–19. DOI: 10.1016/j.csbj.2019.11.002.

Giri, V.P., Shukla, P., Tripathi, A., Verma, P., Kumar, N. *et al.* (2023) A review of sustainable use of biogenic nanoscale agro-materials to enhance stress tolerance and nutritional value of plants. *Plants* 12, 815. DOI: 10.3390/plants12040815.

Gu, B., Zhang, X., Lam, S.K., Yu, Y., Grinsven, H.J. Van *et al.* (2023) Cost-effective mitigation of nitrogen pollution from global croplands. *Nature* 613, 77–84. DOI: 10.1038/s41586-022-05481-8.

Heslot, N., Akdemir, D., Sorrells, M.E. and Jannink, J.L. (2014) Integrating environmental covariates and crop modeling into the genomic selection framework to predict genotype by environment interactions. *Theoretical and Applied Genetics* 127, 463–480. DOI: 10.1007/s00122-013-2231-5.

Javaid, A., Hameed, S., Li, L., Zhang, Z., Zhang, B. and -Rahman, M.U. (2024) Can nanotechnology and genomics innovations trigger agricultural revolution and sustainable development? *Functional & Integrative Genomics* 24, 216. DOI: 10.1007/s10142-024-014.

Kudapa, H., Ghatak, A., Barmukh, R., Chaturvedi, P., Khan, A. *et al.* (2024) Integrated multi-omics analysis reveals drought stress response mechanism in chickpea (Cicer arietinum L.). *The Plant Genome* 17, e20337. DOI: 10.1002/tpg2.20337.

Li, B., Sun, C., Li, J. and Gao, C. (2024) Targeted genome-modification tools and their advanced applications in crop breeding. *Nature Reviews Genetics* 25, 603–622. DOI: 10.1038/s41576-024-00720-2.

Li, G., An, L., Yang, W., Yang, L., Wei, T. *et al.* (2025) Integrated biotechnological and AI innovations for crop improvement. *Nature* 643, 925–937. DOI: 10.1038/s41586-025-09122-8.

Liu, H., Li, J. and Singh, B.K. (2024) Harnessing co-evolutionary interactions between plants and Streptomyces to combat drought stress. *Nature Plants* 10, 1159–1171. DOI: 10.1038/s41477-024-01749-1.

Milner, S.G., Jost, M., Taketa, S., Mazón, E.R., Himmelbach, A. *et al.* (2019) Genebank genomics highlights the diversity of a global barley collection. *Nature Genetics* 51, 319–326. DOI: 10.1038/s41588-018-0266-x.

Nabuurs, G.-J., Mrabet, R., Abu Hatab, A., Bustamante, M., Clark, H. *et al.* (2022) Agriculture, Forestry and Other Land Uses (AFOLU). In: Shukla, P.R., Skea, J., Slade, R., Al Khourdajie, A., van Diemen, R. *et al.* (eds) *IPCC, 2022: Climate Change 2022: Mitigation of Climate Change. Contribution of Working Group III to the Sixth Assessment Report of the Intergovernmental Panel on Climate Change*. Cambridge University Press, Cambridge, UK and New York, NY, USA. DOI: 10.1017/9781009157926.009.

Prado, M., Famoso, A., Guidry, K. and Fritsche-Neto, R. (2024) Optimizing multi-environment trials in the Southern US Rice belt via smart-climate-soil prediction-based models and economic importance. *Frontiers in Plant Science* 15, 1458701. DOI: 10.3389/fpls.2024.1458701.

Ray, D.K., Mueller, N.D., West, P.C. and Foley, J.A. (2013) Yield trends are insufficient to double global crop production by 2050. *PLoS One* 8, e66428. DOI: 10.1371/journal.pone.0066428.

Redhead, J.W., Brown, M., Price, J., Robinson, E., Nicholls, R.J., Warren, R. and Pywell, R.F. (2025) National horizon scanning for future crops under a changing UK climate. *Climate Resilience and Sustainability* 4, e70007. DOI: 10.1002/cli2.70007.

Roorkiwal, M., Bharadwaj, C., Barmukh, R., Dixit, G.P., Thudi, M. *et al.* (2020) Integrating genomics for chickpea improvement: Achievements and opportunities. *Theoretical and Applied Genetics* 133, 1703–1720. DOI: 10.1007/s00122-020-03584-2.

Schiermeier, Q. (2018) Climate as culprit. *Nature* 560, 20–22. DOI: 10.1038/d41586-018-05849-9.

Stokes, C. and Howden, M. (eds) (2010). In: *Adapting Agriculture to Climate Change: Preparing Australian Agriculture, Forestry and Fisheries for the Future*. CSIRO Publishing, Australia.

Sun, C., Li, H., Liu, Y., Li, Y., Gao, R. *et al.* (2025) Iterative recombinase technologies for efficient and precise genome engineering across kilobase to megabase scales. *Cell* 188, 4693–4710. DOI: 10.1016/j.cell.2025.07.011.

Vannier, N., Mesny, F., Getzke, F., Chesneau, G., Dethier, L. *et al.* (2023) Genome-resolved metatranscriptomics reveals conserved root colonization determinants in a synthetic microbiota. *Nature Communications* 14, 8274. DOI: 10.1038/s41467-023-43688-z.

Varshney, R.K., Bohra, A., Roorkiwal, M., Barmukh, R., Cowling, W.A. *et al.* (2021a) Fast-forward breeding for a food-secure world. *Trends in Genetics* 37, 1124–1136. DOI: 10.1016/j.tig.2021.08.002.

Varshney, R.K., Roorkiwal, M., Sun, S., Bajaj, P., Chitikineni, A. *et al.* (2021b) A chickpea genetic variation map based on the sequencing of 3,366 genomes. *Nature* 599, 622–627. DOI: 10.1038/s41586-021-04066-1.

Wang, P., Lehti-Shiu, M.D., Lotreck, S., Segura Abá, K., Krysan, P.J. and Shiu, S.H. (2024) Prediction of plant complex traits via integration of multi-omics data. *Nature Communications* 15, 6856. DOI: 10.1038/s41467-024-50701-6.

Zahnow, R., Yousefnia, A.R., Hassankhani, M. and Cheshmehzangi, A. (2025) Climate change inequalities: A systematic review of disparities in access to mitigation and adaptation measures. *Environmental Science & Policy* 165, 104021. DOI: 10.1016/j.envsci.2025.104021.

Zhang, H., Zhu, J., Gong, Z. and Zhu, J.K. (2022) Abiotic stress responses in plants. *Nature Reviews Genetics* 23, 104–119. DOI: 10.1038/s41576-021-00413-0.

2 Climate Change Projections: Regional Variations and Emission Scenario Dependencies

Jatin Kala[1] and Kerryn Hawke[2,*]

[1]*School of Environmental and Conservation Sciences, Murdoch University, 6150, Western Australia;* [2]*School of Environmental and Conservation Sciences, Murdoch University, 6150, Western Australia*

Abstract

This chapter explores the history of climate-change emission scenarios, including the Special Report on Emission Scenarios (SRES), Representative Concentration Pathways (RCPs) and Shared Socioeconomic Pathways (SSPs), highlighting their evolution and underlying assumptions. It then examines recent research on crop impacts under different RCPs and SSPs, drawing on both global and regional studies. Key uncertainties in projecting future climate impacts on agricultural crops are also discussed, including those from climate and crop models. The chapter also explores how different possible future climate scenarios may affect water availability and the resulting implications for irrigation in major agricultural regions worldwide.

2.1 The History of Climate Change Emission Scenarios

The creation of climate change projections using climate models invariably requires assumptions be made about future greenhouse gas emissions, which depend on a very large number of factors, e.g. population and associated economic growth and technological developments. Future climate scenarios as used in different Assessment Reports (AR) of the IPCC have been developed as part of the various Coupled Model Intercomparison Project (CMIP) phases, with the different phases of CMIP historically designed to align with IPCC assessment reports. This provides the scientific foundation of the assessments of climate change science. For example, the third and 4th Assessment reports of the IPCC (AR3 and AR4) relied heavily on CMIP3; the fifth assessment report (AR5) on CMIP5; and, the sixth assessment report (AR6) on CMIP6. Over the past 10–20 years, these future emission scenarios and associated climate projections have been used to better understand how climate change might affect agriculture across the IPCC AR cycle. This section provides an overview of these different greenhouse gas emission scenarios, as well as context on the history and evolution of these scenarios from CMIP3 to CMIP6, and what is being planned as part of CMIP7 and beyond.

*Corresponding authors. Email: J.Kala@murdoch.edu.au; Kerryn.Hawke@murdoch.edu.au.

 Genomic Solutions for Breeding Climate-resilient Crops (R.K. Varshney, R. Barmukh and R.J. Harper)
DOI: 10.1079/9781800627185.0002

2.1.1 The Special Report on Emissions Scenarios (SRES)

The Special Report on Emissions Scenarios (SRES) were developed for the 3rd and 4th assessment reports of the IPCC as part of CMIP3 and CMIP4, respectively. The SRES is comprised of four narrative storylines (A1, A2, B1, and B2), with each describing different future worlds based on varying assumptions about economic growth, technological development, and population growth. SRES scenarios do not include specific climate policies but cover a wide range of possible futures (IPCC, 2000):

- **A1**: Rapid economic growth, global population peaking mid-century, and rapid technological advancement. The A1 scenario family develops into a further three groups that are distinguishable from each other based on alternative technological emphasis: fossil-intensive (A1FI), non-fossil energy sources (A1T), and balance across all sources (A1B).
- **A2**: Heterogeneous world with high population growth, slower economic development and technological change.
- **B1**: Convergent world with the same population growth as A1 but with rapid changes towards a service and information economy.
- **B2**: Local solutions to economic, social and environmental sustainability with moderate population growth and intermediate economic development. Represents a world with increasing global population rates, but lower than the A2 scenario.

Given computational and data restraints, climate impact studies typically do not consider all 4 SRES scenarios, but rather many studies have predominantly used the SRES A2 scenario as a conservative approach to examine impacts of future climate change on indices relevant to crop production in key agricultural regions of the globe (e.g. Olesen *et al.*, 2007; Mihailović *et al.*, 2015; Andrys *et al.*, 2017). Although these emissions scenarios are currently considered out-of-date, the scientific findings and methodologies used are still relevant, and critical to better understanding emissions scenarios used in subsequent IPCC reports.

2.1.2 Representative Concentration Pathways (RCPs)

Following on from CMIP3 and CMIP4, CMIP5 involved the use of Representative Concentration Pathways (RCPs) as used in the 5th Assessment Report. While the SRES scenarios were narrative-based and focused on socio-economic storylines, the RCPs are concentration-based focusing on specific radiative forcing targets by 2100. In other words, an RCP is a greenhouse gas concentration pathway that is not related to a trajectory of emissions and socioeconomic factors. Another key difference is that while the SRES did not include climate policies, the RCPs included scenarios with explicit mitigation efforts to reduce greenhouse gas emissions. Taken together, SRES scenarios and RCPs are complementary to each other. There are four most commonly evaluated RCPs named after their radiative forcing values in 2100 (i.e. 2.6, 4.5, 6.0, 8.5 W/m^2) (IPCC, 2019):

- **RCP2.6**: Strong mitigation scenario aiming to keep global warming below 2°C. This represents a "best-case scenario" with near-term global warming (2031–2050) estimated at 1.1–2.0°C and end-of-century warming (2081–2100) estimated at 0.9–2.4°C.
- **RCP4.5**: Stabilization scenario where emissions peak around 2040 and then decline. This is considered a scenario with an intermediate level of mitigation (for example via emissions reduction and carbon sequestration), with near-term global warming estimated at 1.3–2.2°C and end-of-century warming estimated at 1.7–3.3°C.
- **RCP6.0**: Stabilization scenario with emissions peaking around 2080. Similar to RCP4.5, this is also considered an "intermediate" scenario between the best (RCP2.6) and worst case (RCP8.5), but with higher radiative forcing than RCP4.5, with near-term global warming estimated at 1.2–2.0°C and end-of-century global warming estimated at 2.0–3.8°C.
- **RCP8.5**: High emissions scenario with continued increases in greenhouse gas emissions. This represents the "worst-case scenario" with near-term global warming estimated at 1.5–2.4°C and end-of-century warming estimated at 3.2–5.4°C.

Studies using regional climate models at high spatial resolutions to better represent agricultural land-use as part of the Coordinated Reginal Climate Downscaling Experiment (CORDEX) have typically made use of RCP4.5 as a "mid-range" scenario and RCP8.5 as the "worst-case" scenario (e.g. Bukovsky and Mearns, 2020; Jacob *et al.*, 2020; Evans *et al.*, 2021) given computational and data limitations associated with high spatial resolution regional climate projections.

2.1.3 Shared Socioeconomic Pathways (SSPs)

Following on from CMIP5, in CMIP6, Shared Socioeconomic Pathways (SSPs) were developed by building on RCPs by essentially adding five storylines pathways based on different socio-economic factors. Specifically, SSPs address two key limitations from SRES and RCPs. The SRES scenarios, which examined four different scenarios in the 1990s, had quickly become out-of-date due to significant changes to society and the economy that occurred over the subsequent couple of decades. While the RCPs were more up-to-date and included climate mitigation, these purposefully did not include any socioeconomic narratives to go alongside them. These two gaps are addressed with SSPs and five main narratives (Riahi *et al.*, 2017):

- **SSP1:** Sustainability – Taking the Green Road (Low challenges to mitigation and adaptation). The world shifts gradually, but pervasively, toward a more sustainable path, emphasizing more inclusive development that respects perceived environmental boundaries. Management of the global commons slowly improves, educational and health investments accelerate the demographic transition, and the emphasis on economic growth shifts toward a broader emphasis on human wellbeing. Driven by an increasing commitment to achieving development goals, inequality is reduced both across and within countries. Consumption is oriented toward low material growth and lower resource and energy intensity.
- **SSP2:** Middle of the Road (Medium challenges to mitigation and adaptation). The world follows a path in which social, economic, and technological trends do not shift markedly from historical patterns. Development and income growth proceeds unevenly, with some countries making relatively good progress while others fall short of expectations. Global and national institutions work toward but make slow progress in achieving sustainable development goals (SDGs). Environmental systems experience degradation, although there are some improvements and overall, the intensity of resource and energy use declines. Global population growth is moderate and levels off in the second half of the century. Income inequality persists or improves only slowly and challenges to reducing vulnerability to societal and environmental changes remain.
- **SSP3:** Regional Rivalry – A Rocky Road (High challenges to mitigation and adaptation). A resurgent nationalism, concerns about competitiveness and security, and regional conflicts push countries to increasingly focus on domestic or, at most, regional issues. Policies shift over time to become increasingly oriented toward national and regional security issues. Countries focus on achieving energy and food security goals within their own regions at the expense of broader-based development. Investments in education and technological development decline. Economic development is slow, consumption is material-intensive and inequalities persist or worsen over time. Population growth is low in industrialized and high in developing countries. A low international priority for addressing environmental concerns leads to strong environmental degradation in some regions.
- **SSP4:** Inequality – A Road Divided (Low challenges to mitigation, high challenges to adaptation). Highly unequal investments in human capital, combined with increasing disparities in economic opportunity and political power, lead to increasing inequalities and stratification both across and within countries. Over time, a gap widens between an internationally connected society that contributes to knowledge- and capital-intensive sectors of the global economy, and a fragmented collection of lower-income, poorly educated societies that work in a labor intensive, low-tech economy. Social

cohesion degrades and conflict and unrest become increasingly common. Technology development is high in the high-tech economy and sectors. The globally connected energy sector diversifies, with investments in both carbon-intensive fuels like coal and unconventional oil, but also low-carbon energy sources. Environmental policies focus on local issues around middle- and high-income areas.

- **SSP5:** Fossil-fueled Development – Taking the Highway (High challenges to mitigation, low challenges to adaptation). This world places increasing faith in competitive markets, innovation, and participatory societies to produce rapid technological progress and development of human capital as the path to sustainable development. Global markets are increasingly integrated. There are also strong investments in health, education, and institutions to enhance human and social capital. At the same time, the push for economic and social development is coupled with the exploitation of abundant fossil fuel resources and the adoption of resource- and energy-intensive lifestyles around the world. All these factors lead to rapid growth of the global economy, while global population peaks and declines in the 21st century. Local environmental problems such as air pollution are successfully managed. There is faith in the ability to effectively manage social and ecological systems, including by geo-engineering if necessary.

SSPs are developed using Integrated Assessment Models (IAMs), incorporating insights from economics, land use, and climate science to provide an overall holistic view of future emission scenarios. By exploring a range of possible different futures that include different technological developments and policy interventions (e.g. mitigation, CO_2 removal, carbon capture and storage), they help policymakers assess the potential outcomes of various policy choices, including mitigation and adaptation (Skea *et al.*, 2021). SSPs have two numbers. For example, SSP3-7.0 is an emission scenario that follows the SSP3 narrative and results in a radiative forcing of 7.0 W/m^2 by 2100, i.e. SSP3-7.0 is related to RCP7.0, but has a storyline attached. The link between SSPs and RCP is summarized in Figure 2.1. Of note in this figure is that an 8.5 W/m^2 radiative forcing by 2100 is only achievable under the SSP5 storyline, i.e. SSP5-8.5, whereas lower levels of radiative forcing between 3.4 and 6 W/m^2 are achieved under all SSPs. Hence, SSP5-8.5 is generally considered a less plausible future scenario as compared with others.

2.1.4 The Distinction Between "Worst Case Scenarios" and "Business as Usual"

Over the past decades, many researchers have used the highest emission scenarios, e.g. RCP8.5 from CMIP5 and SSP5-8.5 from CMIP6 as a "business as usual" scenario, synonymous with a "worst case" scenario of unchecked global warming driven by exploitation of fossil fuels globally (e.g. Zittis *et al.*, 2021). However, there is disagreement within the climate science community as to whether this is correct. Christensen *et al.* (2018) argue that RCP8.5 may in fact underestimate future concentrations of greenhouse gases with a 35% probability that emissions will exceed those assumed under RCP8.5. However, others have argued that greenhouse gas concentrations under RCP8.5, although plausible, remain highly unlikely given updated information on global coal reserves, and recent trends in the energy sector towards increased renewables (e.g. Ritchie and Dowlatabadi, 2017). Rather than referring to a single "business-as-usual" scenario being the scenario with the highest level of warming, it is now recognized that it is far more useful to consider a range of plausible futures. For example, the IPCC 5th assessment report states that "*Scenarios without additional efforts to constrain emissions ('baseline scenarios') lead to pathways ranging between RCP6.0 and RCP8.5*" (IPCC, 2014). The CORDEX initiative, which aims to dynamically downscale global climate models from CMIP6 to higher resolutions that are relevant for agriculture, recommends that modeling groups prioritize the downscaling of SSP3-7.0 and SSP1-2.6 first, to represent the upper and lower estimates respectively, rather than downscale only SSP5-8.5 as the upper estimate. As a result, regional climate modeling groups, for example in Australia, are providing regional climate projections under SSP1-2.6 and SSP3-7.0 (Di Virgilio *et al.*, 2025),

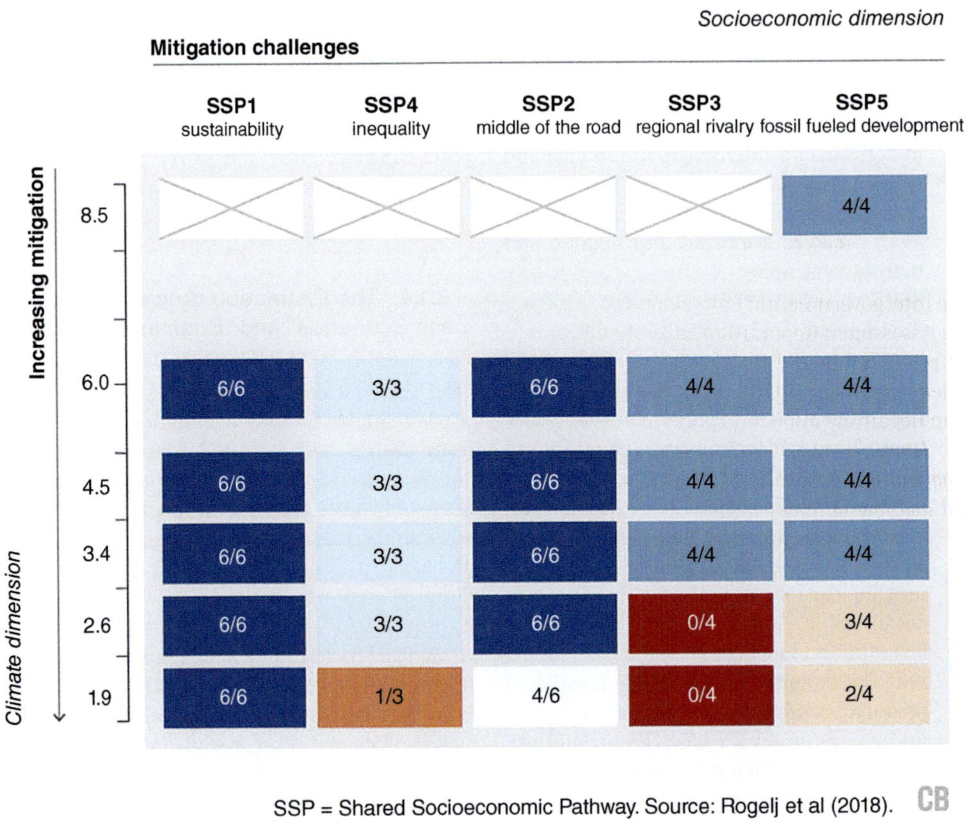

Figure 2.1. Combination of SSP and RCP model runs in the SSP database, with RCPs listed in order of increasing mitigation and SSPs in the (rough) order of increasing mitigation difficulty. Ratios in cells indicate the number of models that succeeded in making the scenario "work" out of the total number of models available for the SSP. Figure chart by Carbon Brief, adapted from Figure S1 in Rogelj *et al.* (2018), used in an article by Hausfather (2018) (see references). Image used with permission.

and this information will be used to better understand impacts of future climate change on crops in Australia.

2.1.5 Future Emission Scenarios Beyond CMIP6

All phases of CMIP up to CMIP6 have used global climate model simulations that are driven by atmospheric greenhouse gas concentrations that are obtained from IAMs. The greenhouse gas concentrations from IAMs that are then used in the global earth system models (commonly referred to as global climate models), represent the net carbon emissions, after accounting for various processes such as mitigation, CO_2 removal etc. While this approach has been very useful to better understand impacts of climate change under multiple future scenarios, there are very large uncertainties in the carbon cycle of these earth system models as processes such as mitigation and CO_2 removal are not represented. It has been argued the earth systems models used as part of CMIP7 and beyond should "*explicitly represent complete cycles of major greenhouse gases, including natural processes and anthropogenic influences*" and that "*that CMIP7 efforts prioritize simulations driven by CO_2 emissions from fossil fuel use and projected deployment of carbon dioxide removal technologies, as well as land use and management, using the process resolution allowed by state-of-the-art ESMs to resolve carbon–climate feedbacks*" (Sanderson *et al.*, 2024). Such simulations will provide an invaluable source of information to better understand possible future impacts of climate change on crops.

2.2 Regional Variations and Emission Scenario Dependencies

2.2.1 Background

The Intergovernmental Panel on Climate (IPCC) 6th Assessment Report (Bezner Kerr *et al.*, 2022) (https://www.ipcc.ch/report/ar6/wg2/) makes it clear that human-induced global warming has been negatively impacting agriculture overall:

Human-induced warming has slowed growth of agricultural productivity over the past 50 years in mid and low latitudes (medium confidence). Crop yields are compromised by surface ozone (high confidence). Methane emissions have negatively impacted crop yields by increasing temperatures and surface ozone concentrations (medium confidence). Warming is negatively affecting crop and grassland quality and harvest stability (high confidence), p. 717.

And there is high confidence that this will worsen under future climate scenarios:

Climate change will make some current food production areas unsuitable (high confidence). Current global crop and livestock areas will increasingly become climatically unsuitable under a high emission scenario (high confidence) (e.g. 10% by 2050, over 30% by 2100 under SSP-8.5 versus below 8% by 2100 under SSP1-2.6), p. 717.

This in turn has cascading impacts on food production and nutrition as illustrated in Figure 2.2.

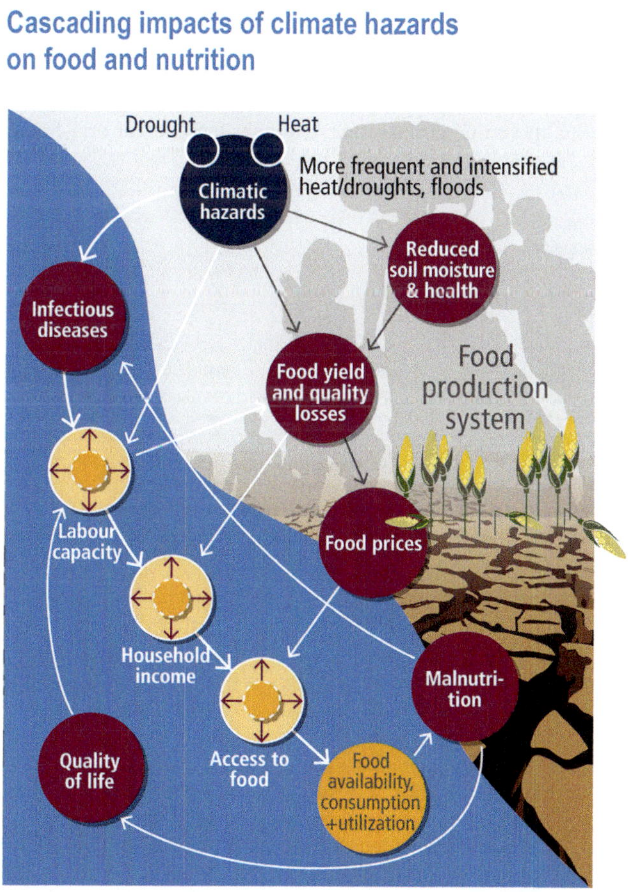

Figure 2.2. Cascading impacts of climate hazards on food and nutrition. The factors involved the impacts on crop production and prices (black arrows) and interaction among food and health (white arrows). Figure from *Bezner Kerr et al.* (2022). Adapted and revised from Phalkey *et al.* (2015).

The observed impacts of climate change vary regionally and for different crops. Figure 2.3, from the IPCC 6th Assessment Report (Bezner Kerr *et al.*, 2022), shows a summary synthesis of observed impacts of climate change on different crop types in key regions, illustrating that although there are regions which have benefited from climate change or have experienced mixed impacts, averaged across the globe, the impacts tend to be negative overall for most crops. Factors that affect crops are both biotic (e.g. pests and diseases) and abiotic (e.g. drought and heat stress). For soybeans and wheat, these vary markedly across the globe, with wheat being much more susceptible to heat stress and aridity (Figure 2.4).

2.2.2 Scenario Dependencies

How crops will be affected by climate change under different future climate scenarios is inherently difficult to quantify, as it is very difficult to accurately consider technological advances and the CO_2 fertilization effect, i.e. more CO_2 enhancing photosynthesis everything else being equal. To better understand the influence of change in climate change alone, it is often useful to exclude these two factors, as illustrated in Figure 2.5 showing projected changes in yield under different RCPs with respect to a baseline period of 2001–2010, without considering adaptation and CO_2 fertilization effects. While there are large variations between the different crop types, RCPs and time-periods, there are some common features. Namely, differences diverge markedly from the mid-century onwards and become more pronounced under RCP8.5. Given the issues around the interpretation of RCP8.5 as "business as usual", as discussed above, it is important to focus on the changes between RCP4.5 and RCP8.5, rather than focus solely on RCP8.5. Of all crop types, changes for maize are most pronounced while changes for soybean show the largest uncertainty and rice yields show the least variation, likely because the simulations assume irrigated conditions (Bezner Kerr *et al.*, 2022). The regional variations are equally very large, notably for Central and South America, Asia,

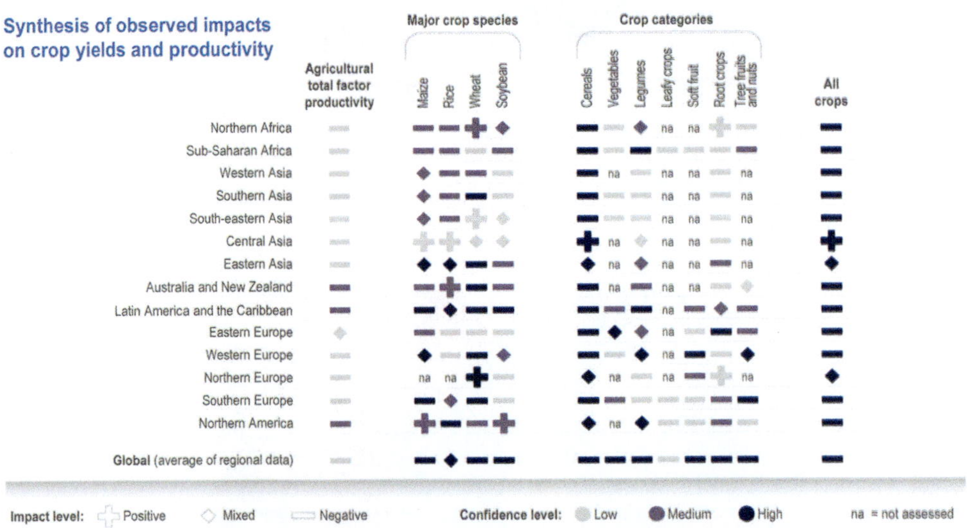

Figure 2.3. Synthesis of literature on observed impacts of climate change on productivity by crop type and region. The figure draws on >150 articles categorized by: agriculture total factor productivity including literature estimating all agricultural outputs in a region; major crop species including literature assessing yield changes in the four major crops; crop categories including productivity changes (yield, quality and other perceived changes) in a range of crops with different growth habits. The assessment uses literature published since AR5, although the timespan often extends prior to 2014. The direction of the effect and the confidence are based on the reported impacts and attribution, and on the number of articles. Figure from *Bezner Kerr et al.* (2022). Image used with permission.

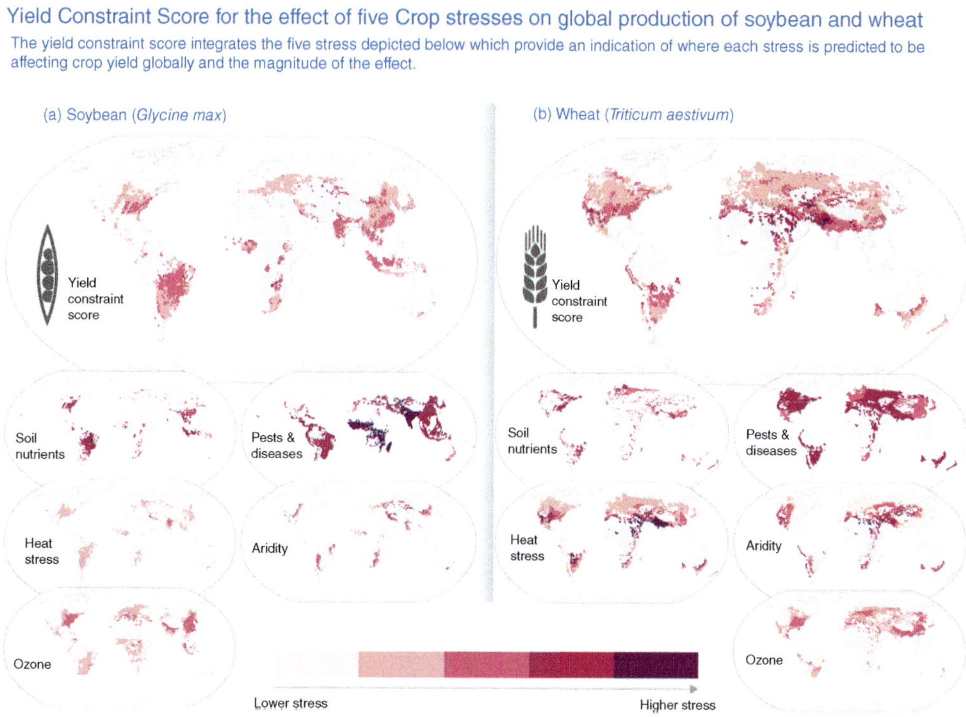

Figure 2.4. The global effects of five biotic and abiotic stresses on soybean and wheat. All data are presented for the 1 × 1° (latitude and longitude) grid squares where the mean production of soybean or wheat was >500 tonnes (0.0005 Tg). The effect of each stress on yield is presented as a Yield Constraint Score (YCS) on a scale of 1–5, where 5 is the highest level of stress from ozone, pests and diseases, heat stress and aridity (*Mills et al.*, 2018). Figure from *Bezner Kerr et al.* (2022). Data are available at Sharps *et al.* (2020). Image used with permission.

and Africa for several crops under RCP8.5. It should be noted that most of these yield change simulations do not incorporate the effects of pest, diseases, and changes in soil conditions for long time-frames (Bezner Kerr *et al.*, 2022), hence these projections should be interpreted with care.

2.2.3 Impacts of Future Climate Change on Crops Under CMIP6 SSPs and CMIP5 RCPs

Numerous studies have examined crop impacts under CMIP6 SSPs and CMIP5 RCPs. Some provide a global perspective (e.g. Jägermeyr *et al.*, 2021; Müller *et al.*, 2021; Li *et al.*, 2025), while others have focused on particular crops or regions of agricultural importance around the world, e.g. the Phichit province in central northern Thailand (Arunrat *et al.*, 2022); the Yellow River Basin (Wang *et al.*, 2017; Guo *et al.*, 2022; X. Li *et al.*, 2024; Xin *et al.*, 2025), rice yields on the Nile River Delta in Egypt (Elsadek *et al.*, 2024); and the vulnerability of barley and *Jatropha curcas* production to climate change in South America (de Oliveira Aparecido *et al.*, 2024, 2025). Many studies focus on evaluating the various models' performance as an important step to understanding the limitations of the models and the quality of the model output.

2.2.4 Ensembles and Uncertainty in Future Climate Projections

With all models being imperfect, many climate projection studies use ensembles to quantify

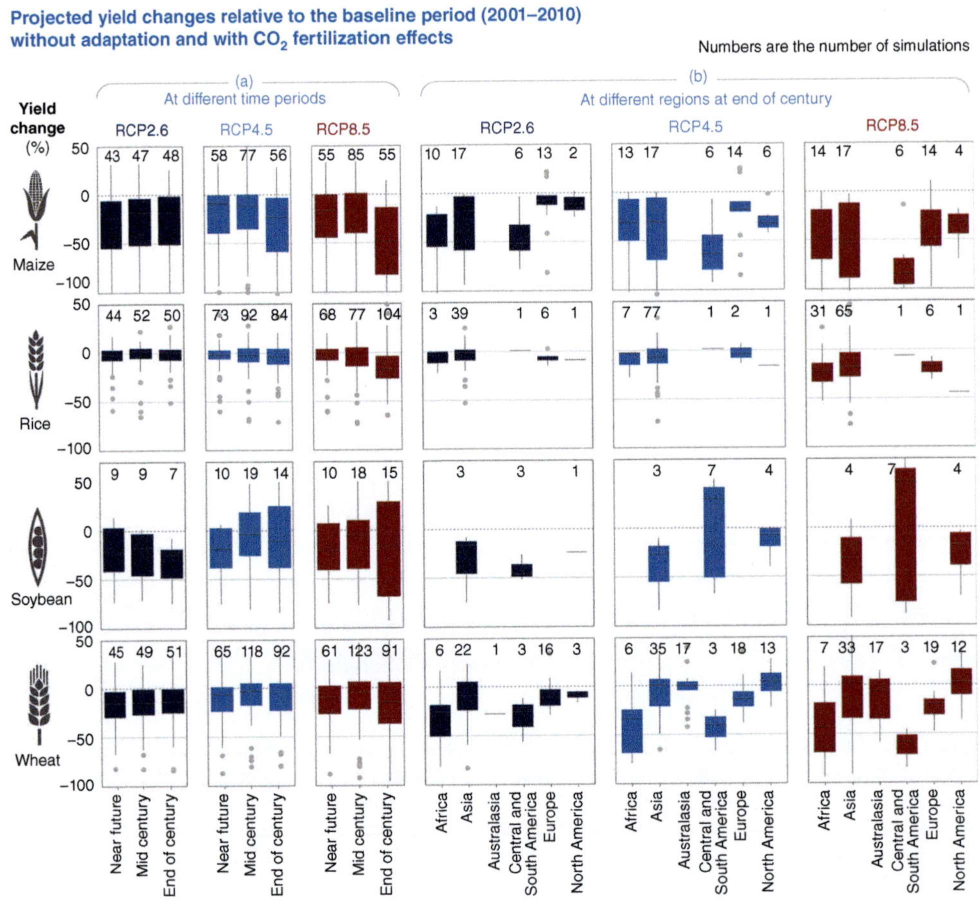

Figure 2.5. Projected yield changes relative to the baseline period (2001–2010) without adaptation and with CO_2 fertilization effects (*Hasegawa et al.*, 2022). The box is the interquartile range (IQR), and the middle line in the box represents the median. The upper and lower end of whiskers are median 1.5 × IQR ± median. Open circles are values outside the 1.5 × IQR. (a) At different time periods (near future, NF, baseline to 2039; mid-century, MC, 2040–2069; end-century, EC, 2070–2100) under three RCPs, and (b) at different regions at EC. Figure from *Bezner Kerr et al.* (2022). Image used with permission.

uncertainty, where an ensemble is simply a group of models (Wilcke and Bärring, 2016). For example, the shading in Figure 2.6 (Jägermeyr *et al.*, 2021) shows the interquartile range or spread of the individual models' change in global productivity of maize and wheat (Figure 2.6a and b, respectively), calculated using data from an ensemble of CMIP6 models under the SSP1-2.6 (green) and the SSP5-8.5 (yellow) scenarios as input to a crop model. The green shading is much more compact than the yellow which illustrates higher uncertainty in the SSP5-8.5 scenario, especially in the latter part of the 21st century. In other words, we can be relatively confident that global maize productivity will decline by the end of the 21st century under the SSP1-2.6 scenario, but some models are predicting a decline of as much as around 10% while others show little change (Figure 2.3a). The model spread is even greater under the SSP5-8.5 scenario for both maize and wheat, with global productivity in maize expected to decrease more around 28% according to the multi-model means. Where there is a greater uncertainty models differ substantially, with individual models predicting a modest decline of

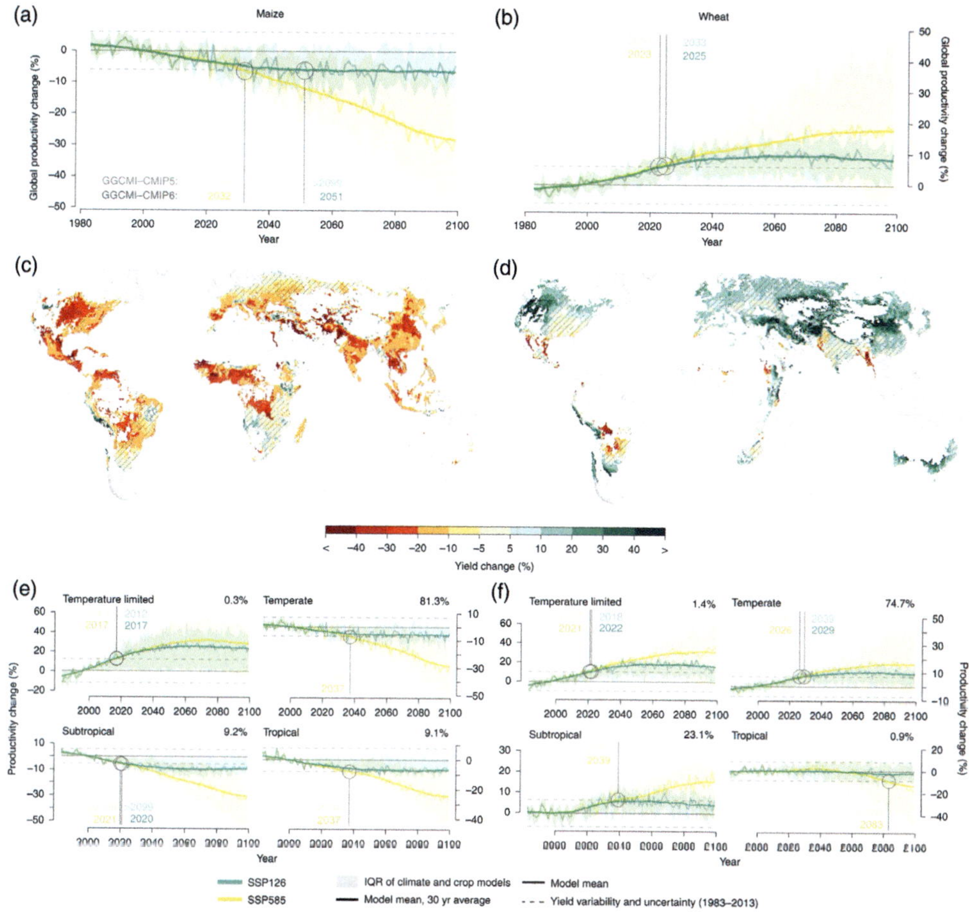

Figure 2.6. Time series of productivity changes for (a) maize and (b) wheat relative to the 1983–2013 reference period under SSP1-26 (green) and SSP5-85 (yellow). Shaded ranges represent the interquartile range of all climate-crop model combinations (5 GCMs × 12 GGCMs). The solid line shows the median response with a 25-year moving average. Horizontal dashed lines mark the standard deviation of historical yield variability and model uncertainty (the 'noise' from individual climate–crop model combinations) and open circles highlight the climate change response emergence (the year in which the smoothed climate change response emerges from the noise). Maps (c) and (d) show median yield changes (2069–2099) for maize (c) and wheat (d) under SSP5-85, with hatching indicating areas where <70% of the climate–crop model combinations agree on the sign of impact. Regional productivity time series for maize (e) and wheat (f) are stratified for the four major Koeppen–Geiger climate zones (temperature limited, temperate/humid, subtropical and tropical), with each zone's contribution to total global production (%) noted in the insets. Figure from *Jägermeyr et al.* (2021). Image used with permission.

~15% but others nearly 50%. Conversely, global wheat productivity (Figure 2.6b) is expected to increase, although the interquartile ranges are similar under the two scenarios.

Some analyses primarily focus on evaluating the source of these uncertainties and whether using the more up-to-date CMIP6 models decreases the extent of these uncertainties. For example, the aforementioned Jägermeyr *et al.* (2021) study examined the projected time of climate impact emergence, finding that CMIP6 models generally project global maize crops to be negatively impacted earlier than was predicted by CMIP5 models. Conversely, wheat yields are expected to be positively impacted as a result of the crop's higher capacity to benefit from elevated CO_2 levels (Kimball, 2016; Jägermeyr *et al.*, 2021). Müller *et al.* (2021) expand on Jägermeyr *et al.* (2021) by evaluating the various configurations or ensembles of both the crop models and the climate data from CMIP5 and CMIP6 models used as input to the crop models in Jägermeyr *et al.* (2021). They found that uncertainties arising from the various crop models are around the same magnitude as the uncertainty arising from the climate models. This highlights the need to not only consider uncertainties from climate models, but also from the crop models used to predict yields based on changes in climate.

Given the value of accurate predictions of future climate impacts on crop yield, studies such as Ramirez-Villegas *et al.* (2013) discuss the importance of GCM skill for predicting the impacts of future climate change on agricultural production and food security. They noted improvements in CMIP5 model skill for mean temperatures (5–15%), diurnal range (2–5%), and precipitation (1–2%) compared with the older CMIP3 suite of models. However, they estimated that achieving substantial improvements in the ability of GCMs to predict regional changes in temperature could take 5–30 years, while precipitation might require 30–50 years before models reach sufficient accuracy in some regions (Ramirez-Villegas *et al.*, 2013). The authors concluded that significant errors persist in CMIP5 GCMs, particularly with respect to seasonal and regional differences in model skill, necessitating a cautious approach to the application of CMIP5 data in agricultural impact studies.

2.2.5 CMIP5 Regional Studies

During the 2010s, many regional studies utilized CMIP5 GCM outputs to explore potential changes in agricultural productivity under various future climate scenarios. For example, Xin *et al.* (2025) statistically downscaled model output from 10 CMIP5 models to find out how agricultural water demand in the Yellow River Basin, an important grain-producing region in China, might change in the future. They found that agricultural irrigation water demand varied significantly across the basin, with greater demand in the middle and lower parts and a smaller demand in the hilly regions. However, in the upper reaches where there are fewer crops, most models indicated a decreasing trend in agricultural water demand from 2020–2070 followed by an increasing trend from 2070–2100. Additionally, there were significant differences between individual models and in their response to the different RCP scenarios, indicative of uncertainties within the GCMs and the RCPs.

Another study that used CMIP5 projections focused on identifying potential areas of agricultural drought around the world and evaluating the uncertainties (Lu *et al.*, 2019). This is illustrated in Figure 2.7, which shows the percentage change in annual mean surface soil moisture across the globe from 2071–2100, compared with the period from 1976–2005 for the four RCP scenarios. Based on the output from multiple CMIP5 models, the authors indicate regions where statistically significant changes are apparent (the stippling on the maps). Statistically significant large-scale drying is apparent over most of South and North America, Australia, Europe, Central Asia and southern Africa, and a few smaller regions of statistically significant wetting in parts of eastern Africa, India, and Central Asia. However, large regions of central and northern Africa, the Middle East, Central Asia, along with regions in Alaska, Canada and Argentina in particular show changes which are not statistically significant, with the authors finding that while regions where the greatest increase in long-term drought is linked with the highest radiative forcing scenarios, the multi-model ensembles have a range that is much larger than the change itself (Lu *et al.*, 2019). They found that more than 80% of the uncertainty comes from differences among models, particularly in how soil moisture, clouds, and precipitation are represented. They concluded that there is a pressing need for future GCMs to improve their ability to simulate surface hydrological processes including soil moisture, to decrease uncertainties in agricultural drought projections.

It is notable that a number of studies examining drought predictions using the newer CMIP6 GCMs have found similar robust increases in drought risk to those using CMIP5 (e.g. Cook *et al.*, 2020; Ukkola *et al.*, 2020). However, model uncertainties are similar in the CMIP6 model suite as found in the CMIP5 models (e.g. Cook

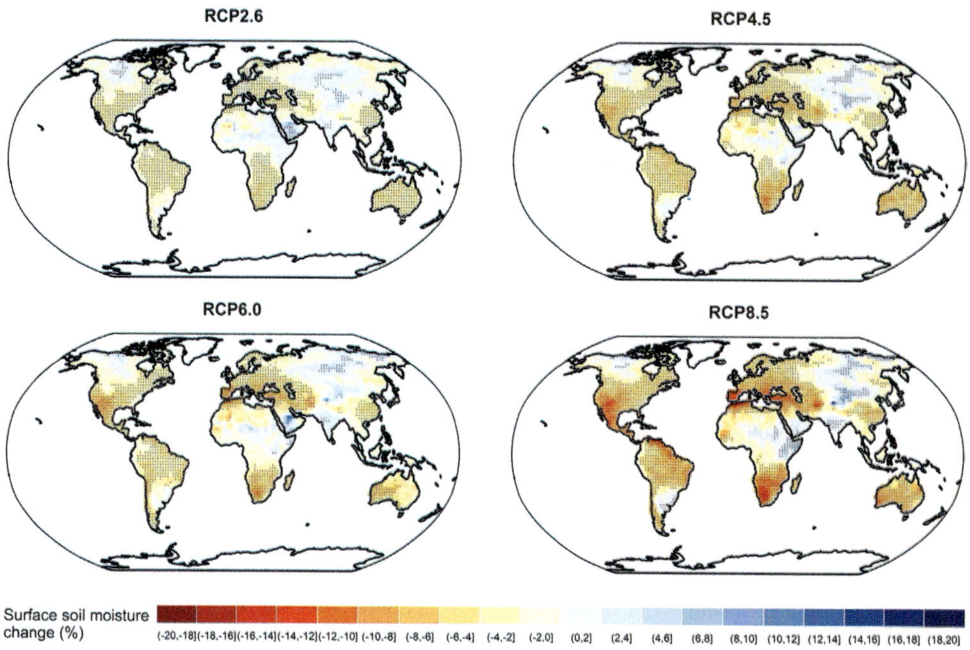

Figure 2.7. Global multi-model mean percentage change in annual mean surface soil moisture from 2071–2100 (RCP forcing) compared with 1976–2005 (historical forcing) based on CMIP5 multi-model ensembles under four scenarios: RCP2.6, RCP4.5, RCP6.0, and RCP8.5. Stippling indicates statistical significance ($p < 0.05$). Figure from *Lu et al.* (2019; CC BY 4.0).

et al., 2020), indicating there is still significant work to be done in this area, particularly with respect to precipitation predictability (e.g. Ukkola *et al.*, 2020).

2.2.6 CMIP6 Regional Studies

Following the release of the CMIP6 suite of models from 2019, numerous studies have emerged that utilize CMIP6 model outputs to assess the risk of climate change on agricultural yield. For example, Arunrat *et al.* (2022) analysed output from 27 CMIP models under SSP2-4.5 and SSP5-8.5 scenarios to evaluate expected changes in agricultural yields in Phichit province, a major agricultural region in central northern Thailand. Their findings indicate that projected increases in rainfall variability, along with increased frequencies and intensities of both drought and floods, coupled with rising temperatures leading to increased evapotranspiration are likely to adversely affect agricultural yields in the region (Arunrat *et al.*, 2022). They noted that the yield from a second rice crop was substantially lower under the SSP5-8.5 scenario compared with the lower (SSP2-4.5) emissions scenario. They recommended that, regardless of SSP scenario, transitioning from multiple annual rice crops to a single rice crop with maize, mung bean or soybean crops during the dry season could mitigate some negative climate impacts, as these alternatives have a smaller water footprint compared with rice.

It is important to note that, while many areas are projected to have decreased crop yields in the future, e.g. Phichit province, Thailand (Arunrat *et al.*, 2022) and the Mediterranean region (Palmer *et al.*, 2021; Mirgol *et al.*, 2024), there are regions where yields for certain crops may increase in response to climate change, leading to region-specific recommendations for the adoption of alternative crops. For example, Karan *et al.* (2022) found that crops such as soybean and cotton are projected to be increasingly vulnerable to climate change in the Pravara

River Basin in India, whereas sorghum and wheat crops would be more resilient to predicted changes in soil moisture regimes in the region. Conversely, another study, focusing on wheat yield in Pakistan, found that projected increasing temperatures and decreasing rainfall during the wheat growing season are likely to result in decreasing wheat yield throughout the 21st century as a result of shortened growing seasons and increased heat stress (Ishaque *et al.*, 2023). The differing responses of different crops in different regions highlight the necessity of region-specific studies and the benefit of utilizing Regional Climate Models (RCMs) that downscale CMIP6 GCM data, allowing for more targeted projections tailored to specific regions.

Another example of a region where adverse climate change impacts are expected is the Mediterranean, where CMIP6 studies have found that increases in temperatures, along with increasing drought intensities and frequencies of dry, hot days are expected to adversely impact crops (Palmer *et al.*, 2021; Mirgol *et al.*, 2024). Mirgol *et al.* (2024) do note, however, that CMIP6 projections indicate that fewer frosts and cold spells in winter can be expected in the southern Mediterranean region, and this may offer some mitigation potential by extending growing seasons in places where frost currently reduces yields due to increased sensitivity and reduced capacity for cold adaptation of plants during their reproductive phase. Similarly, CMIP6 model analysis indicates that winter wheat production in the Huang-Huai-Hai Plain in China, which includes the Yellow River region, can be expected to increase throughout the 21st century as a result of a lessening of the primary constraint of low winter temperatures across much of the region (Xu *et al.*, 2025).

2.2.7 Water Availability and Irrigation

Finally, with increasing pressure on water resources observed in many parts of the world today and given the critical role of water availability in effective crop management, it is essential to understand how varying climate scenarios influence the accessibility of irrigation water and, consequently, agricultural productivity. One significant study by Zhao and Wang (2020) examined future trends of water resources and their implications for agriculture on the Huaibei (Yellow River) Plain in China, utilizing CMIP5 models to analyse the effects of climate change on agricultural productivity and water availability. The study highlighted that precipitation changes varied markedly between RCPs, with an increase under RCP2.6 and RCP4.5 but a decrease under RCP8.5 in the latter part of the century. This indicates that, while they found that there is potential for increased agricultural productivity around the Huaibei Plain region, it is dependent on which scenario unfolds. In particular, under the RCP8.5 scenario, a marked decline in water resources is expected, which would adversely affect crop yields and agricultural sustainability in the region (Zhao and Wang, 2020). This aligns with other studies from the same region such as X. Li *et al.* (2022), who assessed the need for irrigation in the same region under CMIP5 and CMIP6 future climate scenarios. They also found that future precipitation might be expected to increase, with a corresponding decreasing need for irrigation in the region (Figure 2.8). Other studies focus on the projected change in droughts and their effect on agriculture. For example, another CMIP5 study from the same region examined the effects of drought in the Huaibei Plain of China and differences between the various RCP scenarios using a single CMIP5 model. They found that climate change has exacerbated drought conditions, thereby threatening agricultural outputs in this critical farming area (Badji *et al.*, 2023).

Some studies focusing on the southern hemisphere have taken a different approach, by examining changes in barley suitability across South America with and without irrigation, due to thermal excess/deficiency, and due to soil clay content/topography (de Oliveira Aparecido *et al.*, 2024). They noted barley's sensitivity to extreme temperature fluctuations which can significantly reduce crop yields, with 18.62% of the region already unsuitable for barley. By analysing future projections using the four CMIP6 SSP scenarios from one CMIP6 model, they concluded that there would be a reduction in suitable areas around South America under future climate scenarios, with the area suitable for barley

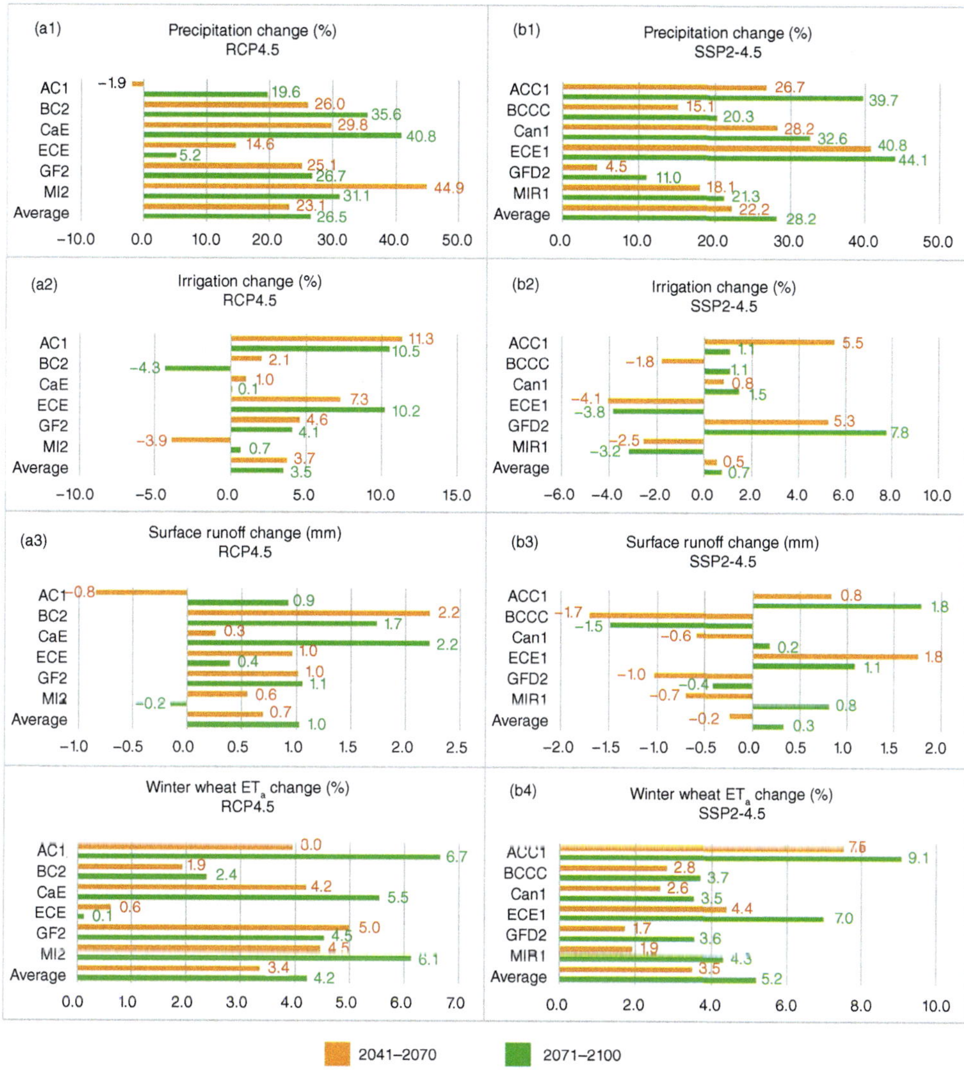

Figure 2.8. Changes in (1) annual precipitation, (2) irrigation, (3) surface runoff, and (4) average actual evapotranspiration (ETa) during the winter wheat growing period simulated by six GCMs under the RCP4.5 (a) and SSP2-4.5 (b) scenarios for the period of 2041–2070 and 2071–2100 compared with the historical period (1971–2000). Figure from X. *Li et al.* (2022). Image used with permission.

decreasing to only 1.48% under the very high emission (SSP5-8.5) scenario. This is illustrated in Figure 2.9, showing the difference in land area that is suitable for barley crops at the moment (the current map) compared with that which is expected to be suitable by the period 2081–2100 under a very high emission (SSP5-8.5) scenario (the SSP5-8.5 2081–2100 map). A significant change is evident, particularly in countries that are currently suitable for barley crops, e.g. Brazil, Paraguay, Uruguay, and Argentina, with much of the currently suitable or restricted areas becoming restricted or unsuitable due to thermal excess (high temperatures). While a small increase in the amount of land expected to be suitable with irrigation (1.04%), it is notable

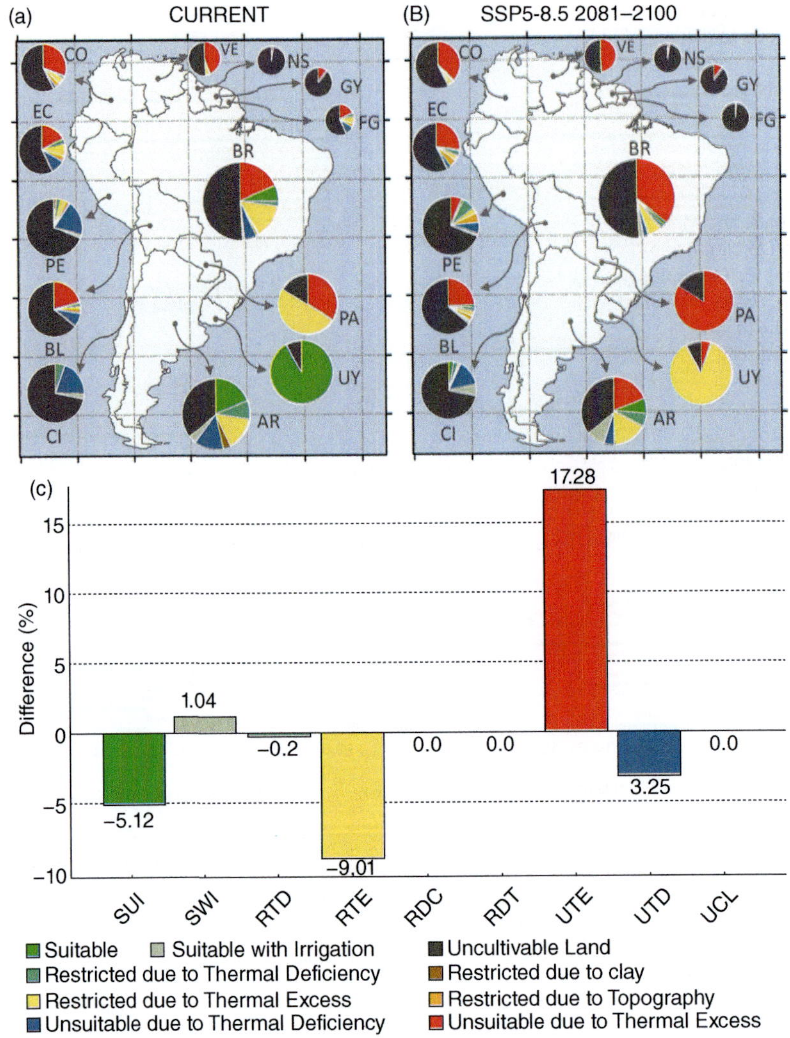

Figure 2.9. Variation of climate classes for barley (*Hordeum vulgare*) in the current scenario and the more drastic climate change scenario (SSP5-8.5) for the period 2081–2100. Figure from *de Oliveira Aparecido et al.* (2024). Image used with permission.

that the amount of suitable land without irrigation is expected to decrease by 5.12% under the SSP5-8.5 scenario indicating that an expansion in the amount of unsuitable land can be expected to co-occur with an increasing need for irrigation.

References

Andrys, J., Kala, J. and Lyons, T.J. (2017) Regional climate projections of mean and extreme climate for the southwest of Western Australia (1970–1999 compared to 2030–2059). *Climate Dynamics* 48(5–6), 1723–1747. DOI: 10.1007/s00382-016-3169-5.

Arunrat, N., Sereenonchai, S., Chaowiwat, W. and Wang, C. (2022) Climate change impact on major crop yield and water footprint under CMIP6 climate projections in repeated drought and flood areas in Thailand. *Science of the Total Environment* 807, 150741. DOI: 10.1016/j.scitotenv.2021.150741.

Badji, O., Zhu, Y., Lü, H., Guédé, K.G., Chen, T. *et al.* (2023) The effects of drought in the Huaibei plain of China due to climate change. *Atmosphere* 14(5), 860. DOI: 10.3390/atmos14050860.

Bezner Kerr, R., Hasegawa, T., Lasco, R., Bhatt, I., Deryng, D. *et al.* (2022) Food, fibre and other ecosystem products. In: Pörtner, H.-O., Roberts, D.C., Tignor, M., Poloczanska, E.S., Mintenbeck, K., Alegría, A., Craig, M., Langsdorf, S., Löschke, S., Möller, V., Okem, A. and Rama, B. (eds) *Climate Change 2022 – Impacts, Adaptation and Vulnerability. Contribution of Working Group II to the Sixth Assessment Report of the Intergovernmental Panel on Climate Change*. Cambridge University Press, Cambridge, UK and New York, NY, USA, pp. 713–906. DOI: 10.1017/9781009325844.007.

Bukovsky, M.S. and Mearns, L.O. (2020) Regional climate change projections from NA-CORDEX and their relation to climate sensitivity. *Climatic Change* 162(2), 645–665. DOI: 10.1007/s10584-020-02835-x.

Christensen, P., Gillingham, K. and Nordhaus, W. (2018) Uncertainty in forecasts of long-run economic growth. *Proceedings of the National Academy of Sciences of the United States of America* 115(21), 5409–5414. DOI: 10.1073/pnas.1713628115.

Cook, B.I., Mankin, J.S., Marvel, K., Williams, A.P., Smerdon, J.E. and Anchukaitis, K.J. (2020) Twenty-first century drought projections in the CMIP6 forcing scenarios. *Earth's. Future* 8(6), e2019EF001461. DOI: 10.1029/2019EF001461.

de Oliveira Aparecido, L.E., Torsoni, G.B., Lorençone, P.A., Lorençone, J.A. and Souza Rolim, G. de (2024) Barley vulnerability to climate change: Perspectives for cultivation in South America. *International Journal of Biometeorology* 69, 123–136. DOI: 10.1007/s00484-024-02798-z.

de Oliveira Aparecido, L.E., Torsoni, G.B., Costa Baratti, A.C., Melo, F. de, Rossi, M. *et al.* (2025) Agroecological vulnerability and climate resilience of *Jatropha curcas* under anthropogenic climate extremes in South America. *Biomass and Bioenergy* 194, 107654. DOI: 10.1016/j.biombioe.2025.107654.

Di Virgilio, G., Evans, J.P., Ji, F., Tam, E., Kala, J. *et al.* (2025) Design, evaluation, and future projections of the NARCliM2.0 CORDEX-CMIP6 Australasia regional climate ensemble. *Geoscientific Model Development* 18(3), 671–702. DOI: 10.5194/gmd-18-671-2025.

Elsadek, E.A., Zhang, K., Hamoud, Y.A., Mousa, A., Awad, A. *et al.* (2024) Impacts of climate change on rice yields in the Nile River Delta of Egypt: A large-scale projection analysis based on CMIP6. *Agricultural Water Management* 292, 108673. DOI: 10.1016/j.agwat.2024.108673.

Evans, J.P., Virgilio, G. Di, Hirsch, A.L., Hoffmann, P., Remedio, A.R. *et al.* (2021) The CORDEX-Australasia ensemble: Evaluation and future projections. *Climate Dynamics* 57(5–6), 1385–1401. DOI: 10.1007/s00382-020-05459-0.

Guo, Y., Yu, X., Xu, Y.P., Wang, G., Xie, J. and Gu, H. (2022) A comparative assessment of CMIP5 and CMIP6 in hydrological responses of the Yellow River Basin China. *Hydrology Research* 53(6), 867–891. DOI: 10.2166/nh.2022.001.

Hasegawa, T., Wakatsuki, H., Ju, H., Vyas, S., Nelson, G.C. *et al.* (2022) A global dataset for the projected impacts of climate change on four major crops. *Scientific Data* 9(1), 58. DOI: 10.1038/s41597-022-01150-7.

Hausfather, Z. (2018) Explainer: How 'Shared Socioeconomic Pathways' explore future climate change. *Carbon Brief.* Available at: https://www.carbonbrief.org/explainer-how-shared-socioeconomic-pathways-explore-future-climate-change/ (accessed 30 January 2025).

IPCC (2000) *Emissions Scenarios. Summary for Policymakers: A Special Report of IPCC Working Group III.* Intergovernmental Panel on Climate Change. Cambridge, UK and New York, NY, USA. Available at: https://www.ipcc.ch/site/assets/uploads/2018/03/sres-en.pdf (accessed 10 February 2025).

IPCC (2014). In: Core Writing Team, Pachauri, R.K. and Meyer, L.A. (eds) *Climate Change 2014: Synthesis Report. Contribution of Working Groups I, II and III to the Fifth Assessment Report of the Intergovernmental Panel on Climate Change*. Intergovernmental Panel on Climate Change. Geneva, Switzerland. Available at: https://www.ipcc.ch/report/ar5/syr/ (accessed 26 February 2025).

IPCC (2019) Summary for policymakers. In: Pörtner, H.-O., Roberts, D.C., Masson-Delmotte, V., Zhai, P., Tignor, M., Poloczanska, E., Mintenbeck, K., Alegría, A., Nicolai, M., Okem, A., Petzold, J., Rama, B. and Weye, N.M. (eds) *Special Report on the Ocean and Cryosphere in a Changing Climate*. Hamish Pritchard, Cambridge, UK and New York, NY, USA.

Ishaque, W., Osman, R., Hafiza, B.S., Malghani, S., Zhao, B., Xu, M. and Ata-Ul-Karim, S.T. (2023) Quantifying the impacts of climate change on wheat phenology, yield, and evapotranspiration under irrigated and rainfed conditions. *Agricultural Water Management* 275, 108017. DOI: 10.1016/j.agwat.2022.108017.

Jacob, D., Teichmann, C., Sobolowski, S., Katragkou, E., Anders, I. *et al.* (2020) Regional climate downscaling over Europe: Perspectives from the EURO-CORDEX community. *Regional Environmental Change* 20(2), 51. DOI: 10.1007/s10113-020-01606-9.

Jägermeyr, J., Müller, C., Ruane, A.C., Elliott, J., Balkovic, J. *et al.* (2021) Climate impacts on global agriculture emerge earlier in new generation of climate and crop models. *Nature Food* 2(11), 873–885. DOI: 10.1038/s43016-021-00400-y.

Karan, K., Singh, D., Singh, P.K., Bharati, B., Singh, T.P. and Berndtsson, R. (2022) Implications of future climate change on crop and irrigation water requirements in a semi-arid river basin using CMIP6 GCMs. *Journal of Arid Land* 14(11), 1234–1257. DOI: 10.1007/s40333-022-0081-1.

Kimball, B.A. (2016) Crop responses to elevated CO_2 and interactions with H_2O, N, and temperature. *Current Opinion in Plant Biology* 31, 36–43. DOI: 10.1016/j.pbi.2016.03.006.

Li, X., Tan, L., Li, Y., Qi, J., Feng, P. *et al.* (2022) Effects of global climate change on the hydrological cycle and crop growth under heavily irrigated management – A comparison between CMIP5 and CMIP6. *Computers and Electronics in Agriculture* 202, 107408. DOI: 10.1016/j.compag.2022.107408.

Li, X., Tan, J., Wang, X., Han, G., Qian, Z. *et al.* (2024) Responses of spring wheat yield and growth period to different future climate change models in the Yellow River irrigation area based on CMIP6 and WOFOST models. *Agricultural and Forest Meteorology* 353, 110071. DOI: 10.1016/j.agrformet.2024.110071.

Li, C., Camac, J., Robinson, A. and Kompas, T. (2025) Predicting changes in agricultural yields under climate change scenarios and their implications for global food security. *Scientific Reports* 15, 2858. DOI: 10.1038/s41598-025-87047-y.

Lu, J., Carbone, G.J. and Grego, J.M. (2019) Uncertainty and hotspots in 21st century projections of agricultural drought from CMIP5 models. *Scientific Reports* 9(1). DOI: 10.1038/s41598-019-41196-z.

Mihailović, D.T., Lalić, B., Drešković, N., Mimić, G., Djurdjević, V. and Jančić, M. (2015) Climate change effects on crop yields in Serbia and related shifts of Köppen climate zones under the SRES-A1B and SRES-A2. *International Journal of Climatology* 35(11), 3320–3334. DOI: 10.1002/joc.4209.

Mills, G., Sharps, K., Simpson, D., Pleijel, H., Frei, M. *et al.* (2018) Closing the global ozone yield gap: Quantification and cobenefits for multistress tolerance. *Global Change Biology* 24(10), 4869–4893. DOI: 10.1111/gcb.14381.

Mirgol, B., Dieppois, B., Northey, J., Eden, J., Jarlan, L. *et al.* (2024) Future changes in agrometeorological extremes in the southern Mediterranean region: When and where will they affect croplands and wheatlands? *Agricultural and Forest Meteorology* 358, 110232. DOI: 10.1016/j.agrformet.2024.110232.

Müller, C., Franke, J., Jägermeyr, J., Ruane, A.C., Elliott, J. *et al.* (2021) Exploring uncertainties in global crop yield projections in a large ensemble of crop models and CMIP5 and CMIP6 climate scenarios. *Environmental Research Letters* 16(3), 034040. DOI: 10.1088/1748-9326/abd8fc.

Olesen, J.E., Carter, T.R., Díaz-Ambrona, C.H., Fronzek, S., Heidmann, T. *et al.* (2007) Uncertainties in projected impacts of climate change on European agriculture and terrestrial ecosystems based on scenarios from regional climate models. *Climatic Change* 81(Suppl. 1), 123–143. DOI: 10.1007/s10584-006-9216-1.

Palmer, T.E., Booth, B.B.B. and McSweeney, C.F. (2021) How does the CMIP6 ensemble change the picture for European climate projections? *Environmental Research Letters* 16(9), 094042. DOI: 10.1088/1748-9326/ac1ed9.

Ramirez-Villegas, J., Challinor, A.J., Thornton, P.K. and Jarvis, A. (2013) Implications of regional improvement in global climate models for agricultural impact research. *Environmental Research Letters* 8(2), 024018. DOI: 10.1088/1748-9326/8/2/024018.

Riahi, K., Vuuren, D.P. van, Kriegler, E., Edmonds, J., O'Neill, B.C. *et al.* (2017) The Shared Socioeconomic Pathways and their energy, land use, and greenhouse gas emissions implications: An overview. *Global Environmental Change* 42, 153–168. DOI: 10.1016/j.gloenvcha.2016.05.009.

Ritchie, J. and Dowlatabadi, H. (2017) The 1000 GtC coal question: Are cases of vastly expanded future coal combustion still plausible? *Energy Economics* 65, 16–31. DOI: 10.1016/j.eneco.2017.04.015.

Sanderson, B.M., Booth, B.B.B., Dunne, J., Eyring, V., Fisher, R.A. *et al.* (2024) The need for carbon-emissions-driven climate projections in CMIP7. *Geoscientific Model Development* 17(22), 8141–8172. DOI: 10.5194/gmd-17-8141-2024.

Skea, J., Shukla, P., Khourdajie, A. Al and McCollum, D. (2021) Intergovernmental Panel on Climate Change: Transparency and integrated assessment modeling. *Wiley Interdisciplinary Reviews: Climate Change* 12(5), e727. DOI: 10.1002/wcc.727.

Ukkola, A.M., Kauwe, M.G. De, Roderick, M.L., Abramowitz, G. and Pitman, A.J. (2020) Robust future changes in meteorological drought in CMIP6 projections despite uncertainty in precipitation. *Geophysical Research Letters* 47(11), 4922. DOI: 10.1029/2020GL087820.

Wang, G., Zhang, J., Jin, J., Weinberg, J., Bao, Z. *et al.* (2017) Impacts of climate change on water resources in the Yellow River basin and identification of global adaptation strategies. *Mitigation and Adaptation Strategies for Global Change* 22(1), 67–83. DOI: 10.1007/s11027-015-9664-x.

Wilcke, R.A.I. and Bärring, L. (2016) Selecting regional climate scenarios for impact modelling studies. *Environmental Modelling and Software* 78, 191–201. DOI: 10.1016/j.envsoft.2016.01.002.

Xin, J., Xin, Y., Wu, H. and Zhou, S. (2025) Predicting the spatiotemporal evolution characteristics of future agricultural water demand in the Yellow River Basin under climate-change conditions. *Water (Switzerland)* 17(1), 31. DOI: 10.3390/w17010031.

Xu, Y., Li, T., Xu, M., Tan, L. and Shen, S. (2025) Assessing climate change effects on winter wheat production in the 3H plain: Insights from bias-corrected CMIP6 projections. *Agriculture (Switzerland)* 15(3), 2858. DOI: 10.3390/agriculture14030469.

Zhao, J. and Wang, Z. (2020) Future trends of water resources and influences on agriculture in China. *PLoS One* 15(4), e0231671. DOI: 10.1371/journal.pone.0231671.

Zittis, G., Hadjinicolaou, P., Almazroui, M., Bucchignani, E., Driouech, F. *et al.* (2021) Business-as-usual will lead to super and ultra-extreme heatwaves in the Middle East and North Africa. *npj Climate and Atmospheric Science* 4(1), 20. DOI: 10.1038/s41612-021-00178-7.

3 Impacts of Climate Change on Crop Production

Nicholas A. George[1], Caitlin E. Moore[2,3] and Sarita Jane Bennett[1,*]
[1]*The School of Molecular and Life Sciences, Curtin University;* [2]*The School of Agriculture and Environment, The University of Western Australia;* [3]*Centre for Water and Spatial Science, The University of Western Australia*

Abstract
This chapter synthesizes evidence on how anthropogenic climate change is reshaping global crop production. Meeting projected food demand by 2050 is estimated to require a doubling of output, even as climate pressures intensify. Weather, especially temperature and rainfall, already accounts for up to half of yield variability. Warming in major breadbaskets already exceeds historical inter-annual ranges. Water stress is increasing as droughts become more frequent, intense, and abrupt in onset, while rising atmospheric vapor-pressure deficit (VPD) further suppresses stomatal conductance and photosynthesis. Shifts toward more extreme rainfall and temperature events compound risks through flooding, hail, frost loss and heat damage at critical stages. The overall trajectory is declining productivity and greater volatility, with significant implications for food security. Building resilience will demand coordinated adaptation, including stress-tolerant germplasm and diversified systems, improved water management and soil health, climate-smart agronomy and early-warning systems, and broader, transformative changes across food systems.

3.1 Introduction: The Growing Threat of Climate Change to Global Crop Production

As reviewed in Chapter 1, anthropogenic climate change, caused chiefly by greenhouse gas emissions and land use change, has resulted in a rise in mean global temperatures, changing rainfall patterns, and more frequent and extreme weather events. Meteorological variables, notably temperature and rainfall, have been shown to explain up to half of the yield variance in crop production worldwide (Masson-Delmotte *et al.*, 2019; Vogel *et al.*, 2019; Moore *et al.*, 2021; Ortiz-Bobea *et al.*, 2021; Bernacchi *et al.*, 2025). These studies demonstrate how agricultural production and food security are already strongly influenced by climate, and change climates can be expect to strongly influence agriculture and food security in the future. Adding to this challenge, it has been estimated that global crop production needs to double by 2050 to meet projected demand for food from population increases, changing dietary patterns and increased biofuel demand (Long and Ort, 2010; Ray *et al.*, 2013). However, growth in the productivity of major grain crops is already declining and is below the 1% required to meet future

*Corresponding author. Email: Sarita.Bennett@curtin.edu.au

 Genomic Solutions for Breeding Climate-resilient Crops (R.K. Varshney, R. Barmukh and R.J. Harper)
DOI: 10.1079/9781800627185.0003

population growth (Hemathilake and Gunathilake, 2022). Due to factors such as extreme weather and socio-political events, the number of people facing chronic hunger rose to 735 million in 2022, pushing the aim of achieving zero hunger by 2030 even further out of reach (FAO IFAD UNICEF WFP and WHO, 2023). Without mitigation, climate change is expected to further decrease agricultural production by exacerbating many of the abiotic and biotic stresses experienced by crops, as well as challenge farm management due to changes in growing season length and timing, and increased incidences of extreme weather events that impede regular management practices (Lobell, 2014; Anderson *et al.*, 2020). The growing impacts of climate change on global food security poses a grave threat to human wellbeing (Ripple *et al.*, 2021; Ripple *et al.*, 2023). This chapter aims to review and discuss the current and projected impacts of climate change on global food systems and introduce potential ways for agricultural systems to become more climate resilient, which will be discussed in more detail in later chapters. A concept diagram summarizing the drivers of climate change, the agroecosystem response and the subsequent potential impacts on global crop production is shown in Figure 3.1.

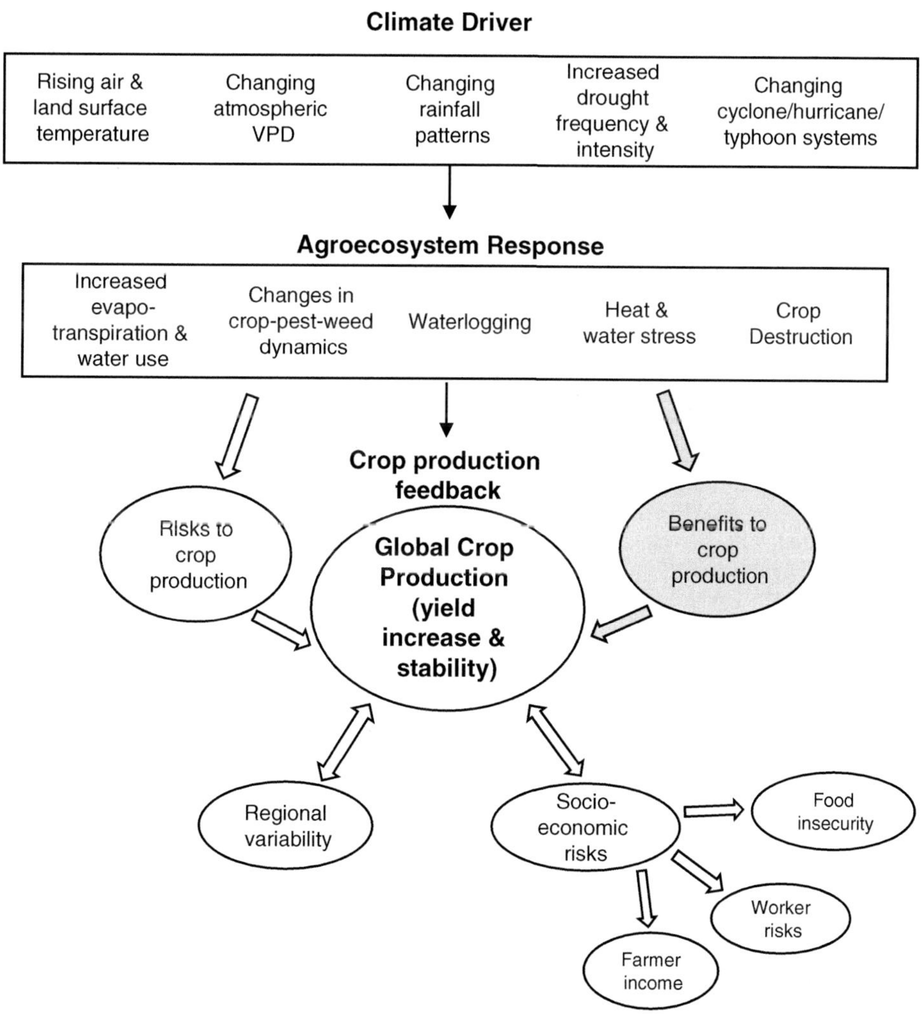

Figure 3.1. Concept diagram of the multiple drivers of climate change, the agroecosystem response and subsequent potential impacts on global crop production. The grey shaded area highlights that there are some potential benefits, but that the majority of impacts (white areas) are all negatives. VPD: vapor-pressure deficit (Author's own image).

3.2 The Main Climate-Related Impacts on Crop Production

3.2.1 Rising Temperatures and Extreme Temperature Events

Temperature is a key factor influencing crop adaptation and growth, affecting plant processes from the enzyme to ecosystem scale (Moore *et al.*, 2021). Temperature plays a particularly critical role in regulating the enzyme-driven photosynthetic reactions in plants, with most reactions having a thermal optimum that results in a decline in productivity once exceeded (Bernacchi *et al.*, 2001; Moore *et al.*, 2021). All crops experience reduced photosynthesis when temperatures exceed critical thresholds, with irreversible damage typically above 40–45°C (Bernacchi *et al.*, 2025). This decline in productivity subsequently reduces biomass accumulation, accelerates development rates, and limits the potential for radiation interception by plant canopies (Pratley, 2003; Stern *et al.*, 2006; Fatima *et al.*, 2020).

Observed land surface warming in major global cropping regions is already exceeding long-term, inter-annual variability by more than two standard deviations in some cases (Herold *et al.*, 2018; Ebi *et al.*, 2021b; Arrighi *et al.*, 2024; Jeong *et al.*, 2025; Lobell and Di Tommaso, 2025). Long-term observations indicate that these warming-induced trends and their impacts on crop growth are often unprecedented compared with recorded observations over the past several centuries (Aono and Saito, 2010; Aono, 2015; Labbé *et al.*, 2019; EMG, 2025; Jeong *et al.*, 2025). For example, Labbé *et al.* (2019) showed that grape harvests from the vine region of Beaune in Burgundy, France, have shifted on average 13 days earlier the last 31 years, in association with record temperatures and drought conditions linked to the Trans-Eurasian heatwave-drought.

In addition to rising global temperatures, the frequency and intensity of extreme heat events have increased globally, while cold extremes have decreased (Easterling *et al.*, 2000; Di Luca *et al.*, 2020; IPCC, 2021; Clarke *et al.*, 2022). Although in some regions, the occurrence of frost events is increasing due to decreased cloud cover during colder winter months, and subsequently greater radiative cooling (Herold *et al.*, 2018). Gradual exposure to high temperatures can trigger protective plant responses, whilst sudden exposure generally does not (Bernacchi *et al.*, 2025). Higher temperatures and extreme heat events have been linked to declines in crop productivity globally, or have limited the gains in yield that would otherwise have resulted from genetic improvement and enhanced management (Lobell *et al.*, 2011; Lesk *et al.*, 2016; Hochman *et al.*, 2017; Zhao *et al.*, 2017; Morgounov *et al.*, 2018; Anderson *et al.*, 2020; Ortiz-Bobea *et al.*, 2021; Habib-ur-Rahman *et al.*, 2022; Lobell and Di Tommaso, 2025).

It is projected that increasing occurrence, intensity, and duration of extreme heat events, particularly during sensitive phenological stages, will negatively affect global crop production (Motha, 2011; Herold *et al.*, 2018; Arnell *et al.*, 2019; Zhang *et al.*, 2024; Martinez-Villalobos *et al.*, 2025). In the future, large areas of agricultural land are likely to become susceptible to at least one type of climate extreme and to experience more rapid temperature fluctuations (Biess *et al.*, 2024; Wu *et al.*, 2025). In the absence of further genetic improvement or adaptive management strategies, continued warming and the increasing frequency of extreme heat events will reduce attainable yields of major grain crops across a significant fraction of current cropland (Stokes and Howden, 2010; Pugh *et al.*, 2016; Zhao *et al.*, 2017; Rezaei *et al.*, 2023). It is important to note that the most significant negative impact of rising temperatures may be in the middle latitudes, rather than in the hottest regions of the world (Hultgren *et al.*, 2025). This is because the hottest regions are, within limits, already adapted to warmer temperatures, whilst the coldest agricultural regions are likely to benefit from increased temperatures (Hultgren *et al.*, 2025). The suitability of certain regions for cultivating crops with chilling or vernalization requirements may be negatively impacted (Stokes and Howden, 2010; Pathak *et al.*, 2018; Marklein *et al.*, 2020).

3.2.2 Decreased Rainfall and Changing Rainfall Patterns

Water stress in plants is a key factor that reduces photosynthesis and, therefore, productivity. A well-watered crop can maintain photosynthesis under elevated temperatures for longer than crops grown under water stress (Fitzgerald *et al.*, 2016). There is a strong correlation between yield and rainfall, most notably in dryland farming systems with little or no irrigation (Rovira, 1992; Stephens and Lyons, 1998; Hay and Porter, 2006; Samarah, 2016; Barrett-Lennard *et al.*, 2024). Two-thirds of the Earth's land surface experienced drier conditions in the three decades before 2020 compared with the preceding 30-year period, and approximately 8% of global land has shifted from less arid to more arid conditions (Vicente-Serrano *et al.*, 2024; Chen *et al.*, 2025). This has resulted in decreased productivity gains in some globally important grain production regions, with some areas becoming unprofitable to farm, and others becoming opportunistically cropped with fallow years required between cropping years to build up stored soil moisture (Sprigg *et al.*, 2014; Ortiz-Bobea *et al.*, 2021).

Drought events are projected to increase globally under future climate change scenarios (Herold *et al.*, 2018; Arnell *et al.*, 2019; Biess *et al.*, 2024), and the transitions between drought and wet conditions have already become more abrupt in many regions, heightening risks for agriculture (Qing *et al.*, 2023; Swain *et al.*, 2025). Greater intra-seasonal variability in rainfall and shifts in rainfall patterns can have a greater negative impact on the long-term viability of agricultural production than mean reductions in total annual rainfall alone (Fishman, 2016; Shortridge, 2019; Scanlon and Doncon, 2020). Future declines in rainfall and increased rainfall variability are therefore likely to further reduce crop productivity and increase risks for growers.

Declining rainfall and changing rainfall patterns will likely increase demand for irrigation, whilst also negatively impacting irrigation water supplies. It is predicted that the majority of irrigated regions will be negatively affected by future climate change (Döll, 2002). Even accounting for adaptation and CO_2 fertilization (which increase crop water use efficiency), areas currently reliant on irrigation, such as the western USA, China, and West, South, and Central Asia, will face challenges from increasing water scarcity and be forced to transition to rainfed agriculture, with concomitant loss of productivity (Elliott *et al.*, 2014; Fader *et al.*, 2016; Elnashar and Elyamany, 2023).

3.2.3 Co-occurrence of Heat and Drought Events

Biotic stresses rarely occur in isolation, and a changing climate is further increasing the risk of multiple stressors affecting crop production simultaneously. Drought and heat stress often co-occur, and when they do, their combined impact on crops is typically more severe than the effect of either stress alone (Anderson *et al.*, 2020). For example, in the Midwest USA, a drought in 2012 combined with a hotter than average summer growing season caused widespread yield declines in maize and soybean yields compared with other years (Lobell *et al.*, 2014; Moore *et al.*, 2022). Human-induced climate change has likely already increased the probability of such compounded extreme events, and this trend is expected to continue as global temperatures rise (IPCC, 2021). These interacting climate stresses pose a significant and growing threat to agricultural productivity.

3.2.4 Extreme Rainfall Events

A warmer atmosphere can hold more water, resulting in a climate feedback that increases the intensity of rainfall and cyclone events, which are becoming more frequent in many regions of the world (IPCC, 2021). More frequent and intense rainfall and cyclone events can lead to issues such as waterlogging (Salinger *et al.*, 2000; Arnell *et al.*, 2019; Martel *et al.*, 2021), inundation/flooding (Mullen and Barrett-Lennard, 2010); and land degradation processes such as soil erosion (Hermans and McLeman, 2021; Talukder *et al.*, 2021) and soil salinisation (Bennett and Barrett-Lennard, 2013). The impact of these under a changing climate can be positive and negative depending on the region, and anthropogenic influences on the landscape.

Waterlogging occurs when the crop root zone becomes inundated with water, leading to anaerobic and anoxic conditions, reducing crop respiration (Bennett *et al.*, 2009). When the crop root zone becomes waterlogged, it can trigger a cascade of physiological and biochemical conditions that are detrimental to crop growth (Colmer and Greenway, 2011; Tian *et al.*, 2021). Inundation occurs when water sits or travels over the soil surface, covering the shoots with water and thus affecting both the shoots and roots (Voesenek *et al.*, 2004). The severity and frequency of inundation is expected to be particularly severe for south-east Asia, for e.g., as monsoon rainfall increases in intensity (Douglas, 2009). The impacts of inundation are still poorly understood, but impact plant growth through a reduction in gaseous exchange, primarily oxygen, between plant leaves and the atmosphere (Colmer, 2003; Voesenek *et al.*, 2004). Soil salinisation is an escalating global threat and occurs due to saltwater intrusion of coastal agricultural lands (Su *et al.*, 2025), groundwater extraction with reduced freshwater inflow due to drought conditions (Su *et al.*, 2025) and anthropogenic land clearance leading to a rise in a saline watertable (Bennett *et al.*, 2009). The later is referred to as secondary salinity and has been estimated to occur over 76.6 million hectares globally, although this is thought to be an under-estimate (Mullen and Barrett-Lennard, 2010). Climate projections also show that hail events, which can be highly damaging to crops, are likely to increase in frequency and intensity (Raupach *et al.*, 2021). Increasing precipitation intensity due to climate change is therefore likely to result in negative consequences for crop production in some regions.

3.2.5 Increased Evapotranspiration and Water Use

Studies on the impact of a changing climate tend to focus on the impact of increasing temperatures and decreasing precipitation, but other climatic variables can also be agriculturally important. Atmospheric vapor pressure deficit is a critical factor influencing plant photosynthesis, and the negative correlation between agricultural production and high temperatures can often be explained by elevated vapor pressure deficits increasing evapotranspiration (Bernacchi and VanLoocke, 2015; Yuan *et al.*, 2019; Lobell and Di Tommaso, 2025).

As identified in the previous section, a warmer atmosphere can hold more water, meaning a stronger gradient of water vapor deficit can occur in the atmosphere to drive land surface water losses via evaporation and plant transpiration processes. Vapor pressure deficit is rising in many cropping regions outside the tropics, which has already resulted in reduced vegetative growth, biomass production, and crop productivity (Yuan *et al.*, 2019; Wanniarachchi and Sarukkalige, 2022; Lobell and Di Tommaso, 2025). DeLucia *et al.* (2019) showed that despite being projected to experience increased annual rainfall, the Midwestern USA is likely to reach a water limit due to increased atmospheric evaporative demand linked with rising temperatures. Even small changes in evapotranspiration can have significant consequences for plant growth in arid climates (Moratiel *et al.*, 2010). Farming systems in arid and semi-arid regions appear particularly vulnerable to increasing evapotranspiration, given that crops are already potentially water-stressed (Dimitriadou and Nikolakopoulos, 2021).

In addition, evapotranspiration represents a major consumptive use of both irrigation water and rainfall in agriculture, and irrigation demands are already rising due to rising temperatures (Wanniarachchi and Sarukkalige, 2022; Moyers *et al.*, 2024). Higher evapotranspiration under future climate conditions is likely to further contribute to reduced crop yields and increased crop water requirements. Increasing use of irrigation to meet crop water needs also heightens the risk of soil salinity (Talukder *et al.*, 2021).

3.2.6 Shifting Agricultural Production Environments

A changing climate is altering the geographic suitability of current agricultural regions, reshaping where different crops can be viably produced. In some cases, climate change is enabling the cultivation of species in previously unsuitable regions, however, it is also reducing suitability in areas where those crops were traditionally grown (Daly *et al.*, 2012; Hambly, 2022; Gasch *et al.*, 2023). For example, yields from temperate crops, as well as land areas suitable for production of these crops, have increased in northern Europe over the past several decades, while yields in other more southern regions of Europe have decreased (Moore and Lobell, 2015; Anderson *et al.*, 2020).

Many established farming regions are expected to experience temperatures that regularly exceed the physiological limits of key grain and horticultural crops, forcing changes in the current geospatial extent of major agricultural zones (King *et al.*, 2018; Anderson *et al.*, 2020; Marklein *et al.*, 2020; Huang *et al.*, 2022; Biess *et al.*, 2024). In some regions, there may be no viable alternative locations for relocating current farming industries, and rainfed agriculture or other adaptation measures may be insufficient to offset the impacts of the changing climate (Thomson *et al.*, 2014; Marklein *et al.*, 2020; AAS, 2021; Torbenson *et al.*, 2024). Sustaining global food production under these conditions will likely require large-scale shifts in land use (Pugh *et al.*, 2016), as growers are required to change both farming practice and the crops they traditionally grow. These changes are expected to produce widespread disruption to established farming systems and supply chains, with significant economic and social consequences.

3.2.7 Risks to Crop Wild Relatives and Sources of Genetic Diversity

Traditional crop varieties and crop wild relatives play a critical role in safeguarding global food security by serving as essential sources of genetic diversity for crop improvement. They offer valuable traits that can be used in breeding programs to enhance resistance to biotic and abiotic stresses (Thorman and Engles, 2015; Swarup *et al.*, 2021; Bohra *et al.*, 2022). However, both crop genetic diversity and wild relatives are increasingly threatened by a range of anthropogenic pressures, including habitat loss, land use change and climate change (Castañeda-Álvarez *et al.*, 2016; FAO, 2019). Protecting these genetic resources against the threat of climate change is therefore essential to ensure the resilience and adaptability of agricultural systems.

3.2.8 Changes in Crop–Pest Dynamics

The dynamics of agricultural pests – including vertebrates, invertebrates and microorganisms – are being reshaped by changing climates. The distribution of plant pests has already changed in response to shifting climatic conditions (Anderson *et al.*, 2020). Some crop pests are projected to benefit from future climate change, potentially exacerbating yield losses in vulnerable systems (Schneider *et al.*, 2022). For example, insect physiology is highly temperature-sensitive, with metabolic rates roughly doubling per 10°C increase, leading to faster consumption, development, and movement that can alter population dynamics by affecting reproduction, survival, life cycle duration, population size and distribution (Bale *et al.*, 2002; Dukes *et al.*, 2009). Changing rainfall patterns, high air humidity and increased soil moisture favor the development and spread of many microorganisms that cause crop diseases (Anderson *et al.*, 2020). Climate change may also favor shifts in species composition and facilitate the introduction of new species, with potentially significant ecological and agronomic consequences (Peters *et al.*, 2014). For example, fall armyworm (*Spodoptera frugiperda*) is a damaging insect pest on a range of important crops. It has invaded most of sub-Saharan Africa, Asia, and parts of Australia, and the global area for the species is expected to increase by nearly 50% under climate warming scenarios (Zacarias, 2020). These emerging pest dynamics will require proactive monitoring and adaptive management strategies to safeguard agricultural productivity.

3.2.9 Socio-economic Risks

Extreme heat events have significant negative impacts on human health and economies, and can be especially severe for agricultural workers and rural communities. This is further exacerbated in developing nations where working conditions can be sub-standard and potentially associated with poor air quality that is commonly associated with agricultural burning (Ebi *et al.*, 2021a; El Khayat *et al.*, 2022; Romanello *et al.*, 2023; Dewi *et al.*, 2024; Huang *et al.*, 2025). A wet-bulb temperature of 35°C has been proposed as the theoretical upper limit for human thermoregulation, although some recent research suggests the threshold may be as low as 26°C (Vecellio *et al.*, 2023; Wu *et al.*, 2025). Future climate change is likely to increase the exposure of populations to unprecedented extreme heat conditions with high wet-bulb temperatures, and may result in large areas of the globe routinely exceeding safe working limits, making outdoor activities required for agriculture increasingly difficult or dangerous at certain times of year (Herold *et al.*, 2018; Ebi *et al.*, 2021b; Ripple *et al.*, 2023; Grant *et al.*, 2025; Wu *et al.*, 2025). Other types of extreme weather events, such as flooding and storms, also disproportionately affect the health and wellbeing of farmers and their communities, as livelihoods are more directly impacted (Hughes *et al.*, 2016; Clarke *et al.*, 2022; Dewi *et al.*, 2024). The social impacts of extreme weather events are therefore likely to have negative consequences for viable food production.

Agricultural infrastructure, postharvest systems, and food supply chains are also at significant risk under a changing climate. Rising temperatures and an increase in extreme weather events threaten rural infrastructure and disrupt agricultural logistics, with lower-income nations particularly vulnerable (Davis *et al.*, 2021; Chirisa and Nel, 2022; Rojas-Reyes *et al.*, 2024; Tchonkouang *et al.*, 2024). For example, higher temperatures contribute to increased food loss and waste, in part by reducing the quality of stored grains and vegetables and by intensifying pest pressures in storage environments, which can lead to damage, contamination and increased postharvest costs (Moretti *et al.*, 2010; Moses *et al.*, 2015; Masson-Delmotte *et al.*, 2019; Gerken and Morrison III, 2022). These changes result in product losses and reduced income for producers. Climate change also threatens food safety at multiple points along the supply chain, increasing the risks of mycotoxin and biotoxin contamination, environmental residues, and food- and water-borne diseases (Misiou and Koutsoumanis, 2022). Collectively, such disruptions compromise the resilience, safety and economic viability of agricultural systems under a changing climate.

3.3 Are There Benefits to Agriculture from Climate Change?

Although climate change presents serious challenges to agriculture, it may also create new opportunities and benefits for food production systems under certain conditions. Elevated atmospheric CO_2 can enhance the efficiency of resource use by plants and, under non-stress conditions, increase yields in C_3 crops such as wheat and rice (Long *et al.*, 2004; Ainsworth and Long, 2021; Dimitriadou and Nikolakopoulos, 2021). However, field experiments have shown that elevated atmospheric CO_2 does not always result in increased productivity, especially when interacting with other stressors such as high temperatures and atmospheric vapor pressure deficit (Long and Ort, 2010; Fitzgerald *et al.*, 2016; Ruiz-Vera *et al.*, 2018; Anderson *et al.*, 2020; Wang *et al.*, 2020; Habib-ur-Rahman *et al.*, 2022; Rezaei *et al.*, 2023; Lobell and Di Tommaso, 2025). C_4 crops such as sorghum and maize are not expected to show yield gains due to elevated CO_2 unless water is limiting (Lobell and Burke, 2009). In some regions, a warmer climate may enable double cropping within a season or expand the geographic range of subtropical and tropical crops into areas currently considered marginal (Stokes and Howden, 2010). Higher temperatures and increased precipitation could also boost agricultural productivity and crop diversity in high-latitude regions, such as northern Europe, where yields and suitable production areas have already expanded in recent decades (Pugh *et al.*, 2016; Redhead *et al.*, 2025). Relocating croplands to more climatically optimal locations and allowing natural ecosystems to regenerate in abandoned areas may provide environmental co-benefits (Beyer *et al.*, 2022). Warmer temperatures may also benefit some perennial species in cooler climates, although annual crops and those in already hot regions may be negatively affected (Stokes and Howden, 2010). Overall, while some regions and crops may benefit, the impact of climate change on agricultural productivity remains highly heterogeneous and context-dependent (Zhao *et al.*, 2017; Lobell and Di Tommaso, 2025). Scenarios involving high levels of warming are projected to result in substantial declines in global agricultural output and pose significant risks to food security (IPCC, 2021).

3.4 Ways Forward

In the absence of the massive relocation of croplands to more climatically suitable areas, developing new crop species and varieties capable of withstanding a wide range of biotic and abiotic stresses will be central to climate adaptation. While elevated atmospheric CO_2 and longer growing seasons may increase yields in some areas and circumstances, climate change will intensify stressors such as drought, flooding, heat and pest pressure, requiring crops to maintain high productivity under more challenging and variable conditions (Anderson *et al.*, 2020).

Conventional molecular breeding approaches focusing on single abiotic stress-resistance genes have often fallen short in the field due to the complexity and diversity of real-world conditions (Parmar *et al.*, 2017; Prado *et al.*, 2025). Stress responses in plants involve multiple pathways and trade-offs, such as allocating resources toward survival rather than yield under drought stress (Hay and Porter, 2006; Lambers *et al.*, 2008; Samarah, 2016). As such, strategies for improving tolerance to future growing conditions must consider physiological traits (e.g. root architecture, aerenchyma formation), envirotyping, genotype-by-environment (G×E) interactions and genotype-by-management (G×M) interactions (Lobell, 2014; Kaur *et al.*, 2020; Xiong *et al.*, 2021). Future breeding may also need to target controlled-environment agriculture, orphan crops, *de novo* domestication, and the use of genomics and microbiome manipulation to expand adaptive potential (Kromdijk *et al.*, 2016; Cavanagh *et al.*, 2022; Benitez-Alfonso *et al.*, 2023; George *et al.*, 2023; Phillips *et al.*, 2025; Prado *et al.*, 2025).

Adapting agriculture to climate change will require a broad suite of holistic strategies, including soil and water management, crop rotation, and system diversification (IPCC, 2019). At the local and regional levels, farmers can adopt improved crop varieties, agronomic practices and water conservation techniques, supported by knowledge transfer, training and weather forecasting systems (Grigorieva *et al.*, 2023). In northern latitudes, new varieties may be needed to take advantage of extended growing seasons, while in lower latitudes, declining yields may require shifting to more heat-tolerant or alternative crops (Anderson *et al.*, 2020). New crops and improved controlled-environment systems offer further adaptation options (George *et al.*, 2023, Phillips *et al.*, 2025, Prado *et al.*, 2025). Critically, increasing consumer acceptance of new crop varieties and food sources must also play a role in the success of adaptation measures (Stokes and Howden, 2010). Climate adaptation in agriculture will therefore require integrated, context-specific and socially informed responses to ensure food system resilience under uncertain future conditions.

The scale and complexity of climate change impacts on agriculture demand systemic and transformational responses. Adaptation may involve not only technological solutions such as climate-resilient crops, precision agriculture and diversification, but also more disruptive changes to land use and land distribution (Rickards and Howden, 2012; AAS, 2021). These shifts may offer new opportunities, but they are also likely to be costly and disruptive to existing agricultural systems and rural communities. Failure to act may, however, be more expensive in the long term. The ability and desire of farmers to adapt is one component of climate change impact prediction and modeling that is rarely addressed (Hultgren *et al.*, 2025), with an assumption that farmers will always aim to optimize yield (Lobell and Asseng, 2017). However, farmers typically seek to optimize profit rather than yield, as this is more sustainable for the farm business (Tey and Brindal, 2015). Furthermore, the concept of *optimum sustainable yield* underscores that maximizing economic profit occurs at less than maximum biological yield (Balmford *et al.*, 2018). In Europe, for example, agricultural profits may increase slightly under climate change if farmers implement appropriate adaptation strategies, but profits could decline substantially without adaptation (Moore and Lobell, 2014).

Transformational adaptation will require a nuanced understanding of local and regional realities, proactive planning to mitigate maladaptation, and consideration of farmer livelihoods and long-term sustainability (Hellin *et al.*, 2023). Broader support at national and international levels – including financial schemes, insurance and agricultural extension services – will be vital in enabling effective and equitable responses (Grigorieva *et al.*, 2023). Addressing climate change in agriculture will necessitate a coordinated and interdisciplinary approach that moves beyond conventional research models and embraces the complexity of future challenges (Rickards and Howden, 2012). Utilizing crop genetics will continue to be an essential component of this multi-faceted approach, as the development of improved crop varieties provides a continual adaptation method that farmers already utilize, and can continue to use, that does not require the complexity and cost-benefit analysis of other adaptation strategies.

References

AAS (2021) *The Risks to Australia of a 3°C Warmer World*. Australian Academy of Science, p. 98. https://science.org.au/our-work/resources-reports/reports-publications/risks-australia-3degc-warmer-world.

Ainsworth, E.A. and Long, S.P. (2021) 30 years of free-air carbon dioxide enrichment (FACE): What have we learned about future crop productivity and its potential for adaptation? *Global Change Biology* 27(1), 27–49. DOI: 10.1111/gcb.15375.

Anderson, R., Bayer, P.E. and Edwards, D. (2020) Climate change and the need for agricultural adaptation. *Current Opinion in Plant Biology* 56, 197–202. DOI: 10.1016/j.pbi.2019.12.006.

Aono, Y. (2015) Cherry blossom phenological data since the seventeenth century for Edo (Tokyo), Japan, and their application to estimation of March temperatures. *International Journal of Biometeorology* 59, 427–434. DOI: 10.1007/s00484-014-0854-0.

Aono, Y. and Saito, S. (2010) Clarifying springtime temperature reconstructions of the medieval period by gap-filling the cherry blossom phenological data series at Kyoto, Japan. *International Journal of Biometeorology* 54, 211–219. DOI: 10.1007/s00484-009-0272-x.

Arnell, N.W., Lowe, J.A., Challinor, A.J. and Osborn, T.J. (2019) Global and regional impacts of climate change at different levels of global temperature increase. *Climatic Change* 155, 377–391. DOI: 10.1007/s10584-019-02464-z.

Arrighi, J., Otto, F.E.L., Marghidan, C.P., Philip, S., Singh, R.P. *et al.* (2024) *Climate Change and the Escalation of Global Extreme Heat: Assessing and Addressing the Risks*. Climate Central, Red Cross Red Crescent Climate Centre, World Weather Attribution, p. 14. https://www.climatecentral.org/report/climate-change-and-the-escalation-of-global-extreme-heat-2025.

Bale, J.S., Masters, G.J., Hodkinson, I.D., Awmack, C., Bezemer, T.M. *et al.* (2002) Herbivory in global climate change research: Direct effects of rising temperature on insect herbivores. *Global Change Biology* 8(1), 1–16. https://doi.org/10.1046/j.1365-2486.2002.00451.x.

Balmford, A., Amano, T., Bartlett, H., Chadwick, D., Collins, A. *et al.* (2018) The environmental costs and benefits of high-yield farming. *Nature Sustainability* 1(9), 477–485. PMID: 30450426; PMCID: PMC6237269.

Barrett-Lennard, E.G., George, N., D'Antuono, M., Holmes, K.W. and Ward, P.R. (2024) Rain and potential evapotranspiration are the main drivers of yield for wheat and barley in southern Australia: Insights from 12 years of National Variety Trials. *Crop and Pasture Science* 75(6), CP23320. https://doi.org/10.1071/CP23320.

Benitez-Alfonso, Y., Soanes, B.K., Zimba, S., Sinanaj, B., German, L. *et al.* (2023) Enhancing climate change resilience in agricultural crops. *Current Biology* 33(23), R1246–R1261. https://doi.org/10.1016/j.cub.2023.10.028.

Bennett, S.J., Barrett-Lennard, E. and Colmer, T. (2009) Salinity and waterlogging as constraints to saltland pasture production: A review. *Agriculture, Ecosystems & Environment* 129(4), 349–360. https://doi.org/10.1016/j.agee.2008.10.013.

Bernacchi, C.J. and VanLoocke, A. (2015) Terrestrial ecosystems in a changing environment: A dominant role for water. *Annual Review of Plant Biology* 66, 599–622. DOI: 10.1146/annurev-arplant-043014-114834.

Bernacchi, C., Singsaas, E., Pimentel, C., Portis Jr A. and Long, S.P. (2001). Improved temperature response functions for models of RUBISCO-limited photosynthesis. *Plant, Cell & Environment* 24(2), 253–259. https://doi.org/10.1111/j.1365-3040.2001.00668.x.

Bernacchi, C.J., Long, S.P. and Ort, D.R. (2025) Safeguarding crop photosynthesis in a rapidly warming world. *Science* 388(6752), 1153–1160. DOI: 10.1126/science.adv5413.

Beyer, R.M., Hua, F., Martin, P.A., Manica, A. and Rademacher, T. (2022) Relocating croplands could drastically reduce the environmental impacts of global food production. *Communications Earth & Environment* 3(1), 49. https://doi.org/10.1038/s43247-022-00360-6.

Biess, B., Gudmundsson, L., Windisch, M.G. and Seneviratne, S.I. (2024) Future changes in spatially compounding hot, wet or dry events and their implications for the world's breadbasket regions. *Environmental Research Letters* 19(6), 064011. DOI 10.1088/1748-9326/ad4619.

Bohra, A., Kilian, B., Sivasankar, S., Caccamo, M., Mba, C., McCouch, S.R. and Varshney, R.K. (2022) Reap the crop wild relatives for breeding future crops. *Trends in Biotechnology* 40(4), 412–431. DOI: 10.1016/j.tibtech.2021.08.009.

Castañeda-Álvarez, N.P., Khoury, C.K., Achicanoy, H.A., Bernau, V., Dempewolf, H. *et al.* (2016) Global conservation priorities for crop wild relatives. *Nature Plants* 2(4), 1–6. DOI: 10.1038/nplants.2016.22.

Cavanagh, A.P., South, P.F., Bernacchi, C.J. and Ort, D.R. (2022) Alternative pathway to photorespiration protects growth and productivity at elevated temperatures in a model crop. *Plant Biotechnology Journal* 20(4), 711–721. https://doi.org/10.1111/pbi.13750.

Chen, L., Brun, P., Buri, P., Fatichi, S., Gessler, A. *et al.* (2025) Global increase in the occurrence and impact of multiyear droughts. *Science* 387(6731), 278–284. DOI: 10.1126/science.ado4245.

Chirisa, I. and Nel, V. (2022) Resilience and climate change in rural areas: A review of infrastructure policies across global regions. *Sustainable and Resilient Infrastructure* 7(5), 380–390. https://doi.org/10.1080/23789689.2020.1871538.

Clarke, B., Otto, F., Stuart-Smith, R. and Harrington, L. (2022) Extreme weather impacts of climate change: An attribution perspective. *Environmental Research: Climate* 1(1), 012001. DOI 10.1088/2752-5295/ac6e7d.

Colmer, T.D. (2003) Long-distance transport of gases in plants: a perspective on internal aeration and radial oxygen loss from roots. *Plant, Cell and Environment* 26, 17–36. https://doi.org/10.1046/j.1365-3040.2003.00846.x.

Colmer, T.D. and Greenway, H. (2011) Ion transport in seminal and adventitious roots of cereals during O_2 deficiency. *Journal of Experimental Botany* 62(1), 39–57. https://doi.org/10.1093/jxb/erq271.

Daly, C., Widrlechner, M.P., Halbleib, M.D., Smith, J.I. and Gibson, W.P. (2012) Development of a new USDA plant hardiness zone map for the United States. *Journal of Applied Meteorology and Climatology* 51(2), 242–264. DOI: 10.1175/2010JAMC2536.1.

Davis, K.F., Downs, S. and Gephart, J.A. (2021) Towards food supply chain resilience to environmental shocks. *Nature Food* 2(1), 54–65. https://doi.org/10.1038/s43016-020-00196-3.

DeLucia, E.H., Chen, S., Guan, K., Peng, B., Li, Y. *et al.* (2019) Are we approaching a water ceiling to maize yields in the United States? *Ecosphere* 10(6), e02773. https://doi.org/10.1002/ecs2.2773.

Dewi, S.P., Kasim, R., Sutarsa, I.N. and Dykgraaf, S.H. (2024) A scoping review of the impact of extreme weather events on health outcomes and healthcare utilization in rural and remote areas. *BMC Health Services Research* 24(1), 1–16. DOI: 10.1186/s12913-024-11695-5.

Di Luca, A., Elía, R. de, Bador, M. and Argüeso, D. (2020) Contribution of mean climate to hot temperature extremes for present and future climates. *Weather and Climate Extremes* 28, 100255. https://doi.org/10.1016/j.wace.2020.100255.

Dimitriadou, S. and Nikolakopoulos, K.G. (2021) Evapotranspiration trends and interactions in light of the anthropogenic footprint and the climate crisis: A review. *Hydrology* 8(4), 163. https://doi.org/10.3390/hydrology8040163.

Döll, P. (2002) Impact of climate change and variability on irrigation requirements: A global perspective. *Climatic Change* 54(3), 269–293. https://doi.org/10.1023/A:1016124032231.

Douglas, I. (2009) Climate change, flooding and food security in south Asia. *Food Security* 1, 127–136. https://doi.org/10.1007/s12571-009-0015-1.

Dukes, J.S., Pontius, J., Orwig, D., Garnas, J.R., Rodgers, V.L. *et al.* (2009) Responses of insect pests, pathogens, and invasive plant species to climate change in the forests of northeastern North America: What can we predict? *Canadian Journal of Forest Research* 39(2), 231–248. http://dx.doi.org/10.1139/X08-171.

Easterling, D.R., Meehl, G.A., Parmesan, C., Changnon, S.A., Karl, T.R. and Mearns, L.O. (2000) Climate extremes: Observations, modeling, and impacts. *Science* 289(5487), 2068–2074. DOI: 10.1126/science.289.5487.2068.

Ebi, K.L., Capon, A., Berry, P., Broderick, C., de Dear, R. *et al.* (2021a) Hot weather and heat extremes: Health risks. *The Lancet* 398(10301), 698–708. DOI: 10.1016/S0140-6736(21)01208-3.

Ebi, K.L., Vanos, J., Baldwin, J.W., Bell, J.E., Hondula, D.M. *et al.* (2021b) Extreme weather and climate change: Population health and health system implications. *Annual Review of Public Health* 42(1), 293–315. https://doi.org/10.1146/annurev-publhealth-012420-105026.

El Khayat, M., Halwani, D. A., Hneiny, L., Alameddine, I., Haidar, M. A., & Habib, R. R. (2022) Impacts of climate change and heat stress on farmworkers' health: a scoping review. *Frontiers in Public Health* 10, 78281. https://doi.org/10.3389/fpubh.2022.782811.

Elliott, J., Deryng, D., Müller, C., Frieler, K., Konzmann, M. *et al.* (2014) Constraints and potentials of future irrigation water availability on agricultural production under climate change. *Proceedings of the National Academy of Sciences USA* 111(9), 3239–3244. https://doi.org/10.1073/pnas.1222474110.

Elnashar, W. and Elyamany, A. (2023) Managing risks of climate change on irrigation water in arid regions. *Water Resources Management* 37(6), 2429–2446. https://doi.org/10.1007/s11269-022-03267-1.

EMG (2025) *Ecological Meterology Group, Osaka Prefecture University Graduate School of Life and Environmental Sciences Department of Landscape and Environmental Sciences*. Available at: http://atmenv.envi.osakafu-u.ac.jp/ (accessed March 2025).

Fader, M., Shi, S., Bloh, W. von, Bondeau, A. and Cramer, W. (2016) Mediterranean irrigation under climate change: More efficient irrigation needed to compensate for increases in irrigation water requirements. *Hydrology and Earth System Sciences* 20(2), 953–973. https://doi.org/10.5194/hess-20-953-2016.

FAO (2019) The state of the world's biodiversity for food and agriculture. In: Bélanger, J. and Pilling, D. (eds) *Commission on Genetic Resources for Food and Agriculture Assessments*. FAO, Rome, p. 572. Available at: http://www.fao.org/3/CA3129EN/CA3129EN.pdf.

FAO IFAD UNICEF WFP and WHO (2023) *The State of Food Security and Nutrition in the World 2023: Urbanization, Agrifood Systems Transformation and Healthy Diets across the Rural–Urban Continuum*. FAO, Rome. https://doi.org/10.4060/cc3017en.

Fatima, Z., Ahmed, M., Hussain, M., Abbas, G., Ul-Allah, S. *et al.* (2020) The fingerprints of climate warming on cereal crops phenology and adaptation options. *Scientific Reports* 10(1), 18013. https://doi.org/10.1038/s41598-020-74740-3.

Fishman, R. (2016) More uneven distributions overturn benefits of higher precipitation for crop yields. *Environmental Research Letters* 11(2), 024004. DOI 10.1088/1748-9326/11/2/024004.

Fitzgerald, G.J., Tausz, M., O'Leary, G., Mollah, M.R., Tausz-Posch, S. *et al.* (2016) Elevated atmospheric [CO_2] can dramatically increase wheat yields in semi-arid environments and buffer against heat waves. *Global Change Biology* 22(6), 2269–2284. DOI: 10.1111/gcb.13263.

Gasch, C., Hasan, J., Friedrichsen, C., Shipka, M., Anderson, J., Anderson, J.V. and Archer, D. (2023) *Alaskan Agriculture and the Collaborative Food Production Systems Research Project*. ASA, CSSA, SSSA International Annual Meeting, St. Louis, MO.

George, N.A., Coorey, R., Dixon, K. and Bennett, S.J. (2023) A review of the de novo domestication and cultivation of edible Australian native plants as food crops. *Journal of Agricultural Science* 161(6), 778–793. https://doi.org/10.1017/S0021859624000078.

Gerken, A.R. and Morrison, III W.R. (2022) Pest management in the postharvest agricultural supply chain under climate change. *Frontiers in Agronomy* 4, 918845. https://doi.org/10.3389/fagro.2022.918845.

Grant, L., Vanderkelen, I., Gudmundsson, L., Fischer, E., Seneviratne, S.I. and Thiery, W. (2025) Global emergence of unprecedented lifetime exposure to climate extremes. *Nature* 641(8062), 374–379. DOI: 10.1038/s41586-025-08907-1.

Grigorieva, E., Livenets, A. and Stelmakh, E. (2023) Adaptation of agriculture to climate change: A scoping review. *Climate* 11(10), 202. https://doi.org/10.3390/cli11100202.

Habib-ur-Rahman, M., Ahmad, A., Raza, A., Hasnain, M.U., Alharby, H.F. *et al.* (2022) Impact of climate change on agricultural production; issues, challenges, and opportunities in Asia. *Frontiers in Plant Science* 13, 925548. https://doi.org/10.3389/fpls.2022.925548.

Hambly, E. (2022) Climate change: Avocados and exotic plants grow in hot UK summer. Available at: https://www.bbc.com/news/science-environment-62665053 (accessed 14 May 2025).

Hay, R.K. and Porter, J.R. (2006) *The Physiology of Crop Yield*. Cambridge, UK: Cambridge University Press.

Hellin, J., Fisher, E., Taylor, M., Bhasme, S. and Loboguerrero, A.M. (2023) Transformative adaptation: From climate-smart to climate-resilient agriculture. *CABI Agriculture and Bioscience* 4(1), 30. https://doi.org/10.1186/s43170-023-00172-4.

Hemathilake, D. and Gunathilake, D. (2022) Agricultural productivity and food supply to meet increased demands. In R. Bhat (ed.) *Future Foods – Global Trends. Opportunities, and Sustainability Challenges*. Elsevier, pp. 539–553.

Hermans, K. and McLeman, R. (2021) Climate change, drought, land degradation and migration: Exploring the linkages. *Current Opinion in Environmental Sustainability* 50, 236–244. https://doi.org/10.1016/j.cosust.2021.04.013.

Herold, N., Ekström, M., Kala, J., Goldie, J. and Evans, J. (2018) Australian climate extremes in the 21st century according to a regional climate model ensemble: Implications for health and agriculture. *Weather and Climate Extremes* 20, 54–68. https://doi.org/10.1016/j.wace.2018.01.001.

Hochman, Z., Gobbett, D.L. and Horan, H. (2017) Climate trends account for stalled wheat yields in Australia since 1990. *Global Change Biology* 23(5), 2071–2081. https://doi.org/10.1111/gcb.13604.

Huang, N., Song, Y., Wang, J., Zhang, Z., Ma, S., Jiang, K. and Pan, Z. (2022) Climatic threshold of crop production and climate change adaptation: A case of winter wheat production in China. *Frontiers in Ecology and Evolution* 10, 1019436. https://doi.org/10.3389/fevo.2022.1019436.

Huang, T., Li, Y., Li, J., Sung, J.J.Y. and Yim, S.H.L. (2025) PM2.5-associated premature mortality attributable to hot-and-polluted episodes and the inequality between the global north and the global South. *GeoHealth* 9(5), e2024GH001290. https://doi.org/10.1029/2024GH001290.

Hughes, L., Rickards, L., Steffen, W., Stock, P. and Rice, M. (2016) *On the Frontline: Climate Change and Rural Communities*. Climate Council of Australia.

Hultgren, A., Carleton, T., Delgado, M., Gergel, D.R., Greenstone, M. *et al.* (2025) Impacts of climate change on global agriculture accounting for adaptation. *Nature* 642(8068), 644–652. DOI: 10.1038/s41586-025-09085-w.

IPCC (2019) *Climate Change and Land: An IPCC Special Report on Climate Change, Desertification, Land Degradation, Sustainable Land Management, Food Security, and Greenhouse Gas Fluxes in Terrestrial Ecosystems*. Intergovernmental Panel on Climate Change, University Printing House, Cambridge CB2 8BS, United Kingdom. DOI: 10.1017/9781009157988.

IPCC (2021). In: Masson-Delmotte, V., Zhai, P., Pirani, A. *et al.* (eds) *Climate Change 2021.The Physical Science Basis. Contribution of Working Group I to the Sixth Assessment Report of the Intergovernmental Panel on Climate Change*. Cambridge, United Kingdom and New York, NY, USA. DOI: 10.1017/9781009157896.

Jeong, J.-H., Kim, M.-S., Yoon, J.-H., Kim, H., Wang, S.-Y.S., Woo, S.-H. and Linderholm, H.W. (2025) Emerging trans-Eurasian heatwave-drought train in a warming climate. *Science Advances* 11(18), eadr7320. DOI: 10.1126/sciadv.adr7320.

Kaur, G., Singh, G., Motavalli, P.P., Nelson, K.A., Orlowski, J.M. and Golden, B.R. (2020) Impacts and management strategies for crop production in waterlogged or flooded soils: A review. *Agronomy Journal* 112(3), 1475–1501. https://doi.org/10.1002/agj2.20093.

King, M., Altdorff, D., Li, P., Galagedara, L., Holden, J. and Unc, A. (2018) Northward shift of the agricultural climate zone under 21st-century global climate change. *Scientific Reports* 8(1), 7904. https://doi.org/10.1038/s41598-018-26321-8.

Kromdijk, J., Głowacka, K., Leonelli, L., Gabilly, S.T., Iwai, M., Niyogi, K.K. and Long, S.P. (2016) Improving photosynthesis and crop productivity by accelerating recovery from photoprotection. *Science* 354(6314), 857–861. DOI: 10.1126/science.aai8878.

Labbé, T., Pfister, C., Brönnimann, S., Rousseau, D., Franke, J. and Bois, B. (2019) The longest homogeneous series of grape harvest dates, Beaune 1354–2018, and its significance for the understanding of past and present climate. *Climate of the Past* 15(4), 1485–1501. https://doi.org/10.5194/cp-15-1485-2019.

Lambers, H., Chapin III F.S. and Pons, T.L. (2008) *Plant Physiological Ecology*. Springer Science & Business Media, New York.

Lesk, C., Rowhani, P. and Ramankutty, N. (2016) Influence of extreme weather disasters on global crop production. *Nature* 529(7584), 84–87. DOI: 10.1038/nature16467.

Lobell, D.B. (2014) Climate change adaptation in crop production: Beware of illusions. *Global Food Security* 3(2), 72–76. https://doi.org/10.1016/j.gfs.2014.05.002.

Lobell, D.B. and Asseng, S. (2017) Comparing estimates of climate change impacts from process-based and statistical crop models. *Environmental Research Letters* 12(1), 015001. DOI: 10.1088/1748-9326/015001.

Lobell, D.B. and Burke, M. (eds) (2009). *Climate Change and Food Security: Adapting Agriculture to a Warmer World*. Springer, Netherlands.

Lobell, D.B. and Tommaso, S. Di (2025) A half-century of climate change in major agricultural regions: Trends, impacts, and surprises. *Proceedings of the National Academy of Sciences USA* 122(20), e2502789122. https://doi.org/10.1073/pnas.2502789122.

Lobell, D.B., Schlenker, W. and Costa-Roberts, J. (2011) Climate trends and global crop production since 1980. *Science* 333(6042), 616–620. DOI: 10.1126/science.1204531.

Lobell, D.B., Roberts, M.J., Schlenker, W., Braun, N., Little, B.B., Rejesus, R.M. and Hammer, G.L. (2014) Greater sensitivity to drought accompanies maize yield increase in the US Midwest. *Science* 344(6183), 516–519. DOI: 10.1126/science.1251423.

Long, S.P. and Ort, D.R. (2010) More than taking the heat: Crops and global change. *Current Opinion in Plant Biology* 13(3), 240–247. DOI: 10.1016/j.pbi.2010.04.008.

Long, S.P., Ainsworth, E.A., Rogers, A. and Ort, D.R. (2004) Rising atmospheric carbon dioxide: Plants FACE the future. *Annual Review of Plant Biology* 55(1), 591–628. DOI: 10.1146/annurev.arplant.55.031903.141610.

Marklein, A., Elias, E., Nico, P. and Steenwerth, K. (2020) Projected temperature increases may require shifts in the growing season of cool-season crops and the growing locations of warm-season crops. *Science of the Total Environment* 746, 140918. DOI: 10.1016/j.scitotenv.2020.140918.

Martel, J.-L., Brissette, F.P., Lucas-Picher, P., Troin, M. and Arsenault, R. (2021) Climate change and rainfall intensity–duration–frequency curves: Overview of science and guidelines for adaptation. *Journal of Hydrologic Engineering* 26(10), 03121001. https://doi.org/10.1061/(ASCE)HE.1943-5584.0002122.

Martinez-Villalobos, C., Fu, D., Loikith, P.C. and Neelin, J. D. (2025). Accelerating increase in the duration of heatwaves under global warming. *Nature Geoscience* 18(8), 716–723. DOI: 10.1038/s41561-025-01737-w.

Masson-Delmotte, V., Pörtner, H., Skea, J., Buendía, E., Zhai, P. and Roberts, D. (2019) *Climate Change and Land*. IPCC Report.

Misiou, O. and Koutsoumanis, K. (2022) Climate change and its implications for food safety and spoilage. *Trends in Food Science & Technology* 126, 142–152. DOI: 10.1016/j.tifs.2021.03.031.

Moore, F.C. and Lobell, D.B. (2014) Adaptation potential of European agriculture in response to climate change. *Nature Climate Change* 4(7), 610–614. DOI: 10.1038/nclimate2228.

Moore, F.C. and Lobell, D.B. (2015) The fingerprint of climate trends on European crop yields. *Proceedings of the National Academy of Sciences USA* 112(9), 2670–2675. https://doi.org/10.1073/pnas.1409606112.

Moore, C.E., Meacham-Hensold, K., Lemonnier, P., Slattery, R.A., Benjamin, C. *et al.* (2021) The effect of increasing temperature on crop photosynthesis: From enzymes to ecosystems. *Journal of Experimental Botany* 72(8), 2822–2844. https://doi.org/10.1093/jxb/erab090.

Moore, C.E., Gibson, C.D., Miao, G., Dracup, E.C., Gomez-Casanovas, N. *et al.* (2022) Substantial carbon loss respired from a corn–soybean agroecosystem highlights the importance of careful management as we adapt to changing climate. *Environmental Research Letters* 17(5), 054029. DOI 10.1088/1748-9326/ac661a.

Moratiel, R., Durán, J. and Snyder, R.L. (2010) Responses of reference evapotranspiration to changes in atmospheric humidity and air temperature in Spain. *Climate Research* 44(1), 27–40.

Moretti, C., Mattos, L., Calbo, A. and Sargent, S. (2010) Climate changes and potential impacts on postharvest quality of fruit and vegetable crops: A review. *Food Research International* 43(7), 1824–1832. https://doi.org/10.1016/j.foodres.2009.10.013.

Morgounov, A., Sonder, K., Abugalieva, A., Bhadauria, V., Cuthbert, R.D. *et al.* (2018) Effect of climate change on spring wheat yields in North America and Eurasia in 1981–2015 and implications for breeding. *PLoS ONE* 13(10), e0204932. https://doi.org/10.1371/journal.pone.0204932.

Moses, J., Jayas, D. and Alagusundaram, K. (2015) Climate change and its implications on stored food grains. *Agricultural Research* 4, 21–30. DOI: 10.5073/jka.2018.463.022.

Motha, R.P. (2011) The impact of extreme weather events on agriculture in the United States. In: Attri, S.D., Rathore, L.S., Sivakumar, M.V.K. and Dash, S.K. (eds) *Challenges and Opportunities in Agrometeorology*. Springer Nature, Berlin Heidelberg, pp. 397–408.

Moyers, K., Abatzoglou, J.T., Escriva-Bou, A., Medellín-Azuara, J. and Viers, J.H. (2024) An invisible water surcharge: Climate warming increases crop water demand in the San Joaquin Valley's groundwater-dependent irrigated agriculture. *PLOS Water* 3(3), e0000184. https://doi.org/10.1371/journal.pwat.0000184.

Mullen, D.J. and Barrett-Lennard, E.G. (2010) Breeding crops for tolerance to salinity, waterlogging and salinity. In: Climate change and crop production (Ed. M.P. Reynolds). CABI, Oxfordshire, UK.

Ortiz-Bobea, A., Ault, T.R., Carrillo, C.M., Chambers, R.G. and Lobell, D.B. (2021) Anthropogenic climate change has slowed global agricultural productivity growth. *Nature Climate Change* 11(4), 306–312. https://doi.org/10.1038/s41558-021-01000-1.

Parmar, N., Singh, K.H., Sharma, D., Singh, L., Kumar, P. *et al.* (2017). Genetic engineering strategies for biotic and abiotic stress tolerance and quality enhancement in horticultural crops: a comprehensive review. *3 Biotech* 7, 1–35. DOI: 10.1007/s13205-017-0870-y.

Pathak, T.B., Maskey, M.L., Dahlberg, J.A., Kearns, F., Bali, K.M. and Zaccaria, D. (2018) Climate change trends and impacts on California agriculture: A detailed review. *Agronomy* 8(3), 25. https://doi.org/10.3390/agronomy8030025.

Peters, K., Breitsameter, L. and Gerowitt, B. (2014) Impact of climate change on weeds in agriculture: A review. *Agronomy for Sustainable Development* 34, 707–721. https://doi.org/10.1007/s13593-014-0245-2.

Phillips, A., Schultz, C. J. andBurton, R. A. (2025) New crops on the block: effective strategies to broaden our food, fibre, and fuel repertoire in the face of increasingly volatile agricultural systems. *Journal of Experimental Botany* 76(8), 2043–2063. https://doi.org/10.1093/jxb/eraf023.

Prado, K., Holland, B.L., Gardener, B.M.S., Lundquist, P.K., Santiago, J.P., VanBuren, R. and Rhee, S.Y. (2025) Building climate-resilient crops: Genetic, environmental, and technological strategies for heat and drought stress tolerance. *Journal of Experimental Botany* 76, 4395–4413. DOI: 10.1093/jxb/eraf111.

Pratley, J.E. (ed.) (2003). *Principles of Field Crop Production*. Oxford University Press, South Melbourne. Available at: https://www.csu.edu.au/research/grahamcentre/publications/e-books

Pugh, T., Müller, C., Elliott, J., Deryng, D., Folberth, C. *et al.* (2016) Climate analogues suggest limited potential for intensification of production on current croplands under climate change. *Nature Communications* 7(1), 1–8. https://doi.org/10.1038/ncomms12608.

Qing, Y., Wang, S., Yang, ZL. *et al.* Soil moisture–atmosphere feedbacks have triggered the shifts from drought to pluvial conditions since 1980. *Commun Earth Environ* 4, 254(2023). https://doi.org/10.1038/s43247-023-00922-2.

Raupach, T.H., Martius, O., Allen, J.T., Kunz, M., Lasher-Trapp, S. *et al.* (2021) The effects of climate change on hailstorms. *Nature Reviews Earth & Environment* 2(3), 213–226. https://doi.org/10.1038/s43017-020-00133-9.

Ray, D.K., Mueller, N.D., West, P.C. and Foley, J.A. (2013) Yield trends are insufficient to double global crop production by 2050. *PLoS ONE* 8(6), e66428. https://doi.org/10.1371/journal.pone.0066428.

Redhead, J.W., Brown, M., Price, J., Robinson, E., Nicholls, R.J., Warren, R. and Pywell, R.F. (2025) National horizon scanning for future crops under a changing UK climate. *Climate Resilience and Sustainability* 4(1), e70007. https://doi.org/10.1002/cli2.70007.

Rezaei, E.E., Webber, H., Asseng, S., Boote, K., Durand, J.L. *et al.* (2023) Climate change impacts on crop yields. *Nature Reviews Earth & Environment* 4(12), 831–846. DOI: 10.1038/s43017-023-00491-0.

Rickards, L. and Howden, S.M. (2012) Transformational adaptation: Agriculture and climate change. *Crop and Pasture Science* 63(3), 240–250. DOI: 10.1071/CP11172.

Ripple, W.J., Wolf, C., Newsome, T.M., Gregg, J.W., Lenton, T.M. *et al.* (2021) World scientists' warning of a climate emergency 2021. *Bioscience* 71(9), 894–898. https://doi.org/10.1093/biosci/biab079.

Ripple, W.J., Wolf, C., Gregg, J.W., Rockström, J., Newsome, T.M., Law, B.E., ... & King, S.D.A. (2023) The 2023 state of the climate report: Entering uncharted territory. *Bioscience* 73(12), 841–850. https://doi.org/10.1093/biosci/biad080.

Rojas-Reyes, J.J., Rivera-Cadavid, L. and Peña-Orozco, D.L. (2024) Disruptions in the food supply chain: A literature review. *Heliyon* 10(14), e34730. https://doi.org/10.1016/j.heliyon.2024.e34730.

Romanello, M., Napoli, C. Di, Green, C., Kennard, H., Lampard, P. *et al.* (2023) The 2023 report of The Lancet countdown on health and climate change: The imperative for a health-centred response in a world facing irreversible harms. *The Lancet* 402(10419), 2346–2394. DOI: 10.1016/S0140-6736(23)01859-7.

Rovira, A. (1992) Dryland mediterranean farming systems in Australia. *Australian Journal of Experimental Agriculture* 32(7), 801–809. http://hdl.handle.net/102.100.100/248237?index=1.

Ruiz-Vera, U.M., Siebers, M.H., Jaiswal, D., Ort, D.R. and Bernacchi, C.J. (2018) Canopy warming accelerates development in soybean and maize, offsetting the delay in soybean reproductive development by elevated CO_2 concentrations. *Plant, Cell & Environment* 41(12), 2806–2820. DOI: 10.1111/pce.13410.

Salinger, M., Stigter, C. and Das, H. (2000) Agrometeorological adaptation strategies to increasing climate variability and climate change. *Agricultural and Forest Meteorology* 103(1–2), 167–184. https://doi.org/10.1016/S0168-1923(00)00110-6.

Samarah, N.H. (2016). In: Hossain, M.A., Wani, S.H., Bhattacharjee, S., Burritt, D.J. and Tran, L.-S.P. (eds) *Introduction. Drought Stress Tolerance in Plants, Vol. 2: Molecular and Genetic Perspectives*. Springer Nature, Switzerland, pp. 3–37. https://doi.org/10.1007/978-3-319-32423-4.

Scanlon, T.T. and Doncon, G. (2020) Rain, rain, gone away: Decreased growing-season rainfall for the dryland cropping region of the south-west of Western Australia. *Crop and Pasture Science* 71(2), 128–133. DOI: 10.1071/CP19294.

Schneider, L., Rebetez, M. and Rasmann, S. (2022) The effect of climate change on invasive crop pests across biomes. *Current Opinion in Insect Science* 50, 100895. https://doi.org/10.1016/j.cois.2022.100895.

Shortridge, J. (2019) Observed trends in daily rainfall variability result in more severe climate change impacts to agriculture. *Climatic Change* 157(3), 429–444. DOI: 10.1007/s10584-019-02555-x.

Sprigg, H., Belford, R., Milroy, S., Bennett, S.J. and Bowran, D. (2014) Adaptations for growing wheat in the drying climate of Western Australia. *Crop and Pasture Science* 65(7), 627–644. https://doi.org/10.1071/CP13352.

Stephens, D. and Lyons, T. (1998) Rainfall-yield relationships across the Australian wheatbelt. *Australian Journal of Agricultural Research* 49(2), 211–224. DOI: 10.1071/A96139.

Stern, K.R., Bidlack, J.E., Jansky, S. and Uno, G. (2006). In: Bidlack, S.H. and Jansky, G.E. (eds) *Introductory Plant Biology/Kingsley R. Stern; with Contributions by James E.* McGraw-Hill Higher Education, Boston.

Stokes, C. and Howden, M. (eds) (2010). *Adapting Agriculture to Climate Change: Preparing Australian Agriculture, Forestry and Fisheries for the Future*. CSIRO Publishing, Collingwood, Victoria, Australia.

Su, Q., Kambale, R.D., Tzeng, J.-H., Amy, G.L., Ladner, D.A. and Karthikeyan, R. (2025) The growing trend of saltwater intrusion and its impact on coastal agriculture: Challenges and opportunities. *Science of The Total Environment* 966, 178701. https://doi.org/10.1016/j.scitotenv.2025.178701.

Swain, D.L., Prein, A.F., Abatzoglou, J.T., Albano, C.M., Brunner, M. *et al.* (2025) Hydroclimate volatility on a warming earth. *Nature Reviews Earth & Environment* 6(1), 35–50. DOI: 10.1038/s43017-024-00624-z.

Swarup, S., Cargill, E.J., Crosby, K., Flagel, L., Kniskern, J. and Glenn, K.C. (2021) Genetic diversity is indispensable for plant breeding to improve crops. *Crop Science* 61(2), 839–852. https://doi.org/10.1002/csc2.20377.

Talukder, B., Ganguli, N., Matthew, R., VanLoon, G.W., Hipel, K.W. and Orbinski, J. (2021) Climate change-triggered land degradation and planetary health: A review. *Land Degradation & Development* 32(16), 4509–4522. https://dx.doi.org/10.1002/ldr.4056.

Tchonkouang, R.D., Onyeaka, H. and Nkoutchou, H. (2024) Assessing the vulnerability of food supply chains to climate change-induced disruptions. *Science of the Total Environment* 920, 171047. https://doi.org/10.1016/j.scitotenv.2024.171047.

Tey, Y.S. and Brindal, M. (2015) Factors influencing farm profitability. *Sustainable Agriculture Reviews*, 235–255. https://link.springer.com/chapter/10.1007/978-3-319-09132-7_5.

Thomson, G., McCaskill, M., Goodwin, I., Kearney, G. and Lolicato, S. (2014) Potential impacts of rising global temperatures on Australia's pome fruit industry and adaptation strategies. *New Zealand Journal of Crop and Horticultural Science* 42(1), 21–30. https://doi.org/10.1080/01140671.2013.838588.

Thorman, I. and Engles, J.M.M. (2015). Genetic diversity and erosion – a global perspective. In: Ahuja, M. and Mohan Jain, S. (eds) *Genetic Diversity and Erosion in Plants*, Springer, Switzerland, pp. 263–294. DOI: 10.1007/978-3-319-25954-3.

Tian, L.-X., Zhang, Y.-C., Chen, P.-I., Zhang, F.-F., Li, J. *et al.* (2021) How does the waterlogging regime affect crop yield? A global meta-analysis. *Frontiers in Plant Science* 12, 634898. https://doi.org/10.3389/fpls.2021.634898.

Torbenson, M., Esper, J., Brázdil, R., Büntgen, U., Olesen, J. *et al.* (2024) Past and future climate-driven changes of agricultural land in Central Europe. *Geophysical Research Letters* 51(24), e2024GL112363. https://doi.org/10.1029/2024GL112363.

Vecellio, D.J., Kong, Q., Kenney, W.L. and Huber, M. (2023) Greatly enhanced risk to humans as a consequence of empirically determined lower moist heat stress tolerance. *Proceedings of the National Academy of Sciences* 120(42), e2305427120. https://doi.org/10.1073/pnas.2305427120.

Vicente-Serrano, S.M., Pricope, N.G., Toreti, A., Morán-Tejeda, E., Spinoni, J. *et al.* (2024) *The Global Threat of Drying Lands: Regional and Global Aridity Trends and Future Projections. A Report of the Science–Policy Interface*. United Nations Convention to Combat Desertification (UNCCD), Bonn, Germany, p. 156.

Voesenek, L.A.C.J., Rijnders, J.H.G.M., Peeters, A.J.M., van de Steeg, H.M. and de Kroon, H. (2004) Plant hormones regulate fast shoot elongation under water: from genes to communities. *Ecology* 85, 16–27. https://doi.org/10.1890/02-740.

Vogel, E., Donat, M.G., Alexander, L.V., Meinshausen, M., Ray, D.K. *et al.* (2019) The effects of climate extremes on global agricultural yields. *Environmental Research Letters* 14(5), 054010. DOI: 10.1088/1748-9326/ab154b.

Wang, W., Cai, C., He, J., Gu, J., Zhu, G. *et al.* (2020) Yield, dry matter distribution and photosynthetic characteristics of rice under elevated CO_2 and increased temperature conditions. *Field Crops Research* 248, 107605. https://doi.org/10.1016/j.fcr.2019.107605.

Wanniarachchi, S. and Sarukkalige, R. (2022) A review on evapotranspiration estimation in agricultural water management: Past, present, and future. *Hydrology* 9(7), 123. https://doi.org/10.3390/hydrology9070123.

Wu, S., Luo, M., Lau, G.N.-C., Zhang, W., Wang, L. *et al.* (2025) Rapid flips between warm and cold extremes in a warming world. *Nature Communications* 16(1), 3543. DOI: 10.1038/s41467-025-58544-5.

Xiong, W., Reynolds, M.P., Crossa, J., Schulthess, U., Sonder, K. *et al.* (2021) Increased ranking change in wheat breeding under climate change. *Nature Plants* 7(9), 1207–1212. DOI: 10.1038/s41477-021-00988-w.

Yuan, W., Zheng, Y., Piao, S., Ciais, P., Lombardozzi, D. *et al.* (2019) Increased atmospheric vapor pressure deficit reduces global vegetation growth. *Science Advances* 5(8),eaax1396. DOI: 10.1126/sciadv.aax1396.

Zacarias, D.A. (2020) Global bioclimatic suitability for the fall armyworm, *Spodoptera frugiperda* (Lepidoptera: Noctuidae), and potential co-occurrence with major host crops under climate change scenarios. *Climatic Change* 161(4), 555–566. DOI: 10.1007/s10584-020-02722-5.

Zhang, H., Yu, X., Qiao, F. and Li, C. (2024) Increasing concurrent exposure of global breadbaskets to reproductive heat extremes. *Environmental Research Letters* 19(8), 084046. DOI: 10.1088/1748-9326/ad6467.

Zhao, C., Liu, B., Piao, S., Wang, X., Lobell, D.B. *et al.* (2017) Temperature increase reduces global yields of major crops in four independent estimates. *Proceedings of the National Academy of Sciences USA* 114(35), 9326–9331. https://doi.org/10.1073/pnas.1701762114.

4 Genomic Interventions for Future-proofing Crops

Mahendar Thudi[1,2,*], Yogesh Dashrath Naik[3,4], Anil Namdev Kale[3], Sagar Krushnaji Rangari[3,5], Ashish Kumar[6], Uday Jha[7], Murukarthick Jayakodi[8,9], Annapurna Chitikineni[10], Stuti Sharma[6], Reyazul Rouf Mir[10], Venugopal Mendu[11], Manish Pandey[5], Naveen Puppala[4], Vinay Kumar Sharma[3] and Somashekhar Punnuri[1,*]

[1]College of Agriculture, Family Sciences and Technology, 1005 State University Dr, Fort Valley State University, Fort Valley, GA, USA; [2]Centre for Crop Health and School of Agriculture and Environmental Science, University of Southern Queensland, Toowoomba, Australia; [3]Department of Agricultural Biotechnology and Molecular Biology, Dr. Rajendra Prasad Central Agricultural University, Pusa, Bihar, India; [4]New Mexico State University, Agricultural Science Center, Clovis, 88101, USA; [5]International Crops Research Institute for the Semi-Arid Tropics (ICRISAT), Patancheru, Hyderabad, Telangana, India; [6]Jawaharlal Nehru Agricultural University, Jabalpur, Madhya Pradesh, India; [7]Indian Council for Agricultural Research-Indian Institute of Pulses Research, Kanpur, Uttar Pradesh, India; [8]Texas A&M AgriLife Research Center, Dallas, TX, USA; [9]Department of Soil & Crop Sciences, Texas A&M University, College Station, TX, USA; [10]WA State Agricultural Biotechnology Centre, Centre for Crop and Food Innovation, Murdoch University, Murdoch, WA 6150, Australia; [11]Department of Agriculture, Agribusiness and Environmental Sciences, Texas A&M University-Kingsville, Texas, USA

Abstract

The rising global challenges of food insecurity demand innovative agricultural solutions. Addressing global food security requires a proactive approach, including the need to future-proof crops against adverse environmental conditions. Climate change has intensified abiotic stresses such as drought, salinity, heat and flooding, resulting in substantial yield losses in globally important food and feed crops. This chapter explores the advantages offered by the integration of advanced genomic tools and multidisciplinary approaches in enhancing crop resilience against the dual challenges of food insecurity and climate change. Genomic approaches such as genome-wide association studies, genomic selection and pangenomics have accelerated the discovery and improvement of complex traits, including yield and stress tolerance. The utilization of genome editing technologies, plant metabolic engineering and synthetic biology has enabled the development of crops with tailored traits and optimized metabolic pathways. For instance, CRISPR-Cas9 has been used to enhance drought tolerance in rice by targeting the *OsPYL* gene, and metabolic engineering has improved provitamin A content in Golden Rice. Similarly, synthetic biology approaches have been applied to optimize nitrogen use efficiency in maize and to engineer C4 photosynthetic traits into rice for higher productivity. Precision agriculture supported with artificial intelligence

*Corresponding authors. Email: mahendar.thudi@fvsu.edu; punnuris@fvsu.edu

 Genomic Solutions for Breeding Climate-resilient Crops (R.K. Varshney, R. Barmukh and R.J. Harper)
DOI: 10.1079/9781800627185.0004

and machine learning further complements these scientific and technological advancements by optimizing resource use and decision-making. While challenges persist, these technologies collectively offer tremendous potential to address global agricultural demands and environmental sustainability.

4.1 Introduction

Global hunger has been steadily rising during the last decade with the situation worsening in recent years. In 2022, an estimated 691–783 million people worldwide faced hunger, representing 9.2% of the global population, with an increase from 7.9% in 2019. Factors such as the COVID-19 pandemic and economic downturns have exacerbated food insecurity, making the goal of eradicating hunger by 2030 increasingly challenging (FAO, IFAD, UNICEF, WFP and WHO, 2023). Compounding this crisis, climate change threatens to further destabilize food systems as the food demand is projected to rise by 30–62% with hunger risks increasing up to 30% (van Dijk *et al.*, 2021). In the context of climate change, drought stress has become a growing concern. Projections indicate that by the 2030s, the population at risk of drought could rise by 21.96–25.95%, escalating to 36.64–45.40% by the 2050s due to rapid population growth and increasing drought hazards (Wang and Sun, 2023). Since 1880, average global temperatures have risen by approximately 1°C (1.7°F). By 2050, global temperatures are expected to increase by about 1.5°C (2.7°F) and could rise further by 2–4°C (3.6–7.2°F) by 2100, depending on the extent of greenhouse gas emissions increase (https://www.ipcc.ch/sr15/). Without effective adaptation measures, a 1°C increase in global mean temperature could reduce global yields by 6.0% for wheat (*Triticum aestivum*), 3.2% for rice (*Oryza sativa*), 7.4% for maize (*Zea mays*), and 3.1% for soybeans (*Glycine max*) (Zhao *et al.*, 2017). Approximately 58% of global land is expected to face increased drought hazards, disproportionately affecting arid and semi-arid regions such as Africa, Central and West Asia, and Australia, more particularly in less-developed economies. These regions are highly vulnerable to drought, with extreme events anticipated to cause significant yield losses up to 54.15% for rice, 29.30% for wheat, and 50.66% for maize (Prodhan *et al.*, 2022). In the face of these challenges, genomic innovations offer transformative potential to enhance crop resilience, safeguard yields and stabilize food production. By accelerating the identification and deployment of stress-tolerant genes, genome-enabled breeding and biotechnology can directly mitigate the adverse effects of climate-induced stressors. This makes genomics an indispensable tool in the global effort to adapt agriculture to a changing climate.

The adverse effects of climate change are already evident with heat stress, drought, flooding, and salinity leading to substantial yield losses in major global crops (Table 4.1). Heatwaves and rising temperatures accelerate crop maturation and impair grain filling, which has the potential to cause yield reductions of up to 50% under severe conditions. Drought further exacerbates water scarcity, disrupting photosynthesis and nutrient uptake (Bailey-Serres *et al.*, 2019). To minimize the adverse effects of these factors, genetic advancements such as the *Submergence 1A* (*SUB1A*) and *Deeper Rooting 1* (*DRO1*) genes in rice provide partial solutions (Soe *et al.*, 2024). Similarly, flooding intensified by torrential rains damages low-lying crops such as rice, causing poor survival, mechanical damage and nutrient losses. Although genetic improvements by utilization of the *SUB1A* gene enhance submergence tolerance, these interventions are limited in scope (Lin *et al.*, 2019). Soil salinity, driven by inappropriate irrigation practices and rising sea levels, further challenges productivity. The genetically conferred solution such as the *HKT1* gene in wheat offer potent promise in this context but requires broader implementation (Mao *et al.*, 2023).

The advancement of quantitative genetics in modern plant breeding represents a transformative shift from classical assumptions and considerations, such as Hardy–Weinberg equilibrium and independent segregation of loci, to advanced methodologies that utilize molecular markers and mixed-model analysis for precise genotypic value prediction (Bernardo, 2020). In recent years, the integration of genomic technologies, computational tools and simulation platforms has enabled breeders to overcome complex challenges and design more effective

Table 4.1. Summary of climate change impact and yield penalty reported in major field crops. (Authors' own table.)

Crop	Stress	Developmental stage	Impact	Yield loss (%)	References
Rice (*Oryza sativa* L.)	Drought	Vegetative phase	Photosynthesis	21–50.6%	Zhang *et al.* (2018)
		Reproductive stage	Chalky kernel and chalkiness	24.00%	Yang *et al.* (2019)
		Pre-Anthesis	–	65–85%	Kumar *et al.* (2008)
	Heat	Grain-filling stage	Seed setting	39.60%	Xiong *et al.* (2017)
		Tillering	Yellowing of tillers	35%	Soda *et al.* (2018).
		Grain filling	Short grain-filling duration	50%	Sreenivasulu *et al.* (2015)
	Salinity	Seedling stage	–	58.30%	Xu *et al.* (2024)
Wheat (*Triticum aestivum* L.)	Drought	Tillering, booting, anthesis, and grain filling	Reduces seedling vigor	69%	Khan *et al.* (2023)
		Grain-filling stage	Biomass and grain size	50–60%	Zhao *et al.* (2020)
	Heat	Anthesis	Reduces grain number	29%	Djanaguiraman *et al.* (2020)
		Grain filling		44%	
	Salinity	Anthesis	Reduces grain number	39.10%	El Sabagh *et al.* (2021)
		Early booting		24.30%	
		Mid grain filling	–	13.40%	
Maize (*Zea mays* L.)	Drought	Silking	Kernel abortion	50%	Liang *et al.* (2020)
		Grain-filling stages	Kernel abortion	21%	Liang *et al.* (2020)
	Heat	Silking	Pollen viability	13%	Siebers *et al.* (2017)
		Grain filling	Reduces maize kernel number	31%	Yang *et al.* (2017)
	Salinity	–	–	34%	Cucci *et al.* (2019)
Soyabean (*Glycine max* L.)	Drought	Flowering and pod-setting	Reduce the pod number	73–82%	Wei *et al.* (2018)
	Heat	Complete life cycle	–	51%	Jumrani *et al.* (2022)
	Salinity	Complete life cycle	–	40%	Hoang *et al.* (2024)
Sorghum (*Sorghum bicolor*)	Drought	Vegetative and reproductive	–	36% and 55%	Assefa *et al.* (2010)
		Grain filling	Alters intra-panicle grain number	53%	Adotey *et al.* (2021)
	Heat	PMC stage	Non-viable pollen grains	100% non-viable pollen	Smith *et al.* (2023)
	Salt	After 2 weeks of planting still growth phase	Chlorophyll loss	10–23%	Rajabi *et al.* (2023)
Barley (*Hordeum vulgare*)	Drought	Flowering	Restricts the growth of roots	20.16%	Li *et al.* (2020)
		Grain filling	–	49–87%	Samarah *et al.* (2009)

Continued

Table 4.1. Continued.

Crop	Stress	Developmental stage	Impact	Yield loss (%)	References
	Heat	Grain filling	Decreasing the number and size of grains	10-40%	Dawood *et al.* (2020)
Chickpea (*Cicer arietinum*)	Drought	Reproductive stage	Loss of flowers	30–40%	Rani *et al.* (2020)
		Flowering and seed filling	–	40–45%	Rani *et al.* (2020)
	Heat	Reproductive stage	Reduce pollen grain and stigma viability	63–66%	Kaushal *et al.* (2013)
		Reproductive stage		39%	Than *et al.* (2007)
Pigeonpea (*Cajanus cajan*)	Drought	Fowering stage	–	40–55%	Nam *et al.* (2001)
		Flowering and early pod development	Reduced number of pods	37%	Lopez *et al.* (1996)

breeding programs. This chapter examines the application of genomic innovations aimed at developing climate-resilient crops and ensuring food and nutritional security for future generations (Figure 4.1). This chapter is organized into three broad categories: (1) Key technologies; (2) Enabling systems; and (3) Outcomes and future perspectives.

4.2 Key Technologies for Future-proofing Crops

4.2.1 Innovations in Trait Discovery and Crop Improvement

Before the advent of genome sequencing technologies, trait discovery and mapping in major cereals and legumes relied heavily on the use of molecular markers, marking a paradigm shift in plant breeding approaches among key cereals and legumes. Early approaches in the 1980s and 1990s involved restriction fragment length polymorphisms (RFLPs), which provided the first genetic linkage maps for crops such as maize and soybean (Botstein *et al.*, 1980; Tanksley *et al.*, 1989). The introduction of PCR-based markers, such as random amplified polymorphic DNA (RAPD), simple sequence repeats (SSRs) and amplified fragment length polymorphisms (AFLP) enhanced the efficiency of marker development and trait mapping (Williams *et al.*, 1990). These molecular markers facilitated the identification of major genes and QTLs associated with yield, disease resistance and abiotic stress tolerance in cereals such as rice and wheat (McCouch *et al.*, 1997). Similarly, in legumes, foundational studies on chickpea (*Cicer arietinum*) and common bean (*Phaseolus vulgaris* L.) utilized SSR markers to map key agronomic traits (Burr *et al.*, 1988; Miklas *et al.*, 2006; Nayak *et al.*, 2010). This era established the foundation for marker-assisted selection (MAS), a transformative approach that linked phenotypic variability with genetic information (Varshney *et al.*, 2014a). MAS significantly accelerated breeding outcomes by enabling precise introgression of traits such as disease resistance (Mannur *et al.*, 2019; Ellur *et al.*, 2025), drought tolerance (Bharadwaj *et al.*, 2021) and improved quality (Singer *et al.*, 2023; Yang *et al.*, 2023), thereby reducing the number of breeding cycles required for cultivar development. While powerful, these approaches were often constrained by low marker density, limited genome coverage and reliance on biparental populations. The pre-genome sequencing era thus represented an essential transitional phase, which bridged classical genetics and modern genomics, setting the stage for high-throughput methods and genome-wide association studies

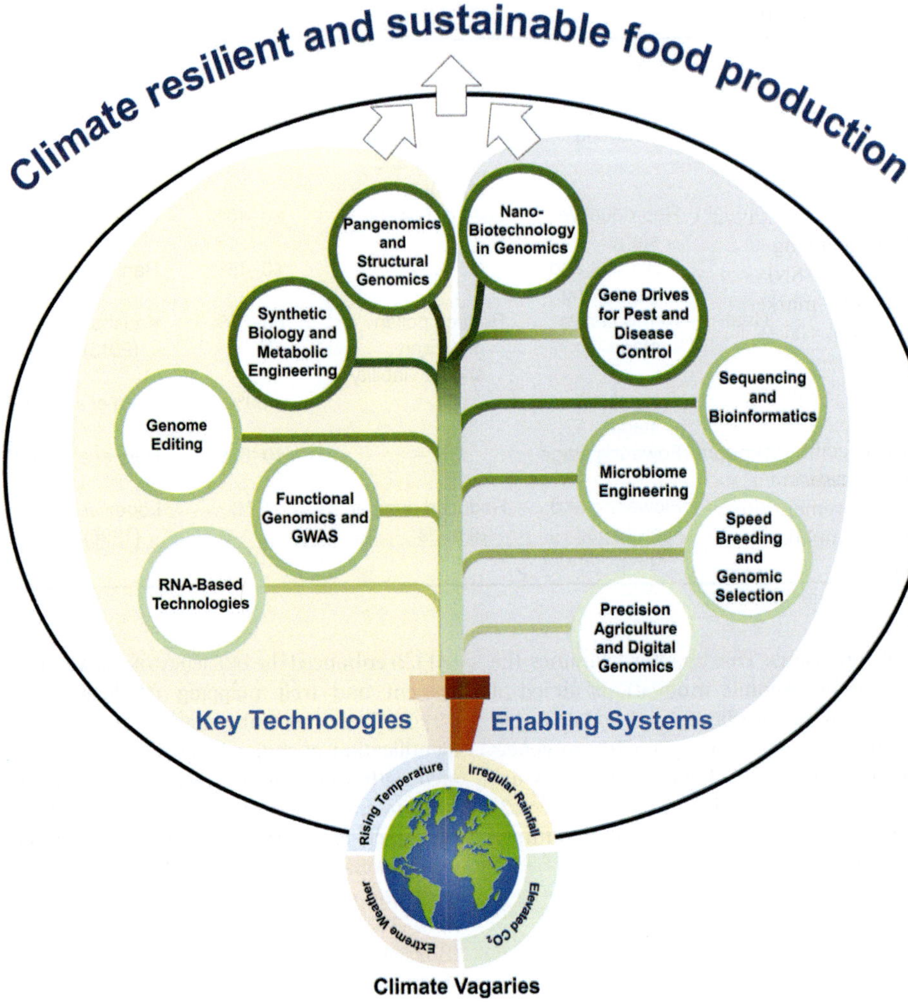

Figure 4.1. Conceptual framework illustrating the integration of key genomic and biotechnological innovations with enabling systems to support climate-resilient and sustainable food production. (Authors' own image.)

(GWAS) (Uffelmann *et al.*, 2021; Gangurde *et al.*, 2022; Clauw *et al.*, 2024).

4.2.2 Advanced Genotyping Platforms

The advent of advanced genotyping platforms during the post-genome sequencing era has significantly transformed the landscape of trait discovery, gene mapping and genetic improvement in major cereals and legumes. These platforms have evolved in scale and precision, catering to diverse breeding objectives and contributing to crop improvement strategies. Early efforts focused on low-throughput technologies such as SSRs and AFLPs. These markers were pivotal in constructing linkage maps and mapping traits such as disease resistance and yield in crops such as rice, soybean and common bean (Williams *et al.*, 1990; McCouch *et al.*, 1997; Miklas *et al.*, 2006; Thudi *et al.*, 2011). However, their limited throughput and low genome coverage restricted their applicability to large-scale breeding programs. With the advent of medium-throughput technologies such as

Diversity Arrays Technology (DArT), genotyping became more cost-effective, enabling simultaneous analysis of hundreds to thousands of loci. This innovation facilitated genetic diversity studies and QTL mapping in crops such as wheat and barley (Jaccoud *et al.*, 2001). The real transformation, however, came with high-throughput platforms such as SNP arrays and genotyping-by-sequencing (GBS, exemplified by the Illumina 50K SNP chip for maize to provide high-density marker data that enhanced the resolution of GWAS for complex traits such as drought tolerance; Gore *et al.*, 2009). In legumes, the Axiom ® CicerSNP array was instrumental in high-throughput genotyping for chickpea, facilitating rapid identification of genomic loci associated with stress tolerance and yield improvement (Roorkiwal *et al.*, 2018). GBS further simplified genotyping, offering a cost-effective, high-resolution alternative that proved invaluable for trait mapping in sorghum (*Sorghum bicolor*) and peanut (*Arachis hypogaea*) (Elshire *et al.*, 2011). The introduction of ultra-high-throughput platforms, such as whole-genome resequencing (WGR) and targeted sequencing panels, further revolutionized crop genomics. WGR allowed the detection of genome-wide SNPs and short insertions/deletions, enabling fine mapping of traits with better precision. For instance, targeted sequencing approaches, such as exome capture arrays, facilitated the discovery of candidate genes in pigeonpea (*Cajanus cajan*), advancing the precision of trait improvement programs (Varshney *et al.*, 2017a).

The availability of genomic resources has remarkably varied between cereals and legumes over the past two decades (Naik *et al.*, 2025a, 2025b). Cereals such as rice, maize and wheat have benefitted from substantial funding and well-curated genomic databases, such as the International Rice Genome Sequencing Project (IRGSP) and MaizeGDB (IRGSP, 2005; Schnable *et al.*, 2009). These resources have enabled detailed structural and functional genomic studies, paving the way for the development of climate-resilient, high-yielding varieties. Contrarily, legumes historically lagged in genomic advancements due to limited funding. However, international collaborations and initiatives, such as the Legume Information System and resequencing projects for chickpea and pigeonpea, have begun bridging this gap (Thudi *et al.*, 2016a, 2016b; Varshney *et al.*, 2017a; Varshney *et al.*, 2018). Further, the draft genomes and the pangenomes that are available in different crop plants are examples of significant strides made in recent years, providing critical insights for trait improvement (Crop genomes and beyond, 2020; Schreiber *et al.*, 2024).

GWAS, leveraging high-throughput SNP arrays, has been instrumental in identifying loci linked to drought tolerance in maize and resistance to *Ascochyta* blight in chickpea (Huang *et al.*, 2010). Similarly, marker-assisted backcrossing (MABC) has been extensively used to introgress key traits, such as the *Sub1A* gene for submergence tolerance in rice and *Xa21* for bacterial blight resistance (Xu *et al.*, 2006). The integration of genomic selection (GS) with high-throughput genotyping has accelerated breeding programs in cereals and legumes by predicting breeding values and enabling rapid selection of superior genotypes (Meuwissen *et al.*, 2001; Crossa *et al.*, 2017).

4.2.3 Genomes, Pangenomes and Super-pangenomes

The past two decades have witnessed appreciable advancements in sequencing technologies, enabling rapid and cost-effective genome sequencing of major crops. The transition from first-generation Sanger sequencing to second-generation (e.g. Illumina) and third-generation platforms (e.g. Pacific Biosciences, Oxford Nanopore) has revolutionized genomics research. These advancements have significantly reduced sequencing costs, with the cost per genome plummeting from millions of dollars to a few hundred, while improving throughput and read accuracy (Naik *et al.*, 2024). Third-generation sequencing technologies offer the additional advantage of generating long reads, which are crucial for resolving complex regions of the genome, including repetitive sequences, tandem duplications and structural variants (van Dijk *et al.*, 2018). Using these cutting-edge tools, the genome sequences of major crops such as rice, wheat, maize, pearl millet (*Pennisetum glaucum*), pigeaonpea, chickpea, soybean and peanut have been decoded, providing invaluable insights into the genetic architecture of their beneficial traits and assisting in crop improvement programs

(IRGSP, 2005; Varshney *et al.*, 2012; Varshney *et al.*, 2013; Varshney *et al.*, 2017b). In addition, several germplasm lines in both cereals and legumes have been resequenced (Thudi *et al.*, 2016a, 2016b; Pankin *et al.*, 2018; Lv *et al.*, 2020; Varshney *et al.*, 2021a; Boatwright *et al.*, 2021; Grzybowski *et al.*, 2023; Thudi *et al.*, 2023). The reference genome serves two key roles in genomics as it offers a consistent framework for reporting findings and reduces computational costs and processing time by providing a reliable scaffold for analysis (Kaye and Wasserman, 2021). However, relying on a single reference genome is insufficient to fully capture the genetic diversity within a species (Tao *et al.*, 2019). Many sequences and genes present in some individuals are absent in the reference genome. For instance, a comparison between two maize inbred lines, B73 and Mo17, showed that over 10% of genes in one genome lacked homologs in the other (Sun *et al.*, 2018). This limitation often results in inefficiencies as millions of unmapped reads are discarded in resequencing studies, and additional sequencing is required to uncover missing sequences in map-based cloning efforts.

Further, the development of pangenomes has evolved alongside rapid advancements in sequencing and computational technologies, offering diverse methods to address specific research and breeding objectives. Pangenomes are now constructed through *de novo* assemblies from representative genotypes. Moving beyond single-species genomic exploration, super-pangenomes extend this concept across wild relatives sharing the same genus for revealing interspecies variations and evolutionary insights that drive cross-species breeding and genomic conservation efforts. Pangenomes primarily focus on intraspecific genetic diversity, identifying core genes shared by all accessions and variable genes specific to certain individuals or ecotypes. In contrast, super-pangenomes extend this framework to capture interspecies variations and evolutionary relationships, enabling cross-species trait discovery. For example, the rice and maize pangenomes have driven trait improvement within these crops, while the citrus and Brassica super-pangenomes have provided insights into cross-species breeding and evolutionary dynamics (Golicz *et al.*, 2016; Zhao *et al.*, 2018). Although computationally less demanding, pangenomes offer species-specific insights, while super-pangenomes provide additional genomic perspectives for applications in conservation and cross-species breeding (Table 4.2).

4.2.3.1 De novo *Assembly-based Methods*

Chromosome-level genome assemblies are constructed for selected individuals, followed by comparative analysis to identify core and variable genes. This approach captures unique genomic regions and large structural variations that are often missed in the single reference method. For instance, the tomato (*Solanum lycopersicum*) pangenome identified rare alleles regulating fruit flavor through *de novo* assembly, providing breeders with richer genetic resources (Gao *et al.*, 2019). Similarly, the barley (*Hordeum vulgare*) pangenome revealed hidden genetic diversity resulting from mutation breeding, highlighting the potential for leveraging this diversity in breeding programs (Jayakodi *et al.*, 2020). Considering the less expensive accurate long read sequencing technology, genome assembly based pangenomes have now become preferential for crop plants.

4.2.3.2 Super-pangenome Construction

Super-pangenomes expand the pangenome concept to genus level, capturing interspecies genomic diversity and evolutionary patterns. This approach combines multiple species' pangenomes to identify conserved and divergent genomic features. For example, the Brassica super-pangenome revealed the impact of genome duplication and divergence on adaptive evolution, providing insights into the genomic mechanisms underlying species diversification (Golicz *et al.*, 2016). While super-pangenomes offer broader insights, they demand higher computational resources and require high-quality genomic data from multiple species. For example, in rice, a graph-based super-pangenome was constructed using 251 accessions from cultivated and wild Asian and African species, identifying genetic variants linked to submergence tolerance, seed shattering and plant architecture, with independent selection shaping adaptation and domestication in both regions (Shang *et al.*, 2022).

Table 4.2. Summary of pangenome studies and genome editing using CRISPR/Cas9 in major crop plants. (Authors' own table.)

Pangenome studies		
Crops	Number of accessions	Reference
Rice	73 genomes of Asian rice subpopulations and two wild relatives	Zhou *et al.* (2023)
	413 accessions	Wang *et al.* (2023)
	74 high-accuracy genomes	Wu *et al.* (2023)
	251 accessions	Shang *et al.* (2022)
	33 accessions	Qin *et al.* (2021)
Wheat	17 cultivars enhancing Eastern Asia diversity	Jiao *et al.* (2024)
	16 cultivars	Bayer *et al.* (2022)
	18 cultivars	Montenegro *et al.* (2017)
Barley	76 long-read and 1,315 short-read sequences	Jayakodi *et al.* (2024)
	300 accessions	Jayakodi *et al.* (2020)
Sorghum	13 accessions	Tao *et al.* (2021)
	354 accessions	Ruperao *et al.* (2021)
	62 accessions	Wang *et al.* (2021)
	394 accessions	Yan *et al.* (2023)
Soybean	7 accessions	Li *et al.* (2014)
	204 accessions	Torkamaneh *et al.* (2021)
Chickpea	Annual *Cicer* wild species	Khan *et al.* (2024)
Pigeonpea	89 accessions	Zhao *et al.* (2020)
Genome editing using CRISPR/Ca9		
Crop	Gene edited	Reference
Rice	*OsPYL9*	Usman et al. (2020)
	OsDjA2 and *OsERF104*	Távora *et al.* (2022)
	OsHPPD and *OsUbi2*	Lv *et al.* (2024)
	OsPIN5b, GS3, and *OsMYB30*	Zeng *et al.* (2020)
Wheat	γ- and ω-gliadin	Sánchez-León *et al.* (2024)
	Receptor-like protein kinase 1	Rahim *et al.* (2024)
	Sal1	Mohr *et al.* (2022)
Barley	D hordein	Li *et al.* (2020)
	HvARE1	Karunarathne *et al.* (2022)
Sorghum	cinnamyl alcohol dehydrogenase (*CAD*) and phytoene desaturase (*PDS*)	Liu *et al.* (2019)
	CCD gene	Hao *et al.* (2023)
Chickpea	Cytokinin dehydrogenase	Mahto *et al.* (2022)
Soybean	*GmF3H1, GmF3H2,* and *GmFNSII-1*	Zhang *et al.* (2020)
Pigeon pea	Phytoene desaturase	Prasad *et al.* (2024)

5.2.4 Utilization of Pangenomes to Identify Genetic Diversity and Structural Variations in Crop Populations

The utilization of pangenomes has revolutionized the study of genetic diversity and structural variations in crop populations by providing a comprehensive framework to capture the broad spectrum of genetic variation within a species (Thudi *et al.*, 2025). Unlike single reference genomes, pangenomes incorporate multiple genomes, enabling the discovery of structural variations that are absent from the reference genome but present in other individuals or subpopulations. This approach has been particularly transformative for major crops, such as wheat, rice, and maize. For instance, the recent construction of a pangenome for *O. sativa* identified over 13,000 genes that were absent from the commonly used reference genome, including genes associated with abiotic stress tolerance and improved grain yield, which can be utilized to breed climate-resilient rice varieties (Zhao *et al.*, 2018). Similarly, the maize pangenome revealed extensive presence-absence variations, encompassing over 10% of the total gene content and linked several of these variations to key agronomic traits, including resistance to northern leaf blight and nitrogen use efficiency (Hufford *et al.*, 2021) (Table 4.2). Advances in sequencing technologies and bioinformatics tools have further facilitated the assembly and analysis of pangenomes, as demonstrated by the soybean pangenome, which uncovered structural variations affecting flowering time, seed protein content and oil composition, providing valuable targets for precision breeding (Liu *et al.*, 2020). By integrating these insights into breeding programs, pangenomes serve as powerful resources to harness genetic diversity and address challenges in global agriculture production.

With the availability of whole-genome sequences for major crops such as rice (IRGSP, 2005), maize (Schnable *et al.*, 2009), soybean (Schmutz *et al.*, 2009), and pearl millet (Varshney *et al.*, 2017b), researchers gained unprecedented access to dense and evenly distributed markers across the genome. The emergence of next-generation sequencing (NGS) technologies enabled GWAS, which use single nucleotide polymorphisms (SNPs) to identify QTLs associated with complex traits such as yield, drought tolerance and disease resistance (Huang *et al.*, 2010; Varshney *et al.*, 2021b). GWAS marked a new milestone, as it allowed the analysis of natural populations instead of biparental populations, increasing the power and resolution of trait mapping. In legumes, genome sequencing of chickpea and pigeonpea accelerated the identification of loci underlying stress tolerance and nutritional traits (Varshney *et al.*, 2012; Varshney *et al.*, 2013). The integration of GS, which combines genotypic data and predictive models to estimate breeding values, further advances crop improvement (Meuwissen *et al.*, 2001; Crossa *et al.*, 2017). This era has also seen the development of pangenomes, providing insights into structural variations and gene presence-absence variations critical for breeding strategies (Zhao *et al.*, 2018).

While traditional genetic analyses relied on SNPs, the discovery of SVs necessitates new approaches to link genetic variation to traits. SVs in genomes arise through different mechanisms. Transposable elements can replicate within a genome and relocate gene sequences, potentially disrupting coding regions. Errors during meiotic recombination, such as unequal recombination and improper DNA repair, also lead to SVs (Della *et al.*, 2021). In plants, presence-absence variations (PAVs) can occur due to genome changes after whole-genome duplication, though their role can vary across species. Compared with single-reference genomes, pangenomes provide a more comprehensive representation of structural variations, including PAVs, duplications, inversions and translocations (Zanini *et al.*, 2022). These structural variants might influence phenotypic diversity and are linked to critical agronomic traits (Torkamaneh *et al.*, 2021; Yan *et al.*, 2023; Khan *et al.*, 2024). Recent research emphasizes the importance of analyzing non-coding regions, regulatory sequences, and repetitive elements within pangenomes, requiring high-quality genome assemblies and advanced tools such as graphical pangenomics (Wang *et al.*, 2023). Despite challenges in studying crops with high repeat content or complex polyploidy, pangenomics offers significant potential for identifying SVs linked to domestication (Cai *et al.*, 2021). For example, a TE insertion near the teosinte branched1 (*tb1*) gene significantly influenced maize domestication (Studer and Doebley, 2011). Continued innovation in data structures and functional analysis will enable deeper insights into plant pangenomes, advancing crop breeding and functional genomics. Understanding SVs is crucial for improving crop traits and guiding future breeding efforts.

4.2.5 Integration of Haplotypes to Design Breeding Strategies

The integration of haplotype-resolved genomes into breeding strategies has significantly advanced the precision and efficiency of crop improvement. By elucidating genetic diversity and structural variations, these genomes enable the identification of favorable alleles and facilitate MAS and GS. In cereal crops such as rice, wheat and maize, haplotype mapping has unraveled key traits including yield, drought tolerance and hybrid vigor, as seen in studies by Walkowiak *et al.* (2020). For instance, haplotype-based GWAS has been successfully used to uncover the genetic architecture of grain yield in wheat, identifying stable associations and epistatic interactions across diverse environments, which are essential for breeding resilient varieties (Sehgal *et al.*,

2020). Similarly, haplotypes have been utilized to pinpoint superior alleles linked to drought tolerance in rice, enabling the development of high-yielding, stress-resilient varieties through haplo-pheno analysis (Singh *et al.*, 2024a). In GS, haplotype blocks reduce redundancy in marker data by summarizing co-inherited features and capturing local epistasis, thereby improving prediction accuracy for complex traits in crops such as maize, wheat, and soybean. However, the impact varies by trait and prediction model (Weber *et al.*, 2023). Moreover, LD-based haplotype tagging enhances the efficiency of GS by optimizing marker pre-selection, which reduces genotyping costs and improves prediction accuracy for traits such as spike length and flag leaf area in wheat (Alemu *et al.*, 2023). These examples underscore the importance of haplotypes in refining genomic methodologies to accelerate crop improvement and address breeding challenges. Overall, haplotype-based breeding represents an effective approach for developing crop varieties tailored to address global food security challenges by optimizing yield, stress tolerance, and quality traits. Similarly, in legumes such as soybean, chickpea, and peanut, haplotype-based genomic approaches have identified critical traits such as disease resistance and nitrogen fixation efficiency, as reported by Liu *et al.* (2020) and Varshney *et al.* (2021). Comparative analyses highlight that while cereals often prioritize yield and abiotic stress resilience, legumes focus on nitrogen fixation and seed quality. The application of these insights, such as the incorporation of drought tolerance haplotypes in wheat and chickpea or disease resistance in soybean and maize, underscores their potential in breeding programs. Future efforts should emphasize expanding these genomic resources to underutilized crops and refining computational tools for haplotype analysis to ensure broader adoption and impact across agricultural systems.

4.2.6 Genome Editing for Crop Improvement

Genome editing involves advanced techniques to precisely modify genome sequences and regulate gene expression, offering significant advancements in agriculture. These methods effectively enable addition of beneficial traits or removal of undesirable ones, enhancing genetic and phenetic architecture of crops with simplicity and precision. Techniques such as zinc-finger nucleases (ZFNs), transcription-activator-like effector nucleases (TALENs) and the widely preferred clustered regularly interspaced short palindromic repeats (CRISPR)/CRISPR-associated protein 9 (Cas9) system have revolutionized plant breeding. CRISPR-Cas9 induces DNA double-strand breaks (DSBs), which can be repaired via three approaches: (a) site-directed nuclease 1 (SDN1) to introduce random mutations through non-homologous end joining (NHEJ) without templates; (b) site-directed nuclease 2 (SDN2) to enable precise edits using synthetic templates via homologous recombination (HR). Both approaches are considered regulatory-exempt as edited plants (Buchholzer and Frommer, 2023). Contrarily, site-directed nuclease 3 (SDN3) involves incorporation of genes through HR, resulting in transgenic crops.

CRISPR-Cas9 has been effectively utilized to reduce the severity of abiotic and biotic stresses by targeting specific genes in crops (Table 4.2). It has been employed to create novel allelic variants that enhance stress tolerance, such as drought-tolerant maize by producing a new allelic variant (Shi *et al.*, 2017). In rice, a CRISPR-Cas9-mediated knockout mutation in the *NIP* gene family significantly reduced arsenic concentration (Singh *et al.*, 2024b). Similarly, the simultaneous modification of two yield-related genes and one cold-tolerance gene in rice improved yield under cold stress conditions (Zeng *et al.*, 2020). CRISPR-Cas9 has also been used to improve salt stress tolerance by knocking out the *OsRR22* gene in rice (Zhang *et al.*, 2019). For biotic stress, CRISPR-Cas9 improved blast resistance in rice by knocking out the *OsERF922* gene (Wang *et al.*, 2016). Multiplex CRISPR-Cas9, a novel technique targeting multiple loci simultaneously, has advanced crop improvement further. For example, a drought-tolerant wheat cultivar was developed by knocking out the *TaSal1* gene, a negative regulator of drought tolerance (Abdallah *et al.*, 2022). Another innovative approach, *de novo* domestication, leverages CRISPR-Cas9 to edit agronomically important genes in wild plant progenitors. This technique enhances traits such as stress tolerance while restoring valuable wild genes. In tomato (*Solanum lycopersicum* L.), *de novo* domestication improved bacterial spot resistance and salt tolerance (Li *et al.*, 2018a, 2018b). These advancements demonstrate CRISPR-Cas9's potential to improve agricultural productivity and stress tolerance in crops.

The integration of CRISPR technology into global agriculture is shaped not only by scientific progress but also by complex regulatory and socioeconomic landscapes. Regulatory frameworks differ markedly across regions, reflecting varying philosophies on biotechnology oversight. In the USA, the USDA has adopted a product-based regulatory approach, exempting many CRISPR-edited crops – particularly those using SDN1 and SDN2 techniques – from traditional GMO regulations if the resulting traits could have been achieved through conventional breeding and do not involve transgenes or pesticidal properties (Ahmad *et al.*, 2023; Ahmad *et al.*, 2024; https://edis.ifas.ufl.edu/publication/HS1334). This facilitates faster commercialization and broader adoption. In contrast, the European Union applies a process-based framework, classifying most CRISPR-edited crops as GMOs, thereby subjecting them to strict approval protocols. Nonetheless, evolving legislative proposals seek to modernize EU regulations by differentiating among gene-editing approaches, which may lead to a more flexible pathway for certain CRISPR applications while preserving rigorous safety evaluations (https://www.isaaa.org/blog/entry/default.asp?BlogDate=5%2F7%2F2025).

Alongside these regulatory variations, socioeconomic barriers further complicate equitable access to CRISPR technologies. A heavily contested patent landscape – dominated by US and European stakeholders – can restrict availability to smaller research institutions and developing nations. For instance, crops such as CRISPR-edited tomatoes enriched in GABA content may face limited deployment in resource-poor regions due to intellectual property constraints (GABA-enriched tomato is first CRISPR-edited food to enter market., 2022). The concentration of CRISPR innovation among a few large biotech corporations risks exacerbating existing inequalities, particularly in staple crops such as maize and wheat that are vital to food security in low-income countries (Molinari *et al.*, 2024). Nations such as India and several African countries are thus challenged to ensure that transformative CRISPR advancements, such as drought-tolerant maize or disease-resistant cassava reach smallholder farmers and not just large agribusinesses (Munawar *et al.*, 2023). As the technology matures, there is a critical need for dynamic, science-based regulatory systems that address safety concerns – such as off-target effects and long-term ecological consequences – through comprehensive risk assessment frameworks. Equally important is fostering collaboration among scientists, regulators and policymakers to ensure that CRISPR is integrated responsibly and inclusively, building public trust and supporting global food and nutritional security.

4.2.7 Synthetic Biology and Metabolic Engineering

Synthetic biology and metabolic engineering have emerged as new disciplines in advancing crop improvement by applying engineering principles to biology. Synthetic biology enables the design of novel genetic circuits and stress-responsive pathways in crops, offering unprecedented opportunities to enhance climate resilience. By engineering plants with traits such as drought tolerance, heat resistance and improved nutrient use efficiency, synthetic biology can help agriculture adapt to increasingly unpredictable environments. These fields enable the design and construction of new biological parts, devices and systems, as well as the re-design of existing biological systems for useful purposes. Recent successes in these areas highlight their potential to address pressing challenges in agriculture. For instance, enhancing photosynthetic efficiency has shown promise in increasing crop yields. South *et al.* (2019) demonstrated a 14–20% increase in biomass in tobacco plants by accelerating recovery from photoprotection, a mechanism that safeguards plants from excessive light, thus serving as a model for improving other crops. Similarly, the microbial production of complex plant metabolites has been achieved through synthetic biology. For example, Levisson *et al.* (2021) engineered yeast to produce bioactive isoflavonoids, compounds with significant health benefits, paving the way for sustainable and scalable production of plant secondary metabolites.

Advancements in enzyme engineering have also accelerated the optimization of biocatalysts, as shown by Huber *et al.* (2023) who developed an automated platform for improving enzyme functions in various metabolic engineering processes, thereby enhancing the production of desired plant compounds. Moreover, modular biosynthesis platforms have enabled the *de novo* production of halogenated tryptophan derivatives in microbes, as demonstrated by Xu *et al.* (2023) showcasing synthetic biology's potential in producing pharmaceutical-grade natural products through engineered biosynthetic pathways. In another breakthrough, engineered eukaryote-like regulatory circuits in prokaryotes expanded artificial regulatory networks to allow more precise control over metabolic pathways and improved the efficiency of bioproduction efforts (Bartsch *et al.*, 2022). These advancements underscore the rapid progress in synthetic biology and metabolic engineering, providing tools for developing crops with enhanced traits, increased resilience, and improved nutritional profiles. Such innovations hold immense potential to contribute to sustainable agriculture and food security globally.

4.2.8 RNA-Based Technologies: RNAi, Small RNAs and Synthetic mRNA Vaccines

RNA based technologies have emerged as innovative tools in plant science, offering sustainable solutions to enhance crop resilience, productivity, and protection against pests and diseases. These technologies, including RNA interference (RNAi), small RNAs (sRNAs), and synthetic mRNA vaccines, have significantly transformed the management of challenges faced by cereals and legumes in modern agriculture. RNAi is a gene-silencing mechanism that utilizes double-stranded RNA (dsRNA) to degrade specific mRNA transcripts, thereby effectively silencing targeted genes. This technology has been extensively used to manage pests and pathogens in crops, providing an eco-friendly alternative to chemical pesticides. In legumes, RNAi has been deployed to control *Helicoverpa armigera*, a major pest in chickpea, by silencing the *Dicer-like* gene critical for pest survival, resulting in population suppression (Baum *et al.*, 2007). Similarly, in cereals, RNAi constructs targeting fungal pathogens such as *Fusarium graminearum* have shown significant success in controlling Fusarium Head Blight in wheat and barley, reducing yield losses and mycotoxin contamination (Moonjely *et al.*, 2023). These examples underscore the potential of RNAi in enhancing crop protection while minimizing environmental impact. sRNAs, including microRNAs (miRNAs) and small interfering RNAs (siRNAs), regulate gene expression at the post-transcriptional level and play a pivotal role in plant growth, development and stress responses. In legumes, miRNAs such as miR2111 regulate root nodulation by targeting TIR1/AFB auxin receptors, optimizing nitrogen fixation under low nitrogen conditions (Wang *et al.*, 2020). Additionally, siRNAs derived from transposable elements in maize enhance gene regulation under environmental stress, further demonstrating the role of sRNAs in crop adaptation (Slotkin *et al.*, 2009). Advances in sRNA profiling and biogenesis have facilitated their application in improving key traits such as grain size, disease resistance and abiotic stress tolerance, making them a valuable tool in modern plant breeding.

Synthetic mRNA vaccines represent a frontier technology in plant protection, leveraging RNA-based mechanisms to combat viral diseases. These vaccines encode viral coat proteins or RNA-dependent RNA polymerases, triggering RNAi like response in plants that suppresses viral replication. In cereals, synthetic mRNA approaches are being explored to manage Barley Yellow Dwarf Virus (BYDV), a significant threat to wheat and barley production (Tenllado *et al.*, 2003). In legumes, mRNA vaccines targeting Bean Common Mosaic Virus (BCMV) have shown promising results, providing a sustainable alternative to conventional virus control methods (Sharma *et al.*, 2022). Although RNA interference (RNAi) holds significant promise in crop protection and improvement, its practical application is hindered by several inherent limitations. For instance, a major constraint is its variable silencing efficiency and specificity, often resulting in inconsistent outcomes. For example, in *Helicoverpa armigera*, RNAi efficacy is reduced due to degradation of dsRNA in the alkaline gut environment, although chitosan nanoparticle formulations have shown some success in enhancing stability and delivery (Kolge *et al.*, 2021). Moreover, delivery remains a core challenge, particularly for exogenous applications such as spray-induced gene silencing (SIGS), where environmental factors such as UV light, temperature, and RNases rapidly degrade naked dsRNA, necessitating complex formulations (Germing *et al.*, 2025). Even with stabilization, cellular uptake can be inefficient, especially in pest species like Lepidoptera, which exhibit lower RNAi sensitivity due to barriers in dsRNA internalization and processing. Additionally, off-target effects remain a concern. Despite bioinformatics tools to predict cross-reactivity, empirical validation is still required to ensure specificity and minimize unintended impacts on non-target organisms or beneficial species (Vatanparast *et al.*, 2024). Regulatory complexity adds another layer of limitation, particularly in regions such as the EU, where RNAi products face classification ambiguities and lack standardized approval pathways. Finally, stable RNAi expression through transgenic plants raises biosafety and public acceptance concerns, while transient approaches often lack durability and consistent inheritance. These cumulative challenges underscore the need for refined delivery systems, improved target selection strategies and harmonized regulatory frameworks to fully realize the potential of RNAi in sustainable agriculture.

4.2.9 Functional Genomics and Genome-wide Association Studies

Functional genomics and GWAS have become indispensable tools for understanding gene function and identifying loci associated with agronomically important traits in crop plants. These approaches enable researchers to dissect complex traits such as drought tolerance, yield and disease resistance, providing valuable insights for crop improvement programs. Below, we discuss the applications and success stories in key crops, including sorghum, pearl millet, rice, chickpea, pigeonpea, peanut and soybean. Rice serves as a model crop for functional genomics due to its well-annotated genome. GWAS has been extensively used to map traits such as submergence tolerance with the identification of the *SUB1A* gene, which enhances survival during seedling stage submergence (Xu *et al.*, 2006). Recent studies have also focused on the genetic basis of yield-related traits, such as grain size and panicle architecture. For instance, the *GW5* gene was identified as a key regulator of grain width, contributing to yield improvement (Weng *et al.*, 2008). In sorghum, GWAS has been pivotal in identifying genes related to drought and heat tolerance. Further, GWAS also identified candidate genes encoding leucine-rich repeats (LRR), Avr proteins, lipoxygenases (LOXs), calmodulins (CAM) dependent protein kinase, WRKY transcription factors, flavonoid biosynthesis genes and 12-oxo-phytodienoic acid reductase associated with Aphid resistance in sorghum (Punnuri *et al.*, 2022; Thudi *et al.*, 2024). Similarly, GWAS identified loci associated with grain quality traits, including tannin content, which is important for nutritional and industrial applications (Chen *et al.*, 2021). Pearl millet, a staple food-grain crop in arid regions, has benefited from GWAS and transcriptomic analyses. In chickpea, GWAS has facilitated the identification of loci associated with drought and heat tolerance. Functional genomics has also revealed the role of the *Dehydrin* family of proteins in conferring drought tolerance, providing targets for genetic improvement. Pigeonpea, known for its resilience to harsh conditions, has seen advancements in trait mapping through GWAS. In peanut, GWAS has identified loci associated with oil content, seed size, and resistance to leaf spot diseases. A major breakthrough was the identification of *AhFAD2*, a gene controlling oleic acid content, which enhances the nutritional quality of peanut oil (Pandey *et al.*, 2014). Functional studies have also elucidated the role of *NBS-LRR* genes in conferring resistance to late leaf spots and rust. Functional genomics has also revealed pathways involved in nitrogen fixation, with key genes such as *Nod* factors playing a central role in symbiosis with rhizobia. The integration of functional genomics with GWAS accelerates the discovery of candidate genes and their validation. Multi-omics approaches, combining transcriptomics, proteomics and metabolomics in legumes, such as chickpea and soybean, are uncovering genes involved in stress responses and nutritional traits. As sequencing technologies advance, the application of pangenomics and haplotype mapping will further enrich functional genomics and GWAS, enabling the development of climate-resilient and high-yielding crop varieties.

4.2.10 Integration of Multi-omics for Crop Improvement

The integration of multi-omics studies, including genomics, transcriptomics, proteomics, and metabolomics, has revolutionized crop research and improvement. These approaches enable a comprehensive understanding of gene function, regulatory networks and metabolic pathways, facilitating targeted breeding strategies. In staple crops such as rice, wheat, maize, pearl millet, sorghum, chickpea, pigeonpea, peanut and soybean, multi-omics integration has been pivotal for improving traits such as yield, stress resilience and nutritional quality, though the extent and success vary among crops.

In rice, multi-omics studies are most advanced due to its well-annotated genome and global research focus. Genomic analyses, combined with transcriptomic profiling, have identified key regulatory networks for traits such as drought and submergence tolerance, including the discovery of the *SUB1A* gene for submergence tolerance (Xu *et al.*, 2006). In wheat (*Triticum aestivum*), the large and complex genome initially limited multi-omics studies, but advancements in sequencing have enabled integrative analyses. For example, combining genomics and transcriptomics revealed loci controlling grain size and yield, while proteomics and metabolomics identified pathways for enhanced gluten content and drought adaptation (Borrill *et al.*, 2018). In maize, multi-omics approaches have been successfully employed to dissect quantitative traits such as nitrogen use efficiency. Comparatively, in pearl millet, multi-omics studies are less extensive. Sorghum, a drought-tolerant cereal, has benefited from multi-omics integration, particularly for understanding the *Stay-Green* trait. Genomic studies have mapped key loci, while transcriptomic analyses have identified transcription factors regulating drought resilience (Mace *et al.*, 2013).

Proteomics and metabolomics further revealed secondary metabolites associated with stress adaptation, such as polyphenols and flavonoids. In legumes such as chickpea, transcriptomics has been instrumental in identifying genes associated with root architecture and nitrogen fixation, critical for drought adaptation (Kudapa *et al.*, 2024). Combining these with metabolomics has highlighted drought-responsive metabolites such as proline and trehalose, enhancing our understanding of legume resilience. Proteomics and metabolomics are still underutilized in this crop, underscoring the need for integrative studies. Similarly, in peanut, metabolomics combined with genomic data has identified biomarkers for high oleic acid content and resistance to leaf spot diseases, while proteomic studies remain sparse (Pandey *et al.*, 2014). Soybean has seen significant advancements in multi-omics integration due to its economic importance. Genomics and transcriptomics have identified key genes regulating flowering time and nodulation, while proteomics has elucidated proteins involved in abiotic stress tolerance (Naik *et al.*, 2025a, 2025b). Metabolomics has further enhanced our understanding of seed composition, identifying metabolites associated with protein and oil content. Compared with cereals, legumes such as pigeonpea and peanut lag in the comprehensive use of multi-omics approaches, but ongoing research is narrowing the gap. Despite the successes, several challenges remain. In crops such as pearl millet and pigeonpea, the lack of high-quality genomic resources limits the full potential of multi-omics integration. Moreover, while genomics and transcriptomics are widely used, proteomics and metabolomics are often underutilized due to technical and cost-related constraints. Comparing crops, cereals such as rice and maize have benefited from robust funding and extensive datasets, enabling advanced multi-omics research. In contrast, legumes require further investment to unlock their full potential. Future research should focus on integrating all omics layers, coupled with advanced computational tools, to fully exploit the genetic and biochemical diversity in these crops for sustainable agriculture.

Although genomics and transcriptomics dominate current strategies in crop improvement, proteomics and metabolomics offer critical insights into post-transcriptional and metabolic regulation that are essential for understanding complex stress responses. These layers remain underutilized, often due to technical complexity and limited integration pipelines. For instance, in wheat, Li *et al.* (2024) identified a kinase fusion protein, *AeIMLKL*, involved in powdery mildew resistance through proteomic analysis – a discovery missed by transcriptomic profiling alone. In another study, Kosová *et al.* (2022) demonstrated that drought-tolerant wheat cultivar Septima maintained higher levels of mitochondrial respiration proteins and redox-regulating enzymes under combined water deficit and aphid stress, highlighting proteomics as a powerful tool to dissect combined abiotic-biotic stress responses. These studies reinforce the value of integrating multi-omics – genomics, transcriptomics, proteomics and metabolomics – into a systems biology framework. Such integrative approaches not only enhance our understanding of stress adaptation but also refine trait selection and accelerate the development of climate-resilient cultivars.

4.3 Enabling Systems

4.3.1 Speed Breeding and Genomic Selection: Combining Speed Breeding Protocols with High-throughput Genomic Data for Rapid Trait Introgression

Recent advancements in plant breeding have demonstrated the efficacy of combining speed breeding protocols with high-throughput genomic data to accelerate trait introgression (Table 4.3). Speed breeding, which involves manipulating environmental conditions to shorten generation times, enables the rapid development of multiple plant generations annually. When integrated with high-throughput genomic technologies, this approach facilitates the swift incorporation of desirable traits into crop varieties (Ghosh *et al.*, 2018). For instance, in wheat, researchers have employed speed breeding alongside GS to expedite the introgression of disease-resistance genes. By utilizing genomic data to identify and select resistance-associated markers, breeders have successfully developed wheat lines with enhanced resistance to stem rust and other pathogens within a reduced timeframe (Watson *et al.*, 2018). This integration of technologies has significantly decreased the breeding cycle duration, allowing for the rapid deployment of improved cultivars to combat emerging diseases (Hickey *et al.*, 2019). Similarly, in pearl millet, the combination of speed breeding and high-throughput phenotyping has been utilized to enhance drought tolerance. By accelerating generation turnover and employing genomic tools to select drought-resilient traits, researchers have developed millet varieties capable of withstanding arid conditions (Voss-Fels *et al.*, 2019). This approach not only shortens the breeding timeline but also ensures the precision of trait incorporation, addressing critical challenges imposed by climate change. These examples underscore the transformative potential of integration of speed breeding protocols with high-throughput genomic data in modern agriculture. By enabling rapid and precise trait introgression, this combined approach holds enormous promise for the genetic enhancement of crop varieties that meet the demands of a growing global population and changing environmental conditions.

The deployment of GS in various crops has revolutionized modern breeding programs by accelerating genetic gains and improving precision in trait selection. In rice, large GS panels, such as 21,945 hybrids (Xu *et al.*, 2014), have facilitated significant advancements in yield stability, pest resistance, and tolerance to abiotic stresses. These improvements are critical for adaptation of rice varieties to diverse and changing environmental conditions (Table 4.3). Similarly, in wheat, GS has reduced breeding cycles and improved selection accuracy for disease resistance and yield traits. Studies involving 4464 breeding lines (Montesinos-López *et al.*, 2024) have demonstrated enhanced resistance to rust diseases and improved performance under diverse agro-climatic conditions, highlighting its role in maintaining food security. In barley, GS has contributed to the development of superior doubled haploid lines, as observed in a study with 980 such lines (Bhatta *et al.*, 2020), enabling improvements in malting quality, drought resilience and biomass yield. Chickpea breeding, based on panels of 315 breeding lines, has also benefited from GS with increased precision in selecting traits such as early flowering, drought tolerance and pod filling (Li *et al.*, 2022). This has expedited the development of climate-resilient varieties suited for semi-arid regions. For peanut, the application of GS using 192 genotypes (Lubanga *et al.*, 2024) has significantly advanced breeding efforts for aflatoxin resistance, disease tolerance and kernel quality, addressing both productivity and market demands. Soybean has similarly seen transformative gains with GS, particularly in developing

Table 4.3. Summary of enabling systems (speed breeding and genomic selection) deployed for future proofing of crops. (Authors' own table.)

Examples of speed breeding		
Crop	Controlled conditions adopted	Reference
Rice	24 h day photoperiod for the initial 15 days	Kabade *et al.* (2024)
	Developed a vertical hydroponic breeding system with LED lighting in a closed plant factory	Liu *et al.* (2024)
	Confirmed that one generation can be achieved within 68–75 days	Sandhu *et al.* (2024)
	Demonstrated that reducing cycle time and refining target environments can enhance genetic gain	Seck *et al.* (2024)
	Gene *hst1* and controlled temperature and day length	Rana *et al.* (2019)
Wheat	Synergized breeding strategy reduced the time to develop a BC_2 near-isogenic line	Cha *et al.* (2024)
	An energy-efficient system reduced breeding time and cut energy use by 80%	Cha *et al.* (2023)
	22 h day/2 h night, 25°C/22°C	Schoen *et al.* (2023)
	Constant temperature (17°C ± 2°C) with 24 h diurnal natural light	Alahmad *et al.* (2018)
	Photoperiod: 22 h/2 h; Temperature: 25°C (light), 22°C (dark)	Chhabra *et al.* (2024)
Barley	22 h of light and 2 h of darkness	Rossi *et al.* (2024)
	Controlled temperature and constant (24 h) light conditions	Hickey *et al.* (2017)
Chickpea	Fresh immature green seed sowing enables F_2–F_5 progression in 1.5 years	Gurumurthy *et al.* (2024)
	Light intensity greatly affected flowering, maturity and plant stress compared with normal conditions	Mitache *et al.* (2024)
	12 h light period from 18:00 to 06:00 still first flower emergence	Samineni *et al.* (2020)
Soybean	10 h light (30°C), 14 h dark (25°C), CO_2 (550 ppm)	Taku *et al.* (2024)
	Shorter day length and higher temperature	Fang *et al.* (2021)
Pigeonpea	13 h:8 h:13 h photoperiod for vegetative, flowering, and pod filling stages	Gangashetty *et al.* (2024)

Genomic selection		
Crop	Composition of genomic selection panel	Reference
Rice	334 synthetic lines	Baertschi *et al.* (2021)
	669 breeding lines	Biswas *et al.* (2023)
	1495 hybrids	Cui *et al.* (2020)
	323 accessions	Huang *et al.* (2019)
	21,945 hybrids	Xu *et al.* (2014)
Wheat	597 lines	Kaushal *et al.* (2024)
	4464 lines	Montesinos-López *et al.* (2024)
	420 accessions	Juliana *et al.* (2020)
	273 breeding lines	He *et al.* (2016)
	254 breeding lines	Poland *et al.* (2012)
Barley	2250 lines	Raffo *et al.* (2025)
	980 doubled haploid lines	Bhatta *et al.* (2020)
	116 and 294	Schmidt *et al.* (2016)
	Parent set with subsequent progeny sets	Sallam *et al.* (2014)
Sorghum	95 genotypes	Behera *et al.* (2024)
	602 hybrids	Sapkota *et al.* (2023)

Continued

Table 4.3. Continued.

Examples of speed breeding		
Crop	Controlled conditions adopted	Reference
Chickpea	315 breeding lines	Li *et al.* (2022)
	132 varieties	Li *et al.* (2018)
	320 breeding lines	Roorkiwal *et al.* (2016, 2018)
Peanut	192 genotypes	Lubanga *et al.* (2024)
	775 accessions	Manley *et al.* (2024)
Soybean	1501 breeding lines	Miller *et al.* (2023a)
	702 breeding lines	Miller *et al.* (2023b)
	301 genotypes	Bandillo *et al.* (2022)
	284 accessions and 180 inbred lines	Qin *et al.* (2022)
Pearl Millet	242 accessions	Yan *et al.* (2024)
	320 hybrids	Jarquin *et al.* (2020)

high-yielding, disease-resistant varieties with improved oil quality. Studies employing 1501 breeding lines (Miller *et al.*, 2023a) have highlighted the increased genetic gains achievable through genomic prediction as compared with traditional methods. These outcomes collectively demonstrate the versatility and scalability of GS across crops, effectively addressing critical agricultural challenges such as climate resilience, biotic and abiotic stress management, and the demand for higher productivity. The integration of GS into breeding programs underscores its potential to future-proof global agriculture by enhancing the efficiency and effectiveness of crop improvement strategies in a rapidly changing world.

4.3.2 Plant Microbiome Engineering

Microbiome engineering has emerged as a critical frontier in genomic interventions aimed at enhancing crop resilience and sustainability. The plant microbiome, encompassing diverse microorganisms such as bacteria, fungi, archaea, viruses and protists, is intimately associated with plant tissues and significantly influences plant health, nutrient uptake and stress tolerance. Advances in metagenomics and sequencing technologies have illuminated the functional capabilities of these microbial communities, allowing researchers to identify key microbes that contribute to plant performance. Beneficial microorganisms such as nitrogen-fixing bacteria (e.g. *Rhizobium*) and phosphate-solubilizing fungi (e.g. *Arbuscular Mycorrhizal Fungi*, AMF) play pivotal roles in nutrient cycling and plant growth promotion. For instance, AMF has demonstrated the ability to enhance phosphorus uptake and confer drought tolerance in crops such as maize and soybean (Bodenhausen *et al.*, 2014; Wubs *et al.*, 2016). These insights provide a foundation for designing microbial-based interventions tailored to counteract biotic and abiotic stresses. In the face of challenges such as drought, salinity and pathogen pressure exacerbated by climate change, microbiome engineering offers a sustainable alternative to agrochemical reliance. By leveraging plant-microbe interactions and optimizing microbial partnerships, crops can be fortified for productivity and resilience.

The synergy between plant genetic engineering and microbiome manipulation is driving a significant change in crop improvement strategies. Genomic tools, particularly CRISPR-Cas9 technology, have enabled researchers to modify plant genomes to actively shape their microbiomes. By engineering plants to produce specific root exudates, researchers can attract beneficial microbes that enhance nutrient acquisition and stress tolerance. For example, root exudates such as coumarins and lignin precursors selectively recruit microbial taxa that improve phosphorus solubilization and pathogen resistance (Beckers and Spoel, 2016; Stringlis *et al.*, 2018). Moreover, advancements in synthetic biology and precision breeding are being integrated with microbiome research to optimize the composition and functionality of plant-associated microbial communities. This approach fosters a holistic crop improvement strategy where plant genetics and microbial partnerships are co-engineered. Additionally, agricultural practices such as crop rotation and intercropping are being optimized to support a diverse and stable microbiome, further enhancing soil health and crop productivity (Berendsen *et al.*, 2012). The concept of "microbiome genes" (M genes) has also gained traction, highlighting host genetic traits that influence microbiome composition. Selective incorporation of M genes into breeding programs offers a pathway to reduce agrochemical dependency while boosting resistance to pathogens and environmental stressors (Escudero-Martinez *et al.*, 2022).

Despite enormous potential, microbiome engineering faces significant challenges that must be addressed to realize its full benefits in agricultural sustainability. The dynamic and context-dependent nature of microbial communities poses a hurdle in ensuring the stability and efficacy of engineered microbiomes across diverse environments. Microbial inoculants often struggle with competition from native microbiota, leading to inconsistent performance. Regulatory and public concerns about the release of genetically engineered plants and their interactions with native ecosystems also require careful consideration. Furthermore, a deeper understanding of the molecular mechanisms governing plant–microbe interactions is essential for refining microbiome engineering approaches. GWAS are crucial for identifying genes that regulate microbiome composition and functionality. For example, endophytic bacteria in rice have been shown to confer complete disease resistance, emphasizing the potential of microbiome-based solutions to address crop vulnerabilities (Matsumoto *et al.*, 2021). Additionally, exploring orphan crops, which often exhibit unique plant–microbe interactions, can provide insights into novel genetic and microbial resources for crop improvement. By addressing these challenges and advancing precision in microbiome engineering, agricultural systems can transition towards sustainable practices that ensure global food security in the face of climate change (Ali *et al.*, 2024).

Microbiome-based technologies are advancing rapidly, offering transformative potential for sustainable agriculture and climate-resilient crop systems. In Brazil, soybean inoculation with nitrogen-fixing *Bradyrhizobium* spp. has become a model for large-scale biological nitrogen fixation, saving over $13 billion annually in fertilizer costs (Patyal *et al.*, 2025). In maize, inoculation with drought-tolerant *Bacillus* spp. significantly improved plant biomass, photosynthetic efficiency, and drought tolerance, demonstrating their utility under limited water conditions (Azeem *et al.*, 2022). Similarly, *Piriformospora indica*, a salt-tolerant endophytic fungus, has been successfully used to enhance ion homeostasis and antioxidant activity in crops such as rice and tomato under salinity stress (Reshna *et al.*, 2022). Recent studies have highlighted the efficacy of microbial consortia in enhancing crop growth under various stress conditions. For instance, a consortium of *Pseudomonas syringae*, *Enterobacter aerogenes*, and *P. fluorescens* improved maize growth and yield under salt-affected field conditions by regulating osmolyte and antioxidant balances (Afzal *et al.*, 2023). In wheat, a multi-strain bacterial consortium demonstrated superior performance in promoting growth under salinity stress compared with single-strain inoculations (Khan *et al.*, 2022). Additionally, the application of a microbial consortium comprising *Azotobacter chroococcum* and *Trichoderma afroharzianum* significantly increased tomato yield and improved plant water relations under reduced water and nitrogen inputs (Cirillo *et al.*, 2023). Furthermore, the concept of M gene breeding – leveraging host genetic variation to enrich beneficial microbiota – has been demonstrated in rice, where lignin precursors like *p*-coumaric acid and scopoletin selectively recruited protective microbial communities (Cernava, 2024). These examples highlight the growing capacity of microbiome engineering to enhance crop productivity, abiotic stress tolerance, and reduce dependence on chemical inputs.

4.3.3 Nanobiotechnology for Genomics Mediated Crop Improvement

A pivotal advancement is the nanobiotechnology field such as development of nanocarriers for gene delivery, which significantly improves the precision and efficiency of genome editing in plants. Traditional methods, such as Agrobacterium-mediated transformation and biolistics, often face limitations in species compatibility and potential tissue damage. In contrast, engineered nanocarriers, including mesoporous silica nanoparticles and carbon nanotubes, offer a non-invasive alternative for delivering genetic material into plant cells. These nanocarriers facilitate the direct delivery of genome-editing tools such as CRISPR-Cas9 ribonucleoproteins, enabling transgene-free genome editing with reduced off-target effects. This method not only enhances editing efficiency but also aligns with regulatory frameworks favoring non-transgenic modifications, thereby accelerating the development of climate-resilient crop varieties (Cunningham *et al.*, 2018). In addition to gene delivery, nanobiotechnology has propelled advancements in plant disease diagnostics through the creation of nano-based diagnostic tools for rapid genotyping. Early and accurate detection of plant pathogens is crucial for effective disease management, especially under climate-induced stress conditions that can exacerbate susceptibility. Nanomaterials, such as gold nanoparticles and quantum dots, have been employed to develop nanosensors capable of detecting specific nucleic acids or proteins associated with pathogens. These nanosensors offer high sensitivity and specificity, enabling the prompt identification of diseases at early stages. For instance, gold nanoparticle-based assays have been utilized for the rapid detection of plant viruses, facilitating timely intervention and mitigating crop losses. The integration of these nano-based diagnostics into agricultural practices enhances the capacity for real-time monitoring and management of plant health, contributing to the sustainability and productivity of crops in changing climates (Kashyap *et al.*, 2017). The convergence of nanobiotechnology with genomics holds immense potential for future-proofing crops against the challenges posed by climate change. By enabling precise genome editing and rapid disease diagnostics, these technologies contribute to the development of crop varieties that are not only resilient to environmental stresses but also capable of maintaining high yields and quality. As research in this domain progresses, the continued refinement and application of nanocarriers and nano-based diagnostics are expected to play a pivotal role in securing global food security in the face of an evolving climate.

Despite their promising potential for targeted delivery and enhanced crop resilience in agriculture, nanocarriers raise significant biosafety and environmental concerns. Key issues include nanoparticle accumulation in plant tissues – sometimes leading to phytotoxicity or entry into food chains; potential toxicity to non-target organisms such as soil microbes and pollinators; and long-term ecological impacts such as persistent soil contamination and groundwater leaching. Currently, there is no globally harmonized regulatory framework for nano-enabled agricultural products; most countries rely on existing chemical or pesticide regulations, which are often inadequate for addressing the unique properties of nanomaterials. In the USA, nano-pesticides are evaluated case-by-case under the Federal Insecticide, Fungicide, and Rodenticide Act (FIFRA), while the European Union applies general chemical frameworks such as REACH that lack nano-specific guidelines. Some countries, such as India and Canada, are developing more tailored approaches, emphasizing risk assessment and life-cycle evaluation. Rigorous, nanomaterial-specific risk assessment protocols and transparent regulatory guidelines are essential for ensuring the safe and sustainable deployment of nanocarriers in agriculture, alongside ongoing research into biodegradable nanocarriers and exposure mitigation strategies. Moreover, access to such cutting-edge nanotechnologies remains limited for smallholder farmers due to high production costs, limited supply chains and the need for technical infrastructure and expertise, potentially widening the technological divide in global agriculture (Kráľová and Jampílek, 2023).

4.3.4 Precision Agriculture and Digital Genomics

Precision agriculture integrates cutting-edge technologies to optimize crop production and resource use efficiency at the field and farm levels. Recently, high-throughput phenotyping approaches have been deployed for trait mapping in both cereals and legumes. For instance, the LeasyScan platform at the International Crops Research Institute for the Semi-Arid Tropics (ICRISAT) employs 3D laser scanning to assess plant canopies, ground-robot and drone-based phenotyping approaches facilitating climate-smart sorghum breeding (Boatwright *et al.*, 2024;

Hivare *et al.*, 2024). While in the case of rice, a novel imaging platform integrates RGB, near-infrared and fluorescence imaging to evaluate drought tolerance traits (Kim *et al.*, 2020). Additionally, the LemnaTec 3D Scanalyzer system has been utilized for non-invasive screening of salinity tolerance in rice (Gill *et al.*, 2022). Robotic technologies have also been integrated into HTP; for example, the Robotanist, a ground-based agricultural robot, enables high-throughput crop phenotyping (Das *et al.*, 2017). Similarly, the Vinobot and Vinoculer robotic platforms facilitate detailed field phenotyping of individual plants and entire fields, respectively (Jin *et al.*, 2017). These technologies enable precise, non-destructive, and rapid phenotyping, accelerating crop improvement efforts across diverse species. Tools such as GPS-guided equipment, remote sensing and IoT-enabled devices have revolutionized farming practices, enabling site-specific crop management. Phenoshpere when complemented by digital genomics, which harnesses genomic information to identify and enhance traits such as drought tolerance, disease resistance, and nutrient use efficiency, offers a holistic approach to addressing environmental challenges and enhancing productivity (Heuermann *et al.*, 2023). The convergence of artificial intelligence (AI) and machine learning (ML) with precision agriculture and genomics has further transformed this landscape. AI algorithms analyze vast datasets from sensors, drones, and genomic studies, providing real-time decision-making capabilities for optimal crop management. For instance, ML models can predict crop yields based on environmental factors and genetic data, offering actionable insights that help farmers maximize productivity (Khan *et al.*, 2021). This synergy creates an ecosystem where data-driven genomics and AI-guided precision farming collectively meet the demands of sustainable agriculture. However, access to these advanced technologies remains uneven; smallholder farmers in low-resource settings often face barriers due to high costs, limited digital infrastructure, and a lack of localized decision-support tools, potentially exacerbating global disparities in agricultural innovation and productivity (Yuan and Sun, 2024).

The application of AI and ML in digital genomics has been significantly advanced by developments in high-throughput sequencing and computational biology. These innovations generate vast genomic datasets, revealing complex relationships between genetic variations and phenotypic traits. AI tools such as deep learning models have been particularly effective in analyzing pan-genomic datasets, uncovering structural variations linked to traits such as heat tolerance and pest resistance (Shook and Edger, 2021). AI-driven genome annotation systems streamline the identification of functional genes, enabling breeders to target specific loci for improvement. Furthermore, ML models are instrumental in predicting gene expression patterns and integrating multi-omics datasets, such as genomics, transcriptomics and proteomics, which underpin precision breeding programs. These integrative approaches have enabled breeders to develop tailored crop varieties suited to specific environments and stress conditions (Hirsch and Springer, 2021). Initiatives such as the Agricultural Genome to Phenome Initiative (AG2PI) emphasize the role of AI and ML in advancing agricultural genomics and phenomics, addressing critical challenges such as global food security (AG2PI Consortium, 2021). Nevertheless, concerns persist around algorithmic transparency, reproducibility, and data bias – particularly when training datasets lack representation from diverse agroecological regions. These issues may limit model accuracy and reinforce inequities unless explicitly addressed through inclusive, explainable AI frameworks (Tzachor *et al.*, 2022; Atapattu *et al.*, 2024).

The combination of AI, ML and genome editing technologies, such as CRISPR-Cas9, has further expanded the scope of precision agriculture. Genome editing allows precise modifications of crop genomes, introducing desirable traits without the drawbacks of traditional breeding methods. AI models enhance the specificity and efficiency of CRISPR systems by predicting off-target effects and optimizing guide RNA design (Dara *et al.*, 2024). Moreover, AI-powered platforms analyze environmental data to recommend optimal crop varieties and farming practices tailored to local conditions, enabling sustainable intensification. Integrating omics data, supported by AI and ML, also generates information pertinent to adaptive crop management strategies that respond dynamically to climate variability. Beyond genomics, ML algorithms are transforming genomic analyses by identifying patterns and predictive features directly from data, making them particularly valuable for analyzing complex datasets (Libbrecht and Noble, 2015).

4.3.5 Gene Drives for Pest and Disease Control: A Revolution in Precision Pest Management

Gene drives represent a powerful technology in precision pest management, designed to control pest populations and mitigate vector-borne diseases by promoting the inheritance of specific genetic traits (Brossard *et al.*, 2019). This approach ensures the rapid spread of modifications through populations, offering precise and sustainable solutions for agriculture. A notable application is the CRISPR-based gene drive targeting *Drosophila suzukii*, a pest of soft fruits (Yadav *et al.*, 2023). By disrupting the *doublesex* gene essential for female fertility, researchers achieved significant population suppression in laboratory trials, offering a model for other pests (Kandul *et al.*, 2019). Beyond insect pests, gene drives show promise in managing fungal pathogens affecting cereals and legumes. By spreading deleterious traits, these drives can reduce pathogen virulence and disease incidence (Gardiner *et al.*, 2020). While offering sustainable alternatives to chemical pesticides, the deployment of gene drives necessitates addressing ecological risks, ethical concerns and regulatory challenges. Integration with precision agriculture, digital genomics, and AI could further optimize their application. By leveraging these advancements, gene drives have the potential to revolutionize pest management, reduce chemical inputs and enhance global food security. However, ensuring their safe and effective use will require collaboration among researchers, policymakers, and stakeholders.

4.4 Climate-smart Crop Varieties

Recent advancements in genomics-assisted breeding have led to the development and release of several climate-resilient crop varieties with enhanced traits, enabling sustainable agricultural production in the face of climate change. For instance, in the case of rice, herbicide-tolerant varieties such as 'Pusa Basmati 1979' and 'Pusa Basmati 1985' enable direct-seeded systems, reducing water and labor needs (chrome-extension://efaidnbmnnnibpcajpcglclefindmkaj/; https://www.iari.res.in/files/Latest-News/Launch_of_Seed_24052024.pdf). Additionally, 'Swarna Sub1', a submergence-tolerant variety, was developed through MAS, providing farmers with a reliable option in flood-prone areas (Singh *et al.*, 2009). 'DRR Dhan 50' was released for its resistance to bacterial blight, while 'CR Dhan 801', developed for coastal saline soils, ensures stable yields under high salinity (chrome-extension://efaidnbmnnnibpcajpcglclefindmkaj/; https://icar-nrri.in/wp-content/uploads/2022/04/RB-34.pdf). In pearl millet, significant strides have been made in addressing downy mildew, a devastating disease. Varieties such as 'HHB 67 Improved', developed through marker-assisted selection, feature enhanced resistance to downy mildew (chrome-extension://efaidnbmnnnibpcajpcglclefindmkaj/; https://oar.icrisat.org/5661/1/HHB67_Flyer_2009.pdf). Additionally, broken resistance to downy mildew in earlier varieties has been successfully reconstituted using genomics-assisted approaches, ensuring long-term durability of resistance and yield stability. Similarly, sorghum varieties with improved drought tolerance and grain quality are vital for food security in arid and semi-arid regions (Mwamahonje *et al.*, 2024). These genomics-driven advancements provide robust solutions to challenges posed by climate change, ensuring resilience and productivity across diverse agro-climatic zones. In chickpea, varieties such as 'Pusa 10216', a drought-tolerant cultivar suitable for dryland farming (Bharadwaj *et al.*, 2021), and 'Super Annigeri 1', which is high-yielding and disease-resistant (Mannur *et al.*, 2019), have been developed. Additionally, 'Pusa Manav', an early-maturing variety suitable for mechanical harvesting, addresses both climate and labor constraints (Bharadwaj *et al.*, 2022a, 2022b). In peanut, high oleic acid varieties like 'Girnar 4' and 'Girnar 5', with enhanced oil quality and shelf life, were developed through genomics-assisted breeding to support cultivation in water-limited conditions (Pandey *et al.*, 2024) (Table 4.4).

Table 4.4. Summary of marker-assisted breeding efforts and developed/released varieties in major crops for enhanced traits. (Authors' own table.)

Crop	Trait	MABC/ GAB	Variety released/ under pipeline	Reference
Rice	Bacterial blight resistance	MABC	Pusa Basmati 1718	Singh *et al.* (2018)
	Bacterial blight resistance	MABC	Pusa Basmati 1728	Singh *et al.* (2017b)
	Bacterial blight resistance	MABC	Improved Samba Mahsuri	Sundaram *et al.* (2008)
	Neck blast resistant	MABC	Pusa Basmati 1637	Singh *et al.* (2017a)
	Tolerance against biotic stresses and drought stress	MAFB and MABC	8 MAFB lines and 12 MABC lines	Ramayya *et al.* (2021)
	Salinity tolerance	MABC	Introgression lines	Yadav *et al.* (2020)
	Drought tolerance	MABC	Introgressed lines	Dhawan *et al.* (2020)
	Herbicide tolerance		PB 1121	Grover *et al.* (2020)
Wheat	Drought tolerance	MABC	HD3411	Kumar *et al.* (2023a)
	Leaf rust resistance	MABC	DWR 162	Yadawad *et al.* (2017)
	Leaf rust and drought tolerance	MABC	HD3471	Sunilkumar *et al.* (2022)
	Triple rust resistance	MABC	HD3407	Mallick *et al.* (2024)
	Fusarium head blight resistance	MABC	Introgressed lines	Hu *et al.* (2023)
	Stripe rust resistance and agronomic performance	MABC	Introgressed lines	Zhang *et al.* (2022)
Chickpea	Drought tolerance	MABC	Pusa 10216	Bharadwaj *et al.* (2021)
	Fusarium wilt resistance	MABC	Super Annigeri 1 and improved JG 74	Mannur *et al.* (2019)
	Fusarium wilt resistance	MABC	Introgressed lines	Varshney *et al.* (2014a)
	Highly resistant to Fusarium wilt and stunt	MABC	BG 4005	Bharadwaj *et al.* (2021)
	High yielding Fusarium wilt resistant	MABC	Pusa Manav/Pusa Chickpea 20211	Bharadwaj *et al.* (2022)
Peanut	Foliar disease resistance and high oleic acid	MABC	Introgression lines	Shasidhar *et al.* (2020)
	High oleic acid content	MABC	Improved GPBD 4 and G 2–52	Jadhav *et al.* (2021)
	High oleic acid content	MABC	IL-23	Vargheese *et al.* (2024)
	Bacterial wilt resistance, high shelling percentage and early maturity	MABC	Yuhua183	Fang *et al.* (2022)
	High oleic acid content	MABC	YH61	Tang *et al.* (2022)
	Higher kernel yield, higher oil, oleic acid, and shelling percentage	MAS	GG 39 (ICGV 16697)	State Variety Release Committee of Gujarat
	High oleic acid content	GAB	Girnar 4 and Girnar 5	Annual Groundnut Workshop held in Andhra University, Vizag during 25th-27th May 2019
Barley	Abiotic stress	MABC	Introgressed lines	Mikołajczak *et al.* (2016)
Soybean	Kunitz trypsin inhibitor free	MABC	NRC 127	Kumar *et al.* (2020)
Sorghum	Striga-resistant	MABC	ASARECA.W2 Striga	BMZ-supported project involving the International Crops Research
	Grain mold resistance	MABC		

Continued

Table 4.4. Continued.

Crop	Trait	MABC/ GAB	Variety released/ under pipeline	Reference
Pearl Millet	Downy mildew resistance	MABC	HHB 67 Improved	https://colostate.pressbooks.pub/pgrsuccessstories/chapter/pearl-millet-hhb-67-improved-downy-mildew-resistance/?utm_source=chatgpt.com
	Tolerance to terminal drought stress	MABC	HHB 226 Improved	Rani *et al.* (2022)

Note: Genomic Assisted Breeding (GAB); Marker Assisted Back Crossing (MABC); marker aided selection (MAS)

4.5 Challenges, Opportunities and Future Perspective

The ever-growing global food demand, compounded by climate change, poses significant challenges to sustainable agriculture. Rising global temperatures, projected to increase by 1.5°C by 2050 and potentially up to 4°C by 2100, threaten crop yields and exacerbate food insecurity (IPCC, 2021). These temperature increases are expected to reduce yields of staple crops such as wheat, maize, rice, and soybeans by 3–7% for every degree of warming, with drought, flooding, and soil salinity intensifying these impacts (Zhao *et al.*, 2017; Prodhan *et al.*, 2022). Developing climate-resilient crops faces technical hurdles, such as the complexity of multigenic traits, the lack of comprehensive genomic datasets and the variable efficiency of genetic interventions across diverse environments. Additionally, integrating microbiome engineering and genome editing technologies, such as CRISPR-Cas9, into breeding programs is challenged by the dynamic nature of microbial communities, regulatory hurdles, and concerns about unintended ecological consequences (Ali *et al.*, 2024). Furthermore, the limited application of precision agriculture tools in developing countries underscores disparities in technology access and capacity building, creating a critical need for equitable dissemination of innovations. Despite these challenges, advancements in genomic technologies, data analytics and precision agriculture present unprecedented opportunities for addressing global agricultural challenges. High-throughput sequencing and pangenome construction have revolutionized crop genomics, uncovering genetic diversity and structural variations essential for breeding climate-resilient crops (Zhao *et al.*, 2018; Hufford *et al.*, 2021). ML and AI are accelerating genomic analyses by identifying complex patterns in large datasets, optimizing genome editing tools and predicting phenotypes from genotypes (Libbrecht and Noble, 2015; Xu and Jackson, 2019). For instance, AI-driven CRISPR platforms enhance the specificity and efficiency of genome edits, enabling precise trait improvement (Dara *et al.*, 2024). In addition, microbiome engineering offers potential for improving nutrient uptake, stress tolerance and disease resistance through targeted plant–microbe interactions, supported by advances in metagenomics and synthetic biology (Beckers and Spoel, 2016; Escudero-Martinez *et al.*, 2022). Integrating these technologies with speed breeding and digital genomics can shorten breeding cycles and enable rapid deployment of improved crop varieties, addressing food security and environmental sustainability simultaneously (Watson *et al.*, 2018; Voss-Fels *et al.*, 2019).

The seamless integration of emerging technologies into holistic crop improvement frameworks further enhances crop improvement greatly. Enhancing genomic resources through pangenomics, super-pangenomes and high-quality genome assemblies will be critical for capturing the full spectrum of genetic variation and enabling precision breeding strategies. Expanding the application of AI and ML across various plant phenotyping, genomic prediction and comparative genomics will further optimize breeding programs and enhance decision-making in precision agriculture (Khan *et al.*, 2021; Shook and Edger, 2021). Addressing regulatory and societal concerns through transparent communication and equitable technology dissemination is imperative for fostering public trust and ensuring global adoption. Future efforts should focus on integrating genomic tools with traditional practices, optimizing microbiome engineering for diverse agro-ecological systems, and leveraging underutilized crops to broaden the genetic base for food production (Ali *et al.*, 2024). By addressing these challenges and harnessing the opportunities, agriculture can transition towards a resilient and equitable system capable of feeding the growing global population under changing environmental conditions.

References

Abdallah, N.A., Elsharawy, H., Abulela, H.A., Thilmony, R., Abdelhadi, A.A. and Elarabi, N.I. (2022) Multiplex CRISPR/Cas9-mediated genome editing to address drought tolerance in wheat. *GM Crops & Food* 16(1), 1–17. DOI: 10.1080/21645698.2022.2120313.

Adotey, R.E., Patrignani, A., Bergkamp, B., Kluitenberg, G., Prasad, P.V. and Jagadish, S.K. (2021). *Crop Science* 61(4), 2680–2695. DOI: 10.1002/csc2.20532.

Afzal, A., Khan, M.Y., Zahir, Z.A., Asghar, H.N., Muhmood, A. *et al.* (2023) Plant growth-promoting bacterial consortia improved the physiology and growth of maize by regulating osmolytes and antioxidants balance under salt-affected field conditions. *Heliyon* 9(7), e17816. DOI: 10.1016/j.heliyon.2023.e17816.

AG2PI Consortium (2021) The Agricultural Genome to Phenome Initiative (AG2PI): Creating a shared vision across crop and livestock research. *Genome Biology* 25(143), 100382. DOI: 10.1186/s13059-021-02570-1.

Ahmad, A., Jamil, A. and Munawar, N. (2023) GMOs or non GMOs? The CRISPR conundrum. *Frontiers in Plant Science* 14, 1232938. DOI: 10.3389/fpls.2023.1232938.

Ahmad, A., Hoffman, N.E., Jones, M.G.K. and Zhang, B. (2024) Editorial: Frontiers in global regulatory landscape of CRISPR-edited plants. *Frontiers in Plant Science* 15, 1367698. DOI: 10.3389/fpls.2024.1367698.

Alahmad, S., Dinglasan, E., Leung, K.M., Riaz, A., Derbal, N. *et al.* (2018) Speed breeding for multiple quantitative traits in durum wheat. *Plant Methods* 14, 1–15. DOI: 10.1186/s13007-018-0302-y.

Alemu, A., Batista, L., Singh, P.K., Ceplitis, A. and Chawade, A. (2023) Haplotype-tagged SNPs improve genomic prediction accuracy for Fusarium head blight resistance and yield-related traits in wheat. *Theoretical and Applied Genetics* 136(4), 92. DOI: 10.1007/s00122-023-04352-8.

Ali, Q., Ali, M., Jing, H., Hussain, A., Manghwar, H. *et al.* (2024) Power of plant microbiome: A sustainable approach for agricultural resilience. *Plant Stress* 100681. DOI: 10.1016/j.stress.2024.100681.

Assefa, Y., Staggenborg, S.A. and Prasad, V.P. (2010) Grain sorghum water requirement and responses to drought stress: A review. *Crop Management* 9(1), 1–11. DOI: 10.1094/CM-2010-1109-01-RV.

Atapattu, A.J., Perera, L.K., Nuwarapaksha, T.D., Udumann, S.S. and Dissanayaka, N.S. (2024) Challenges in achieving artificial intelligence in agriculture. In: Chouhan, S.S., Saxena, A., Singh, U.P., Jain, S. (eds) *Artificial Intelligence Techniques in Smart Agriculture*. Springer, Singapore. DOI: 10.1007/978-981-97-5878-4_2.

Azeem, M., Haider, M.Z., Javed, S., Saleem, M.H. and Alatawi, A. (2022) Drought stress amelioration in maize (*Zea mays* L.) by inoculation of *Bacillus* spp. strains under sterile soil conditions. *Agriculture* 12, 50. DOI: 10.3390/agriculture12010050.

Baertschi, C., Cao, T.V., Bartholomé, J., Ospina, Y., Quintero, C. *et al.* (2021) Impact of early genomic prediction for recurrent selection in an upland rice synthetic population. *G3* 11(12), jkab320. DOI: 10.1093/g3journal/jkab320.

Bailey-Serres, J., Parker, J.E., Ainsworth, E.A., Giles, E.O. and Julian, I.S. (2019) Genetic strategies for improving crop yields. *Nature* 575, 109–118. DOI: 10.1038/s41586-019-1679-0.

Bandillo, N.B., Jarquin, D., Posadas, L.G., Lorenz, A.J. and Graef, G.L. (2023) Genomic selection performs as effectively as phenotypic selection for increasing seed yield in soybean. *The Plant Genome* 16(1), e20285. DOI: 10.1002/tpg2.20285.

Bartsch, R., Beckmann, M. and Weuster-Botz, D. (2022) Engineering eukaryote-like regulatory circuits in prokaryotes to expand artificial regulatory networks. *Communications Biology* 5(1), 1–13. DOI: 10.1038/s42003-022-03070-z.

Baum, J.A., Bogaert, T., Clinton, W., Heck, G.R., Feldmann, P. *et al.* (2007) Control of coleopteran insect pests through RNA interference. *Nature Biotechnology* 25(11), 1322–1326. DOI: 10.1038/nbt1359.

Bayer, P.E., Petereit, J., Durant, É., Monat, C., Rouard, M. *et al.* (2022) Wheat Panache: A pangenome graph database representing presence–absence variation across sixteen bread wheat genomes. *The Plant Genome* 15(3), 20221. DOI: 10.1002/tpg2.20221.

Beckers, G.J. and Spoel, S.H. (2016) Fine-tuning plant defence signalling: Salicylate versus jasmonate. *Frontiers in Plant Science* 8(1), 1–10. DOI: 10.1055/s-2005-872705.

Behera, P.P., Singode, A., Bhat, B.V., Ronda, V., Borah, N. *et al.* (2024) Genetic gains in forage sorghum for adaptive traits for non-conventional area through multi-trait-based stability selection methods. *Frontiers in Plant Science* 15, 1248663. DOI: 10.3389/fpls.2024.1248663.

Berendsen, R.L., Pieterse, C.M. and Bakker, P.A. (2012) The rhizosphere microbiome and plant health. *Trends in Plant Science* 17(8), 478–486. DOI: 10.1016/j.tplants.2012.04.001.

Bernardo, R. (2020) Reinventing quantitative genetics for plant breeding: Something old, something new, something borrowed, something BLUE. *Heredity* 125, 375–385. DOI: 10.1038/s41437-020-0312-1.

Bharadwaj, C., Tripathi, S., Soren, K.R., Thudi, M., Singh, R.K. *et al.* (2021) Introgression of "QTL-hotspot" region enhances drought tolerance and grain yield in three elite chickpea cultivars. *The Plant Genome* 14, e20076. DOI: 10.1002/tpg2.20076.

Bharadwaj, C., Jorben, J., Rao, A., Roorkiwal, M., Patil, B.S. *et al.* (2022a) Development of high yielding Fusarium wilt resistant cultivar by pyramiding of "genes" through marker-assisted backcrossing in chickpea (*Cicer arietinum* L.). *Frontiers in Genetics* 13, 924287. DOI: 10.3389/fgene.2022.924287.

Bharadwaj, C., Jorben, J., Rao, A., Roorkiwal, M., Patil, B.S. *et al.* (2022b) Development of high-yielding Fusarium wilt resistant cultivar by pyramiding of "genes" through marker-assisted backcrossing in chickpea (*Cicer arietinum* L.). *Frontiers in Genetics* 13, 924287. DOI: 10.3389/fgene.2022.924287.

Bhatta, M., Gutierrez, L., Cammarota, L., Cardozo, F., Germán, S. *et al.* (2020) Multi-trait genomic prediction model increased the predictive ability for agronomic and malting quality traits in barley (*Hordeum vulgare* L.). *G3: Genes. Genomes, Genetics* 10, 1113–1124. DOI: 10.1534/g3.119.400968.

Biswas, P.S., Ahmed, M.E., Afrin, W., Rahman, A., Shalahuddin, A.K.M. *et al.* (2023) Enhancing genetic gain through the application of genomic selection in developing irrigated rice for the favorable ecosystem in Bangladesh. *Frontiers in Genetics* 14, 1083221. DOI: 10.3389/fgene.2023.1083221.

Boatwright, J.L., Sapkota, S., Jin, H., Schnable, J.C., Brenton, Z., Boyles, R. and Kresovich, S. (2021) Sorghum association panel whole-genome sequencing establishes pivotal resource for dissecting genomic diversity. *The Plant Journal* 111(3), 888–904. DOI: 10.1111/tpj.15853.

Boatwright, L., Thudi, M., Sangireddy, M.K.R., Coffin, A.W., Tadesse, H.K. *et al.* (2024) GWAS analysis for plant height and stem diameter in sorghum using multiple phenotyping approaches. *The Plant Phenome Journal* 7(1), e70008. DOI: 10.1002/ppj2.70008.

Bodenhausen, N., Bortfeld-Miller, M., Ackermann, M. and Vorholt, J.A. (2014) A synthetic community approach reveals plant genotypes affecting the phyllosphere microbiota. *PLoS Genetics* 10(4), e1004283. DOI: 10.1371/journal.pgen.1004283.

Borrill, P., Harrington, S.A. and Uauy, C. (2018) Applying the latest advances in genomics and phenomics for trait discovery in polyploid wheat. *The Plant Journal* 97(1), 56–72. DOI: 10.1111/tpj.14150.

Botstein, D., White, R.L., Skolnick, M. and Davis, R.W. (1980) Construction of a genetic linkage map in man using restriction fragment length polymorphisms. *American Journal of Human Genetics* 32(3), 314–331. Available at: https://pmc.ncbi.nlm.nih.gov/articles/PMC1686077/ (accessed 20 March 2025).

Brossard, D., Belluck, P., Gould, F. and Wirz, C.D. (2019) Promises and perils of gene drives: Navigating the communication of complex, post-normal science. *Proceedings of the National Academy of Sciences USA* 116(16), 7692–7697. DOI: 10.1073/pnas.1805874115.

Buchholzer, M. and Frommer, W.B. (2023) An increasing number of countries regulate genome editing in crops. *New Phytologist* 237(1), 7–8. DOI: 10.1111/nph.18333.

Burr, B., Burr, F.A., Thompson, K.H., Albertson, M.C. and Stuber, C.W. (1988) Gene mapping with recombinant inbreds in maize. *Genetics* 118, 519–526. DOI: 10.1093/genetics/118.3.519.

Cai, X., Chang, L., Zhang, T., Chen, H., Zhang, L. *et al.* (2021) Impacts of allopolyploidization and structural variation on intraspecific diversification in *Brassica rapa*. *Genome Biology* 22(1), 166. DOI: 10.1186/s13059-021-02383-2.

Cernava, T. (2024) Coming of age for microbiome gene breeding in plants. *Nature Communication* 15, 6623. DOI: 10.1038/s41467-024-50700-7.

Cha, J.K., Park, H., Choi, C., Kwon, Y., Lee, S.M. *et al.* (2023) Acceleration of wheat breeding: Enhancing efficiency and practical application of the speed breeding system. *Plant Methods* 19, 118. DOI: 10.1186/s13007-023-01083-1.

Cha, J.K., Park, H., Kwon, Y., Lee, S.M., Jang, S.G., Kwon, S.W. and Lee, J.H. (2024) Synergizing breeding strategies via combining speed breeding, phenotypic selection, and marker-assisted backcrossing for the introgression of Glu-B1i in wheat. *Frontiers in Plant Science* 15, 1402709. DOI: 10.3389/fpls.2024.1402709.

Chen, L., Guo, Q., Zhu, Z., Wan, H., Qin, Y. and Zhang, H. (2021) Integrated analyses of the transcriptome and small RNA of the hemiparasitic plant *Monochasma savatieri* before and after establishment of parasite-host association. *BMC Plant Biology* 21, 1–20. DOI: 10.1186/s12870-021-02861-6.

Chhabra, B., Thrasu, S., Wallace, S., Schoen, A., Shahoveisi, F. *et al.* (2024) Evaluation of speed breeding conditions for accelerating Fusarium head blight and deoxynivalenol screening in wheat. *Crop Science* 64, 1586–1594. DOI: 10.1002/csc2.21226.

Cirillo, V., Romano, I., Woo, S.L., Stasio, E. Di, Lombardi, N. *et al.* (2023) Inoculation with a microbial consortium increases soil microbial diversity and improves agronomic traits of tomato under water and nitrogen deficiency. *Frontiers in Plant Science* 14, 1304627. DOI: 10.3389/fpls.2023.1304627.

Clauw, P., Ellis, T.J., Liu, H.J. and Sasaki, E. (2024) Beyond the standard GWAS – A guide for plant biologists. *Plant and Cell Physiology* 11, pcae079. DOI: 10.1093/pcp/pcae079.

Crop genomes and beyond (2020). *Nature Genetics* 52, 865. DOI: 10.1038/s41588-020-0691-5.

Crossa, J., Pérez-Rodríguez, P., Cuevas, J., Montesinos-López, O., Jarquín, D. *et al.* (2017) Genomic selection in plant breeding: Methods, models, and perspectives. *Trends in Plant Science* 22(11), 961–975. DOI: 10.1016/j.tplants.2017.08.011.

Cucci, G., Lacolla, G., Boari, F., Mastro, M.A. and Cantore, V. (2019) Effect of water salinity and irrigation regime on maize (*Zea mays* L.) cultivated on clay loam soil and irrigated by furrow in southern Italy. *Agricultural Water Management* 222, 118–124. DOI: 10.1016/j.agwat.2019.05.033.

Cui, Y., Li, R., Li, G., Zhang, F., Zhu, T. *et al.* (2020) Hybrid breeding of rice via genomic selection. *Plant Biotechnology Journal* 18(1), 57–67. DOI: 10.1111/pbi.13170.

Cunningham, F.J., Goh, N.S., Demirer, G.S., Matos, J.L. and Landry, M.P. (2018) Nanoparticle-mediated delivery towards advancing plant genetic engineering. *Trends in Biotechnology* 36(9), 882–897. DOI: 10.1016/j.tibtech.2018.03.009.

Dara, M., Dianatpour, M., Azarpira, N. and Omidifar, N. (2024) Convergence of CRISPR and artificial intelligence: A paradigm shift in biotechnology. *Human Genetics* 2024, 201297. DOI: 10.1016/j.humgen.2024.201297.

Das, A., Alabi, O., and Aghazadeh, F. (2017) The Robotanist: A ground-based agricultural robot for high-throughput crop phenotyping. In *2017 IEEE International Conference on Computational Intelligence and Virtual Environments for Measurement Systems and Applications (CIVEMSA)*, (pp. 106–111) IEEE. Annecy, France.

Dawood, M.F., Moursi, Y.S., Amro, A., Baenziger, P.S. and Sallam, A. (2020) Investigation of heat-induced changes in the grain yield and grains metabolites, with molecular insights on the candidate genes in barley. *Agronomy* 10, 1730. DOI: 10.3390/agronomy10111730.

Della, C.R., Qiu, Y., Ou, S., Hufford, M.B. and Hirsch, C.N. (2021) How the pan-genome is changing crop genomics and improvement. *Genome Biology* 22, 1–19. DOI: 10.1186/s13059-020-02224-8.

Dhawan, G., Kumar, A., Dwivedi, P., Gopala Krishnan, S., Pal, M. *et al.* (2021) Introgression of qDTY1.1 governing reproductive stage drought tolerance into an elite basmati rice variety “Pusa Basmati 1” through marker-assisted backcross breeding. *Agronomy* 11(2), 202. DOI: 10.3390/agronomy11020202.

Djanaguiraman, M., Narayanan, S., Erdayani, E. and Prasad, P.V. (2020) Effects of high temperature stress during anthesis and grain filling periods on photosynthesis, lipids and grain yield in wheat. *BMC Plant Biology* 20, 1–12. DOI: 10.1186/s12870-020-02479-0.

Sabagh, A. El, Islam, M.S., Skalicky, M., Ali Raza, M., Singh, K. *et al.* (2021) Salinity stress in wheat (*Triticum aestivum* L.) in the changing climate: Adaptation and management strategies. *Frontiers in Agronomy* 3, 661932. DOI: 10.3389/fagro.2021.661932.

Ellur, R.K., Khanna, A., Yadav, A.K., Magdum, S., Maurya, S.K. *et al.* (2025) Development of dual disease resistant Basmati rice varieties offer significant economic advantage and impetus to sustainable crop production. *Journal of Advanced Research* 78, 63–77. DOI: 10.1016/j.jare.2025.02.014.

Elshire, R.J., Glaubitz, J.C., Sun, Q., Poland, J.A., Kawamoto, K., Buckler, E.S. and Mitchell, S.E. (2011) A robust, simple genotyping-by-sequencing (GBS) approach for high diversity species. *PLoS ONE* 6(5), e19379. DOI: 10.1371/journal.pone.0019379.

Escudero-Martinez, C., Bulgarelli, D. and Cope, K.R. (2022) Harnessing host genetic control of root microbiota for sustainable crop improvement. *Current Opinion in Microbiology* 5(61), 257–277. DOI: 10.1146/annurev-phyto-021621-121447.

Fang, Y., Wang, L., Sapey, E., Fu, S., Wu, T. *et al.* (2021) Speed-breeding system in soybean: Integrating off-site generation advancement, fresh seeding, and marker-assisted selection. *Frontiers in Plant Science* 12, 717077. DOI: 10.3389/fpls.2021.717077.

Fang, Y., Liu, H., Zhang, Z., Qi, F., Chen, Q. *et al.* (2022) Genomic identification and phenotypic evaluation of yield traits and bacterial wilt resistance in high oleic peanut breeding lines developed by marker-assisted backcrossing. *Euphytica* 218(6), 80. DOI: 10.1007/s10681-022-03032-w.

FAO, IFAD, UNICEF, WFP and WHO (2023) In *Brief to the State of Food Security and Nutrition in the World. Urbanization, Agrifood Systems Transformation and Healthy Diets across the Rural–Urban Continuum*. Rome, FAO. DOI: 10.4060/cc6550en.

GABA-enriched tomato is first CRISPR-edited food to enter market. 2022. Available at: https://www.nature.com/articles/d41587-021-00026-2.epdf?no_ publisher_access=1&r3_referer=nature

Gangashetty, P.I., Belliappa, S.H., Bomma, N., Kanuganahalli, V., Sajja, S.B. *et al.* (2024) Optimizing speed breeding and seed/pod chip-based genotyping techniques in pigeonpea: A way forward for high throughput line development. *Plant Methods* 20(1), 27. DOI: 10.1186/s13007-024-01155-w.

Gangurde, S.S., Xavier, A., Naik, Y.D., Jha, U.C., Rangari, S.K. *et al.* (2022) Two decades of association mapping: Insights on disease resistance in major crops. *Frontiers in Plant Science* 13, 1064059. DOI: 10.3389/fpls.2022.1064059.

Gao, L., Gonda, I., Sun, H., Ma, Q., Bao, K. *et al.* (2019) The tomato pan-genome uncovers new genes and a rare allele regulating fruit flavor. *Nature Genetics* 51(6), 1044–1051. DOI: 10.1038/s41588-019-0410-2.

Gardiner, D.M., Rusu, A., Barrett, L., Hunter, G.C. and Kazan, K. (2020) Natural gene drives offer potential pathogen control strategies in plants. *bioRxiv*. DOI: 10.1101/2020.04.05.026500.

Germing, K., Navarrete, C.A.D., Schiermeyer, A., Hommen, U., Zühl, L., Eilebrecht, S. and Eilebrecht, E. (2025) Crop protection by RNA interference: A review of recent approaches, current state of developments and use as of 2013. *Environmental Sciences Europe* 37(1), 15. DOI: 10.1186/s12302-025-01052-6.

Ghosh, S., Watson, A., Gonzalez-Navarro, O.E., Ramirez-Gonzalez, R.H., Yanes, L. *et al.* (2018) Speed breeding in growth chambers and glasshouses for crop breeding and model plant research. *Nature Protocols* 13(12), 2944–2963. DOI: 10.1038/s41596-018-0072-z.

Gill, T., Gill, S.K., Saini, D.K., Chopra, Y., Koff, J.P. de and Sandhu, K.S. (2022) A comprehensive review of high throughput phenotyping and machine learning for plant stress phenotyping. *Phenomics* 2, 156–183. DOI: 10.1007/s43657-022-00048-z.

Golicz, A.A., Bayer, P.E., Barker, G.C., Edger, P.P., Kim, H. *et al.* (2016) The pangenome of an agronomically important crop plant *Brassica oleracea*. *Nature Communications* 7, 1–8. DOI: 10.1038/ncomms13390.

Gore, M.A., Chia, J.M., Elshire, R.J., Sun, Q., Ersoz, E.S. *et al.* (2009) A first-generation haplotype map of maize. *Science* 326(5956), 1115–1117. DOI: 10.1126/science.1183168.

Grover, N., Kumar, A., Yadav, A.K., Gopala Krishnan, S., Ellur, R.K. *et al.* (2020) Marker-assisted development and characterization of herbicide-tolerant near-isogenic lines of a mega Basmati rice variety, "Pusa Basmati 1121". *Rice* 13, 1–13. DOI: 10.1186/s12284-020-00423-2.

Grzybowski, M.W., Mural, R.V., Xu, G., Turkus, J., Yang, J. and Schnable, J.C. (2023). *The Plant Journal* 113(6), 1109–1121. DOI: 10.1111/tpj.16123.

Gurumurthy, S., Ashu, A., Kruthika, S., Solanke, A.P., Basavaraja, T. *et al.* (2024) An innovative natural speed breeding technique for accelerated chickpea (*Cicer arietinum* L.) generation turnover. *Plant Methods* 20(1), 177. DOI: 10.1186/s13007-024-01299-9.

Hao, J., Yang, Y., Futrell, S., Kelly, E.A., Lorts, C.M. *et al.* (2023) CRISPR/Cas9-mediated mutagenesis of carotenoid cleavage dioxygenase (CCD) genes in sorghum alters strigolactone biosynthesis and plant biotic interactions. *Phytobiomes Journal* 7(3), 339–351. DOI: 10.1094/PBIOMES-08-22-0053-R.

He, S., Schulthess, A.W., Mirdita, V., Zhao, Y., Korzun, V. *et al.* (2016) Genomic selection in a commercial winter wheat population. *Theoretical and Applied Genetics* 129, 641–651. DOI: 10.1007/s00122-015-2655-1.

Heuermann, M.C., Knoch, D., Junker, A. and Altmann, T. (2023) Natural plant growth and development achieved in the IPK PhenoSphere by dynamic environment simulation. *Nature Communications* 14(1), 5783. DOI: 10.1038/s41467-023-41332-4.

Hickey, L.T., Germán, S.E. and Pereyra, S.A. (2017) Speed breeding for multiple disease resistance in barley. *Euphytica* 213, 64. DOI: 10.1007/s10681-016-1803-2.

Hickey, L.T., Robinson, H., Jackson, S.A., Bertioli, S.A., Tester, M. *et al.* (2019) Breeding crops to feed 10 billion. *Nature Biotechnology* 37(7), 744–754. DOI: 10.1038/s41587-019-0152-9.

Hirsch, C.N. and Springer, N.M. (2021) How the pan-genome is changing crop genomics and improvement. *Genome Biology* 22(3), 1–19. DOI: 10.1186/s13059-020-02224-8.

Hivare, V., Sivasakthi, K. and Choudhary, S. (2024) Advancing sorghum improvement through high-throughput phenotyping technologies. In: Habyarimana, E., Nadeem, M.A., Baloch, F.S., Zencirci, N. (eds) *Omics and Biotechnological Approaches for Product Profile-Driven Sorghum Improvement*. Springer, Singapore. DOI: 10.1007/978-981-97-4347-6_4.

Hoang, E. and Stephenson, P. (2024) *Ascophyllum nodosum* SWE enhances root anatomy, but not POD activity in both a salt-tolerant and salt-sensitive soybean (*Glycine max*) variety exposed to salt stress. *MicroPublication Biology.* DOI: 10.17912/micropub.biology.001046.

Hu, W.J., Fu, L.P., Gao, D.R., Li, D.S., Liao, S. and Lu, C.B. (2023) Marker-assisted selection to pyramid *Fusarium* head blight resistance loci *Fhb1* and *Fhb2* in the high-quality soft wheat cultivar Yangmai 15. *Journal of Integrative Agriculture* 22(2), 360–370. DOI: 10.1016/j.jia.2022.08.057.

Huang, S., Zhang, W., Yu, X. and Huang, Q. (2010) Effects of long-term fertilization on corn productivity and its sustainability in an Ultisol of southern China. *Agriculture, Ecosystems and Environment* 138(1–2), 44–50. DOI: 10.1016/j.agee.2010.03.015.

Huang, M., Balimponya, E.G., Mgonja, E.M., McHale, L.K., Luzi-Kihupi, A., Wang, G.L. and Sneller, C.H. (2019) Use of genomic selection in breeding rice (*Oryza sativa* L.) for resistance to rice blast (*Magnaporthe oryzae*). *Molecular Breeding* 39, 1–16. DOI: 10.1007/s11032-019-1023-2.

Huber, F., Büttel, M. and Tissot, L. (2023) An automated in vivo enzyme engineering platform for biocatalyst optimization. *Nature Communications* 14, 5678. DOI: 10.1038/s41467-024-46574-4.

Hufford, M.B., Seetharam, A.S., Woodhouse, M.R., Chougule, K.M., Ou, S. *et al.* (2021) De novo assembly, annotation, and comparative analysis of 26 diverse maize genomes. *Science* 373(6575), 655–662. DOI: 10.1126/science.abg5289.

IPCC (2021) *Special Report on Global Warming of 1.5°C.* Available at: https://www.ipcc.ch/sr15/ (accessed 20 March 2025).

IRGSP (International Rice Genome Sequencing Project) (2005) The map-based sequence of the rice genome. *Nature* 436, 793–800. DOI: 10.1038/nature03895.

Jaccoud, D., Peng, K., Feinstein, D. and Kilian, A. (2001) Diversity arrays: A solid-state technology for sequence information-independent genotyping. *Nucleic Acids Research* 29(4), e25. DOI: 10.1093/nar/29.4.e25.

Jadhav, M.P., Patil, M.D., Hampannavar, M., Venkatesh, D., Dattatreya, P. *et al.* (2021) Enhancing oleic acid content in two commercially released peanut varieties through marker-assisted backcross breeding. *Crop Science* 61(4), 2435–2443. DOI: 10.1002/csc2.20512.

Jarquin, D., Howard, R., Liang, Z., Gupta, S.K., Schnable, J.C. and Crossa, J. (2020) Enhancing hybrid prediction in pearl millet using genomic and/or multi-environment phenotypic information of inbreds. *Frontiers in Genetics* 10, 1294. DOI: 10.3389/fgene.2019.01294.

Jayakodi, M., Padmarasu, S., Haberer, G., Bonthala, V.S., Gundlach, H. *et al.* (2020) The barley pan-genome reveals the hidden legacy of mutation breeding. *Nature* 588(7837), 284–289. DOI: 10.1038/s41586-020-2947-8.

Jayakodi, M., Lu, Q., Pidon, H., Rabanus-Wallace, M.T., Bayer, M. *et al.* (2024) Structural variation in the pangenome of wild and domesticated barley. *Nature* 636, 654–662. DOI: 10.1038/s41586-024-08187-1.

Jiao, C., Xie, X., Hao, C., Chen, L., Xie, Y. *et al.* (2024) Pan-genome bridges wheat structural variations with habitat and breeding. *Nature* 637, 384–393. DOI: 10.1038/s41586-024-08277-0.

Jin, X., Sun, J., Wu, L., Zheng, H. and Scherer, S. (2017) Vinobot and Vinoculer: Two robotic platforms for high-throughput field phenotyping. *Sensors* 17(1), 214. DOI: 10.3390/s17010214.

Juliana, P., Singh, R.P., Braun, H.J., Huerta-Espino, J., Crespo-Herrera, L. *et al.* (2020) Genomic selection for grain yield in the CIMMYT wheat breeding program—Status and perspectives. *Frontiers in Plant Science* 11, 564183. DOI: 10.3389/fpls.2020.564183.

Jumrani, K., Bhatia, V.S., Kataria, S. and Landi, M. (2022) Screening soybean genotypes for high-temperature tolerance by maximin-minimax method based on yield potential and loss. *Agronomy* 12(11), 2854. DOI: 10.3390/agronomy12112854.

Kabade, P.G., Dixit, S., Singh, U.M., Alam, S., Bhosale, S., Kumar, S. and Singh, V.K. (2024) SpeedFlower: A comprehensive speed breeding protocol for indica and japonica rice. *Plant Biotechnology Journal* 22(5), 1051–1066. DOI: 10.1111/pbi.14245.

Kandul, N.P., Liu, J., Sanchez, C.H.M., Wu, S.L., Marshall, J.M. and Akbari, O.S. (2019) Transforming insect population control with precision-guided sterile males with demonstration in flies. *Nature Communications* 10(1), 84. DOI: 10.1038/s41467-018-07964-7.

Karunarathne, S.D., Han, Y., Zhang, X.Q. and Li, C. (2022) CRISPR/Cas9 gene editing and natural variation analysis demonstrate the potential for HvARE1 in improvement of nitrogen use efficiency in barley. *Journal of Integrative Plant Biology* 64(3), 756–770. DOI: 10.1111/jipb.13214.

Kashyap, P.L., Rai, P., Srivastava, A.K. and Kumar, S. (2017) Trichoderma for climate-resilient agriculture. *World Journal of Microbiology and Biotechnology* 33, 1–18. DOI: 10.1007/s11274-017-2319-1.

Kaushal, N., Awasthi, R., Gupta, K., Gaur, P., Siddique, K.H. and Nayyar, H. (2013) Heat-stress-induced reproductive failures in chickpea (*Cicer arietinum*) are associated with impaired sucrose metabolism in leaves and anthers. *Functional Plant Biology* 40(12), 1334–1349. DOI: 10.1071/FP13082.

Kaushal, S., Gill, H.S., Billah, M.M., Khan, S.N., Halder, J. *et al.* (2024) Enhancing the potential of phenomic and genomic prediction in winter wheat breeding using high-throughput phenotyping and deep learning. *Frontiers in Plant Science* 15, 1410249. DOI: 10.3389/fpls.2024.1410249.

Kaye, A.M. and Wasserman, W.W. (2021) The genome atlas: Navigating a new era of reference genomes. *Trends in Genetics* 37(9), 807–818. DOI: 10.1016/j.tig.2020.12.002.

Khan, A., Islam, M.Z. and Hou, X. (2021) Machine learning in precision agriculture: Applications, challenges, and prospects. *Computers and Electronics in Agriculture* 184, 106075. DOI: 10.1016/j.compag.2021.106075.

Khan, M.Y., Nadeem, S.M., Sohaib, M., Waqas, M.R., Alotaibi, F. *et al.* (2022) Potential of plant growth promoting bacterial consortium for improving the growth and yield of wheat under saline conditions. *Frontiers in Microbiology* 13, 958522. DOI: 10.3389/fmicb.2022.958522.

Khan, M.A., Waseem Akram, M., Iqbal, M., Ghulam, M.H., Rehman, A., Arslan, I.M. and Alam, B. (2023) Multivariate and association analyses of quantitative attributes reveal drought tolerance potential of wheat (*Triticum aestivum* L.) genotypes. *Communications in Soil Science and Plant Analysis* 54(2), 178–195. DOI: 10.1080/00103624.2022.2110893.

Khan, A.W., Garg, V., Sun, S., Gupta, S., Dudchenko, O. *et al.* (2024) Cicer super-pangenome provides insights into species evolution and agronomic trait loci for crop improvement in chickpea. *Nature Genetics* 56(6), 1–10. DOI: 10.1038/s41588-024-01760-4.

Kim, S.L., Kim, N., Lee, H., Lee, E., Cheon, K.-S. *et al.* (2020) High-throughput phenotyping platform for analyzing drought tolerance in rice. *Planta* 252(38), 1–18. DOI: 10.1007/s00425-020-03545-6.

Kolge, H., Kadam, K., Galande, S., Lanjekar, V. and Ghormade, V. (2021) New frontiers in pest control: Chitosan nanoparticles-shielded dsRNA as an effective topical RNAi spray for gram podborer biocontrol. *ACS Applied Bio Materials* 4(6), 5145–5157. DOI: 10.1021/acsabm.1c00349.

Kosová, K., Vítámvás, P., Skuhrovec, J., Vítámvás, J., Planchon, S., Renaut, J. and Saska, P. (2022) Proteomic responses of two spring wheat cultivars to the combined water deficit and aphid (*Metopolophium dirhodum*) treatments. *Frontiers in Plant Science* 13, 1005755. DOI: 10.3389/fpls.2022.1005755.

Kráľová, K., and Jampílek, J. (2023) Effects of nanoparticles/nanotubes on plant growth. In *The Impact of Nanoparticles on Agriculture and Soil* (pp. 183–237). Academic Press, England. DOI: 10.1016/B978-0-323-91703-2.00001-4.

Kudapa, H., Ghatak, A., Barmukh, R., Chaturvedi, P., Khan, A. *et al.* (2024) Integrated multi-omics analysis reveals drought stress response mechanism in chickpea (*Cicer arietinum* L.). *Plant. Genome* 17(1), e20337. DOI: 10.1002/tpg2.20337.

Kumar, A., Bernier, J., Verulkar, S., Lafitte, H.R. and Atlin, G.N. (2008) Breeding for drought tolerance: Direct selection for yield, response to selection and use of drought-tolerant donors in upland and lowland-adapted populations. *Field Crops Research* 107(3), 221–231. DOI: 10.1016/j.fcr.2008.02.007.

Kumar, V., Rani, A., Mourya, V. and Rawal, R. (2020) NRC 127: First Kunitz trypsin inhibitor free Indian soybean variety developed through marker assisted backcrossing. *Soybean Research* 5, 98. DOI: 10.5555/20203478621.

Kumar, P.K., Bellundagi, A., Krishna, H., Mallikarjuna, M.G., Thimmappa, R.K. *et al.* (2023) Development of bread wheat (*Triticum aestivum* L.) variety HD3411 following marker-assisted backcross breeding for drought tolerance. *Frontiers in Genetics* 14, 1046624. DOI: 10.3389/fgene.2023.1046624.

Levisson, M., Patin, D. and Voshol, G.P. (2021) Engineering yeast for the production of bioactive isoflavonoids. *Nature Communications* 12, 4342. DOI: 10.1038/s41467-021-26361-1.

Li, Y.H., Zhou, G., Ma, J., Jiang, W., Jin, L.G. *et al.* (2014) De novo assembly of soybean wild relatives for pan-genome analysis of diversity and agronomic traits. *Nature Biotechnology* 32(10), 1045–1052. DOI: 10.1038/nbt.2979.

Li, T., Yang, X., Yu, Y., Si, X., Zhai, X. *et al.* (2018a) Domestication of wild tomato is accelerated by genome editing. *Nature Biotechnology* 36(12), 1160–1163. DOI: 10.1038/nbt.4273.

Li, Y., Ruperao, P., Batley, J., Edwards, D., Khan, T. *et al.* (2018b) Investigating drought tolerance in chickpea using genome-wide association mapping and genomic selection based on whole-genome resequencing data. *Frontiers in Plant Science* 9, 190. DOI: 10.3389/fpls.2018.00190.

Li, Y., Fu, Y. and Li, M. (2020) Sustainable agriculture in the era of omics: Knowledge-driven crop breeding. *Genome Biology* 21(154), 1–5. DOI: 10.1186/s13059-020-02073-8.

Li, Y., Liu, D., Zong, Y., Jiang, L., Xi, X. *et al.* (2020b) New D hordein alleles were created in barley using CRISPR/Cas9 genome editing. *Cereal Research Communications* 48(2), 131–138. DOI: 10.1007/s42976-020-00023-2.

Li, Y., Ruperao, P., Batley, J., Edwards, D., Martin, W., Hobson, K. and Sutton, T. (2022) Genomic prediction of preliminary yield trials in chickpea: Effect of functional annotation of SNPs and environment. *The Plant Genome* 15(1), e20166. DOI: 10.1002/tpg2.20166.

Li, Y., Jin, M., Zhang, Y., Yang, S., Wang, J. *et al.* (2024) AeIMLKL encodes a novel resistance protein with a kinase fusion for powdery mildew resistance. *Nature Communications* 15, 2934. DOI: 10.1038/s41467-024-46814-7.

Liang, L., Geng, D., Yan, J., Qiu, S., Di, L. *et al.* (2020) Estimating crop LAI using spectral feature extraction and the hybrid inversion method. *Remote Sensing* 12(21), 3534. DOI: 10.3390/rs12213534.

Libbrecht, M.W. and Noble, W.S. (2015) Machine learning applications in genetics and genomics. *Nature Reviews Genetics* 16(6), 321–332. DOI: 10.1038/nrg3920.

Lin, C.C., Chao, Y.T., Chen, W.C., Ho, H.Y., Chou, M.Y. *et al.* (2019) Regulatory cascade involving transcriptional and N-end rule pathways in rice under submergence. *Proceedings of the National Academy of Sciences USA* 116(8), 3300–3309. DOI: 10.1073/pnas.1818507116.

Liu, G., Li, J. and Godwin, I.D. (2019) Genome editing by CRISPR/Cas9 in sorghum through biolistic bombardment. *Sorghum: Methods and Protocols*, 169–183. DOI: 10.1007/978-1-4939-9039-9_12.

Liu, X.H., Lyu, Y.S., Yang, W., Yang, Z.T., Lu, S.J. and Liu, J.X. (2020) A membrane-associated NAC transcription factor OsNTL3 is involved in thermotolerance in rice. *Plant Biotechnology Journal* 18(5), 1317–1329. DOI: 10.1111/pbi.13297.

Liu, Y., Li, Z.G., Cheng, H., Yang, X., Li, M.Y. *et al.* (2024) Plant factory speed breeding significantly shortens rice generation time and enhances metabolic diversity. *Engineering* 50, 259–269. DOI: 10.1016/j.eng.2024.09.019.

Lopez, F.B., Johansen, C. and Chauhan, Y.S. (1996) Effects of timing of drought stress on phenology, yield, and yield components of short-duration pigeonpea. *Journal of Agronomy and Crop Science* 177(5), 311–320. DOI: 10.1111/j.1439-037X.1996.tb00251.x.

Lubanga, N., Okaron, V., Gimode, D.M., Persa, R., Mwololo, J. *et al.* (2024) Genomic prediction by modelling genotype by environment interaction for yield in groundnut. *bioRxiv*. DOI: 10.1101/2024.10.30.621054.

Lv, Q., Li, W., Sun, Z. *et al.* (2020) Resequencing of 1,143 indica rice accessions reveals important genetic variations and different heterosis patterns. *Nature Communications* 11, 4778. DOI: 10.1038/s41467-020-18608-0.

Lv, P., Su, F., Chen, F., Yan, C., Xia, D. *et al.* (2024) Genome editing in rice using CRISPR/Cas12i3. *Plant Biotechnology Journal* 22(2), 379–385. DOI: 10.1111/pbi.14192.

Mace, E.S., Tai, S., Gilding, E.K., Li, Y., Prentis, P.J. *et al.* (2013) Whole-genome sequencing reveals untapped genetic potential in Africa's indigenous cereal crop sorghum. *Nature Communications* 4(1), 1–9. DOI: 10.1038/ncomms3320.

Mahto, R.K., Ambika Singh, C., Chandana, B.S., Singh, R.K., Verma, S. *et al.* (2022) Chickpea biofortification for cytokinin dehydrogenase via genome editing to enhance abiotic-biotic stress tolerance and food security. *Frontiers in Genetics* 13, 900324. DOI: 10.3389/fgene.2022.900324.

Mallick, N., Vinod, J, S.K., Raghunandan, K., Choudhary, M.K., Agarwal, P., Singh, M., Kumari, P., Niranjana, M. and Sivasamy, M. (2024) Marker-assisted development of triple rust resistance wheat variety HD3407. *Cereal Research Communications* 52, 1779–1788. DOI: 10.1007/s42976-024-00501-x.

Manley, A., Brown, M., Ravelombola, W., Cason, J. and Pham, H. (2024) Genome-wide association study and genomic selection for plant growth habit in peanuts using the USDA public data. *American Journal of Plant Sciences* 15(9), 811–834. DOI: 10.4236/ajps.2024.159052.

Mannur, D.M., Babbar, A., Thudi, M., Sabbavarapu, M.M., Roorkiwal, M., Sharanabasappa, Y. and Chamarthi, S.K. (2019) Super Annigeri 1 and improved JG 74: Two Fusarium wilt-resistant introgression lines developed using marker-assisted backcrossing approach in chickpea (*Cicer arietinum* L.). *Molecular Breeding* 39(2), 2. DOI: 10.1007/s11032-018-0908-9.

Mao, H., Jiang, C., Tang, C., Nie, X., Du, L. *et al.* (2023) Wheat adaptation to environmental stresses under climate change: Molecular basis and genetic improvement. *Molecular Plant* 16(10), 1564–1589. DOI: 10.1016/j.molp.2023.09.001.

Matsumoto, H., Fan, X., Wang, Y., Kusstatscher, P., Duan, J. *et al.* (2021) Bacterial seed endophyte shapes disease resistance in rice. *Nature Plants* 7(1), 60–72. DOI: 10.1038/s41477-020-00826-5.

McCouch, S.R., Xiuli, C., Olivier, P., Svetlana, T., Yunbi, X. *et al.* (1997) Microsatellite marker development, mapping, and applications in rice genetics and breeding. In: Sasaki, T. and Moore, G. (eds) *Oryza: From Molecule to Plant*. Springer, Dordrecht. DOI: 10.1007/978-94-011-5794-0_9.

Meuwissen, T.H., Hayes, B.J. and Goddard, M.E. (2001) Prediction of total genetic value using genome-wide dense marker maps. *Genetics* 157(4), 1819–1829. DOI: 10.1093/genetics/157.4.1819.

Miklas, P.N., Kelly, J.D., Beebe, S.E. and Blair, M.W. (2006) Common bean breeding for resistance against biotic and abiotic stresses: From classical to MAS breeding. *Euphytica* 147, 105–131. DOI: 10.1007/s10681-006-4600-5.

Mikolajczak, K., Ogrodowicz, P., Surma, M., Adamski, T. and Kuczyñska, A. (2016) Introgression of LTP2 gene through marker assisted backcross in barley (*Hordeum vulgare* L.). *Electronic Journal of Biotechnology* 19(6), 9–11. DOI: 10.1016/j.ejbt.2016.09.003.

Miller, M.J., Song, Q. and Li, Z. (2023a) Genomic selection of soybean (*Glycine max*) for genetic improvement of yield and seed composition in a breeding context. *Plant Genome* 16(4), 20384. DOI: 10.1002/tpg2.20384.

Miller, M.J., Song, Q., Fallen, B. and Li, Z. (2023b) Genomic prediction of optimal cross combinations to accelerate genetic improvement of soybean (*Glycine max*). *Frontiers in Plant Science* 14, 1171135. DOI: 10.3389/fpls.2023.1171135.

Mitache, M., Baidani, A., Bencharki, B. and Idrissi, O. (2024) Exploring the impact of light intensity under speed breeding conditions on the development and growth of lentil and chickpea. *Plant Methods* 20(1), 30. DOI: 10.1186/s13007-024-01156-9.

Mohr, T., Horstman, J., Gu, Y.Q., Elarabi, N.I., Abdallah, N.A. and Thilmony, R. (2022) CRISPR-Cas9 gene editing of the Sal1 gene family in wheat. *Plants* 11(17), 2259. DOI: 10.3390/plants11172259.

Molinari, M.D.C., Pagliarini, R.F., Florentino, L.H., Lima, R.N., Arraes, F.B.M., Abbad, S.V. *et al.* (2024) Navigating the path from lab to market: Regulatory challenges and opportunities for genome editing technologies for agriculture. In Chen, J.-T. and Ahmar, S. (eds) *Plant Genome Editing Technologies: Speed Breeding, Crop Improvement and Sustainable Agriculture*. Springer Nature Singapore, pp. 25–63.

Montenegro, J.D., Golicz, A.A., Bayer, P.E., Hurgobin, B., Lee, H. *et al.* (2017) The pangenome of hexaploid bread wheat. *The Plant Journal* 90(5), 1007–1013. DOI: 10.1111/tpj.13515.

Montesinos-López, A., Crespo-Herrera, L., Dreisigacker, S., Gerard, G., Vitale, P. *et al.* (2024) Deep learning methods improve genomic prediction of wheat breeding. *Frontiers in Plant Science* 15, 1324090. DOI: 10.3389/fpls.2024.1324090.

Moonjely, S., Ebert, M., Paton-Glassbrook, D., Noel, Z.A., Roze, L. *et al.* (2023) Update on the state of research to manage Fusarium head blight. *Fungal Genetics and Biology* 169, 103829. DOI: 10.1016/j.stress.2024.100681.

Munawar, N., Ahsan, K. and Ahmad, A. (2023) CRISPR-edited plants' social, ethical, policy, and governance issues. In: *Global Regulatory Outlook for CRISPRized Plants*. Elsevier, Netherlands, pp. 367–396. DOI: 10.1016/B978-0-443-18444-4.00011-9.

Mwamahonje, A., Mdindikasi, Z., Mchau, D., Mwenda, E., Sanga, D., Garcia-Oliveira, A.L. and Ojiewo, C.O. (2024) Advances in sorghum improvement for climate resilience in the global arid and semi-arid tropics: A review. *Agronomy* 14(12), 3025.

Naik, Y.D., Zhao, C., Channale, S., Nayak, S.N., Bhutia, K.L. *et al.* (2024) Bioinformatics for plant genetics and breeding research. In *Frontier Technologies for Crop Improvement*. Springer Nature Singapore, pp. 35–64. DOI: 10.1007/978-981-99-4673-0_3.

Naik, Y.D., Bahuguna, R.N., García-Caparrós, P., Zwart, R.S., Reddy, M.S. *et al.* (2025a) Exploring the multifaceted dynamics of flowering time regulation in field crops: Insight and intervention approaches. *The Plant Genome* 18(2), e70017. DOI: 10.1002/tpg2.70017.

Naik, Y.D., Rangari, S.K., García-Caparros, P., Jan, F., Gangurde, S.S. *et al.* (2025b) Genomics, pan-genomics, and super pan-genomics of major oilseed crops. In *Breeding Climate Resilient and Future Ready Oilseed Crops*. Springer Nature Singapore, pp. 7–41. DOI: 10.1007/978-981-97-7744-0_2

Nam, N.H., Chauhan, Y.S. and Johansen, C. (2001) Effect of timing of drought stress on growth and grain yield of extra-short-duration pigeonpea lines. *Journal of Agricultural Science* 136(2), 179–189. DOI: 10.1017/S0021859601008607.

Nayak, S.N., Zhu, H., Varghese, N., Datta, S., Choi, H.K. *et al.* (2010) Integration of novel SSR and gene-based SNP marker loci in the chickpea genetic map and establishment of new anchor points with *Medicago truncatula* genome. *Theoretical and Applied Genetics* 120, 1415–1441. DOI: 10.1007/s00122-010-1265-1.

Pandey, M.K., Wang, M.L., Qiao, L., Feng, S., Khera, P. *et al.* (2014) Identification of QTLs associated with oil content and mapping FAD2 genes and their relative contribution to oil quality in peanut (*Arachis hypogaea* L.). *BMC Genetics* 15, 1–14. DOI: 10.1186/s12863-014-0133-4.

Pandey, M.K., Gangurde, S.S., Shasidhar, Y., Sharma, V., Kale, S.M. *et al.* (2024) High-throughput diagnostic markers for foliar fungal disease resistance and high oleic acid content in groundnut. *BMC Plant Biology* 24(1), 262. DOI: 10.1186/s12870-024-04987-9.

Pankin, A., Altmüller, J., Becker, C. and Korff, M. von (2018) Targeted resequencing reveals genomic signatures of barley domestication. *New Phytologist* 218(3), 1247–1259. DOI: 10.1111/nph.15077.

Patyal, U., Bala, R., Kaur, M., Faizan, M. and Alam, P. (2025) Phyto-microbiome engineering: Designing plant–microbe interactions for improved crop performance. *The Microbe* 6, 100272. DOI: 10.1016/j.microb.2025.100272.

Poland, J., Endelman, J., Dawson, J., Rutkoski, J., Wu, S., Manes, Y., Dreisigacker, S., Crossa, J., Sánchez-Villeda, H., Sorrells, M. and Jannink, J.L. (2012) Genomic selection in wheat breeding using genotyping-by-sequencing. *The Plant Genome* 5(3), 103–113. DOI: 10.3835/plantgenome2012.06.0006.

Prasad, K., Gadeela, H., Bommineni, P.R., Reddy, P.S., Tyagi, W. and Yogendra, K. (2024) CRISPR/Cas9-mediated mutagenesis of phytoene desaturase in pigeonpea and groundnut. *Functional and Integrative Genomics* 24(2), 57. DOI: 10.1007/s10142-024-01336-9.

Prodhan, F.A., Zhang, J., Sharma, T.P.P., Nanzad, L., Zhang, D. *et al.* (2022) Projection of future drought and its impact on simulated crop yield over South Asia using ensemble machine learning approach. *Science of the Total Environment* 807, 151029. DOI: 10.1016/j.scitotenv.2021.151029.

Punnuri, S.M., Ayele, A.G., Harris-Shultz, K.R., Knoll, J.E., Coffin, A.W. *et al.* (2022) Genome-wide association mapping of resistance to the sorghum aphid in *Sorghum bicolor*. *Genomics* 114(4), 110408. DOI: 10.1016/j.ygeno.2022.110408.

Qin, P., Lu, H., Du, H., Wang, H., Chen, W. *et al.* (2021) Pan-genome analysis of 33 genetically diverse rice accessions reveals hidden genomic variations. *Cell* 184(13), 3542–3558. DOI: 10.1016/j.cell.2021.04.046.

Qin, J., Wang, F., Zhao, Q., Shi, A., Zhao, T. *et al.* (2022) Identification of candidate genes and genomic selection for seed protein in soybean breeding pipeline. *Frontiers in Plant Science* 13, 882732. DOI: 10.3389/fpls.2022.882732.

Raffo, M.A., Sarup, P., Jensen, J., Guo, X., Jensen, J.D. *et al.* (2025) Genomic prediction for yield and malting traits in barley using metabolomic and near-infrared spectra. *Theoretical and Applied Genetics* 138(1), 24. DOI: 10.1007/s00122-024-04806-7.

Rahim, A.A., Uzair, M., Rehman, N., Fiaz, S., Attia, K.A. *et al.* (2024) CRISPR/Cas9 mediated TaRPK1 root architecture gene mutagenesis confers enhanced wheat yield. *Journal of King Saud University-Science* 36(2), 103063. DOI: 10.1016/j.jksus.2023.103063.

Rajabi Dehnavi, A., Zahedi, M. and Piernik, A. (2024) Understanding salinity stress responses in sorghum: Exploring genotype variability and salt tolerance mechanisms. *Frontiers in Plant Science* 14, 1296286. DOI: 10.3389/fpls.2023.1296286.

Ramayya, J.P., Vinukonda, V.P., Singh, U.M., Alam, S., Venkateshwarlu, C. *et al.* (2021) Marker-assisted forward and backcross breeding for improvement of elite Indian rice variety Naveen for multiple biotic and abiotic stress tolerance. *PLoS One* 16(9), e0256721. DOI: 10.1371/journal.pone.0256721.

Rana, M.M., Takamatsu, T., Baslam, M., Kaneko, K., Itoh, K. *et al.* (2019) Salt tolerance improvement in rice through efficient SNP marker-assisted selection coupled with speed-breeding. *International Journal of Molecular Sciences* 20(10), 2585. DOI: 10.3390/ijms20102585.

Rani, A., Devi, P., Jha, U.C., Sharma, K.D., Siddique, K.H. and Nayyar, H. (2020) Developing climate-resilient chickpea involving physiological and molecular approaches with a focus on temperature and drought stresses. *Frontiers in Plant Science* 10, 1759. DOI: 10.3389/fpls.2019.01759.

Rani, A., Taunk, J., Jangra, S., Yadav, R.C., Yadav, N.R., Yadav, D. and Yadav, H.P. (2022) Development of advanced pearl millet lines tolerant to terminal drought stress using marker-assisted selection. *Vegetos* 35(1), 63–73. DOI: 10.1007/s42535-021-00284-0.

Reshna, O.P., Beena, R., Joy, M., Viji, M.M. and Roy, S. (2022) Elucidating the effect of growth promoting endophytic fungus *Piriformospora indica* for seedling stage salinity tolerance in contrasting rice genotypes. *Journal of Crop Science and Biotechnology* 25, 393–403. DOI: 10.1007/s12892-022-00153-5.

Roorkiwal, M., Rathore, A., Das, R.R., Singh, M.K., Jain, A. *et al.* (2016) Genome-enabled prediction models for yield-related traits in chickpea. *Frontiers in Plant Science* 7, 1666. DOI: 10.3389/fpls.2016.01666.

Roorkiwal, M., Jarquin, D., Singh, M.K., Gaur, P.M., Bharadwaj, C. *et al.* (2018) Genomic-enabled prediction models using multi-environment trials to estimate the effect of genotype × environment interaction on prediction accuracy in chickpea. *Scientific Reports* 8(1), 11701. DOI: 10.1038/s41598-018-30027-2.

Rossi, N., Powell, W., Mackay, I.J., Hickey, L., Maurer, A. *et al.* (2024) Investigating the genetic control of plant development in spring barley under speed breeding conditions. *Theoretical and Applied Genetics* 137(5), 1–11. DOI: 10.1007/s00122-024-04618-9.

Ruperao, P., Thirunavukkarasu, N., Gandham, P., Selvanayagam, S., Govindaraj, M. *et al.* (2021) Sorghum pan-genome explores the functional utility for genomic-assisted breeding to accelerate the genetic gain. *Frontiers in Plant Science* 12, 666342. DOI: 10.3389/fpls.2021.666342.

Sallam, A.H., Endelman, J.B., Jannink, J.L. and Smith, K.P. (2015) Assessing genomic selection prediction accuracy in a dynamic barley breeding population. *The Plant Genome* 8(1), 1–15. DOI: 10.3835/plantgenome2014.05.0020.

Samarah, N.H., Alqudah, A.M., Amayreh, J.A. and McAndrews, G.M. (2009) The effect of late-terminal drought stress on yield components of four barley cultivars. *Journal of Agronomy and Crop Science* 195(6), 427–441. DOI: 10.1111/j.1439-037X.2009.00387.x.

Samineni, S., Sen, M., Sajja, S.B. and Gaur, P.M. (2020) Rapid generation advance (RGA) in chickpea to produce up to seven generations per year and enable speed breeding. *Crop Journal* 8(1), 164–169. DOI: 10.1016/j.cj.2019.08.003.

Sánchez-León, S., Marín-Sanz, M., Guzmán-López, M.H., Gavilán-Camacho, M., Simón, E. and Barro, F. (2024) CRISPR/Cas9-mediated multiplex gene editing of gamma and omega gliadins: Paving the way for gliadin-free wheat. *Journal of Experimental Botany* 75(22), 7079–7095. DOI: 10.1093/jxb/erae376.

Sandhu, N., Singh, J., Pruthi, G., Verma, V.K., Raigar, O.P. *et al.* (2024) SpeedyPaddy: A revolutionized cost-effective protocol for large-scale offseason advancement of rice germplasm. *Plant Methods* 20(1), 109. DOI: 10.1186/s13007-024-01235-x.

Sapkota, S., Boatwright, J.L., Kumar, N., Myers, M., Cox, A. *et al.* (2023) Genomic prediction of hybrid performance for agronomic traits in sorghum. *G3: Genes, Genomes, Genetics* 13(4), 21. DOI: 10.1093/g3journal/jkac311.

Schmidt, M., Kollers, S., Maasberg-Prelle, A., Großer, J., Schinkel, B. *et al.* (2016) Prediction of malting quality traits in barley based on genome-wide marker data to assess the potential of genomic selection. *Theoretical and Applied Genetics* 129, 203–213. DOI: 10.1007/s00122-015-2639-1.

Schmutz, J., Cannon, S.B., Schlueter, J., Ma, J., Mitros, T. *et al.* (2009) Genome sequence of the palaeopolyploid soybean. *Nature* 463(7278), 178–183. DOI: 10.1038/nature08670.

Schnable, P.S., Ware, D., Fulton, R.S., Stein, J.C., Wei, F. *et al.* (2009) The B73 maize genome: Complexity, diversity, and dynamics. *Science* 326(5956), 1112–1115. DOI: 10.1126/science.1178534.

Schoen, A., Wallace, S., Holbert, M.F., Brown-Guidera, G., Harrison, S. *et al.* (2023) Reducing the generation time in winter wheat cultivars using speed breeding. *Crop Science* 63(4), 2079–2090. DOI: 10.1002/csc2.20989.

Schreiber, M., Jayakodi, M. and Stein, N. (2024) Plant pangenomes for crop improvement, biodiversity and evolution. *Nature Reviews Genetics* 25, 563–577. DOI: 10.1038/s41576-024-00691-4.

Seck, F., Prakash, P.T., Covarrubias-Pazaran, G., Gueye, T., Diédhiou, I. *et al.* (2024) Stochastic simulation to optimize rice breeding at IRRI. *Frontiers in Plant Science* 15, 1488814. DOI: 10.3389/fpls.2024.1488814.

Sehgal, D., Mondal, S., Crespo-Herrera, L., Velu, G., Juliana, P. *et al.* (2020) Haplotype-based, genome-wide association study reveals stable genomic regions for grain yield in CIMMYT spring bread wheat. *Frontiers in Genetics* 11, 589490. DOI: 10.3389/fgene.2020.589490.

Shang, L., Li, X., He, H., Yuan, Q., Song, Y. *et al.* (2022) A super pan-genomic landscape of rice. *Cell Research* 32(10), 878–896. DOI: 10.1038/s41422-022-00685-z.

Sharma, A., Oda, G. and Holodniy, M. (2022) Effectiveness of messenger RNA-based vaccines during the emergence of the severe acute respiratory syndrome coronavirus 2 omicron variant. *Clinical Infectious Diseases* 75(12), 2186–2192. DOI: 10.1093/cid/ciac325.

Shasidhar, Y., Variath, M.T., Vishwakarma, M.K., Manohar, S.S., Gangurde, S.S. *et al.* (2020) Improvement of three popular Indian groundnut varieties for foliar disease resistance and high oleic acid using SSR markers and SNP array in marker-assisted backcrossing. *The Crop Journal* 8(1), 1–15. DOI: 10.1016/j.cj.2019.07.001.

Shi, J., Gao, H., Wang, H., Lafitte, H.R., Archibald, R.L. *et al.* (2017) ARGOS 8 variants generated by CRISPR-Cas9 improve maize grain yield under field drought stress conditions. *Plant Biotechnology Journal* 15(2), 207–216. DOI: 10.1111/pbi.12603.

Shook, J.M. and Edger, P.P. (2021) Advances in pan-genomics: Driving phenotypic and functional genomic discoveries in plants. *Nature Reviews Genetics* 22(5), 324–338. DOI: 10.1038/s41576-021-00330-3.

Siebers, M.H., Slattery, R.A., Yendrek, C.R., Locke, A.M., Drag, D. *et al.* (2017) Simulated heat waves during maize reproductive stages alter reproductive growth but have no lasting effect when applied during vegetative stages. *Agricultura, Ecosystems & Environment* 240, 162–170. DOI: 10.1016/j.agee.2016.11.008.

Singer, W.M., Lee, Y.C., Shea, Z., Vieira, C.C., Lee, D. *et al.* (2023) Soybean genetics, genomics, and breeding for improving nutritional value and reducing antinutritional traits in food and feed. *The Plant Genome* 16, 20415. DOI: 10.1002/tpg2.20415.

Singh, S., Mackill, D.J. and Ismail, A.M. (2009) Responses of SUB1 rice introgression lines to submergence in the field: Yield and grain quality. *Field Crops Research* 113, 12–23. DOI: 10.1016/j.fcr.2009.04.003.

Singh, A.K., Gopala Krishnan, S., Nagarajan, M., Bhowmick, P.K., Ellur, R.K. *et al.* (2017a) Basmati rice variety – Pusa Basmati 1637. *Indian Journal of Genetics* 77(4), 583.

Singh, A.K., Gopala Krishnan, S., Ellur, R.K., Bhowmick, P.K., Nagarajan, M. *et al.* (2017b) Pusa Basmati 1728. *Indian Journal of Genetics* 77(4), 584.

Singh, A.K., Ellur, R.K., Gopala Krishnan, S., Bhowmick, P.K., Nagarajan, M. *et al.* (2018) Notification of Basmati rice variety Pusa Basmati 1718. *Indian Journal of Genetics* 78(1), 151.

Singh, P., Sundaram, K.T., Vinukonda, V.P., Venkateshwarlu, C., Paul, P.J. *et al.* (2024a) Superior haplotypes of key drought-responsive genes reveal opportunities for the development of climate-resilient rice varieties. *Communications Biology* 7(1), 89. DOI: 10.1038/s42003-024-05769-7.

Singh, P., Kumar, A., Singh, T., Anto, S., Indoliya, Y. *et al.* (2024b) Targeting OsNIP3; 1 via CRISPR/Cas9: A strategy for minimizing arsenic accumulation and boosting rice resilience. *Journal of Hazardous Materials* 471, 134325. DOI: 10.1016/j.jhazmat.2024.134325.

Slotkin, R.K., Vaughn, M., Borges, F., Tanurdzić, M., Becker, J.D., Feijó, J.A. and Martienssen, R.A. (2009) Epigenetic reprogramming and small RNA silencing of transposable elements in pollen. *Cell* 136(3), 461–472. DOI: 10.1016/j.cell.2008.12.038.

Smith, A., Gentile, B.R., Xin, Z. and Zhao, D. (2023) The effects of heat stress on male reproduction and tillering in *Sorghum bicolor*. *Food and Energy Security* 12(6), 510. DOI: 10.1002/fes3.510.

Soda, N., Gupta, B.K., Anwar, K., Sharan, A., Govindjee, S.-P. and S.L., and Pareek, A. (2018) Rice intermediate filament, OsIF, stabilizes photosynthetic machinery and yield under salinity and heat stress. *Scientific Reports* 8(1), 4072. DOI: 10.1038/s41598-018-22131-0.

Soe, M.S., Hlaing, S.S., Mon, A.S. and Lynn, K.T. (2024) Prevalence of hypertension and factors associated with the utilization of primary health care services for hypertension among hypertensive population aged 40 years and above in Pyin Oo Lwin Township. *Myanmar. PLoS ONE* 19(10), e0312186. DOI: 10.1371/journal.pone.0312186.

South, P.F., Cavanagh, A.P., Liu, H.W. and Ort, D.R. (2019) Synthetic glycolate metabolism pathways stimulate crop growth and productivity in the field. *Science* 363(6422), eaat9077. DOI: 10.1126/science.aat9077.

Sreenivasulu, N., Butardo, V.M., Misra, G., Cuevas, R.P., Anacleto, R. and Kavi Kishor, P.B. (2015) Designing climate-resilient rice with ideal grain quality suited for high-temperature stress. *Journal of Experimental Botany* 66(7), 1737–1748. DOI: 10.1093/jxb/eru544.

Stringlis, I.A., Yu, K., Feussner, K., Jonge, R. de, Bentum, S. Van *et al.* (2018) MYB72-dependent coumarin exudation shapes root microbiome assembly to promote plant health. *Proceedings of the National Academy of Sciences USA* 115(22), E5213–E5222. DOI: 10.1073/pnas.1722335115.

Studer, A.J. and Doebley, J.F. (2011) Do large effect QTL fractionate? A case study at the maize domestication QTL teosinte branched1. *Genetics* 188(3), 673–681. DOI: 10.1534/genetics.111.126508.

Sun, S., Zhou, Y., Chen, J., Shi, J., Zhao, H. *et al.* (2018) Extensive intraspecific gene order and gene structural variations between Mo17 and other maize genomes. *Nature Genetics* 50(9), 1289–1295. DOI: 10.1038/s41588-018-0182-0.

Sundaram, R.M., Vishnupriya, M.R., Biradar, S.K., Laha, G.S., Reddy, A.G. *et al.* (2008) Marker assisted introgression of bacterial blight resistance in Samba Mahsuri, an elite indica rice variety. *Euphytica* 160, 411–422. DOI: 10.1007/s10681-007-9564-6.

Sunilkumar, V.P., Krishna, H., Devate, N.B., Manjunath, K.K., Chauhan, D. *et al.* (2022) Marker assisted improvement for leaf rust and moisture deficit stress tolerance in wheat variety HD3086. *Frontiers in Plant Science* 13, 1035016. DOI: 10.3389/fpls.2022.1035016.

Taku, M., Saini, M., Kumar, R., Debbarma, P., Rathod, N.K.K. *et al.* (2024) Modified speed breeding approach reduced breeding cycle to less than half in vegetable soybean (*Glycine max* (L.) Merr.). *Physiologia Plantarum* 30(9), 1463–1473. DOI: 10.1007/s12298-024-01503-z.

Tang, Y., Qiu, X., Hu, C., Li, J., Wu, L. *et al.* (2022) Breeding of a new variety of peanut with high-oleic-acid content and high-yield by marker-assisted backcrossing. *Molecular Breeding* 42(7), 42. DOI: 10.1007/s11032-022-01313-9.

Tanksley, S.D., Young, N.D., Paterson, A.H. and Bonierbale, M.W. (1989) RFLP mapping in plant breeding: New tools for an old science. *Nature Biotechnology* 7(3), 257–264. DOI: 10.1038/nbt0389-257.

Tao, Y., Jordan, D.R. and Mace, E.S. (2019) Crop genomics goes beyond a single reference genome. *Trends in Plant Science* 24(12), 1072–1074. DOI: 10.1016/j.tplants.2019.10.001.

Távora, F.T., Meunier, A.C., Vernet, A., Portefaix, M., Milazzo, J. *et al.* (2022) CRISPR/Cas9-targeted knockout of rice susceptibility genes OsDjA2 and OsERF104 reveals alternative sources of resistance to *Pyricularia oryzae*. *Rice Science* 29(6), 535–544. DOI: 10.1016/j.rsci.2022.04.001.

Tenllado, F., Martínez-García, B., Vargas, M. and Díaz-Ruíz, J.R. (2003) Crude extracts of bacterially expressed dsRNA can be used to protect plants against virus infections. *BMC Biotechnology* 3(3), 1–11. DOI: 10.1186/1472-6750-3-3.

Than, A.M., Maw, J.B., Aung, T., Gaur, P.M. and Gowda, C.L.L. (2007) Development and adoption of improved chickpea varieties in Myanmar. *Journal of SAT Agricultural Research* 5(1), 1–3.

Thudi, M., Abhishek, B., Spurthi, N.N., Nicy, V., Trushar, M.S. *et al.* (2011) Novel SSR markers from BAC-end sequences, DArT arrays, and a comprehensive genetic map with 1,291 marker loci for chickpea (*Cicer arietinum* L.). *PLoS One* 6(11), e27275. DOI: 10.1371/journal.pone.0027275.

Thudi, M., Chitikineni, A., Liu, X., He, W., Roorkiwal, M. *et al.* (2016a) Recent breeding programs enhanced genetic diversity in both desi and kabuli varieties of chickpea (*Cicer arietinum* L.). *Scientific Reports* 6(1), 38636. DOI: 10.1038/srep38636.

Thudi, M., Khan, A.W., Kumar, V., Gaur, P.M., Katta, K. *et al.* (2016b) Whole genome re-sequencing reveals genome-wide variations among parental lines of 16 mapping populations in chickpea (*Cicer arietinum* L.). *BMC Plant Biology* 16, 53–64. DOI: 10.1186/s12870-015-0690-3.

Thudi, M., Samineni, S., Li, W., Boer, M.P., Roorkiwal, M. *et al.* (2023) Whole genome resequencing and phenotyping of MAGIC population for high-resolution mapping of drought tolerance in chickpea. *The Plant Genome*. 17, e20333. DOI: 10.1002/tpg2.20333.

Thudi, M., Reddy, M.S., Naik, Y.D., Cheruku, V.K.R., Sangireddy, M.K.R. *et al.* (2024) Invasive sorghum aphid: A decade of research on deciphering plant resistance mechanisms and novel approaches in breeding for sorghum resistance to aphids. *Crop Science* 64(5), 2436–2458. DOI: 10.1002/csc2.21301.

Thudi, M., Mascher, M. and Jayakodi, M. (2025) Pangenome charts the genomic path for wheat improvement. *Trends in Plant Science* 30(7), 687–689. DOI: 10.1016/j.tplants.2025.03.002.

Torkamaneh, D., Lemay, M.A. and Belzile, F. (2021). *Plant Biotechnology Journal* 19(9), 1852–1862. DOI: 10.1111/pbi.13600.

Tzachor, A., Devare, M., King, B., Avin, S. and Ó Héigeartaigh, S. (2022) Responsible artificial intelligence in agriculture requires systemic understanding of risks and externalities. *Nature Machine Intelligence* 4, 104–109. DOI: 10.1038/s42256-022-00440-4.

Uffelmann, E., Qin, Q.H., Nchangwi, S.M., Jantina, D.V., Yukinori, O. *et al.* (2021) Genome-wide association studies. *Nature Reviews Methods Primers* 1(1), 1–21. DOI: 10.1038/s43586-021-00056-9.

Usman, B., Nawaz, G., Zhao, N., Liao, S., Liu, Y. and Li, R. (2020) Precise editing of the OsPYL9 gene by RNA-guided Cas9 nuclease confers enhanced drought tolerance and grain yield in rice (*Oryza sativa* L.) by regulating circadian rhythm and abiotic stress responsive proteins. *International Journal of Molecular Sciences* 21(21), 7854. DOI: 10.3390/ijms21217854.

van Dijk, D., Sharma, R., Nainys, J., Yim, K., Kathail, P. *et al.* (2018) Recovering gene interactions from single-cell data using data diffusion. *Cell* 174(3), 716–729.e27. DOI: 10.1016/j.cell.2018.05.061.

van Dijk, M., Morley, T., Rau, M.L. and Saghai, Y. (2021) A meta-analysis of projected global food demand and population at risk of hunger for the period 2010–2050. *Nature Food* 2(7), 494–501. DOI: 10.1038/s43016-021-00322-9.

Vargheese, R.L., Saravanan, S., Hepziba, S.J., Kumari, S.M.P., Pushpam, A.K., Kanagarajan, S. and Pillai, M.A. (2024) Marker-assisted introgression to improve the oleic acid content in the TMV 7 groundnut (*Arachis hypogaea* L.) variety suitable for the oil industry. *BMC Plant Biology* 24(1), 733. DOI: 10.1186/s12870-024-05387-9.

Varshney, R.K., Ribaut, J.M., Buckler, E.S., Tuberosa, R., Rafalski, J.A. and Langridge, P. (2012) Can genomics boost productivity of orphan crops? *Nature Biotechnology* 30(12), 1172–1176. DOI: 10.1038/nbt.2440.

Varshney, R.K., Chi, S., Saxena, R.K., Azam, S., Yu, S. *et al.* (2013) Draft genome sequence of chickpea (*Cicer arietinum*) provides a resource for trait improvement. *Nature Biotechnology* 31(3), 240–246. DOI: 10.1038/nbt.2491.

Varshney, R.K., Terauchi, R. and McCouch, S.R. (2014) Harvesting the promising fruits of genomics: Applying genome sequencing technologies to crop breeding. *PLoS Biology* 12(6), e1001883. DOI: 10.1371/journal.pbio.1001883.

Varshney, R.K., Mohan, S.M., Gaur, P.M., Chamarthi, S.K., Singh, V.K. *et al.* (2014a) Marker-assisted backcrossing to introgress resistance to Fusarium wilt race 1 and Ascochyta blight in C 214, an elite cultivar of chickpea. *The Plant Genome* 7(1), 1–14. DOI: 10.3835/plantgenome2013.10.0035.

Varshney, R.K., Saxena, R.K., Upadhyaya, H.D., Khan, A.W., Yu, Y. *et al.* (2017a) Whole-genome resequencing of 292 pigeonpea accessions identifies genomic regions associated with domestication and agronomic traits. *Nature Genetics* 49(7), 1082–1088. DOI: 10.1038/ng.3872.

Varshney, R.K., Shi, C., Thudi, M., Mariac, C., Wallace, J. *et al.* (2017b) Pearl millet genome sequence provides a resource to improve agronomic traits in arid environments. *Nature Biotechnology* 35(10), 969–976. DOI: 10.1038/nbt.3943.

Varshney, R.K., Thudi, M., Pandey, M.K., Tardieu, F., Ojiewo, C. *et al.* (2018) Accelerating genetic gains in legumes for the development of prosperous smallholder agriculture: Integrating genomics, phenotyping, systems modelling and agronomy. *Journal of Experimental Botany* 69(13), 3293–3312. DOI: 10.1093/jxb/ery088.

Varshney, R.K., Roorkiwal, M., Sun, S., Bajaj, P., Chitikineni, A. *et al.* (2021a) A chickpea genetic variation map based on the sequencing of 3,366 genomes. *Nature* 599(7886), 622–627. DOI: 10.1038/s41586-021-04066-1.

Varshney, R.K., Bohra, A., Yu, J., Graner, A., Zhang, Q. and Sorrells, M.E. (2021b) Designing future crops: Genomics-assisted breeding comes of age. *Trends in Plant Science* 26(6), 631–669. DOI: 10.1016/j.tplants.2021.03.010.

Vatanparast, M., Merkel, L. and Amari, K. (2024) Exogenous application of dsRNA in plant protection: Efficiency, safety concerns and risk assessment. *International Journal of Molecular Sciences* 25(12), 6530. DOI: 10.3390/ijms25126530.

Voss-Fels, K.P., Stahl, A., Wittkop, B., Lichthardt, C., Nagler, S. *et al.* (2019) Breeding improves wheat productivity under contrasting agrochemical input levels. *Nature Plants* 5(7), 706–714. DOI: 10.1038/s41477-019-0445-5.

Walkowiak, S., Gao, L., Monat, C., Haberer, G., Kassa, M.T. *et al.* (2020) Multiple wheat genomes reveal global variation in modern breeding. *Nature* 588(7837), 277–283. DOI: 10.1038/s41586-020-2961-x.

Wang, T. and Sun, F. (2023) Integrated drought vulnerability and risk assessment for future scenarios: An indicator-based analysis. *Science of the Total Environment* 900, 165591. DOI: 10.1016/j.scitotenv.2023.165591.

Wang, F., Wang, C., Liu, P., Lei, C., Hao, W. *et al.* (2016) Enhanced rice blast resistance by CRISPR/Cas9-targeted mutagenesis of the ERF transcription factor gene OsERF922. *PLoS ONE* 11(4), 0154027. DOI: 10.1371/journal.pone.0154027.

Wang, B., Zhong, Z., Wang, X., Han, X., Yu, D., Wang, C. and Zhang, Y. (2020) Knockout of the OsNAC006 transcription factor causes drought and heat sensitivity in rice. *International Journal of Molecular Sciences* 21(7), 2288. DOI: 10.3390/ijms21072288.

Wang, B., Jiao, Y., Chougule, K., Olson, A., Huang, J. *et al.* (2021) Pan-genome analysis in sorghum highlights the extent of genomic variation and sugarcane aphid resistance genes. *BioRxiv*, 2021–2001. DOI: 10.1101/2021.01.03.424980.

Wang, S., Qian, Y.Q., Zhao, R.P., Chen, L.L. and Song, J.M. (2023) Graph-based pan-genomes: Increased opportunities in plant genomics. *Journal of Experimental Botany* 74(1), 24–39. DOI: 10.1093/jxb/erac412.

Watson, A., Ghosh, S., Williams, M.J., Cuddy, W.S., Simmonds, J. *et al.* (2018) Speed breeding is a powerful tool to accelerate crop research and breeding. *Nature Plants* 4(1), 23–29. DOI: 10.1038/s41477-017-0083-8.

Weber, S.E., Frisch, M., Snowdon, R.J. and Voss-Fels, K.P. (2023) Haplotype blocks for genomic prediction: A comparative evaluation in multiple crop datasets. *Frontiers in Plant Science* 14, 1217589. DOI: 10.3389/fpls.2023.1217589.

Wei, Y., Jin, J., Jiang, S., Ning, S. and Liu, L. (2018) Quantitative response of soybean development and yield to drought stress during different growth stages in the Huaibei Plain. *China*. Agronomy 8(7), 97. DOI: 10.3390/agronomy8070097.

Weng, J., Gu, S., Wan, X., Gao, H., Guo, T. *et al.* (2008) Isolation and initial characterization of GW5, a major QTL associated with rice grain width and weight. *Cell Research* 18(12), 1199–1209. DOI: 10.1038/cr.2008.307.

Williams, J.G., Kubelik, A.R., Livak, K.J., Rafalski, J.A. and Tingey, S.V. (1990) DNA polymorphisms amplified by arbitrary primers are useful as genetic markers. *Nucleic Acids Research* 18(22), 6531–6535. DOI: 10.1093/nar/18.22.6531.

Wu, D., Xie, L., Sun, Y., Huang, Y., Jia, L. *et al.* (2023) A syntelog-based pan-genome provides insights into rice domestication and de-domestication. *Genome Biology* 24(1), 179. DOI: 10.1186/s13059-023-03017-5.

Wubs, E.R., Putten, W.H. van der, Bosch, M. and Bezemer, T.M. (2016) Soil inoculation steers restoration of terrestrial ecosystems. *Nature Plants* 11(2), 16107. DOI: 10.1038/nplants.2016.107.

Xiong, D., Ling, X., Huang, J. and Peng, S. (2017) Meta-analysis and dose-response analysis of high temperature effects on rice yield and quality. *Environmental and Experimental Botany* 141, 1–9. DOI: 10.1016/j.envexpbot.2017.06.007.

Xu, C. and Jackson, S.A. (2019) Machine learning and complex biological data. *Genome Biology* 20, 1–4. DOI: 10.1186/s13059-019-1689-0.

Xu, K., Xu, X., Fukao, T., Canlas, P., Heuer, S. *et al.* (2006) *Sub1A* is an ethylene-response-factor-like gene that confers submergence tolerance to rice. *Nature* 442(7103), 705–708. DOI: 10.1038/nature04920.

Xu, S., Zhu, D. and Zhang, Q. (2014) Predicting hybrid performance in rice using genomic best linear unbiased prediction. *Proceedings of the National Academy of Sciences USA* 111(34), 12456–12461. DOI: 10.1073/pnas.1413750111.

Xu, Y., Jin, Z. and Chen, Z. (2023) Modular biosynthesis platform for halogenated tryptophan derivatives. *Nature Communications* 14, 6847. DOI: 10.1038/s41467-024-47387-1.

Xu, Y., Bu, W., Xu, Y., Fei, H., Zhu, Y. *et al.* (2024) Effects of salt stress on physiological and agronomic traits of rice genotypes with contrasting salt tolerance. *Plants* 13(8), 1157. DOI: 10.3390/plants13081157.

Yadav, A.K., Kumar, A., Grover, N., Ellur, R.K., Krishnan, S.G. *et al.* (2020) Marker aided introgression of 'Saltol', a major QTL for seedling stage salinity tolerance into an elite Basmati rice variety 'Pusa Basmati 1509'. *Scientific Reports* 10(1), 13877. DOI: 10.1038/s41598-020-70664-0.

Yadav, A.K., Butler, C., Yamamoto, A., Patil, A.A., Lloyd, A.L. and Scott, M.J. (2023) CRISPR/Cas9-based split homing gene drive targeting *doublesex* for population suppression of the global fruit pest *Drosophila suzukii*. *Proceedings of the National Academy of Sciences USA* 120(25), e2301525120. DOI: 10.1073/pnas.2301525120.

Yadawad, A., Gadpale, A., Hanchinal, R.R., Nadaf, H.L., Desai, S.A., Biradar, S. and Naik, V.R. (2017) Pyramiding of leaf rust resistance genes in bread wheat variety DWR 162 through marker assisted backcrossing. *Indian Journal of Genetics and Plant Breeding* 77(02), 251–257. DOI: 10.5958/0975-6906.2017.00033.5.

Yan, H., Sun, M., Zhang, Z., Jin, Y., Zhang, A. *et al.* (2023) Pangenomic analysis identifies structural variation associated with heat tolerance in pearl millet. *Nature Genetics* 55(3), 507–518. DOI: 10.1038/s41588-023-01302-4.

Yan, H., Jin, Y., Yu, H., Wang, C., Wu, B. *et al.* (2024) Genomic selection for agronomical phenotypes using genome-wide SNPs and SVs in pearl millet. *Theoretical and Applied Genetics* 137(10), 244. DOI: 10.1007/s00122-024-04754-2.

Yang, H., Huang, T., Ding, M., Lu, D. and Lu, W. (2017) High temperature during grain filling impacts on leaf senescence in waxy maize. *Agronomy Journal* 109(3), 906–916. DOI: 10.2134/agronj2016.08.0452.

Yang, X., Wang, B., Chen, L. and Li, and Cao, C. (2019) The different influences of drought stress at the flowering stage on rice physiological traits, grain yield, and quality. *Theoretical and Applied Genetics* 9(1), 37–42. DOI: 10.1038/s41598-019-40161-0.

Yang, X., Pan, Y., Xia, X., Qing, D., Chen, W. *et al.* (2023) Molecular basis of genetic improvement for key rice quality traits in southern China. *Genomics* 115(6), 110745. DOI: 10.1016/j.ygeno.2023.110745.

Yuan, Y. and Sun, Y. (2024) Practices, challenges, and future of digital transformation in smallholder agriculture: Insights from a literature review. *Agriculture (Basel)* 14(12). DOI: 10.3390/agriculture14122193.

Zanini, S.F., Bayer, P.E., Wells, R., Snowdon, R.J., Batley, J. *et al.* (2022) Pangenomics in crop improvement – From coding structural variations to finding regulatory variants with pangenome graphs. *The Plant Genome* 15(1), 20177. DOI: 10.1002/tpg2.20177.

Zeng, Y., Wen, J., Zhao, W., Wang, Q. and Huang, W. (2020) Rational improvement of rice yield and cold tolerance by editing the three genes OsPIN5b, GS3, and OsMYB30 with the CRISPR–Cas9 system. *Frontiers in Plant Science* 10, 1663. DOI: 10.3389/fpls.2019.01663.

Zhang, J., Zhang, S., Cheng, M., Jiang, H., Zhang, X. *et al.* (2018) Effect of drought on agronomic traits of rice and wheat: A meta-analysis. *International Journal of Environmental Research and Public Health* 15(5), 839. DOI: 10.3390/ijerph15050839.

Zhang, A., Liu, Y., Wang, F., Li, T., Chen, Z. *et al.* (2019) Enhanced rice salinity tolerance via CRISPR/Cas9-targeted mutagenesis of the OsRR22 gene. *Molecular Breeding* 39, 1–10. DOI: 10.1007/s11032-019-0954-y.

Zhang, P., Du, H., Wang, J., Pu, Y., Yang, C. *et al.* (2020). *Plant Biotechnology Journal* 18(6), 1384–1395. DOI: 10.1111/pbi.13302.

Zhang, S., Zhang, R., Gao, J., Song, G., Li, J. *et al.* (2021). *Plant Biotechnology Journal* 19(9), 1684. DOI: 10.1111/pbi.13647.

Zhang, M., Fang, T., Zhou, X., Chen, X., Li, X. *et al.* (2022) Combination of marker-assisted backcross selection of Yr59 and phenotypic selection to improve stripe rust resistance and agronomic performance in four elite wheat cultivars. *Agronomy* 12(2), 497. DOI: 10.3390/agronomy12020497.

Zhao, C., Liu, B., Piao, S., Wang, X., Lobell, D.B. *et al.* (2017) Temperature increase reduces global yields of major crops in four independent estimates. *Proceedings of the National Academy of Sciences USA* 114(35), 9326–9331. DOI: 10.1073/pnas.1701762114.

Zhao, Q., Feng, Q., Lu, H., Li, Y., Wang, A. *et al.* (2018) Pan-genome analysis highlights the extent of genomic variation in cultivated and wild rice. *Nature Genetics* 50(2), 278–284. DOI: 10.1038/s41588-018-0041-z.

Zhao, J., Bayer, P.E., Ruperao, P., Saxena, R.K., Khan, A.W. *et al.* (2020) Trait associations in the pangenome of pigeon pea (*Cajanus cajan*). *Plant Biotechnology Journal* 18(9), 1946–1954. DOI: 10.1111/pbi.13354.

Zhou, W., Liang, D., Tang, J., Jia, D., Zhang, Z., Dai, T. and Jiang, D. (2016) Identification of drought responsive proteins in wheat (*Triticum aestivum* L.) using a proteomic approach. *Plant Molecular Biology Reporter* 34, 330–342. DOI: 10.1007/s11105-015-0914-0.

Zhou, Y., Yu, Z., Chebotarov, D., Chougule, K., Lu, Z. *et al.* (2023) Pan-genome inversion index reveals evolutionary insights into the subpopulation structure of Asian rice. *Nature Communications* 14(1), 1567. DOI: 10.1038/s41467-023-37004-y.

5 Genomics-assisted Breeding for Abiotic Stress Tolerance in Crops

Ali Raza[1,*], Xinyue Zhang[2], Cassandria G. Tay Fernandez[2] and Rajeev K. Varshney[2]

[1]*Guangdong Key Laboratory of Plant Epigenetics, College of Life Sciences and Oceanography, Shenzhen University, Shenzhen, China;* [2]*WA State Agricultural Biotechnology Centre, Centre for Crop and Food Innovation, Food Futures Institute, Murdoch University, Murdoch, WA, Australia*

Abstract

Climate change-driven abiotic stresses pose significant threats to global agricultural productivity by reducing crop yields and challenging food security and economic stability. Recent innovations in genomics-assisted breeding (GAB) tools, gene editing, high-throughput phenotyping and artificial intelligence propose fast-forward solutions to alleviate these challenges. Significant advances in GAB, including genome-wide association studies, genomic selection, pangenomics and haplotype-based mapping, have facilitated the identification of key genes, markers and stress-responsive pathways underlying multiple stress tolerance. Integrating GAB with modern breeding methods facilitates the discovery of stress-smart traits, while artificial intelligence-driven models boost the precision of crop improvement efforts. Moreover, combining speed breeding protocols with genomic selection pipelines fast-tracks genetic gains, particularly under controlled stress settings. Therefore, this chapter briefly highlights the economic and agricultural impacts of abiotic stresses and investigates an innovative roadmap of GAB tools for designing stress-smart future crops. By leveraging these tools, biotechnologists or breeders can enhance crop productivity and tolerance against single and combined abiotic stresses.

5.1 Introduction

The increasing and irregular challenge of climate change and the escalation of diverse abiotic stresses (e.g. drought, waterlogging, soil salinity, extreme temperatures, metals toxicity, nutrient deficiencies, ozone, UV radiation, elevated CO_2) pose real threats to universal agricultural yield and food security (Benitez-Alfonso *et al.*, 2023; Janni *et al.*, 2024; Yi Yang *et al.*, 2024b; Raza *et al.*, 2025a). The negative impact of diverse abiotic stresses also significantly obstructs sustainable crop production. These stresses alone or in combination hamper the normal functionality of plant stress-responsive mechanisms at biochemical, physiological, molecular and cellular levels, which ultimately trigger negative impacts on plant health, growth, development and yield (Zandalinas *et al.*, 2021; Zandalinas *et al.*, 2024; Raza *et al.*, 2025b). With reference to continuing climate change events, it is anticipated that the occurrence and strength of abiotic stresses (either single or combined) will continue to increase, and it has become more urgent than ever to mitigate

*Corresponding author. Email: alirazamughal143@gmail.com

 Genomic Solutions for Breeding Climate-resilient Crops (R.K. Varshney, R. Barmukh and R.J. Harper)
DOI: 10.1079/9781800627185.0005

their negative impacts on crops (Janni *et al.*, 2024; Yi Yang *et al.*, 2024b; Raza *et al.*, 2025a). As the world's population continues to grow, safeguarding sustainable food production under gradually irregular climatic conditions has become a main concern for crop scientists and breeders.

Over the years, traditional breeding approaches have achieved sizable achievement in advancing multiple traits, including stress tolerance (Ahmar *et al.*, 2020; Varshney *et al.*, 2021b; Krishna *et al.*, 2023; Raza *et al.*, 2024b). However, these approaches are often slow, count on narrow genetic diversity and face disputes in adopting the multifaceted nature of abiotic stresses. To overcome these bottlenecks, recent improvements in genomics and high-throughput sequencing tools have transformed our understanding of plant stress biology (Varshney *et al.*, 2021c; Varshney *et al.*, 2021b; Garg *et al.*, 2024; Khan *et al.*, 2024; Mishra *et al.*, 2024; Xie *et al.*, 2024; Raza *et al.*, 2025b). The genomic basis of abiotic stress tolerance can guide and deliver unique prospects to discover key stress-responsive genes, regulatory networks and adaptive traits. Modern genomics-assisted breeding (GAB) tools, such as quantitative trait loci (QTL) mapping, genome-wide association studies (GWAS), pangenomics, haplotype-assisted breeding (HAB), genomic selection (GS) and single-cell genomics (sc-genomics), have empowered breeders to discover the genetic groundworks of stress tolerance at a record resolution (Bhat *et al.*, 2021; Varshney *et al.*, 2021b; Cuperus, 2022; Depuydt *et al.*, 2023; Nolan and Shahan, 2023; Alemu *et al.*, 2024; Raza *et al.*, 2024b).

In this chapter, we comprehensively discuss the recent innovations of modern GAB tools to understand and boost abiotic stress tolerance in crop plants. We start the discussion with (1) an appraisal of the impact of abiotic stresses, (2) the contribution and application of modern GAB tools in designing stress-smart future crops and (3) the integration of genomics-driven genome editing (GE), artificial intelligence (AI), and high-throughput phenotyping (HTP) for fast-tracking breeding programs. By leveraging these modern breeding methods, we can fast-track the designing of climate-smart crop varieties capable of thriving under adverse stress conditions that can contribute to safeguarding food security under climate change.

5.2 A Brief Appraisal of the Economic and Agricultural Impact of Abiotic Stresses

Major abiotic stresses can significantly shrink global crop yields by 50–70%, with major impacts on agricultural productivity and economic stability. For instance, drought triggered ~$37 billion in agricultural losses over the past decade, with extreme effects in regions where agriculture is the primary livelihood (https://www.fao.org/interactive/disasters-in-agriculture/en/). The rapid commencement of droughts, now affecting 74% of global regions, further complicate these challenges that fast-track crop collapses and increase the cost of adaptation for farmers and governments (Yuan *et al.*, 2023). Climate-induced yield losses (e.g. 24% in maize) compromise staple food supplies (https://svs.gsfc.nasa.gov/13979). Disturbances in 10 major crops (i.e. rice, wheat, maize, sorghum, barley, cassava, oil palm, rapeseed, soybean and sugarcane) have even decreased consumable food calories by 1%, which challenges food security and increases rural deprivation (Schleussner *et al.*, 2018; Ray *et al.*, 2019). These extensive losses are further demonstrated in FAO report (Chapter I: https://openknowledge.fao.org/items/e5a5771a-6127-4e4a-ae1c-9f26fe8aa770), which illustrates the global distribution of production losses across several stress types and regions that further highlight the scale and complexity of climate disasters.

In addition to yield losses, abiotic stresses erode economic efficiency by increasing input costs, reducing arable land area, and upsetting agricultural supply chains. For instance, salinity alone causes an annual productivity loss of $31 billion, rendering 1.5 million hectares of cropland unproductive each year (https://www.fao.org/global-soil-partnership/areas-of-work/soil-salinity/en/). Prolonged waterlogging, predicted to raise yield consequences to 10–20% by 2080, further impairs this trend and emphasizes the growing liability of agricultural systems (Ke Liu *et al.*, 2023). These losses have stronger economic consequences, such as demoted export revenues, increased food prices and dependence on imports, mainly in low-income regions. Without direct investment in climate-smart policies, these intensifying impacts will continue to challenge both global food security and economic stability (https://www.worldbank.org/en/topic/climate-smart-agriculture).

5.3 Modern Genomics Tools for Designing Stress-smart Future Crops: An Integrative Roadmap

To fully harness the power of GAB tools, we need to design an integrative roadmap that demonstrates their interconnected roles in discovering and applying novel genetic elements for designing stress-smart crop varieties (Figure 5.1). These tools aim to jointly evolve sustainable agriculture by assisting precise trait adjustment and optimizing breeding strategies. In this context, the subsequent sections shed light on how breeders can leverage these GAB tools to fast-track the designing of stress-smart crop plants that can adapt and tolerate stressful field conditions.

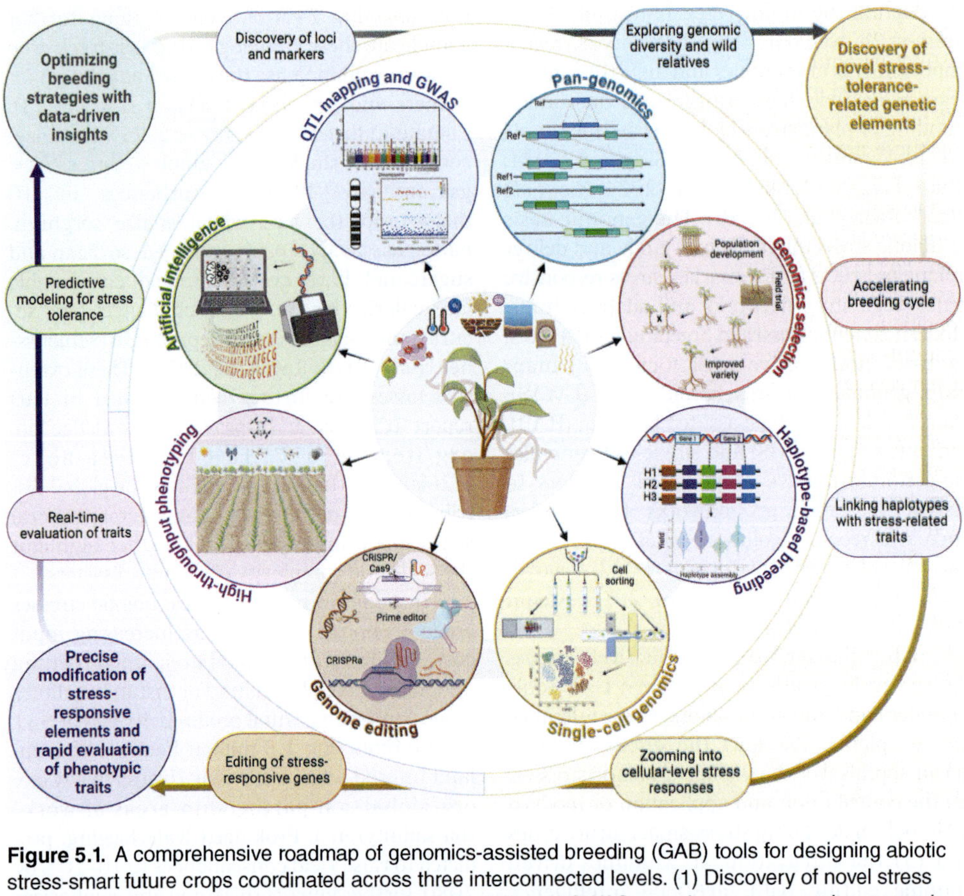

Figure 5.1. A comprehensive roadmap of genomics-assisted breeding (GAB) tools for designing abiotic stress-smart future crops coordinated across three interconnected levels. (1) Discovery of novel stress tolerance-related genetic elements; this level covers QTL and GWAS mapping, pan-genomics, genomic selection, haplotype-based breeding, and single-cell genomics. (2) Precise modification of stress-responsive elements and rapid evaluation of phenotypic traits; this level covers genome editing (e.g., CRISPR/Cas) and high-throughput phenotyping. (3) Optimizing breeding strategies with data-driven insights; this level covers comprehensive artificial intelligence that integrates multi-omics and phenotypic data for predictive modeling and stress-tolerance optimization, enabling data-driven decision-making in breeding programs. This roadmap provides a universal strategy for designing stress-smart crops for sustainable agriculture. Created with BioRender.com.

5.3.1 Identification of Candidate Genes and Genomic Regions via QTL Mapping and GWAS

QTL and GWAS mapping have become core methods in plant GAB, which allows the discovery of novel genes and genomic regions accompanying abiotic stress tolerance. Both methods enable us to examine the genetic basis of complicated traits and deliver novel insights into the stress-tolerance mechanisms in crop plants (Varshney *et al.*, 2021c; Varshney *et al.*, 2021b; Raza *et al.*, 2024b).

QTL mapping utilizes segregating populations derived from crosses between genetically distinct parental lines to identify chromosomal regions linked to phenotypic variation in stress-related traits (Yang Xu *et al.*, 2017). By genotyping these populations with molecular markers and correlating marker data with phenotypic performance, we can locate genomic regions harboring genes that contribute to stress adaptation. For instance, the *qDTY12.1* QTL in rice, associated with enhanced yield under drought conditions, has been widely utilized in breeding drought-tolerant varieties (Henry *et al.*, 2014; Henry *et al.*, 2019; Sanchez *et al.*, 2022). Likewise, the *Saltol* QTL in rice (Yadav *et al.*, 2020; Marè *et al.*, 2023) and *HKT1;5* QTL across plant species (Venkataraman *et al.*, 2021; Rawat *et al.*, 2022) have been extensively introgressed and contributed to breeding salinity-smart crop plants.

GWAS complements QTL mapping by harnessing the natural genetic diversity present in large, diverse populations (e.g. landraces, cultivars and wild relatives). This method provides high-resolution mapping to find causal variants at the single-nucleotide level (Huang and Han, 2014; Alseekh *et al.*, 2021; Cortes *et al.*, 2021). For instance, GWAS has been helpful in discovering loci-associated and candidate genes associated with multiple stress tolerance across various crops. In wheat, GWAS recognized 53 single nucleotide polymorphisms (SNPs) and 44 unique candidate genes (e.g. encoding polygalacturonase, auxilin-related protein 1, peptide deformylase and receptor-like kinase) associated with drought tolerance (Nouraei *et al.*, 2024). Likewise, GWAS and eGWAS analysis of emmer wheat discovered 86 QTLs and 68 eQTL hotspots correlated with drought tolerance, with a key regulator *TtOTS1* aiding root development under drought stress (Guang Yang *et al.*, 2024a). Another GWAS under heat stress in wheat showed 115 QTLs, including 15 stable clusters on Chr-4B harboring candidate genes such as *TaWRKY74-B* and *TaSnRK3.15-B*, which are positively associated with seedling-stage heat stress tolerance (Fu *et al.*, 2024). Moreover, Zhen Zhang *et al.* (2024b) reported 5320 drought-related important SNPs and 27 SNP clusters, with a key hotspot region where *TaPP2C6* gene was found to negatively regulate drought response.

In rice, GWAS identified 10 salinity tolerance-related candidate genes (e.g. *OsWRKY53* and *OsMKK10.2*), coordinating root Na^+ flux and Na^+ homeostasis, suggesting the powerful role of *OsWRKY53-OsHKT1;5* module in salinity tolerance (Yu *et al.*, 2023). Another combined GWAS and linkage mapping investigation in *Japonica* rice discovered a 195-kb candidate interval containing *LOC_Os12g34450* gene for salinity tolerance at the seedling stage (Shanbin Xu *et al.*, 2023). GWAS in rice germplasm identified 11 candidate genes for heat stress tolerance and numerous novel marker-trait associations under combined drought and heat stress (Kommana *et al.*, 2023).

In addition to SNP-based GWAS, emerging methods such as SV-based GWAS, k-mer GWAS, and presence/absence variation (PAV)-based GWAS have expanded the scope of genetic association investigations by capturing previously untagged variants linked to stress-adaptive traits. For instance, in pearl millet, a 716-bp insertion upstream of PMA5G03691.1 (encoding a pollen-related protein) was identified through PAV-based GWAS as a candidate variant associated with grain number per panicle and heat-responsive gene expression. This SV, likely under positive selection during domestication, was more dominant in wild accessions and may contribute to enhanced seed production under heat stress (Haidong Yan *et al.*, 2023b). Similarly, SV-based GWAS and eQTL mapping using pangenome data identified causal variants for grain weight (e.g., *QTN64*) in rice. This study discovered multiple selection events across genes related to seed shattering and submergence tolerance, specifically in African and Asian rice populations (Shang *et al.*, 2022). These studies demonstrate the power of integrating SNPs,

SVs, InDels and gene expression datasets to interpret complex traits shaped by domestication and environmental adaptation.

Moreover, the above-discussed studies demonstrate the power of QTL/GWAS in understanding the genetic basis of abiotic stress tolerance that contributes to advanced marker-assisted breeding (MAB). By spotting stress-responsive loci/markers, both methods facilitate the integration of beneficial alleles into elite varieties via marker-assisted selection/backcrossing. Despite challenges (e.g. accurate phenotyping under flexible environments, detecting minor QTLs and complex gene networks), improvements in multi-environment phenotyping, pangenomics and sc-genomics will support their application by overcoming persisting challenges and accelerate progress towards breeding and designing of stress-smart crop varieties.

5.3.2 Pangenomics: Leveraging Crop Wild Relatives for Stress-smart Crops

A single reference genome is insufficient to capture the full genetic diversity of a species. In this context, pangenomics integrates manifold genomes and allows the examination of a complete set of genes and their variations across accessions. This method helps discover SVs, rare alleles, and stress-responsive genes often lost during domestication. Pangenomes incorporate both the core genome (genes shared by all individuals) and the dispensable genome (genes distinctive to some individuals), which suggests a powerful platform for crop advancements. Moreover, recent innovations in sequencing tools have further evolved the concept into super-pangenomes, which integrate crop wild relatives (CWRs) to capture the full spectrum of genetic diversity and design stress-smart crops (Bohra *et al.*, 2022; Raza *et al.*, 2023a; Khan *et al.*, 2024; Tingting Shi *et al.*, 2024; Jayakodi *et al.*, 2025; Jiao *et al.*, 2025).

CWRs have noteworthy pools of genetic traits that have developed under diverse and extreme climate conditions. By integrating the dispensable genome of CWRs into breeding programs, novel alleles for abiotic stress tolerance can be harnessed. For instance, a super-pangenome of *Populus* genus (comprising 19 genomes) suggested the key role of private genes in species-specific adaptation to climatic stresses (Tingting Shi *et al.*, 2024). Integration of pangenome data with transcriptomic, methylomic, and chromatin accessibility data features the collaboration between gene evolution, epigenetic landscapes (e.g. CG methylation) and abiotic stress responses. This analysis recognizes 142,202 SVs, with private genes indicating a strong association with local climate adaptation (i.e. heat and hypoxia tolerance) in *Populus koreana* (Tingting Shi *et al.*, 2024). Another super-pangenome for wild North American grapes, based on nine genomes, mapped genetic diversity and discovered key loci associated with salinity tolerance. A significant SNP on Chr08, associated with chloride exclusion, and a gene homologous to *Arabidopsis* cation/H^+ exchanger, exhibited the power of CWRs for identifying salinity-tolerant traits (Cochetel *et al.*, 2023). In rice, a super-pangenome incorporating 251 accessions identified SVs and genes (i.e. *Sub-1A/B/C*, *SNORKEL1/2*, *DEC1*, *ACE1* and *ACE1-LIKE1*) associated with submergence tolerance, seed shattering, and plant architecture, that spotted key genetic options for stress adaptation (Shang *et al.*, 2022).

The building of high-quality pangenomes has been empowered by advances in sequencing technologies that assist with the discovery of SVs, copy number variations, and rare alleles associated with stress tolerance. These insights are essential for understanding how genetic change is associated with phenotypic responses to stresses. For instance, the pearl millet pangenome, based on 10 chromosomal genomes, identified 424,085 SVs, including 622,584 presence/absence variations, and suggested the expansion of RWP-RK transcription factor family for heat tolerance (Raza *et al.*, 2023b; Haidong Yan *et al.*, 2023b). In wheat, chromosome-level assemblies of 17 cultivars highlighted 249,976 SVs that signal the breeding history and adaptation to climate change (e.g. extreme temperatures), with evidence for *VRN-A1* gene mutations guiding the shift from spring to winter types (Jiao *et al.*, 2024). These examples highlight the game-changing power of pangenomics for designing stress-smart crops by reintroducing lost genetic diversity from CWRs.

Based on the above arguments, it can be concluded that pangenomics has the ability to bridge the gap between genomic findings and practical breeding goals. MAB and GS approaches can now incorporate pangenome data, which permits breeders to target formerly distant regions of the genome. Furthermore, advances in genomic editing can be used to precisely transfer beneficial alleles from CWRs into elite germplasm without compromising agronomic traits.

5.3.3 Genomics Selection Accelerates the Breeding of Stress-smart, High-yielding Crops

GS has transformed molecular breeding by enabling the prediction of phenotypic performance based on genome-wide marker data, which helps bypass the limitations of traditional marker-trait association methods. This powerful approach leverages statistical models trained on genotypic and phenotypic data, which facilitates the selection of superior lines early in the breeding process (He and Li, 2020; Alemu *et al.*, 2024). For instance, GS application in cowpeas discovered SNP markers associated with drought-tolerance indicators for key agronomic traits, with moderate prediction accuracy and tools for MAS (Ravelombola *et al.*, 2021). Similarly, whole-genome resequencing of chickpeas identified over 144,000 SNPs, including those in auxin-related genes associated with yield under drought stress. Integrating GWAS results into GS models significantly enhanced prediction accuracy, which highlights the power of GS for breeding drought-smart legumes (Yongle Li *et al.*, 2018).

In maize, GS contributed to achieving genetic gains under stress conditions. For instance, two studies exhibited genetic gains for grain yield under drought and waterlogging stresses using rapid-cycle GS using phenotypic selection (PS) and genomic-estimated breeding values (GEBVs) in yellow synthetic populations while retaining genetic diversity without yield consequences under optimal conditions (Das *et al.*, 2020; Das *et al.*, 2021). GEBVs-enabled GS-derived maize populations achieved 31–53% higher yields compared with the F_2 generation, outperforming PS (Vivek *et al.*, 2017). Likewise, Beyene *et al.* (2015) observed an average gain of 0.086 Mg/ha per cycle under managed drought, while Shikha *et al.* (2017) highlighted the superior prediction accuracy of the Bayes B model in identifying SNPs linked to drought-responsive transcription factors. These findings confirm the effectiveness of GS in breeding drought-smart future maize cultivars.

In rice, GS was used to enhance salinity tolerance that illuminated key genotype-by-environment interactions (Bartholomé *et al.*, 2023). Another study reported yield gains of 16.9% and 12.6% per cycle in *Striga*-infested and optimal environments, respectively, albeit with limited progress under drought due to low genetic variability (Badu-Apraku *et al.*, 2019). These studies collectively highlight the dynamic power of GS in fast-tracking crop improvement under abiotic stresses and suggest new ways for designing stress-smart, high-yielding crop varieties.

Despite the massive potential in transforming breeding methods by enabling early and accurate prediction of genetic potential, GS adoption faces challenges, particularly in resource-limited breeding programs, due to the high cost of generating genomic and phenotypic data, limited training population diversity, and infrastructure requirements. Addressing these issues will require collaborative efforts, including sharing datasets, open-access GS tools, and capacity-building initiatives. Future improvements, such as integrating GS with pangenomics, sc-genomics and genome editing, hold the potential to improve its impact by enabling the efficient development of stress-smart, high-yielding crops.

5.3.4 Haplotype-assisted Breeding for Enhancing Stress Tolerance

HAB is a transformative approach for enhancing stress tolerance by utilizing haplotypes, which are blocks of closely linked genetic variants that capture the combined effects of multiple loci (Bhat *et al.*, 2021). This method provides deeper insights into complex traits such as abiotic stress tolerance, which are governed by interactions among numerous genes. By identifying and deploying favorable haplotypes, HAB accelerates GAB to pinpoint genomic regions where synergistic genetic combinations enhance plant performance and pave the way for designing stress-smart crop varieties to adapt to challenging environmental conditions (Bhat *et al.*, 2021; Varshney *et al.*, 2021a; Raza *et al.*, 2024b). In this context, the emergence of haplotype-based GS (haplo-GS) and haplotype-based GWAS (haplo-GWAS) has offered novel directions for improving complex traits, including drought (Afzal *et al.*, 2019; Sehgal *et al.*, 2024) and salinity (Patil *et al.*, 2016) stress tolerance.

Haplo-GWAS, distinct from conventional SNP-GWAS, considers the collective influence of haplotype blocks rather than individual SNPs, thus improving mapping resolution for polygenic traits. In wheat, Afzal *et al.* (2019) applied both SNP- and haplotype-based GWAS in a synthetic-derived population (SYN-DER) and found that haplo-GWAS identified 20 additional drought-associated genomic regions missed by SNP-GWAS. Complementary selective sweep analysis further identified favorable loci from synthetic hexaploid wheat associated with key agronomic traits such as grain size, morphology, and drought tolerance. Similarly, Sehgal *et al.* (2024) demonstrated that haplo-GWAS outperformed SNP-GWAS by identifying five major haplotype blocks associated with over 700 kg/ha yield gain under drought in wheat, while SNP-GWAS detected only one such region. Selective sweep analysis further discovered 39 differentiation signatures co-localized with genes for yield and flowering. Together, these studies demonstrate the superior resolution and practical breeding value of haplo-GWAS in uncovering hidden diversity and facilitating targeted introgression for stress-smart, high-yielding cultivars.

Moreover, Zhen Zhang *et al.* (2024b) discovered a favored allele, *TaPP2C6*, positively associated with enhanced drought tolerance which advocates its potential for wheat breeding. Likewise, in a recent study of 399 rice genotypes from the 3000 rice genomes panel, whole-genome resequencing detected superior haplotypes for 95 drought-responsive genes (Singh *et al.*, 2024). Association analysis discovered 69 MTAs for single plant yield under drought stress. In comparison, haplotype-phenotype analysis determined superior haplotypes in seven genes, including *OsGSK1-H4*, *OsDSR2-H3* and *OsDREB1C-H3*. These haplotypes were associated with higher single plant yield under drought, and validation confirmed the presence of the *OsDREB1C* haplotype in drought-tolerant rice varieties but its absence in sensitive ones, highlighting its potential for HBB to develop high-yielding, drought-smart rice (Singh *et al.*, 2024).

In soybean, Patil *et al.* (2016) demonstrated how haplotype-based analysis combined with high-throughput SNP genotyping enabled precise identification of salinity-tolerant cultivars. The study detected SVs and favorable alleles of *GmCHX1*, designed six KASP assays, and validated them in diverse germplasm and biparental populations. With >91% accuracy in phenotype prediction and >98% precision in identifying SVs, this research exemplifies how haplotype-assisted markers can fast-track marker-assisted selection for salinity tolerance. Such findings are further reinforced by haplo-GWAS, which improves the resolution of association studies by utilizing haplotype blocks rather than individual SNPs. This allows the identification of superior haplotypes associated with stress tolerance traits and contributes to overcome the limitations of SNP-based GWAS in detecting polygenic traits.

Bai *et al.* (2024) spotted an elite haplotype, Hap2, of the *SiNLP2* gene in foxtail millet, associated with enhanced response to low-nitrogen stress. Notably, *SiNLP2* regulates key nitrogen pathways, further supporting its role in stress adaptation. This haplotype is being progressively utilized in MAS for stress-smart millet breeding (Bai *et al.*, 2024). In sorghum, there are three unique haplotypes of the *SORBI_3004G304700* gene, which is a negative regulator of salinity stress tolerance. Haplotype analysis discovered that haplotype1 is the most common, possibly suggesting a selective advantage for salinity adaptation for sorghum (Li Mao *et al.*, 2024).

Likewise, Lee *et al.* (2018) identified 40 diverse haplotypes of the *GmSALT3* gene, including eight salinity-tolerant and 32 salinity-sensitive haplotypes. Salinity-tolerant haplotypes displayed higher expression levels of *GmSALT3* than sensitive ones. Molecular markers were developed with 98.8% accuracy in screening soybean accessions for salinity tolerance, delivering new MAS resources (Lee *et al.*, 2018). These studies collectively feature the power of HAB in distinguishing superior haplotypes and featuring them in breeding programs, and they have opened new ways for designing stress-smart crop varieties.

Furthermore, adopting a strategy "haplo-GS" (Varshney *et al.*, 2021a), which combines GS with superior haplotypes and its integration with speed breeding (Watson *et al.*, 2018), offers an influential method to quickly adapt crop varieties with higher performance across manifold adaptive parameters and can also fast-track the development of stress-smart crop varieties. Haplo-GS uses haplotype information to enhance the accuracy of genomic predictions by capturing epistatic interactions and linkage disequilibrium more effectively than traditional GS. This allows breeders to select optimal combinations of haplotypes that are associated with complex traits under stress. For example, the integration of haplo-GS and haplo-GWAS in SYN-DER wheat populations, as shown by Afzal *et al.* (2019), enabled both high-resolution mapping and predictive selection of favorable haplotypes under drought that highlight the importance of this combined approach for stress-smart breeding.

The integration of haplotype data into breeding pipelines is gradually supported by innovative computational tools and statistical models that allow for the accurate association of haplotypes with phenotypic traits and confirmation in breeding populations. MAS and GS now combine haplotype-based markers that can advance selection accuracy and lower linkage drag (He and Li, 2020; Bhat *et al.*, 2021; Alemu *et al.*, 2024). In the near future, combining haplo-GWAS for discovery and haplo-GS for predictive breeding can represent a powerful pipeline to design climate-smart cultivars with higher genetic gain and broader adaptation.

5.3.5 Single-cell Genomics for Stress-smart Breeding

Single-cell genomics (sc-genomics) is a relatively new transformative extension of the GAB pipeline, which enables the dissection of stress responses at cellular resolution and thus improves trait discovery, validation, and precision targeting for stress-smart breeding. Sc-genomics, mainly sc-RNA-seq, has transformed plant research by delivering an extraordinary resolution of gene expression, regulatory networks and phenotypic features at cellular levels. This approach discovers cellular heterogeneity, and empowers the documentation of rare or cell-type-specific stress-responsive genes/pathways that often remain hidden in bulk tissue investigation (Cuperus, 2022; Nolan and Shahan, 2023). By identifying unique cellular targets and regulatory circuits, sc-genomics enables the discovery of high-value alleles and stress-responsive elements that can be introduced into elite cultivars through GAB tools such as CRISPR-based editing, GWAS and marker-assisted selection. Such integration is dynamic for stress-smart breeding, as it enables the discovery of novel targets and guides precision editing and selection strategies for crops with improved stress tolerance.

Under stress conditions, plant cells display very specific responses regulated by their location and purpose. Moreover, sc-RNA-seq permits the partition of these cell-type-specific responses to discover genes, regulatory elements and pathways required for stress adaptation. For instance, Dinneny *et al.* (2008) verified that salinity and iron deficiency prompted highly cell-type- and developmental-stage-specific transcriptional responses in *Arabidopsis* root cells. Correspondingly, Gifford *et al.* (2008) discovered cell-specific regulation of nitrogen-responsive transcripts, connecting these responses to developmental events, e.g. lateral root outgrowth facilitated by microRNA167. These discoveries highlight the power of developmental context in understanding stress responses.

Recent improvements have extended this method to diverse plant systems under stressful conditions. For instance, Zhixin Liu *et al.* (2022) used sc-RNA-seq to explore the expression levels of PIFs in leaf epidermal and guard cells under drought and salinity stresses, which suggests their role in regulating cell morphology and stress tolerance. Sun *et al.* (2022) mapped the transcriptional landscape of Chinese cabbage leaves under heat stress, illuminating cell-type-specific gene expression and alterations in subgenome dominance. These reports highlight the power of sc-genomics to identify stress-tolerant cellular states and transcriptional networks triggering abiotic stress responses.

Single-cell approaches also discover the dynamics of root cell responses under stress. For instance, Pengtao Li *et al.* (2024b) built a single-cell transcriptome atlas of salinity-stressed cotton roots, detecting 10 distinct cell-types and tracking changes in cell populations, including stress-specific roles of outer epidermal and inner endodermal cells. The functional contribution of candidate genes, e.g. *GaGH3.6* was confirmed, showing its role in redox homeostasis and salinity tolerance in cotton (Pengtao Li *et al.*, 2024b). Such gene discoveries, enabled by sc-genomics, can directly feed into the GAB pipeline for functional validation, editing, and downstream exploitation in elite lines.

In C_4 species, sc-RNA-seq has informed essential activities such as photosynthesis. For instance, a recent study characterized bermudagrass leaf cells, recognizing the preferential expression of *NADP-ME* genes in bundle sheath cells and transcription factors accompanying C_4 photosynthesis regulation. These results suggest a platform for exploring photosynthetic efficiency at the cellular level under environmental stresses (Bing Zhang *et al.*, 2024a).

Even in the model organism *Arabidopsis*, sc-RNA-seq has detected cell-specific responses under heat stress, highlighting subtle yet significant differences in gene expression among cell-types (Jean-Baptiste *et al.*, 2019). This study indicates that altering stress tolerance to specific cell-types could reduce pleiotropic effects on plant health and yield. In a GAB context, this suggests that breeders could engineer or select specific traits in targeted cell-types (e.g. guard cells or bundle sheath cells) that can minimize negative impacts on whole-plant physiology while enhancing stress tolerance. Nonetheless, technical challenges persist, e.g. the need for larger datasets and superior experimental conditions to discover non-canonical, cell-type-specific stress responses.

Beyond transcriptomics, single-cell tools such as scATAC-seq increase the benefit of sc-genomics by mapping chromatin accessibility and *cis*-regulatory elements. For example, Marand *et al.* (2021) spotted 92 chromatin accessibility states in maize across 52 cell-types, which explains how transcription factor binding supports cellular states during stress conditions. The integration of such epigenomic maps into GAB allows breeders to target not only coding sequences but also regulatory DNA, and it offers novel platforms for engineering stress tolerance at multiple regulatory layers.

These examples demonstrate that sc-genomics is not just a fundamental research tool but a functional asset in the GAB toolkit, and it can bridge gene discovery, regulatory insight, and precision breeding. These approaches deliver new insights into cellular states, which are crucial for stress adaptation and tolerance and, in turn, guide the breeding of stress-smart crops. Nonetheless, challenges such as cell isolation from stressed tissues, high costs, and computational demands limit its application. Advancing isolation techniques, cost-effective sequencing, and bioinformatics tools can help overcome these challenges. Future investigations should plan to expand single-cell assessments, emphasize diverse stress conditions and leverage these insights to manipulate stress-responsive pathways and cell-types precisely.

5.4 Genomics-driven Genome Editing for Stress Tolerance

Genome editing, driven by advances in genomics, has transformed crop improvement by allowing precise amendments to plant genomes. CRISPR-Cas-based editing systems permit targeted manipulations to improve abiotic stress tolerance by proposing valuable mutations, silencing unwanted genes or fine-tuning gene expression. These systems address the polygenic nature of stress tolerance more efficiently than traditional breeding, suggesting precision, speed and flexibility (Cardi *et al.*, 2023; Tariq *et al.*, 2023; Zaman *et al.*, 2023; Boshu Li *et al.*, 2024a). For instance, CRISPR-Cas9 editing of *TaSal1* genes in wheat resulted in mutants with closed stomata and improved drought tolerance, showcasing the system's ability to enhance key stress-related traits (Lou *et al.*, 2025). Similarly, editing *OsDSG1* in rice improved salinity tolerance by improving root and shoot growth under salinity stress without affecting normal growth conditions, which highlights the utility of CRISPR-Cas9 for breeding stress-smart crops (Ly *et al.*, 2024).

CRISPR-based editing of *OsmiR535* in rice discovered its role in stress response, with knockouts improving drought and salinity tolerance while overexpression reduced stress tolerance (Yue *et al.*, 2020). Likewise, *GmHdz4* editing in soybean discovered that modified root architecture and osmolyte accumulation increased drought tolerance (Zhong *et al.*, 2022). Editing of *CsSGR* in cucumber enhanced cold and salinity tolerance while also conferring resistance to fungal pathogens (Dong *et al.*, 2023). These studies highlight how CRISPR-Cas9 can enhance stress-adaptive traits across crops.

Moreover, editing of regulatory elements, i.e. stress-responsive promoters and enhancers, also contributes to enhanced stress tolerance. Synthetic biology approaches have led to the design of stress-inducible synthetic promoters that outperform native ones. For instance, Yongil Yang *et al.* (2021) developed osmotic and salinity stress-inducible promoters using conserved DNA motifs discovered in poplar transcriptomes. These synthetic promoters presented improved inducibility in *Arabidopsis* under water-deficit and salinity stresses, which explains their effectiveness in advancing stress-responsive gene expression. Likewise, Bhadouriya *et al.* (2024) engineered a salinity-inducible promoter with higher and more sustained activity under salinity, drought, and ABA stress compared with the constitutive CaMV35S promoter.

In a corresponding approach, a recent study has developed high-efficiency prime-editing tools to specifically modify *cis*-regulatory elements (CREs), such as heat-shock elements (HSEs), within promoters to improve stress responsiveness (Lou *et al.*, 2025). In this context, Lou *et al.* (2025) confirmed the targeted insertion of a 10-bp HSE into *CWIN* promoters in rice and tomato, resulting in heat-responsive upregulation and upgraded yields under heat stress by 25% and 33%, correspondingly, while saving up to 41% of heat-induced grain losses in rice deprived of quality drawbacks. This sets a strong method for adjusting source-sink relations and advancing crop tolerance under hotter conditions.

Rui Li *et al.* (2019) reported the vital role of *SlNPR1* in regulating drought tolerance in tomato. Mutants demonstrated reduced antioxidant defense and expression of key stress-responsive genes, suggesting the possibility of targeting such regulators for improved tolerance. Also, the activation of *AREB1* using a CRISPRa dCas9HAT system improved drought tolerance in *Arabidopsis* through upregulation of stress-associated pathways (Roca Paixão *et al.*, 2019). Yingfang Zhu *et al.* (2023) exhibited the role of *SlBBX31* in tomato cold tolerance, relating it to specific promoter variants selected during domestication. Hude Mao *et al.* (2022) emphasized the significance of CREs in wheat drought tolerance, detecting a MYB-binding insertion in the *TaNAC071-A* promoter that increases water-use efficiency and stress tolerance, displaying a useful genetic resource for future breeding.

Genome editing with CRISPR-associated proteins is of great power for designing climate-smart crops through precise genetic modifications (Zaman *et al.*, 2023; Boshu Li *et al.*, 2024a). Despite game-changing power, some challenges persist, i.e. off-target effects, regulatory obstacles and public acceptance. In this regard, innovations in CRISPR design, delivery systems, and transparent interaction can help overcome these challenges. Integration of genome editing with multi-omics and HTP improves its impact, assisting the discovery of key stress-related genes and verifying their roles in natural-field conditions. Coupled with pangenomics and HBB, genome editing can rebuild unique stress-tolerant alleles from CWRs into elite cultivars without upsetting desirable traits (Bhat *et al.*, 2021; Bohra *et al.*, 2022; Raza *et al.*, 2023a; Jayakodi *et al.*, 2025). Moreover, AI and ML can further fast-track gene discovery and editing approaches to guide the rapid development of stress-smart crops.

5.5 Fast-tracking Genomics-assisted Breeding via AI and HTP for Enhanced Stress Tolerance

Integration of AI and HTP is transforming GAB, assisting the rapid development of stress-smart crops by directing the bottlenecks of conventional breeding systems (Yunbi Xu *et al.*, 2022; Farooq *et al.*, 2024; Wanchao Zhu *et al.*, 2025). These tools fast-track the discovery of desired traits, modernize data analysis and boost the accuracy of selection processes. HTP, combined with GWAS and QTL mapping, is improving our insight into abiotic stress tolerance-related traits (Varshney *et al.*, 2021b; Farooq *et al.*, 2024; Raza *et al.*, 2024b; Raza *et al.*, 2024c). Moreover, HTP systems, using advanced sensors and imaging tools, assist in the compilation of huge phenotypic data on plant responses to several stresses. This approach also acts as a scalable and precise alternative to labor-intensive traditional phenotyping methods (Yunbi Xu *et al.*, 2022; Farooq *et al.*, 2024; Wanchao Zhu *et al.*, 2025).

For instance, HTP has recognized root architectural traits linked to phosphorus deficiency tolerance in rapeseed, illuminating key QTL for root length and biomass necessary for nutrient-poor soil adaptation (Lei Shi *et al.*, 2013). Similarly, combining HTP with GWAS has discovered drought tolerance loci in maize (Wu *et al.*, 2021) and wheat (Correia *et al.*, 2022), highlighting significant loci and candidate genes for improving drought stress tolerance. In rapeseed, HTP exhibited 928 image-based traits (i-traits) accompanying salinity tolerance, with candidate genes such as *BnCKX5* and *BnERF3* regulating these responses (Guofang Zhang *et al.*, 2023). In wheat, hyperspectral-imaging via HTP mapped yield-related traits under drought conditions, finding key chromosomal regions associated with heat and drought stresses (Mérida-García *et al.*, 2024). These outcomes highlight the power of merging HTP with genomics to discover the genetic design of stress-tolerant traits and deliver helpful insights for breeding stress-smart crops.

Integration of HTP with AI/ML algorithms can handle vast image-based datasets, discovering complex models that improve breeding accuracy. For example, AI models were used to evaluate 28,392 i-traits in wheat under drought, spotting drought tolerance-related significant SNPs with *TaPP2C6*, a key negative regulator of drought response (Zhen Zhang *et al.*, 2024b). Likewise, ML models, e.g. ridge regression and random forests, were employed to forecast maize kernel Cd levels, and this outcome encourages crop development with low Cd accumulation to prevent soil contamination (Huili Yan *et al.*, 2023a). AI-driven models have also boosted selenium-rich maize cultivation, where an artificial neural network recognized areas appropriate for growing selenium-rich crops in selenium-deprived soils (Ma *et al.*, 2022).

The combination of AI and HTP is especially vital in regions such as South Asia, where drought influences crop yields. An ML collaborative approach using "Random Forest" and "Gradient Boosting Machine" models estimated major yield losses for rice, wheat, and maize under future drought conditions. This approximate delivered high predictive precision, advancing worthy insights for future agricultural policy and drought tolerance (Prodhan *et al.*, 2022). Overall, the AI-HTP duo is redesigning plant breeding by predicting yield and fast-tracking the advancement of stress-smart agriculture.

Nonetheless, the adoption of AI and HTP encounters some challenges, e.g., high setup costs, computational loads, and data interoperability problems, uniquely for supply-limited breeding systems. Overcoming these challenges requires collaboration between research institutions, private companies and funding organizations. In the future, integrating AI and HTP with emerging methods such as sc-genomics, drones, and satellite-based systems could enhance our knowledge of stress responses and outspread these skills to large-scale field trials, that bridge the gap between controlled settings and actual agricultural systems.

5.6 Conclusion and Outlook

The mounting impacts of single or combined abiotic stresses on agriculture and the global economy demand groundbreaking solutions to sustain productivity and food security. Genomics tools, such as MAB, GS, and genome editing, have emerged as game-changing methods to fast-track breeding efficiency and improve stress tolerance. Notably, integrating GAB with speed breeding has proven highly effective in rapidly cycling generations, thereby accelerating the development and release of stress-resilient crop varieties (Watson *et al.*, 2018; Raza *et al.*, 2024a). High-throughput trait-related marker development, cost-effective genotyping and HTP tools are essential for fast-tracking genetic gains via GAB. Access to high-quality reference genomes for both wild and cultivated crops delivers new opportunities for gene discovery, marker development, and breeding products with improved stress tolerance. Moreover, advanced tools such as GS, HBB and AI-driven modeling hold enormous potential to meet market and food supply demands by designing improved crop cultivars for future climatic challenges. Likewise, the integration of speed breeding protocols with GS pipelines further enhances selection accuracy and breeding throughput, particularly under controlled stress conditions (Sinha *et al.*, 2023; Alemu *et al.*, 2024). Integration of *de novo* domestication approaches, leveraging genome editing and introgression breeding, can further balance domesticated traits with enhanced tolerance to abiotic stresses.

In the future, interdisciplinary approaches combining genomic insights, AI-powered models and true phenotyping will be vital to bridging the gap between research advancements and field-level applications. Collaborative efforts among researchers, policymakers and stakeholders will enable the availability and development of adaptive and sustainable agricultural systems. In short, we anticipate that employing an integrated GAB framework, alongside targeted gene editing, can speed up the design of stress-smart crop varieties to withstand single or combined abiotic stresses in a rapidly changing climate era.

Acknowledgment

RKV thanks Grains Research & Development Corporation (GRDC), Australia, and Hort Innovation Australia for supporting research projects on genomics and pre-breeding in wheat (UMU2404-003RTX, WSU2303-001RTX), legumes (UMU2403-009RTX, UMU2303-003RTX) and horticultural crops (AS21006, AS23003).

References

Afzal, F., Li, H., Gul, A., Subhani, A., Ali, A. *et al.* (2019) Genome-wide analyses reveal footprints of divergent selection and drought adaptive traits in synthetic-derived wheats. *G3: Genes, Genomes, Genetics* 9, 1957–1973.

Ahmar, S., Gill, R.A., Jung, K.-H., Faheem, A., Qasim, M.U., Mubeen, M. and Zhou, W. (2020) Conventional and molecular techniques from simple breeding to speed breeding in crop plants: Recent advances and future outlook. *International Journal of Molecular Sciences* 21, 2590.

Alemu, A., Åstrand, J., Montesinos-Lopez, O.A., Sanchez, Y.J.I., Fernandez-Gonzalez, J. *et al.* (2024) Genomic selection in plant breeding: Key factors shaping two decades of progress. *Molecular Plant* 17, 552–578.

Alseekh, S., Kostova, D., Bulut, M. and Fernie, A.R. (2021) Genome-wide association studies: Assessing trait characteristics in model and crop plants. *Cellular and Molecular Life Sciences* 78, 5743–5754.

Badu-Apraku, B., Talabi, A., Fakorede, M., Fasanmade, Y., Gedil, M., Magorokosho, C. and Asiedu, R. (2019) Yield gains and associated changes in an early yellow bi-parental maize population following genomic selection for *Striga* resistance and drought tolerance. *BMC Plant Biology* 19, 1–17.

Bai, Y., Wang, J., Tang, W., Sun, D., Wang, S. *et al.* (2024) Genome-wide identification of *NLP* gene families and haplotype analysis of *SiNLP2* in foxtail millet (*Setaria italica*). *International Journal of Molecular Sciences* 25, 12938.

Bartholomé, J., Frouin, J., Brottier, L., Cao, T.-V., Boisnard, A., Ahmadi, N. and Courtois, B. (2023) Genomic selection for salinity tolerance in *japonica* rice. *PLoS ONE* 18, e0291833.

Benitez-Alfonso, Y., Soanes, B.K., Zimba, S., Sinanaj, B., German, L. *et al.* (2023) Enhancing climate change resilience in agricultural crops. *Current Biology* 33, R1246–R1261.

Beyene, Y., Semagn, K., Mugo, S., Tarekegne, A., Babu, R. *et al.* (2015) Genetic gains in grain yield through genomic selection in eight bi-parental maize populations under drought stress. *Crop Science* 55, 154–163.

Bhadouriya, S.L., Karamchandani, A.N., Nayak, N., Mehrotra, S. and Mehrotra, R. (2024) Artificially designed synthetic promoter for a high level of salt induction using a *cis*-engineering approach. *Scientific Reports* 14, 13657.

Bhat, J.A., Yu, D., Bohra, A., Ganie, S.A. and Varshney, R.K. (2021) Features and applications of haplotypes in crop breeding. *Communications Biology* 4, 1266.

Bohra, A., Kilian, B., Sivasankar, S., Caccamo, M., Mba, C., Mccouch, S.R. and Varshney, R.K. (2022) Reap the crop wild relatives for breeding future crops. *Trends in Biotechnology* 40, 412–431.

Cardi, T., Murovec, J., Bakhsh, A., Boniecka, J., Bruegmann, T. *et al.* (2023) CRISPR/Cas-mediated plant genome editing: Outstanding challenges a decade after implementation. *Trends in Plant Science* 28, 1144–1165.

Cochetel, N., Minio, A., Guarracino, A., Garcia, J.F., Figueroa-Balderas, R. *et al.* (2023) A super-pangenome of the North American wild grape species. *Genome Biology* 24, 290.

Correia, P.M., Cairo Westergaard, J., Bernardes Da Silva, A., Roitsch, T., Carmo-Silva, E. and Marques Da Silva, J. (2022) High-throughput phenotyping of physiological traits for wheat resilience to high temperature and drought stress. *Journal of Experimental Botany* 73, 5235–5251.

Cortes, L.T., Zhang, Z. and Yu, J. (2021) Status and prospects of genome-wide association studies in plants. *The Plant Genome* 14, e20077.

Cuperus, J.T. (2022) Single-cell genomics in plants: Current state, future directions, and hurdles to overcome. *Plant Physiology* 188, 749–755.

Das, R.R., Vinayan, M., Patel, M.B., Phagna, R.K., Singh, S. *et al.* (2020) Genetic gains with rapid-cycle genomic selection for combined drought and waterlogging tolerance in tropical maize (*Zea mays* L.). *The Plant Genome* 13, e20035.

Das, R.R., Vinayan, M.T., Seetharam, K., Patel, M., Phagna, R.K. *et al.* (2021) Genetic gains with genomic versus phenotypic selection for drought and waterlogging tolerance in tropical maize (*Zea mays* L.). *The Crop Journal* 9, 1438–1448.

Depuydt, T., Rybel, B. De and Vandepoele, K. (2023) Charting plant gene functions in the multi-omics and single-cell era. *Trends in Plant Science* 28, 283–296.

Dinneny, J.R., Long, T.A., Wang, J.Y., Jung, J.W., Mace, D. *et al.* (2008) Cell identity mediates the response of *Arabidopsis* roots to abiotic stress. *Science* 320, 942–945.

Dong, S., Li, C., Tian, H., Wang, W., Yang, X. *et al.* (2023) Natural variation in *STAYGREEN* contributes to low-temperature tolerance in cucumber. *Journal of Integrative Plant Biology* 65, 2552–2568.

Farooq, M.A., Gao, S., Hassan, M.A., Huang, Z., Rasheed, A. *et al.* (2024) Artificial intelligence in plant breeding. *Trends in Genetics* 40, 891–908.

Fu, C., Zhou, Y., Liu, A., Chen, R., Yin, L., Li, C. and Mao, H. (2024) Genome-wide association study for seedling heat tolerance under two temperature conditions in bread wheat (*Triticum aestivum* L.). *BMC Plant Biology* 24, 430.

Garg, V., Bohra, A., Mascher, M., Spannagl, M., Xu, X. *et al.* (2024) Unlocking plant genetics with telomere-to-telomere genome assemblies. *Nature Genetics* 56, 1788–1799.

Gifford, M.L., Dean, A., Gutierrez, R.A., Coruzzi, G.M. and Birnbaum, K.D. (2008) Cell-specific nitrogen responses mediate developmental plasticity. *Proceedings of the National Academy of Sciences USA* 105, 803–808.

He, T. and Li, C. (2020) Harness the power of genomic selection and the potential of germplasm in crop breeding for global food security in the era with rapid climate change. *The Crop Journal* 8, 688–700.

Henry, A., Dixit, S., Mandal, N.P., Anantha, M., Torres, R. and Kumar, A. (2014) Grain yield and physiological traits of rice lines with the drought yield QTL qDTY12. 1 showed different responses to drought and soil characteristics in upland environments. *Functional Plant Biology* 41, 1066–1077.

Henry, A., Stuart-Williams, H., Dixit, S., Kumar, A. and Farquhar, G. (2019) Stomatal conductance responses to evaporative demand conferred by rice drought-yield quantitative trait locus qDTY12. 1. *Functional Plant Biology* 46, 660–669.

Huang, X. and Han, B. (2014) Natural variations and genome-wide association studies in crop plants. *Annual Review of Plant Biology* 65, 531–551.

Janni, M., Maestri, E., Gullì, M., Marmiroli, M. and Marmiroli, N. (2024) Plant responses to climate change, how global warming may impact on food security: A critical review. *Frontiers in Plant Science* 14, 1297569.

Jayakodi, M., Shim, H. and Mascher, M. (2025) What are we learning from plant pangenomes? *Annual Review of Plant Biology* 76, 663–686.

Jean-Baptiste, K., Mcfaline-Figueroa, J.L., Alexandre, C.M., Dorrity, M.W., Saunders, L. *et al.* (2019) Dynamics of gene expression in single root cells of *Arabidopsis thaliana*. *The Plant Cell* 31, 993–1011.

Jiao, C., Xie, X., Hao, C., Chen, L., Xie, Y. *et al.* (2025) Pan-genome bridges wheat structural variations with habitat and breeding. *Nature* 637, 384–393.

Khan, A.W., Garg, V., Sun, S., Gupta, S., Dudchenko, O. *et al.* (2024) *Cicer* super-pangenome provides insights into species evolution and agronomic trait loci for crop improvement in chickpea. *Nature Genetics* 56, 1225–1234.

Kommana, M., Reddy, D.M., Amarnath, K., Naik, M.V.K., Withanawasam, D. *et al.* (2023) Identification of genomic regions governing moisture and heat stress tolerance employing association mapping in rice (*Oryza sativa* L.). *Molecular Biology Reports* 50, 1499–1515.

Krishna, T.P.A., Veeramuthu, D., Maharajan, T. and Soosaimanickam, M. (2023) The era of plant breeding: Conventional breeding to genomics-assisted breeding for crop improvement. *Current Genomics* 24, 24–35.

Lee, S., Kim, J.-H., Sundaramoorthy, J., Park, G.T., Lee, J.-D. *et al.* (2018) Identification of GmSALT3 haplotypes and development of molecular markers based on their diversity associated with salt tolerance in soybean. *Molecular Breeding* 38, 1–11.

Li, Y., Ruperao, P., Batley, J., Edwards, D., Khan, T. *et al.* (2018) Investigating drought tolerance in chickpea using genome-wide association mapping and genomic selection based on whole-genome resequencing data. *Frontiers in Plant Science* 9, 190.

Li, R., Liu, C., Zhao, R., Wang, L., Chen, L. *et al.* (2019) CRISPR/Cas9-mediated *SlNPR1* mutagenesis reduces tomato plant drought tolerance. *BMC Plant Biology* 19, 1–13.

Li, B., Sun, C., Li, J. and Gao, C. (2024a) Targeted genome-modification tools and their advanced applications in crop breeding. *Nature Reviews Genetics* 25, 603–622.

Li, P., Liu, Q., Wei, Y., Xing, C., Xu, Z. *et al.* (2024b) Transcriptional landscape of cotton roots in response to salt stress at single-cell resolution. *Plant Communications* 5, 100740.

Liu, Z., Guo, C., Wu, R., Wang, J., Zhou, Y. *et al.* (2022) Identification of the regulators of epidermis development under drought-and salt-stressed conditions by single-cell RNA-seq. *International Journal of Molecular Sciences* 23, 2759.

Liu, K., Harrison, M.T., Yan, H., Liu, D.L., Meinke, H. *et al.* (2023) Silver lining to a climate crisis in multiple prospects for alleviating crop waterlogging under future climates. *Nature Communications* 14, 765.

Lou, H., Li, S., Shi, Z., Zou, Y., Zhang, Y. *et al.* (2025) Engineering source-sink relations by prime editing confers heat-stress resilience in tomato and rice. *Cell* 188, 530–549.

Ly, L.K., Ho, T.M., Bui, T.P., Nguyen, L.T., Phan, Q. *et al.* (2024) CRISPR/Cas9 targeted mutations of *OsDSG1* gene enhanced salt tolerance in rice. *Functional & Integrative Genomics* 24, 70.

Ma, X., Yang, Z., Yu, T. and Guan, D.-X. (2022) Probability of cultivating Se-rich maize in Se-poor farmland based on intensive field sampling and artificial neural network modelling. *Chemosphere* 309, 136690.

Mao, H., Li, S., Chen, B., Jian, C., Mei, F. *et al.* (2022) Variation in cis-regulation of a NAC transcription factor contributes to drought tolerance in wheat. *Molecular Plant* 15, 276–292.

Mao, L., Huazhuan, H., Haiyan, G., Wangqi, H., Qizhe, C. *et al.* (2024) Identification, function validation and haplotype analysis of salt-tolerant genes of lectin receptor kinase gene family in sorghum (*Sorghum bicolor* L.). *Frontiers in Genetics* 15, 1464537.

Marand, A.P., Chen, Z., Gallavotti, A. and Schmitz, R.J. (2021) A cis-regulatory atlas in maize at single-cell resolution. *Cell* 184(3041–3055), e21.

Marè, C., Zampieri, E., Cavallaro, V., Frouin, J., Grenier, C. *et al.* (2023) Marker-assisted introgression of the salinity tolerance locus *Saltol* in temperate *Japonica* rice. *Rice* 16(2).

Mérida-García, R., Gálvez, S., Solís, I., Martínez-Moreno, F., Camino, C. *et al.* (2024) High-throughput phenotyping using hyperspectral indicators supports the genetic dissection of yield in durum wheat grown under heat and drought stress. *Frontiers in Plant Science* 15, 1470520.

Mishra, S., Srivastava, A.K., Khan, A.W., Tran, L.-S.P. and Nguyen, H.T. (2024) The era of panomics-driven gene discovery in plants. *Trends in Plant Science* 29, 995–1005.

Nolan, T.M. and Shahan, R. (2023) Resolving plant development in space and time with single-cell genomics. *Current Opinion in Plant Biology* 76, 102444.

Nouraei, S., Mia, M.S., Liu, H., Turner, N.C. and Yan, G. (2024) Genome-wide association study of drought tolerance in wheat (*Triticum aestivum* L.) identifies SNP markers and candidate genes. *Molecular Genetics and Genomics* 299, 22.

Patil, G., Do, T., Vuong, T.D., Valliyodan, B., Lee, J.-D. *et al.* (2016) Genomic-assisted haplotype analysis and the development of high-throughput SNP markers for salinity tolerance in soybean. *Scientific Reports* 6, 19199.

Prodhan, F.A., Zhang, J., Sharma, T.P.P., Nanzad, L., Zhang, D. *et al.* (2022) Projection of future drought and its impact on simulated crop yield over South Asia using ensemble machine learning approach. *Science of the Total Environment* 807, 151029.

Ravelombola, W., Shi, A. and Huynh, B.-L. (2021) Loci discovery, network-guided approach, and genomic prediction for drought tolerance index in a multi-parent advanced generation intercross (MAGIC) cowpea population. *Horticulture Research* 8, 24.

Rawat, N., Wungrampha, S., Singla-Pareek, S.L., Yu, M., Shabala, S. and Pareek, A. (2022) Rewilding staple crops for the lost halophytism: Toward sustainability and profitability of agricultural production systems. *Molecular Plant* 15, 45–64.

Ray, D.K., West, P.C., Clark, M., Gerber, J.S., Prishchepov, A.V. and Chatterjee, S. (2019) Climate change has likely already affected global food production. *PLoS ONE* 14, e0217148.

Raza, A., Bohra, A., Garg, V. and Varshney, R.K. (2023a) Back to wild relatives for future breeding through super-pangenome. *Molecular Plant* 16, 1363–1365.

Raza, A., Bohra, A. and Varshney, R.K. (2023b) Pan-genome for pearl millet that beats the heat. *Trends in Plant Science* 28, 857–860.

Raza, A., Bashir, S., Khare, T., Karikari, B., Copeland, R.G. *et al.* (2024a) Temperature-smart plants: A new horizon with omics-driven plant breeding. *Physiologia Plantarum* 176, e14188.

Raza, A., Chen, H., Zhang, C., Zhuang, Y., Sharif, Y. *et al.* (2024b) Designing future peanut: The power of genomics-assisted breeding. *Theoretical and Applied Genetics* 137, 1–30.

Raza, A., Salehi, H., Bashir, S., Tabassum, J., Jamla, M. *et al.* (2024c) Transcriptomics, proteomics, and metabolomics interventions prompt crop improvement against metal (loid) toxicity. *Plant Cell Reports* 43, 80.

Raza, A., Khare, T., Zhang, X., Rahman, M.M., Hussain, M. *et al.* (2025a) Novel strategies for designing climate-smart crops to ensure sustainable agriculture and future food security. *Journal of Sustainable Agriculture and Environment* 4, e70048.

Raza, A., Li, Y., Prakash, C.S. and Hu, Z. (2025b) Panomics to manage combined abiotic stresses in plants. *Trends in Plant Science* 30, 1079–1084.

Roca Paixão, J.F., Gillet, F.-X., Ribeiro, T.P., Bournaud, C., Lourenço-Tessutti, I.T. *et al.* (2019) Improved drought stress tolerance in Arabidopsis by CRISPR/dCas9 fusion with a Histone AcetylTransferase. *Scientific Reports* 9, 8080.

Sanchez, J., Kaur, P.P., Pabuayon, I.C., Karampudi, N.B.R., Kitazumi, A. *et al.* (2022) DECUSSATE network with flowering genes explains the variable effects of qDTY12. 1 to rice yield under drought across genetic backgrounds. *The Plant Genome* 15, e20168.

Schleussner, C.-F., Deryng, D., Müller, C., Elliott, J., Saeed, F. *et al.* (2018) Crop productivity changes in 1.5 C and 2 C worlds under climate sensitivity uncertainty. *Environmental Research Letters* 13, 064007.

Sehgal, D., Rathan, N.D., Özdemir, F., Keser, M., Akin, B. *et al.* (2024) Genomic wide association study and selective sweep analysis identify genes associated with improved yield under drought in Turkish winter wheat germplasm. *Scientific Reports* 14, 8431.

Shang, L., Li, X., He, H., Yuan, Q., Song, Y. *et al.* (2022) A super pan-genomic landscape of rice. *Cell Research* 32, 878–896.

Shi, L., Shi, T., Broadley, M.R., White, P.J., Long, Y. *et al.* (2013) High-throughput root phenotyping screens identify genetic loci associated with root architectural traits in *Brassica napus* under contrasting phosphate availabilities. *Annals of Botany* 112, 381–389.

Shi, T., Zhang, X., Hou, Y., Jia, C., Dan, X. *et al.* (2024) The super-pangenome of Populus unveils genomic facets for its adaptation and diversification in widespread forest trees. *Molecular Plant* 17, 725–746.

Shikha, M., Kanika, A., Rao, A.R., Mallikarjuna, M.G., Gupta, H.S. and Nepolean, T. (2017) Genomic selection for drought tolerance using genome-wide SNPs in maize. *Frontiers in Plant Science* 8, 550.

Singh, P., Sundaram, K.T., Vinukonda, V.P., Venkateshwarlu, C., Paul, P.J. *et al.* (2024) Superior haplotypes of key drought-responsive genes reveal opportunities for the development of climate-resilient rice varieties. *Communications Biology* 7, 89.

Sinha, D., Maurya, A.K., Abdi, G., Majeed, M., Agarwal, R. *et al.* (2023) Integrated genomic selection for accelerating breeding programs of climate-smart cereals. *Genes* 14, 1484.

Sun, X., Feng, D., Liu, M., Qin, R., Li, Y. *et al.* (2022) Single-cell transcriptome reveals dominant subgenome expression and transcriptional response to heat stress in Chinese cabbage. *Genome Biology* 23, 262.

Tariq, A., Mushtaq, M., Yaqoob, H., Bhat, B.A., Zargar, S.M. *et al.* (2023) Putting CRISPR-Cas system in action: A golden window for efficient and precise genome editing for crop improvement. *GM Crops & Food* 14, 1–27.

Varshney, R.K., Barmukh, R., Roorkiwal, M., Qi, Y., Kholova, J. *et al.* (2021a) Breeding custom-designed crops for improved drought adaptation. *Advances in Genetics* 2, e202100017.

Varshney, R.K., Bohra, A., Yu, J., Graner, A., Zhang, Q. and Sorrells, M.E. (2021b) Designing future crops: Genomics-assisted breeding comes of age. *Trends in Plant Science* 26, 631–649.

Varshney, R.K., Roorkiwal, M., Sun, S., Bajaj, P., Chitikineni, A. *et al.* (2021c) A chickpea genetic variation map based on the sequencing of 3,366 genomes. *Nature* 599, 622–627.

Venkataraman, G., Shabala, S., Véry, A.-A., Hariharan, G.N., Somasundaram, S. *et al.* (2021) To exclude or to accumulate? Revealing the role of the sodium HKT1; 5 transporter in plant adaptive responses to varying soil salinity. *Plant Physiology and Biochemistry* 169, 333–342.

Vivek, B., Krishna, G.K., Vengadessan, V., Babu, R., Zaidi, P. *et al.* (2017) Use of genomic estimated breeding values results in rapid genetic gains for drought tolerance in maize. *The Plant Genome* 10, plantgenome2016.07.0070.

Watson, A., Ghosh, S., Williams, M.J., Cuddy, W.S., Simmonds, J. *et al.* (2018) Speed breeding is a powerful tool to accelerate crop research and breeding. *Nature Plants* 4, 23–29.

Wu, X., Feng, H., Wu, D., Yan, S., Zhang, P. *et al.* (2021) Using high-throughput multiple optical phenotyping to decipher the genetic architecture of maize drought tolerance. *Genome Biology* 22, 1–26.

Xie, L., Gong, X., Yang, K., Huang, Y., Zhang, S. *et al.* (2024) Technology-enabled great leap in deciphering plant genomes. *Nature Plants* 10, 551–566.

Xu, Y., Li, P., Yang, Z. and Xu, C. (2017) Genetic mapping of quantitative trait loci in crops. *The Crop Journal* 5, 175–184.

Xu, Y., Zhang, X., Li, H., Zheng, H., Zhang, J. *et al.* (2022) Smart breeding driven by big data, artificial intelligence, and integrated genomic-enviromic prediction. *Molecular Plant* 15, 1664–1695.

Xu, S., Cui, J., Cao, H., Liang, S., Ma, T. *et al.* (2023) Identification of candidate genes for salinity tolerance in *japonica* rice at the seedling stage based on genome-wide association study and linkage mapping. *Frontiers in Plant Science* 14, 1184416.

Yadav, A.K., Kumar, A., Grover, N., Ellur, R.K., Krishnan, S.G. *et al.* (2020) Marker aided introgression of '*Saltol*', a major QTL for seedling stage salinity tolerance into an elite Basmati rice variety 'Pusa Basmati 1509'. *Scientific Reports* 10, 13877.

Yan, H., Guo, H., Xu, W., Dai, C., Kimani, W. *et al.* (2023a) GWAS-assisted genomic prediction of cadmium accumulation in maize kernel with machine learning and linear statistical methods. *Journal of Hazardous Materials* 441, 129929.

Yan, H., Sun, M., Zhang, Z., Jin, Y., Zhang, A. *et al.* (2023b) Pangenomic analysis identifies structural variation associated with heat tolerance in pearl millet. *Nature Genetics* 55, 507–518.

Yang, Y., Lee, J.H., Poindexter, M.R., Shao, Y., Liu, W. *et al.* (2021) Rational design and testing of abiotic stress-inducible synthetic promoters from poplar *cis*-regulatory elements. *Plant Biotechnology Journal* 19, 1354–1369.

Yang, G., Pan, Y., Pan, W., Song, Q., Zhang, R. *et al.* (2024a) Combined GWAS and eGWAS reveals the genetic basis underlying drought tolerance in emmer wheat (*Triticum turgidum* L.). *New Phytologist* 242, 2115–2131.

Yang, Y., Tilman, D., Jin, Z., Smith, P., Barrett, C.B. *et al.* (2024b) Climate change exacerbates the environmental impacts of agriculture. *Science* 385, eadn3747.

Yu, J., Zhu, C., Xuan, W., An, H., Tian, Y. *et al.* (2023) Genome-wide association studies identify *OsWRKY53* as a key regulator of salt tolerance in rice. *Nature Communications* 14, 3550.

Yuan, X., Wang, Y., Ji, P., Wu, P., Sheffield, J. and Otkin, J.A. (2023) A global transition to flash droughts under climate change. *Science* 380, 187–191.

Yue, E., Cao, H. and Liu, B. (2020) OsmiR535, a potential genetic editing target for drought and salinity stress tolerance in *Oryza sativa*. *Plants* 9, 1337.

Zaman, Q.U., Raza, A., Lozano-Juste, J., Chao, L., Jones, M.G., Wang, H.-F. and Varshney, R.K. (2023) Engineering plants using diverse CRISPR-associated proteins and deregulation of genome-edited crops. *Trends in Biotechnology* 42, 560–574.

Zandalinas, S.I., Fritschi, F.B. and Mittler, R. (2021) Global warming, climate change, and environmental pollution: Recipe for a multifactorial stress combination disaster. *Trends in Plant Science* 26, 588–599.

Zandalinas, S.I., Peláez-Vico, M.Á., Sinha, R., Pascual, L.S. and Mittler, R. (2024) The impact of multifactorial stress combination on plants, crops, and ecosystems: How should we prepare for what comes next? *The Plant Journal* 117, 1800–1814.

Zhang, G., Zhou, J., Peng, Y., Tan, Z., Zhang, Y. *et al.* (2023) High-throughput phenotyping-based quantitative trait loci mapping reveals the genetic architecture of the salt stress tolerance of *Brassica napus*. *Plant, Cell & Environment* 46, 549–566.

Zhang, B., Ma, Z., Guo, H., Chen, S. and Liu, J. (2024a) Single-cell RNA-sequencing provides new insights into the cell-specific expression patterns and transcriptional regulation of photosynthetic genes in bermudagrass leaf blades. *Plant Physiology and Biochemistry* 213, 108857.

Zhang, Z., Qu, Y., Ma, F., Lv, Q., Zhu, X. *et al.* (2024b) Integrating high-throughput phenotyping and genome-wide association studies for enhanced drought resistance and yield prediction in wheat. *New Phytologist* 243, 1758–1775.

Zhong, X., Hong, W., Shu, Y., Li, J., Liu, L. *et al.* (2022) CRISPR/Cas9 mediated gene-editing of *GmHdz4* transcription factor enhances drought tolerance in soybean (*Glycine max* [L.] Merr.). *Frontiers in Plant Science* 13, 988505.

Zhu, Y., Zhu, G., Xu, R., Jiao, Z., Yang, J. *et al.* (2023) A natural promoter variation of SlBBX31 confers enhanced cold tolerance during tomato domestication. *Plant Biotechnology Journal* 21, 1033–1043.

Zhu, W., Li, W., Zhang, H. and Li, L. (2025) Big data and artificial intelligence-aided crop breeding: Progress and prospects. *Journal of Integrative Plant Biology* 67, 722–739.

6 Genomic Solutions for Pest and Disease Management in Wheat

Arielís Morales-Salvá, Vijay Tiwari and Nidhi Rawat*
Department of Plant Sciences and Landscape Architecture, University of Maryland, Maryland, USA

Abstract

Rapid advances in sequencing technologies and genomics pipelines have opened new avenues for applications in crop improvement. In this chapter, we have summarized the progress made in pathogen and pest management in wheat driven by genetic and genomic approaches, including genetic mapping, marker-assisted selection (MAS), positional cloning, genome sequencing, association mapping, genotyping-by-sequencing (GBS), genomic selection, mutagenesis, TILLING/MutRenSeq, gene editing and RNA interference (RNAi). Examples of each of these aspects in wheat have been presented, along with the appropriate techniques used to achieve the specific goals. The chapter highlights the potential of these technologies to improve global food security by promoting the sustainable management of pathogens and pests.

6.1 Introduction

Wheat is a staple food for 35% of the world population (Ajijah *et al.*, 2024). With a production of 1.81 billion bushels from 37.3 million acres in 2023/24 alone, it is the food crop with maximum acreage globally (USDA, 2024). Wheat contributes to at least 20% of the daily calories and proteins consumed by humans. However, wheat production is challenged by a number of diseases and pests. Annually an approximate 21.5% of wheat production is lost to diseases (Singh *et al.*, 2023) and 10–50% to pests (Chai *et al.*, 2022). These losses are predicted to exacerbate even further with a changing climate (Iqbal *et al.*, 2024). With the world's population expected to reach nearly 10 billion by 2050, keeping up with these challenges is critical to ensure global food security (United Nations, 2012). Genomic solutions are expediting the tools for crop improvement to safeguard wheat production against the never-ending threats of pests and diseases.

Bread wheat (*Triticum aestivum*) is an allohexaploid crop, composed of three sub-genomes (ABD). The high similarity of the constituent sub-genomes creates genetic redundancy and complexity in its ~17 Gb large genome, making it uniquely challenging to study and manipulate it toward genetic investigations. The size and intricacy of the wheat genome were the factors leading to slower advances in the genome sequencing and genomics resource development in wheat as compared with other crops such as rice and maize (Shewry *et al.*, 2022). Figure 6.1 shows the chronological

*Corresponding author. Email: nidhirwt@umd.edu

DOI: 10.1079/9781800627185.0006

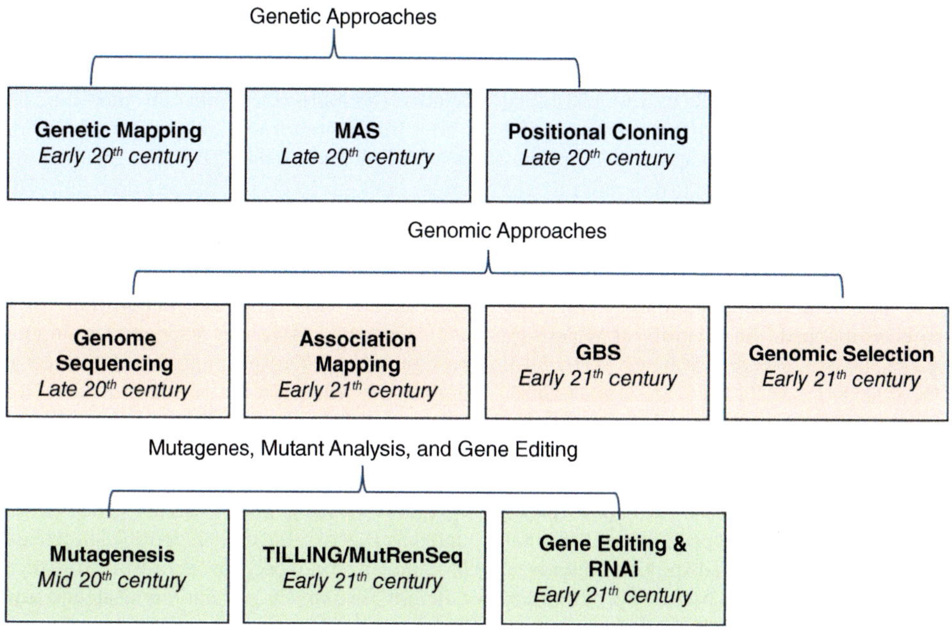

Figure 6.1. Genetic, genomic, and functional genomics techniques in chronological order used for disease and pest management. (Authors' own image.)

advances in genetic and genomics technologies for wheat improvement.

6.2 Genetic Approaches

Genetic approaches, including homology-based searches, genetic mapping, marker-assisted selection and positional cloning help researchers to identify and target specific genes associated with resistance to pathogens. Genetic mapping locates these resistance genes in the genome, while marker-assisted selection allows for the rapid selection of plants with these desirable traits in breeding programs. Further, positional cloning helps in the isolation of individual genes that are responsible for resistance. Together, these approaches play a critical role in developing resilient crops.

6.2.1 Homology-based Searches

Homology-based searches have been a foundational approach in identifying disease resistance genes across plant species. By leveraging conserved genetic sequences among related species, researchers can predict and isolate genes involved in plant defense mechanisms. One notable example is the identification and cloning of the *Mlo* gene in barley using homology to rice (*Oriza sativa*) and wheat, as described by Elliott *et al.* (2002). For stem rust resistance, genes such as *Sr33* and *Sr35* were also identified through sequence similarity to resistance genes in barley and brachypodium, respectively (Periyannan *et al.*, 2013; Saintenac *et al.*, 2013). Similarity, *Lr67*, a multi-pathogen resistance gene effective against leaf and stripe rust, was discovered through homology to sugar transporters involved in plant defense (Moore *et al.*, 2015). Hessian fly resistance genes, such as *H13* and *H26*, were mapped through comparative analyses with insect resistance genes in barley and other cereals (Cox and Hatchett, 1994; Dweikat *et al.*, 1997). As sequencing technologies continue to improve and more reference genomes are assembled, this technique is less prevalent, but still very effective.

6.2.2 Genetic Mapping

Genetic mapping is the first step toward identification for a gene. Numerous resistance genes for diseases such as leaf rust, stem rust, stripe rust, powdery mildew, Fusarium Head blight (FHB), tan spot and pests such as Hessian fly have been mapped in wheat since the 20th century (McIntosh *et al.*, 1995, 2020). Rust resistance genes have been classified into race-specific and broad-spectrum (race-non-specific) categories depending on their response to different races of the rust pathogens (Periyannan *et al.*, 2017). Wild relatives of wheat have been important sources of resistance genes (Wulff and Moscou, 2014). For stem rust approximately 60 genes have been mapped, the majority of which are race-specific. *Sr5, Sr6, Sr7, Sr8, Sr9, Sr10, Sr15, Sr16, Sr18, Sr19, Sr20, Sr23, Sr28, Sr29, Sr30, Sr41, Sr42, Sr48, Sr49, Sr54, SrCad, SrTmp* are examples of race-specific resistance genes (Karelov *et al.*, 2022). More than 80 genes for leaf rust resistance (~50% derived from wild relative species of wheat) have been mapped in wheat (Kumar *et al.*, 2022). *Lr34* is a broad-spectrum resistance gene and has been used in breeding programs for several decades (Borlaug, 1968). More than 80 stripe rust resistance genes have been mapped in wheat, most of which are race-specific and prone to be ineffective for the new and emerging races of the pathogen (Chen *et al.*, 2021). Historically, stripe rust was considered a disease of cool and wet climates, but with the emergence of new high temperature tolerant races, the search for and mapping of new broad-spectrum stripe rust genes resistant to these pathotypes is important.

Fusarium Head blight (FHB) is a major disease of wheat causing not only yield losses, but also mycotoxin contamination of grains. Genetic resistance against FHB is quantitative and more than 500 quantitative trait loci (QTLs) with varying effects have been reported in wheat (Buerstmayr *et al.*, 2009). Among the QTLs identified, a few have shown major effects and have been extensively studied. The most well-known and significant QTL, *Fhb1*, located on chromosome 3BS is associated primarily with Type II resistance, which restricts pathogen spread within the spike (Waldron *et al.*, 1999). *Fhb1* has been integrated into many breeding programs due to its effectiveness and stability across different environments. Another notable QTL, *Fhb2*, located on chromosome 6BS, has also shown a substantial role in improving FHB resistance, particularly in reducing disease spread (Jiang *et al.*, 2007). Additionally, *Fhb5*, situated on chromosome 5AS, contributes to reducing both disease spread and mycotoxin levels in infected grains (Cuthbert *et al.*, 2006). *Fhb7* (Guo *et al.*, 2015) and *Fhb9* (Zhang *et al.*, 2024) are other major effect QTLs sourced from wild wheat relative *Thinopyrum ponticum* (chromosome 7el) and Chinese landraces (chromosome 2DL), respectively. These major QTLs, along with many minor ones, play a crucial role in improving FHB resistance in wheat, though their effectiveness can vary depending on environmental factors and genetic backgrounds, making the development of resistant cultivars a complex but essential task.

Hessian fly (*Mayetiola destructor*) is a major pest of wheat, causing significant economic impact on wheat production worldwide. This pest primarily targets the stem of wheat plants, leading to stunted growth, reduced yield, and in severe infestations, the death of the plant. To date, over 30 Hessian fly resistance genes, designated as *H* genes, have been identified in wheat, providing a broad spectrum of resistance against different Hessian fly populations (Ratcliffe and Hatchett, 1997). Among these resistance genes, a few have shown major effects and have been widely utilized in wheat breeding. The *H13* gene, located on chromosome 6D, provides strong resistance against several Hessian fly biotypes and has been an important source of resistance in wheat breeding programs (Dweikat *et al.*, 1997). Another critical gene, *H26*, found on chromosome 4AL, has demonstrated durable resistance, particularly against more virulent Hessian fly biotypes (Cox and Hatchett, 1994). *H30*, located on chromosome 5BS, is another key gene that has contributed to Hessian fly resistance from wild relative *Aegilops triuncialis* especially in regions where other resistance genes have shown reduced effectiveness due to evolving fly biotypes (Martín-Sánchez *et al.*, 2003). These major resistance genes, together with minor QTLs, play an essential role. However, the effectiveness of these genes can vary depending on the biotypes and environment, underscoring the need for continuous breeding efforts to incorporate new sources of resistance.

6.2.3 Marker-Assisted Selection (MAS)

MAS is a powerful tool in plant breeding, created in the late 20th century, that uses molecular markers to increase the precision and efficiency for selecting desirable traits, such as disease resistance in crops. Traditional breeding methods primarily rely on phenotypic selection, whereas MAS utilizes DNA markers linked to specific traits, facilitating the selection of plants carrying desirable genes early in the breeding process. This technique accelerates the development of varieties, also allowing breeders to screen large populations for traits such as disease and pest resistance while the plants are still in the seedling stage. MAS is being widely used in wheat breeding to incorporate resistance to major diseases such as leaf rust, stripe rust, stem rust and FHB. The distinct advantage of MAS is the ease of combining and pyramiding resistance QTLs without extensive phenotyping at each step of breeding (Possamai *et al.*, 2024). For example, traditional breeding for FHB is challenging due to the quantitative nature of the resistance, which is why MAS was a promising alternative. Zhang *et al.* (2021c) utilized this technique in order to perform pyramiding of resistant genes *Fhb1*, *Fhb4* and *Fhb5*. MAS proved efficient in improving FHB resistance and breaking unfavorable linkages between resistance genes and undesirable traits. The lines developed using MAS demonstrated high resistance levels and maintained desirable agronomic performance. A similar experiment was conducted by Hu *et al.* (2023) utilizing resistant loci *Fhb1* and *Fhb2*, with favorable results. Another study by Guo *et al.* (2015) used MAS to pyramid two key resistant genes, *Fhb1* and *Fhb7*, into new wheat lines, enhancing resistance. This study identified new molecular markers closely linked to *Fhb7* and created wheat lines. These lines were tested for resistance, demonstrating significant reductions in the spread of the disease.

Sunilkumar *et al.* (2022) utilized a form of MAS, marker-assisted backcrossing, to improve wheat variety HD3086 by introducing leaf rust resistance (*Lr24*) and moisture deficit stress tolerance traits. The improved lines, tested in multiple generations, showed enhanced resistance to leaf rust pathotypes 77-5 and 77-9 as well as superior agronomic performance under restricted irrigation.

MAS has been successfully applied to incorporate resistance genes against various diseases, including wheat rust (*Lr34*, *Yr36*), eyespot (*Pch1*), barley yellow mosaic viruses (*rym4*/*rym5*), barley powdery mildew (*mlo*) and Fusarium head blight (*Fhb1*, *Qfhs.ifa-5A*). Despite its advantages, MAS faces challenges, like the need for closely linked diagnostic markers and the complexity of QTLs that are affected by environmental interactions. These limitations will be overcome with advancements in areas such as GS and advancements in high-throughput genotyping platforms (Miedaner and Korzun, 2012). Marker-assisted selection is crucial for accelerating the breeding process, permitting the incorporation of desirable traits and enhancing crop resilience.

6.2.4 Positional Cloning of Genes

Positional cloning refers to the identification of a gene underlying a trait based on its location in the genome, and is crucial for deploying disease- and pest-resistant genes in crops. Gene cloning is vital for advancing our understanding of plant resistance mechanisms and developing robust strategies to manage pests and diseases. Leaf rust resistance genes *Lr10* and *Lr21* were the first genes to be positionally cloned in wheat (Feuillet *et al.*, 2003; Huang *et al.*, 2003). Both *Lr10* and *Lr21* encode nucleotide-binding leucine-rich repeat proteins (NBS-LRR) and their positional cloning demonstrated the feasibility of cloning disease resistance genes in the complex wheat genome. Since then numerous disease resistance genes including *Lr34*, *Lr67*, *Lr14a*, *Lr42* (for leaf rust) *Sr21*, *Sr22*, *Sr31*, *Sr33* (for stem rust), *Yr5*, *Yr10*, *Yr15* (for stripe rust), *Pm1*, *Pm2*, *Pm60* (for powdery mildew), *Fhb1*, *Fhb7* (for FHB), *Stb6* (for Septoria tricitici blotch), *Snn 1* (Septoria nodorum blotch) and *Tsn1* (for tan spot) have been cloned in wheat (Faris *et al.*, 2010). This list is by no means exhaustive and the number of cloned resistance genes is increasing very rapidly because of the advances in genomics and sequencing approaches. Cloning of these genes not only is leading to advances in understanding of how their protein products interact with the pathogens to confer resistance but at the same time it highlights the avenues for transgenic utilization of these genes. Luo *et al.* (2021) transgenically expressed a five-gene cassette containing *Sr45*, *Sr55*, *Sr50*, *Sr35* and *Sr22*, which was found to confer broad-spectrum resistance to rust diseases. Table 6.1 provides a list of major disease resistance genes cloned in wheat to date. The majority of the disease resistance genes cloned for rust diseases encode standard NLR proteins (*Lr10*, *Lr21*, *Lr1*, *Lr22a*, *Lr13*, *Lr42*). However, several new diverse

Table 6.1. Wheat (*Triticum aestivum*) cloned genes for FHB, powdery mildew, rusts, Septoria, leaf tip necrosis, tan spot and Orange wheat blossom midge.

Protein	Disease	Gene	Author
Nucleotide-binding Leucine-rich Repeat (NLR)	Leaf rust	*Lr10*, *Lr21*, *Lr1*, *Lr22a*, *Lr13*, *Lr42*, *Lr47*, *Lr85*	Feuillet *et al.* (2003),Cloutier *et al.* (2007), Thind *et al.* (2017), Hewitt *et al.* (2021a), Lin *et al.* (2022), Li *et al.* (2023), Sharma *et al.* (2024)
NLR	Powdery mildew	*Pm3a*, *Pm3b*, *Pm3d*, *Pm3f, Pm8*, *Pm2a*, *Pm21*, *Pm60*, *MlWE18*, *Pm17*, *Pm5e*, *Pm41*, *Pm1a*, *Pm69*	Srichumpa *et al.* (2005), Yahiaoui *et al.* (2004), Hurni *et al.* (2013), Sánchez-Martín *et al.* (2016), He *et al.* (2017), Xing *et al.* (2018), Zou *et al.* (2018), Singh *et al.* (2018), Xie *et al.* (2020), Li *et al.* (2020), Hewitt *et al.* (2021b), Li *et al.* (2024)
NLR	Stem rust	*Sr33*, *Sr35*, *Sr50*, *Sr22a*, *Sr22b*, *Sr45*, *Sr13*, *Sr21*, *Sr46*, *Sr26*, *Sr61*, *Sr27*, *SrTA1662*, *Sr43*	Periyannan *et al.* (2013), Saintenac *et al.* (2013), Mago *et al.* (2015), Steuernagel *et al.* (2016), Zhang *et al.* (2017), Chen *et al.* (2018), Arora *et al.* (2019), Zhang *et al*. (2021a), Upadhyaya *et al.* (2021), Gaurav *et al.* (2022)
NLR	Stripe rust	*Yr10*, *YrAS2388R*, *Yr27*, *TaYRG1*, *TaSTP3*	Liu *et al.* (2014), Zhang *et al.* (2019), Athiyannan *et al.* (2022), Zhang *et al.* (2022), Huai *et al.* (2022)
Transmembrane domains	Powdery mildew	*TaMlo-B1*, *TaMlo-A1*, *TaMlo-D1*	Elliott *et al.* (2002), Várallyay *et al.* (2012)
Abscisic acid transporter	Leaf rust, stripe rust, stem rust, powdery mildew, leaf tip necrosis	*Lr34*, *Yr18*, *Sr57*, *Pm38*, *Ltn1*	Krattinger *et al.* (2009)
START Kinase	Stripe rust	*Yr36*	Fu *et al.* (2009)
S/TPK-NLR	Septoria nodorum blotch, tan spot	*Tsn1*	Faris *et al.* (2010)
Hexose transporter	Leaf rust, stripe rust, stem rust, powdery mildew, leaf tip necrosis	*Lr67*, *Yr46*, *Sr55*, *Pm46*, *Ltn3*	Moore *et al.* (2015)
Chimeric lectin	Fusarium head blight	*Fhb1*	Rawat *et al.* (2016)
Wall-associated kinase (WAK)	Septoria nodorum blotch	*Snn1*	Shi *et al.* (2016)
WAK-like protein	Septoria tritici blotch resistance	*Stb6*	Saintenac *et al.* (2018)
Tandem kinase-pseudokinase	Stripe rust	*Yr15*	Klymiuk *et al.* (2018)
Tandem kinase	Leaf rust	*Lr9*	Wang *et al.* (2023)
	Powdery mildew	*Pm24*, *Pm57*, *Pm36*	Lu *et al.* (2020), Zhao *et al.* (2024), Li *et al.* (2024)
	Stem rust	*Sr60*, *Sr62*	Chen *et al.* (2020), Yu *et al.* (2022)

BED-NLR	Stripe rust	*Yr5*, *YrSP*, *Yr7*	Marchal *et al.* (2018)
ANK-NLR-WRKY	Stripe rust	*YrU1*	Wang *et al.* (2020b)
NLR-kinase-MSP domains	Orange wheat blossom midge	*Sm1*	Walkowiak *et al.* (2020)
Glutathione S-transferase	Fusarium head blight	*Fhb7*	Wang *et al.* (2020a)
Ankyrin transmembrane domain protein	Leaf rust	*Lr14a*	Kolodziej *et al.* (2021)
PK-MSP	Septoria nodorum blotch	*Snn3-D1*	Zhang *et al.* (2021d)
Cysteine-rich receptor-like kinase	Septoria tritici blotch	*Stb16q*	Saintenac *et al.* (2021)
Multiple C2-domain and transmembrane region(s) kinase protein	Powdery mildew	*Pm4a*, *Pm4b*, *Pm4d*, *Pm4f*, *Pm4g*, *Pm4h*	Sánchez-Martín *et al.* (2021)
Protein disulfide isomerase-like	Wheat yellow mosaic virus	*TaPDIL5-1-4A*, *TaPDIL5-1-4B*, *TaPDIL5-1-4D*	Kan *et al.* (2022)
Sugar transporter	Stripe rust	*TaSTP3*	Huai *et al.* (2022)
NB-ARC	Powdery mildew	*TaRPP13L1-3D*	Zhang *et al.* (2022)
Mixed lineage kinase domain-like protein	Powdery mildew	*Pm13*	Li *et al.* (2024)
Coiled-coil nucleotide-binding site leucine-rich repeat	Powdery mildew	*Pm55*	Lu *et al.* (2024)
Kinase fusion protein	Wheat blast	*Rmg8*	Asuke *et al.* (2024),Yu *et al.* (2023)
	Stem rust	*Sr43*	

genes including Tandem-kinases (*Lr9*, Wang *et al.*, 2023), integrated domain-NLRs, chimeric lectin, Glutathione-S transferase, abscisic acid transporters, hexose transporters, ankyrin transmembrane domain proteins, among others, have been cloned (Table 6.1). Diseases such as leaf rust (*Puccinia triticina*), stem rust (*Puccinia graminis*), stripe rust (*Puccinia striiformis*) and powdery mildew are biotrophic pathogens that lead to reduced photosynthesis, stunted growth and lower crop yields.

Various genes have been cloned for stem rust (Table 6.1). In 1999 a new race of stem rust named UG99 was reported, for which a number of deployed genes were found ineffective. Positional cloning of *Sr33*, *Sr35*, *Sr50*, *Sr13* from *Aegilops tauschii* (Periyannan *et al.*, 2013), *T. monococcum* (Saintenac *et al.*, 2013), rye (Mago *et al.*, 2015), and *T. durum* (Zhang *et al.*, 2017), respectively uncovered genes conferring resistance to UG99 race from diverse sources. With the emergence of high-temperature-tolerant races of stripe rust, its geographic distribution and intensity are increasing. With further climate change, stripe rust in wheat is expected to cause severe epidemics. Several genes that confer different types of resistance have been identified and cloned. *Yr15*, *Yr5*, *Yr7*, *YrU1* and *YrAS2388* are some of these genes, encoding to tandem kinase domain protein (Klymiuk *et al.*, 2018), BED-NLR (Marchal *et al.*, 2018), CC-NB-LRR with N-terminal ankyrin-repeat and C-terminal WRKY domains (Wang *et al.*, 2020b), and atypical NLR domain (Zhang *et al.*, 2019), respectively. However, newer, highly virulent races, such as the High-Temperature Adult Plant (HTAP), have rendered the resistance of Yr7 ineffective, along with other genes (Rani *et al.*, 2019).

FHB is another major fungal disease of wheat that affects yield and safety of the grain. *Fhb1* gene from Chinese wheat cultivar Sumai 3 is the most extensively used major effect QTL for FHB resistance. Using a map-based cloning approach, Rawat *et al.* sequenced overlapping BAC clones from Sumai 3, and annotated 13 genes in the region. Out of these, the pore-forming toxin-like (PFT) gene was identified as the key gene underlying Fhb1 using mutation analysis, RNAi, and overexpression studies (Rawat *et al.*, 2016). Further, Su *et al.* (2019) and Li *et al.* (2019) reported a histidine-rich calcium-binding protein-encoding gene to be also involved with the *Fhb1* locus with contrasting functional validation results. It is possible that multiple genes in the genomic region work together to determine the overall phenotype and further work on mechanistic investigation is needed to pinpoint the role of individual genes in resistance. This case also highlights the complications of gene cloning for quantitative disease resistance traits in the complex wheat genome. The *Fhb7* gene encoding Glutathione-S-transferase that has been cloned for FHB resistance from *Thinopyrum elongatum* is a gene conferring FHB resistance. This gene was reported to be a case of horizontal gene transfer from fungal endophytic *Epichloë* species (Wang *et al.*, 2020a).

Septoria nodorum blotch and tan spot are other important foliar diseases of wheat caused by necrotrophic fungal pathogens. Faris *et al.* (2010) reported *Tsn1* gene, that encodes a serine/threonine protein kinase nucleotide-binding domain leucine-rich repeat (S/TPK-NLR) commonly associated with disease resistance but here linked to susceptibility. Necrotrophic pathogens use an inverse gene-for-gene interaction with such genes, making the plant–pathogen interaction compatible.

6.3 Genomic Technologies

Genomic Technologies have been advancing crop improvement in various ways. For example, genome sequencing allows researchers to explore entire genomes, uncovering variations linked to disease resistance. Association mapping helps identify correlations between genetic markers and phenotypic traits, offering novel targets for disease resistance. Genotyping-by-sequencing (GBS) accelerates the process of identifying genetic variants across large populations, while genomic selection enables the prediction of plant performance based on genetic data, facilitating the breeding of more resilient crops. Together, these methods pave the way for more efficient and targeted strategies to combat plant diseases and pests in agriculture.

6.3.1 Genome Sequencing: Reference Genomes

Genome sequences are important to map genetic regions, develop molecular markers, and for gene cloning, in addition to addressing genome-scale questions. The wheat genome, on account of its polyploidy and high repeat content, is very complex. For around two decades, wheat researchers relied on homologous genes from rice to be able to perform any molecular analysis such as gene cloning. In 2018, The International Wheat Genome Sequencing Consortium (IWGSC) *et al.* (2018) developed a reference genome for the bread wheat reference variety Chinese Spring, changing and facilitating wheat research and breeding. This high-quality, fully annotated reference genome was organized into 21 chromosomes, including approximately 108,000 genes and 500 transposable element families. This annotated genome drastically helps with the identification of important agronomic trait genes and facilitates breeding and techniques such as genome editing. Subsequently, Walkowiak *et al.* (2020) sequenced the genomes of 10 wheat varieties from global breeding programs into pseudomolecule level and five varieties with scaffold assemblies to explore the genomic diversity among wheat lines. Athiyannan *et al.* (2022) generated a chromosome-scale assembly of the wheat variety Kariega by combining high-fidelity long reads, optical mapping, and chromosome conformation capture. The resulting assembly is an order of magnitude more contiguous than previous wheat assemblies.

6.3.2 Pangenomes

Although the development of a reference genome has significantly advanced breeding and research efforts, a single genome sequence cannot capture the full spectrum of genetic diversity within wheat. To address this limitation, the concept of a *pangenome*, a comprehensive collection of genes across multiple wheat varieties, has emerged as a powerful tool for genetic studies (Gao *et al.*, 2023). Pangenomes have two main components: (1) the *core genome*, which consists of genes conserved across all wheat varieties and (2) the *dispensable genome*, which includes genes unique to specific accessions.

For disease resistance research, the pangenome approach enables the identification of novel resistance genes and susceptible genes that might be absent in the reference genome. The dispensable genome is particularly relevant, as it may harbor genes that provide race-specific or environmental-dependent resistance, which traditional breeding programs have overlooked (Montenegro *et al.*, 2017). By integrating pangenomic data with GWAS and mutagenesis approaches, researchers can prioritize candidate genes for functional validation and deployment in breeding programs.

6.3.3 Wild Relatives' Genomes

The genomes of wheat's wild relatives, including *Triticum urartu* (A genome donor), *Aegilops speltoides* (close to the B genome donor) and *Ae. tauschii* (D genome donor), represent a critical reservoir of genetic variation that has been largely eroded in modern cultivars due to breeding bottlenecks (Dubcovsky and Dvorak, 2007). These wild species possess alleles conferring broad-spectrum resistance to major wheat pathogens such as rusts, powdery mildew, and FHB, many of which have not been fully exploited in wheat improvement efforts (Wang *et al.*, 2020a; Zhang *et al.*, 2021b; Li *et al.*, 2022a; Steed *et al.*, 2022; Yu *et al.*, 2023).

Recent advancements in sequencing have enabled the assembly of high-quality reference genomes for these species (Ling *et al.*, 2018; Cavalet-Giorsa *et al.*, 2024), facilitating the identification of resistance loci that were previously inaccessible. By integrating these genomic resources with modern gene-editing approaches such as CRISPR-Cas9 and RNAi, researchers can introduce resistance alleles from wild relatives into elite wheat varieties while minimizing linkage drag. These efforts underscore the crucial role of wild relative genomics in expanding the genetic foundation for wheat disease resistance and ensuring the long-term sustainability of global wheat production.

6.3.4 Association Mapping

Association mapping, also known as linkage disequilibrium (LD) mapping, is a valuable tool for identifying genetic variations associated with traits of interest within populations. This technique differs from traditional linkage mapping, which relies on controlled crosses between parents, because association mapping utilizes natural populations and the historical recombination events within them. This approach involves scanning the genome with molecular markers or genomic sequence tags (k-mers) to associate genetic variants with phenotypic traits, such as pest and disease resistance. This technique offers high resolution, enabling the identification of candidate genes with greater precision.

Gaurav *et al.* used k-mer GWAS on a set of 242 *Ae. tauschii* accessions, and identified a 50-kb linkage disequilibrium block containing two candidate genes for the stem rust resistance gene SrTA1662. Further gene complementation was done to validate SrTA1662 to be an NLR. Cavalet-Giorsa *et al.* (2024) utilized the *Ae. tauschii* genome to perform k-mer-based genome-wide association mapping to pinpoint the location of *Lr39* (leaf rust) and identify candidate genes demonstrating its crucial role in leaf rust defense.

In a study Alemu *et al.* (2021) focused on the identification of genetic loci associated with resistance to stripe rust in Ethiopia durum wheat. They performed a genome-wide association study (GWAS) on a diverse panel of 300 accessions. The study identified several significant SNPs associated with stripe rust resistance. Tomar *et al.* (2021) also identified through GWAS eight QTLs associated with spot blotch, caused by the fungus *Bipolaris sorokiniana*, on 139 advanced spring wheat lines evaluated in India and Bangladesh. These QTLs can be utilized to develop wheat varieties with improved resistance to this disease and contribute to sustainable wheat production in affected regions.

Other studies have been performed on a variety of pests. Joukhadar *et al.* (2013) focused on identifying loci associated with resistance to the Hessian fly, Russian wheat aphid, Sunn pest, wheat stem sawfly, and cereal leaf beetle through GWAS, finding 40 markers in total. Association mapping is essential for identifying genetic variations linked to important agronomic traits, facilitating the development of superior crop varieties.

6.3.5 Genotyping-by-sequencing

The first article describing genotyping-by-sequencing (GBS) was published by Elshire *et al.* (2011). This study detailed an effective method for SNP discovery and genotyping, especially in high-diversity species. To reduce genome complexity, GBS utilizes restriction enzymes that target lower copy regions, therefore avoiding repetitive regions and simplifying sequence alignment. This method enables high-throughput sequencing by using DNA barcodes to mix multiple samples that are later sequenced using next-generation sequencing, such as Ilumina platforms. This obtains as a final result effective SNP discovery and genetic mapping without the need for a reference genome.

High-density genetic maps can be constructed with GBS. Li *et al.* (2015) uncovered 28,644 GBS markers distributed across the wheat genome. Eighteen QTL regions were associated with adult plant resistance to stem, yellow, and leaf rusts, across 13 chromosomes. Jamil *et al.* (2018) also identified regions associated with spot blotch; Arruda *et al.* (2016) identified regions associated with FHB resistance; and Li *et al.* (2015) aimed to map a major gene conferring resistance to the Hessian fly (*Mayetiola destructor*), a significant pest of wheat. These studies highlighted the effectiveness of GBS in identifying high-density markers for complex traits. GBS is important for providing insights into genetic diversity, helping with the discovery of genetic markers and therefore enhancing the efficiency of breeding programs. These techniques can be utilized in wheat breeding programs to enhance disease resistance and improve wheat varieties.

6.3.6 Genomic Selection

Genomic selection consists of utilizing a group of individuals with known genotypes and phenotypes to create a statistical model that will predict the relationship between genetic markers and traits (Meuwissen *et al.*, 2001; Crossa *et al.*, 2017). Three statistical approaches were compared: Least Squares, Best Linear Unbiased Prediction and Bayesian methods utilizing simulated data (Lourenço *et al.*, 2024). Genomic selection helps reduce the time and cost required to phenotype entire populations by making the breeding cycle shorter.

Traditional breeding programs for cereals, such as wheat and maize, often require 8–12 years to develop and release a new variety due to the need for extensive phenotypic evaluations across multiple environments (Heffner *et al.*, 2009). In contrast, genomic selection can reduce this timeline to approximately 4–6 years by enabling early selection based on genetic markers rather than waiting for full phenotypic assessment (Battenfield *et al.*, 2016). This efficiency gain has made genomic selection increasingly attractive in cereal breeding programs, particularly for improving disease resistance and other agronomic traits (Rutkoski *et al.*, 2014; Gill *et al.*, 2021).

The adoption of genomic selection in the cereal industry has accelerated in recent years. Major breeding institutions, including the International Maize and Wheat Improvement Center (CIMMYT) and national breeding programs, now incorporate genomic selection as a core strategy for developing high-yielding, disease-resistant varieties (Dreisigacker *et al.*, 2021). As genotyping costs continue to decrease and machine learning techniques refine predictive models, genomic selection is poised to become the dominant method for accelerating cereal crop improvement.

This technique is particularly important in resource-limited regions, as described by Santantonio *et al.* (2020). One important and challenging aspect is to build an appropriate training set based on the relevant data and needs while using breeding informatics systems, as well as having all of the appropriate accessions genotyped. This is key to having better predictions and success.

This method has been tested to predict diseases such as FHB with variable success. Arruda *et al.* (2015) utilized GBS to identify SNPs that were used as markers to develop a GS model. Additionally, if disease resistance in GS-selected lines fails, identifying the responsible resistance gene(s) can be difficult since GS does not rely on selecting specific loci but rather leverages genome-wide marker effects. Unlike traditional marker-assisted selection, where specific resistance genes can be traced back to known markers, GS predictions are often based on aggregate effects across many loci. This lack of direct gene targeting complicates post-selection validation and may require additional fine mapping or functional validation studies to confirm resistance mechanisms (Crossa *et al.*, 2017; Juliana *et al.*, 2019).

Other traits that can be more complicated and influenced by the environment can be more difficult to predict with GS. For example, one of the most important traits is grain yield, but this particular trait can be heavily influenced by the environment (genotype × environment), therefore making it one of the hardest traits to improve. Given this complexity GS would look like an appealing method to predict yield. To address this, scientists have compared the benefits of GS over traditional pedigree-based selection methods. For example, Juliana *et al.* (2020) evaluated the potential integration of GS for grain yield in bread wheat and observed that genomic prediction accuracies within nurseries were substantially higher than across-nursery accuracies, highlighting the challenges posed by genotype × environment interactions.

However, results have not consistently shown significant differences between GS and traditional breeding selections. For instance, a study by Cerrudo *et al.* (2018) demonstrated that GS outperformed marker-assisted selection for grain yield and physiological traits in maize across water treatments, suggesting that GS can be more effective in certain contexts. Conversely, other studies have reported that the advantages of GS over traditional methods are moderate, particularly in populations without full-sibs or when within-family selections involve elite families with minimal Mendelian sampling variance.

Furthermore, Bonnett *et al.* (2022) investigated early-generation GS in wheat and found that while GS showed potential, its effectiveness varied depending on the statistical models used and the specific breeding context. These mixed results indicate that the effectiveness of GS relative to traditional breeding methods can vary depending on factors such as the genetic architecture of the trait, the breeding population structure and environmental interactions. Therefore, while GS holds promise, its advantages over traditional selection methods may not be universally significant across all scenarios.

6.3.7 Mutagenesis, Mutant Analysis and Gene Editing

On the basis of the process of finding or evaluating genes of interest for diseases, we can divide techniques into two classes: Forward and Reverse genetics. Forward genetics strategies start with a phenotype of interest. In the case of disease resistance, the severity of symptoms of the disease are recorded and then the genes causing these symptoms are discovered. Under this category different techniques of mutagenesis are employed such as chemical mutagenesis with Ethyl methanesulfonate (EMS), sodium azide, or with physical mutagens such as Gamma radiation. Reverse genetics starts with a known gene and then verifies what phenotype will arise when that gene is altered. Techniques used here are gene editing to create knockouts, RNA interference, and other techniques such as Targeting Induced Local Lesions in Genomes (TILLING) and MutRenSeq. Together, forward and reverse genetics provide the necessary tools to discover the relationship between phenotype and genotype.

6.3.8 Forward Genetics: EMS and Gamma Radiation

EMS is a chemical mutagen that induces point mutations in the DNA by alkylating guanine bases, causing mispairing during DNA replication, leading to a single base pair change. EMS-mutagenized populations are used to identify variants in phenotypes such as diseases. One example was done for FHB in Jagger wheat where 74 mutant lines were found to exhibit resistance to FHB spread, with 30 of these also having low deoxynivalenol (DON) content (Chhabra *et al.*, 2021). On the other hand, gamma radiation will create random deletions in the chromosomes. These deletions can occur in susceptible genes for different diseases and create a resistant phenotype. Hassine *et al.* (2023) utilized three different doses of gamma radiation (100, 150, and 250 Gy) to induce mutations in durum wheat (*Triticum turgidum* ssp. *durum*). Diseases such as Septoria tritici blotch (STB), tan spot (TS), stripe rust (SR) and leaf rust (LR) reduced their severity by 80%, 67%, 56% and 40%, respectively. With these techniques, we can accelerate the breeding process and introduce other desirable traits that can improve crop yield and quality.

6.3.9 Reverse Genetics: TILLING and MutRenSeq

TILLING is a technique that consists of chemically mutagenizing seeds with EMS and later screening the resulting populations for mutations in genes using PCR and DNA sequencing. This method allowed us to identify and analyze point mutations that could confer beneficial traits, such as disease resistance, among other phenotypic characteristics, without introducing foreign DNA, making it a non-GMO approach to crop improvement.

TILLING has been used to develop powdery mildew resistant wheat (Acevedo-Garcia *et al.*, 2017). The authors induced mutations in orthologs of the barley *Mlo* gene, *TaMlo*, and identified mutations that conferred resistance to powdery mildew. Several wheat lines showed mutations across *TaMlo-A1*, *B1*, and *D1* genes, showing resistance without pleiotropic effects. However, this resistance was accompanied by enhanced susceptibility to wheat blast, as reported by Gruner *et al.* (2020). Unlike barley, where natural loss-of-fuction mutations in *Mlo* lead to durable resistance against *Blumeria graminis* f. sp. *hordei*, no naturally occurring *mlo* mutants have been identified in wheat (Wang *et al.*, 2014; Acevedo-Garcia *et al.*, 2017). This absence is likely due to the polyploid nature of wheat, where functional redundancy among homoeologous mutations confers resistance (Kusch and Panstruga, 2017). This study demonstrated the potential of TILLING to create wheat varieties that are non-transgenic and disease resistant.

Another technique is Mutational Resistance Gene Enrichment Sequencing (MutRenSeq) designed to identify resistant genes in plants rapidly. This technique utilizes mutagenesis with sequence capture technology and selects enriched genomic regions containing resistant genes (R genes) from the mutagenized populations. After creating the mutants they use sequence capture probes designed from known resistant gene sequences to isolate and sequence these regions. By comparing the sequences from resistant and susceptible mutants this allows for the identification of functional resistant genes against various pathogens. Steuernagel *et al.* (2016) introduced this technique by focusing on cloning the stem rust R genes *Sr22* and *Sr45* from hexaploid wheat. This method is faster and cheaper compared with traditional map-based cloning, making it extremely useful for large and complex genomes such as wheat. The successful application of MutRenSeq in the cloning of these two genes demonstrated the potential in improving disease resistance in wheat and other crops.

The identification and characterization of genes such as *Yr63* in wheat, that confers resistance to stripe rust, was obtained by RNASeq and MutRenSeq (Mackenzie *et al.*, 2023). MutRenSeq has been very important in the identification and cloning of disease resistant genes. This method is used only for NLR genes, as it relies on sequence capture of conserved NLR motifs, making it less effective for resistance genes that do not belong to this class (Jupe *et al.*, 2013). Despite this limitation, MutRenSeq has significantly accelerated the discovery of key resistance genes, contributing to the development of resilient crops and supporting global food security. Other rapid techniques such as AgRenSeq, k-mer GWAS, and MutChromSeq help in the identification of candidate genes without the need for developing and screening high-resolution mapping populations, which is a labor-intensive step required in traditional cloning techniques (Mackenzie *et al.*, 2023). MutRenSeq and related approaches have significantly reduced the time required for gene cloning. Traditional map-based cloning often takes several years due to the need for fine-mapping, positional cloning, and validation steps (Jaganathan *et al.*, 2020). In contrast, MutRenSeq has been shown to reduce this process to a matter of months by integrating high-throughput sequencing with mutant population screening (Steuernagel *et al.*, 2016). Similarly, AgReSeq allows the rapid discovery of resistance genes from wild germplasm in as little as two years, whereas traditional methods could take over a decade (Arora *et al.*, 2019). These advancements have revolutionized resistance gene discovery, making crop improvement efforts more efficient and responsive to emerging diseases.

6.3.10 Gene Editing and RNAi

Gene editing and RNAi are reverse genetic techniques that employ biotechnological approaches. The benefit of gene editing is that the transgene can be segregated out from the edited gene, and thus can be used in agriculture. Gene editing has revolutionized agricultural biotechnology, permitting very precise alterations to the genome in order to improve yield, disease resistance and climate adaptability.

Gene editing using Clustered Regularly Interspaced Short Palindromic Repeats (CRISPR) has been used to generate gene-knockouts by error-prone repair mechanisms of the cells after creating cuts in the gene by an endonuclease (Cas9/ Cas12a/ Cas13/ Cas14). This is mainly used to study the role of a gene by observing the effects caused by their absence. By knocking out genes, science can gain insight into gene function, identify targets for crop improvement, and develop plants with agronomic desirable traits such as disease resistance. Creation of mutants like *Tamlo-R32* in wheat (Li *et al.*, 2022b) provides resistance to powdery mildew without the growth penalties that usually accompany similar genetic changes. By deleting a portion of the *MLO-B1* gene, the researchers were able to trigger the activation of another gene, *TaTMT3B*m which compensates for the negative effects typically associated with *MLO* loss. This breakthrough allows the development of high-yield, disease-resistant wheat varieties. The findings demonstrate the potential of genome editing in crop improvement, highlighting a method to balance resistance and crop performance.

CRISPR-Cas12a system can be used as a new diagnostic method for detecting diseases such as FHB (Mu *et al.*, 2022). This system, by combining PCR amplification with CRISPR-Cas12a detection, enhances specificity and sensitivity. The dual recognition helps distinguish FHB from other Fusarium pathogens. The assay can detect as low as 1 fg/μL of *Fusarium graminearum* DNA after only four days of infection. This method can provide early diagnosis allowing for timely intervention.

On the other hand, RNAi consists of introducing into a cell a double-stranded RNA (dsRNA) where it gets recognized and cleaved by an enzyme called Dicer into small interfering RNAs (siRNA). These are then incorporated into the RNA-induced silencing complex (RISC), guiding it to complementary mRNA sequences. Upon binding, RISC degrades the target mRNA, preventing the translation of the corresponding protein. In wheat, RNAi can be engineered to silence targeted genes essential for the survival of pathogens. Researchers can enhance wheat's resistance to diseases by designing dsRNA molecules specific to these harmful genes.

Shuai *et al.* (2024) explored the use of two methods of RNAi (Spray-induced gene silencing (SIGS) and host-induced gene silencing (HIGS)) to enhance wheat resistance to FHB. The results showed for SIGS that direct application of small interfering (si) RNAs reduced the expression of target genes and suppressed disease symptoms. The combination of all five siRNAs was more effective than individual siRNAs. For HIGS, transgenic wheat lines expressing dsRNA constructs showed significant resistance to FHB and reduced DON levels. These approaches did not impact growth or yield, therefore making them potential techniques to integrate into crop protection strategies.

A critical aspect of gene editing in agriculture is its regulatory classification, particularly its distinction from genetically modified organisms (GMOs). Unlike transgenic approaches that introduce foreign DNA, gene editing technologies such as CRISPR can create targeted mutations without inserting exogenous genetic material. This allows gene-edited crops to be considered non-GM in certain regulatory frameworks, accelerating their acceptance in commercial agriculture (Lassoued *et al.*, 2021). Countries such as the USA, Argentina and Japan have established guidelines that exempt gene-edited crops from strict GMO regulations if no foreign DNA is introduced (Turnbull *et al.*, 2021). The European Union, however, currently classifies gene-edited crops under GMO regulations, limiting their rapid adoption (Schmidt *et al.*, 2020).

In wheat improvement, this distinction is crucial for integrating gene editing into breeding programs. Non-GM classification enables faster market adoption and consumer acceptance while facilitating regulatory approval processes (Ahmad *et al.*, 2023). If gene editing is to play a significant role in wheat improvement, clear regulatory frameworks and public acceptance will be essential for its successful implementation in global agriculture.

6.4 Conclusion

Genomic solutions for pest and disease management in crops offer a diversity of approaches to make agriculture resilient. Marker-based genetic analysis helps in the identification and selection of resistant traits, making possible precise breeding strategies. Genomic analysis deepens our understanding by unraveling genetic architectures and pathways involved in resistance mechanisms. On the other hand, mutagenesis and mutant analysis introduce novel genetic variations that will help expand the pool of resistant genes. Furthermore, more studies in genomic solutions for pests need to be done. The majority of efforts have been around diseases. The increasing demand for food in the world is the best example that we need to incorporate these advanced genomic methodologies for developing crops resilient to pests and diseases to ensure sustainable global food security.

References

Acevedo-Garcia, J., Spencer, D., Thieron, H., Reinstädler, A., Hammond-Kosack, K. *et al.* (2017) mlo-based powdery mildew resistance in hexaploid bread wheat generated by a non-transgenic TILLING approach. *Plant Biotechnology Journal* 15, 367–378. DOI: 10.1111/pbi.12631.

Ahmad, A., Jamil, A. and Munawar, N. (2023) GMOs or non-GMOs? The CRISPR Conundrum. *Frontiers in Plant Science* 14, 1232938. DOI: 10.3389/fpls.2023.1232938.

Ajijah, N., Fiodor, A., Kazimierczuk, K., Urbaniak, M., Enow, E. *et al.* (2024) *Pseudomonas protegens* ML15 and *Trichoderma koningiopsis* Tr21 co-culture: A potent strategy for suppressing *Fusarium cerealis* infections in wheat through augmented antifungal metabolite production. *Biological Control* 198, 105621. DOI: 10.1016/j.biocontrol.2024.105621.

Alemu, S.K., Huluka, A.B., Tesfaye, K., Haileselassie, T. and Uauy, C. (2021) Genome-wide association mapping identifies yellow rust resistance loci in Ethiopian durum wheat germplasm. *PLoS ONE* 16, e0243675. DOI: 10.1371/journal.pone.0243675.

Arora, S., Steuernagel, B., Gaurav, K., Chandramohan, S., Long, Y. *et al.* (2019) Resistance gene cloning from a wild crop relative by sequence capture and association genetics. *Nature Biotechnology* 37, 139–143. DOI: 10.1038/s41587-018-0007-9.

Arruda, M.P., Brown, P.J., Lipka, A.E., Krill, A.M., Thurber, C. *et al.* (2015) Genomic selection for predicting *Fusarium* head blight resistance in a wheat breeding program. *The Plant Genome* 8, plantgenome2015.01.0003. DOI: 10.3835/plantgenome2015.01.0003.

Arruda, M.P., Brown, P., Brown-Guedira, G., Krill, A.M., Thurber, C. *et al.* (2016) Genome-wide association mapping of Fusarium head blight resistance in wheat using genotyping-by-sequencing. *The Plant Genome* 9, plantgenome2015.04.0028. DOI: 10.3835/plantgenome2015.04.0028.

Asuke, S., Morita, K., Shimizu, M., Abe, F., Terauchi, R. *et al.* (2024) Evolution of wheat blast resistance gene Rmg8 accompanied by differentiation of variants recognizing the powdery mildew fungus. *Nature Plants* 10, 971–983. DOI: 10.1038/s41477-024-01711-1.

Athiyannan, N., Abrouk, M., Boshoff, W.H.P., Cauet, S., Rodde, N. *et al.* (2022) Long-read genome sequencing of bread wheat facilitates disease resistance gene cloning. *Nature Genetics* 54, 227–231. DOI: 10.1038/s41588-022-01022-1.

Battenfield, S.D., Guzmán, C., Gaynor, R.C., Singh, R.P., Peña, R.J. *et al.* (2016) Genomic selection for processing and end-use quality traits in the CIMMYT spring bread wheat breeding program. *The Plant Genome* 9, plantgenome2016.01.0005. DOI: 10.3835/plantgenome2016.01.0005.

Bonnett, D., Li, Y., Crossa, J., Dreisigacker, S., Basnet, B. *et al.* (2022) Response to early generation genomic selection for yield in wheat. *Frontiers in Plant Science* 12, 718611. DOI: 10.3389/fpls.2021.718611.

Borlaug, N.E. (1968) Wheat breeding and its impact on world food supply. *Proceedings of the 3rd International Wheat Genetics Symposium*, Canberra, 5–9 August 1968, 1–36.

Buerstmayr, H., Ban, T. and Anderson, J.A. (2009) QTL mapping and marker-assisted selection for Fusarium head blight resistance in wheat: A review. *Plant Breeding* 128, 1–26. DOI: 10.1111/j.1439-0523.2008.01550.x.

Cavalet-Giorsa, E., González-Muñoz, A., Athiyannan, N., Holden, S., Salhi, A. *et al.* (2024) Origin and evolution of the bread wheat D genome. *Nature* 633, 848–855. DOI: 10.1038/s41586-024-07808-z.

Cerrudo, D., Cao, S., Yuan, Y., Martinez, C., Suarez, E.A. *et al.* (2018) Genomic selection outperforms marker assisted selection for grain yield and physiological traits in a maize doubled haploid population across water treatments. *Frontiers in Plant Science* 9, 366. DOI: 10.3389/fpls.2018.00366.

Chai, Y., Senay, S., Horvath, D. and Pardey, P. (2022) Multi-peril pathogen risks to global wheat production: A probabilistic loss and investment assessment. *Frontiers in Plant Science* 13, 1034600. DOI: 10.3389/fpls.2022.1034600.

Chen, S., Zhang, W., Bolus, S., Rouse, M.N. and Dubcovsky, J. (2018) Identification and characterization of wheat stem rust resistance gene Sr21 effective against the Ug99 race group at high temperature. *PLoS Genetics* 14, e1007287. DOI: 10.1371/journal.pgen.1007287.

Chen, S., Rouse, M.N., Zhang, W., Zhang, X., Guo, Y. *et al.* (2020) Wheat gene *Sr60* encodes a protein with two putative kinase domains that confers resistance to stem rust. *New Phytologist* 225, 948–959. DOI: 10.1111/nph.16169.

Chen, S., Hegarty, J., Shen, T., Hua, L., Li, H. *et al.* (2021) Stripe rust resistance gene Yr34 (synonym Yr48) is located within a distal translocation of *Triticum monococcum* chromosome 5AmL into common wheat. *Theoretical and Applied Genetics* 134, 2197–2211. DOI: 10.1007/s00122-021-03816-z.

Chhabra, B., Singh, L., Wallace, S., Schoen, A., Dong, Y. *et al.* (2021) Screening of an ethyl methane sulfonate mutagenized population of a wheat cultivar susceptible to Fusarium head blight identifies resistant variants. *Plant Disease* 105, 3669–3676. DOI: 10.1094/PDIS-03-21-0670-RE.

Cloutier, S., McCallum, B.D., Loutre, C., Banks, T.W., Wicker, T. *et al.* (2007) Leaf rust resistance gene Lr1, isolated from bread wheat (*Triticum aestivum* L.) is a member of the large psr567 gene family. *Plant Molecular Biology* 65, 93–106. DOI: 10.1007/s11103-007-9201-8.

Cox, T.S. and Hatchett, J.H. (1994) Hessian Fly-resistance gene *H26* transferred from *Triticum tauschii* to common wheat. *Crop Science* 34, 958–960. DOI: 10.2135/cropsci1994.0011183X003400040023x.

Crossa, J., Pérez-Rodríguez, P., Cuevas, J., Montesinos-López, O., Jarquín, D. *et al.* (2017) Genomic selection in plant breeding: Methods, models, and perspectives. *Trends in Plant Science* 22, 961–975. DOI: 10.1016/j.tplants.2017.08.011.

Cuthbert, P.A., Somers, D.J., Thomas, J., Cloutier, S. and Brulé-Babel, A. (2006) Fine mapping Fhb1, a major gene controlling Fusarium head blight resistance in bread wheat (*Triticum aestivum* L.). *Theoretical and Applied Genetics* 112, 1465–1472. DOI: 10.1007/s00122-006-0249-7.

Dreisigacker, S., Crossa, J., Pérez-Rodríguez, P., Montesinos-L, O.A., Mondal, S. *et al.* (2021) Implementation of genomic selection in the CIMMYT Global Wheat Program, findings from the past 10 years. *Crop Breeding Genetics and Genomics* 3(2). DOI: 10.20900/cbgg20210005

Dubcovsky, J. and Dvorak, J. (2007) Genome plasticity a key factor in the success of polyploid wheat under domestication. *Science* 316, 1862–1866. DOI: 10.1126/science.1143986.

Dweikat, I., Ohm, H., Patterson, F. and Cambron, S. (1997) Identification of RAPD markers for 11 Hessian fly resistance genes in wheat. *Theoretical and Applied Genetics* 94, 419–423. DOI: 10.1007/s001220050431.

Elliott, C., Zhou, F., Spielmeyer, W., Panstruga, R. and Schulze-Lefert, P. (2002) Functional conservation of wheat and rice *Mlo* orthologs in defense modulation to the powdery mildew fungus. *Molecular Plant-Microbe Interactions* 15, 1069–1077. DOI: 10.1094/MPMI.2002.15.10.1069.

Elshire, R.J., Glaubitz, J.C., Sun, Q., Poland, J.A., Kawamoto, K. *et al.* (2011) A robust, simple genotyping-by-sequencing (GBS) approach for high diversity species. *PLoS ONE* 6, e19379. DOI: 10.1371/journal.pone.0019379.

Faris, J.D., Zhang, Z., Lu, H., Lu, S., Reddy, L. *et al.* (2010) A unique wheat disease resistance-like gene governs effector-triggered susceptibility to necrotrophic pathogens. *Proceedings of the National Academy of Sciences USA* 107, 13544–13549. DOI: 10.1073/pnas.1004090107.

Feuillet, C., Travella, S., Stein, N., Albar, L., Nublat, A. *et al.* (2003) Map-based isolation of the leaf rust disease resistance gene Lr10 from the hexaploid wheat (*Triticum aestivum* L.) genome. *Proceedings of the National Academy of Sciences USA* 100, 15253–15258. DOI: 10.1073/pnas.2435133100.

Fu, D., Uauy, C., Distelfeld, A., Blechl, A., Epstein, L. *et al.* (2009) A kinase-START gene confers temperature-dependent resistance to wheat stripe rust. *Science* 323, 1357–1360. DOI: 10.1126/science.1166289.

Gao, Z., Bian, J., Lu, F., Jiao, Y. and He, H. (2023) Triticeae crop genome biology: An endless frontier. *Frontiers in Plant Science* 14, 1222681. DOI: 10.3389/fpls.2023.1222681.

Gaurav, K., Arora, S., Silva, P., Sánchez-Martín, J., Horsnell, R. *et al.* (2022) Population genomic analysis of *Aegilops tauschii* identifies targets for bread wheat improvement. *Nature Biotechnology* 40, 422–431. DOI: 10.1038/s41587-021-01058-4.

Gill, H.S., Halder, J., Zhang, J., Brar, N.K., Rai, T.S. *et al.* (2021) Multi-trait multi-environment genomic prediction of agronomic traits in advanced breeding lines of winter wheat. *Frontiers in Plant Science* 12, 709545. DOI: 10.3389/fpls.2021.709545.

Gruner, K., Esser, T., Acevedo-Garcia, J., Freh, M., Habig, M. *et al.* (2020) Evidence for allele-specific levels of enhanced susceptibility of wheat mlo mutants to the Hemibiotrophic fungal pathogen *Magnaporthe oryzae* pv. *Triticum*. *Genes* 11, 517. DOI: 10.3390/genes11050517.

Guo, J., Zhang, X., Hou, Y., Cai, J., Shen, X. *et al.* (2015) High-density mapping of the major FHB resistance gene Fhb7 derived from *Thinopyrum ponticum* and its pyramiding with Fhb1 by marker-assisted selection. *Theoretical and Applied Genetics* 128, 2301–2316. DOI: 10.1007/s00122-015-2586-x.

Hassine, M., Baraket, M., Marzougui, N. and Slim-Amara, H. (2023) Screening of the effect of mutation breeding on biotic stress tolerance and quality traits of durum wheat. *Gesunde Pflanz.* 75, 837–846. DOI: 10.1007/s10343-022-00750-y.

He, H., Zhu, S., Ji, Y., Jiang, Z., Zhao, R. *et al.* (2017) Map-Based Cloning of the Gene *Pm21* That Confers Broad Spectrum Resistance to Wheat Powdery Mildew. Preprint, bioRxiv, https://doi.org/10.1101/177857.

Heffner, E.L., Sorrells, M.E. and Jannink, J. (2009) Genomic selection for crop improvement. *Crop Science* 49, 1–12. DOI: 10.2135/cropsci2008.08.0512.

Hewitt, T., Müller, M.C., Molnár, I., Mascher, M., Holušová, K. *et al.* (2021a) A highly differentiated region of wheat chromosome 7AL encodes a *Pm1a* immune receptor that recognizes its corresponding *AvrPm1a* effector from *Blumeria graminis*. *New Phytologist* 229, 2812–2826. DOI: 10.1111/nph.17075.

Hewitt, T., Zhang, J., Huang, L., Upadhyaya, N., Li, J. *et al.* (2021b) Wheat leaf rust resistance gene Lr13 is a specific Ne2 allele for hybrid necrosis. *Molecular Plant* 14, 1025–1028. DOI: 10.1016/j.molp.2021.05.010.

Hu, W., Fu, L., Gao, D., Li, D., Liao, S. *et al.* (2023) Marker-assisted selection to pyramid Fusarium head blight resistance loci Fhb1 and Fhb2 in the high-quality soft wheat cultivar Yangmai 15. *Journal of Integrative Agriculture* 22, 360–370. DOI: 10.1016/j.jia.2022.08.057.

Huai, B., Yuan, P., Ma, X., Zhang, X., Jiang, L. *et al.* (2022) Sugar transporter TaSTP3 activation by TaWRKY19 /61/82 enhances stripe rust susceptibility in wheat. *New Phytologist* 236, 266–282. DOI: 10.1111/nph.18331.

Huang, L., Brooks, S.A., Li, W., Fellers, J.P., Trick, H.N. *et al.* (2003) Map-based cloning of leaf rust resistance gene *Lr21* from the large and polyploid genome of bread wheat. *Genetics* 164, 655–664. DOI: 10.1093/genetics/164.2.655.

Hurni, S., Brunner, S., Buchmann, G., Herren, G., Jordan, T. *et al.* (2013) Rye *P m8* and wheat *P m3* are orthologous genes and show evolutionary conservation of resistance function against powdery mildew. *The Plant Journal* 76, 957–969. DOI: 10.1111/tpj.12345.

Iqbal, N., Shahzad, M.U., Sherif, E.-S.M., Tariq, M.U., Rashid, J. *et al.* (2024) Analysis of wheat-yield prediction using machine learning models under climate change scenarios. *Sustainability* 16, 6976. DOI: 10.3390/su16166976.

Jaganathan, D., Bohra, A., Thudi, M. and Varshney, R.K. (2020) Fine mapping and gene cloning in the post-NGS era: Advances and prospects. *Theoretical and Applied Genetics* 133, 1791–1810. DOI: 10.1007/s00122-020-03560-w.

Jamil, M., Ali, A., Gul, A., Ghafoor, A., Ibrahim, A.M.H. *et al.* (2018) Genome-wide association studies for spot blotch (*Cochliobolus sativus*) resistance in bread wheat using genotyping-by-sequencing. *Phytopathology* 108, 1307–1314. DOI: 10.1094/PHYTO-02-18-0047-R.

Jiang, G.-L., Shi, J. and Ward, R.W. (2007) QTL analysis of resistance to fusarium head blight in the novel wheat germplasm CJ 9306. I. Resistance to fungal spread. *Theoretical and Applied Genetics* 116, 3–13. DOI: 10.1007/s00122-007-0641-y.

Joukhadar, R., El-Bouhssini, M., Jighly, A. and Ogbonnaya, F.C. (2013) Genome-wide association mapping for five major pest resistances in wheat. *Molecular Breeding* 32, 943–960. DOI: 10.1007/s11032-013-9924-y.

Juliana, P., Poland, J., Huerta-Espino, J., Shrestha, S., Crossa, J. *et al.* (2019) Improving grain yield, stress resilience and quality of bread wheat using large-scale genomics. *Nature Genetics* 51, 1530–1539. DOI: 10.1038/s41588-019-0496-6.

Juliana, P., Singh, R.P., Braun, H.-J., Huerta-Espino, J., Crespo-Herrera, L. *et al.* (2020) Genomic selection for grain yield in the CIMMYT wheat breeding program—Status and perspectives. *Frontiers in Plant Science* 11, 564183. DOI: 10.3389/fpls.2020.564183.

Jupe, F., Witek, K., Verweij, W., Śliwka, J., Pritchard, L. *et al.* (2013) Resistance gene enrichment sequencing (Ren Seq) enables reannotation of the NB - LRR gene family from sequenced plant genomes and rapid mapping of resistance loci in segregating populations. *The Plant Journal* 76, 530–544. DOI: 10.1111/tpj.12307.

Kan, J., Cai, Y., Cheng, C., Jiang, C., Jin, Y. *et al.* (2022) Simultaneous editing of host factor gene *TaPDIL5-1* homoeoalleles confers wheat yellow mosaic virus resistance in hexaploid wheat. *New Phytologist* 234, 340–344. DOI: 10.1111/nph.18002.

Karelov, A., Kozub, N., Sozinova, O., Pirko, Y., Sozinov, I. *et al.* (2022) Wheat genes associated with different types of resistance against stem rust (*Puccinia graminis* Pers.). *Pathogens* 11, 1157. DOI: 10.3390/pathogens11101157.

Klymiuk, V., Yaniv, E., Huang, L., Raats, D., Fatiukha, A. *et al.* (2018) Cloning of the wheat Yr15 resistance gene sheds light on the plant tandem kinase-pseudokinase family. *Nature Communications* 9, 3735. DOI: 10.1038/s41467-018-06138-9.

Kolodziej, M.C., Singla, J., Sánchez-Martín, J., Zbinden, H., Šimková, H. *et al.* (2021) A membrane-bound ankyrin repeat protein confers race-specific leaf rust disease resistance in wheat. *Nature Communications* 12, 956. DOI: 10.1038/s41467-020-20777-x.

Krattinger, S.G., Lagudah, E.S., Spielmeyer, W., Singh, R.P., Huerta-Espino, J. *et al.* (2009) A putative ABC transporter confers durable resistance to multiple fungal pathogens in wheat. *Science* 323, 1360–1363. DOI: 10.1126/science.1166453.

Kumar, K., Jan, I., Saripalli, G., Sharma, P.K., Mir, R.R. *et al.* (2022) An update on resistance genes and their use in the development of leaf rust resistant cultivars in wheat. *Frontiers in Genetics* 13, 816057. DOI: 10.3389/fgene.2022.816057.

Kusch, S. and Panstruga, R. (2017) *Mlo* -based resistance: An apparently universal "weapon" to defeat powdery mildew disease. *Molecular Plant-Microbe Interactions* 30, 179–189. DOI: 10.1094/MPMI-12-16-0255-CR.

Lassoued, R., Phillips, P.W.B., Macall, D.M., Hesseln, H. and Smyth, S.J. (2021) Expert opinions on the regulation of plant genome editing. *Plant Biotechnology Journal* 19, 1104–1109. DOI: 10.1111/pbi.13597.

Li, H., Vikram, P., Singh, R.P., Kilian, A., Carling, J. *et al.* (2015) A high density GBS map of bread wheat and its application for dissecting complex disease resistance traits. *BMC Genomics* 16, 216. DOI: 10.1186/s12864-015-1424-5.

Li, G., Zhou, J., Jia, H., Gao, Z., Fan, M. *et al.* (2019) Mutation of a histidine-rich calcium-binding-protein gene in wheat confers resistance to Fusarium head blight. *Nature Genetics* 51, 1106–1112. DOI: 10.1038/s41588-019-0426-7.

Li, M., Dong, L., Li, B., Wang, Z., Xie, J. *et al.* (2020) A CNL protein in wild emmer wheat confers powdery mildew resistance. *New Phytologist* 228, 1027–1037. DOI: 10.1111/nph.16761.

Li, L.-F., Zhang, Z.-B., Wang, Z.-H., Li, N. *et al.* (2022a) Genome sequences of five Sitopsis species of Aegilops and the origin of polyploid wheat B subgenome. *Molecular Plant* 15, 488–503. DOI: 10.1016/j.molp.2021.12.019.

Li, S., Lin, D., Zhang, Y., Deng, M., Chen, Y. *et al.* (2022b) Genome-edited powdery mildew resistance in wheat without growth penalties. *Nature* 602, 455–460. DOI: 10.1038/s41586-022-04395-9.

Li, H., Hua, L., Zhao, S., Hao, M., Song, R. *et al.* (2023) Cloning of the wheat leaf rust resistance gene Lr47 introgressed from *Aegilops speltoides*. *Nature Communications* 14, 6072. DOI: 10.1038/s41467-023-41833-2.

Li, H., Men, W., Ma, C., Liu, Q., Dong, Z. *et al.* (2024) Wheat powdery mildew resistance gene Pm13 encodes a mixed lineage kinase domain-like protein. *Nature Communications* 15, 2449. DOI: 10.1038/s41467-024-46814-7.

Lin, G., Chen, H., Tian, B., Sehgal, S.K., Singh, L. *et al.* (2022) Cloning of the broadly effective wheat leaf rust resistance gene Lr42 transferred from *Aegilops tauschii*. *Nature Communications* 13, 3044. DOI: 10.1038/s41467-022-30784-9.

Ling, H.-Q., Ma, B., Shi, X., Liu, H., Dong, L. *et al.* (2018) Genome sequence of the progenitor of wheat A subgenome *Triticum urartu*. *Nature* 557, 424–428. DOI: 10.1038/s41586-018-0108-0.

Liu, W., Frick, M., Huel, R., Nykiforuk, C.L., Wang, X. *et al.* (2014) The stripe rust resistance gene Yr10 encodes an evolutionary-conserved and unique CC-NBS-LRR sequence in wheat. *Molecular Plant* 7, 1740–1755. DOI: 10.1093/mp/ssu112.

Lourenço, V.M., Ogutu, J.O., Rodrigues, R.A.P., Posekany, A. and Piepho, H.-P. (2024) Genomic prediction using machine learning: A comparison of the performance of regularized regression, ensemble, instance-based and deep learning methods on synthetic and empirical data. *BMC Genomics* 25, 152. DOI: 10.1186/s12864-023-09933-x.

Lu, P., Guo, L., Wang, Z., Li, B., Li, J. *et al.* (2020) A rare gain of function mutation in a wheat tandem kinase confers resistance to powdery mildew. *Nature Communications* 11, 680. DOI: 10.1038/s41467-020-14294-0.

Lu, C., Du, J., Chen, H., Gong, S., Jin, Y. *et al.* (2024) Asuke. *Nature Communications* 15, 503. DOI: 10.1038/s41467-024-44796-0.

Luo, M., Xie, L., Chakraborty, S., Wang, A., Matny, O. *et al.* (2021) A five-transgene cassette confers broad-spectrum resistance to a fungal rust pathogen in wheat. *Nature Biotechnology* 39, 561–566. DOI: 10.1038/s41587-020-00770-x.

Mackenzie, A., Norman, M., Gessese, M., Chen, C., Sørensen, C. *et al.* (2023) Wheat stripe rust resistance locus YR63 is a hot spot for evolution of defence genes – A pangenome discovery. *BMC Plant Biology* 23, 590. DOI: 10.1186/s12870-023-04576-2.

Mago, R., Zhang, P., Vautrin, S., Šimková, H., Bansal, U. *et al.* (2015) The wheat Sr50 gene reveals rich diversity at a cereal disease resistance locus. *Nature Plants* 1, 15186. DOI: 10.1038/nplants.2015.186.

Marchal, C., Zhang, J., Zhang, P., Fenwick, P., Steuernagel, B. *et al.* (2018) BED-domain-containing immune receptors confer diverse resistance spectra to yellow rust. *Nature Plants* 4, 662–668. DOI: 10.1038/s41477-018-0236-4.

Martín-Sánchez, J.A., Gómez-Colmenarejo, M., Moral, J. Del, Sin, E., Montes, M.J. *et al.* (2003) A new hessian fly resistance gene (H30) transferred from the wild grass *Aegilops triuncialis* to hexaploid wheat. *Theoretical and Applied Genetics* 106, 1248–1255. DOI: 10.1007/s00122-002-1182-z.

McIntosh, R.A., Plant Breeding Institute, The University of Sydney, Wellings, C.R. and Park, R.F. (1995) *Wheat Rusts: An Atlas Resistance Genes*. CSIRO, Melbourne.

McIntosh, R.A., Dubcovsky, J., Rogers, W.J., Xia, X.C. and Raupp, W.J. (2020) Catalogue of gene symbols for wheat: 2020 supplement. Annual Wheat Newsletter, Volume 66, pages 19–28.

Meuwissen, T.H.E., Hayes, B.J. and Goddard, M.E. (2001) Prediction of Total genetic value using genome-wide dense marker maps. *Genetics* 157, 1819–1829. DOI: 10.1093/genetics/157.4.1819.

Miedaner, T. and Korzun, V. (2012) Marker-assisted selection for disease resistance in wheat and barley breeding. *Phytopathology* 102, 560–566. DOI: 10.1094/PHYTO-05-11-0157.

Montenegro, J.D., Golicz, A.A., Bayer, P.E., Hurgobin, B., Lee, H. *et al.* (2017) The pangenome of hexaploid bread wheat. *The Plant Journal* 90, 1007–1013. DOI: 10.1111/tpj.13515.

Moore, J.W., Herrera-Foessel, S., Lan, C., Schnippenkoetter, W., Ayliffe, M. *et al.* (2015) A recently evolved hexose transporter variant confers resistance to multiple pathogens in wheat. *Nature Genetics* 47, 1494–1498. DOI: 10.1038/ng.3439.

Mu, K., Ren, X., Yang, H., Zhang, T., Yan, W. *et al.* (2022) CRISPR-Cas12a-based diagnostics of wheat fungal diseases. *Journal of Agricultural and Food Chemistry* 70, 7240–7247. DOI: 10.1021/acs.jafc.1c08391.

Periyannan, S., Moore, J., Ayliffe, M., Bansal, U., Wang, X. *et al.* (2013) The gene *Sr33*, an Ortholog of barley *Mla* genes, encodes resistance to wheat stem rust race Ug99. *Science* 341, 786–788. DOI: 10.1126/science.1239028.

Periyannan, S., Milne, R.J., Figueroa, M., Lagudah, E.S. and Dodds, P.N. (2017) An overview of genetic rust resistance: From broad to specific mechanisms. *PLoS Pathogens* 13, e1006380. DOI: 10.1371/journal.ppat.1006380.

Possamai, T., Scota, L., Velasco, R. and Migliaro, D. (2024) A sustainable strategy for marker-assisted selection (MAS) applied in grapevine (Vitis spp.) breeding for resistance to downy (*Plasmopara viticola*) and powdery (*Erysiphe necator*) mildews. *Plants* 13, 2001. DOI: 10.3390/plants13142001.

Rani, R., Singh, R. and Yadav, N.R. (2019) Evaluating stripe rust resistance in Indian wheat genotypes and breeding lines using molecular markers. *Comptes Rendus Biologies* 342, 154–174. DOI: 10.1016/j.crvi.2019.04.002.

Ratcliffe, R. and Hatchett, J. (1997) *Biology and Genetics of the Hessian Fly and Resistance in Wheat* [WWW Document]. Available at: https://www.ars.usda.gov/research/publications/publication/?seqNo115=77479 (accessed 11 June 24).

Rawat, N., Pumphrey, M.O., Liu, S., Zhang, X., Tiwari, V.K. *et al.* (2016) Wheat Fhb1 encodes a chimeric lectin with agglutinin domains and a pore-forming toxin-like domain conferring resistance to Fusarium head blight. *Nature Genetics* 48, 1576–1580. DOI: 10.1038/ng.3706.

Rutkoski, J.E., Poland, J.A., Singh, R.P., Huerta-Espino, J., Bhavani, S. *et al.* (2014) Genomic selection for quantitative adult plant stem rust resistance in wheat. *The Plant Genome* 7, plantgenome2014.02.0006. DOI: 10.3835/plantgenome2014.02.0006.

Saintenac, C., Zhang, W., Salcedo, A., Rouse, M.N., Trick, H.N. *et al.* (2013) Identification of wheat gene *Sr35* that confers resistance to Ug99 stem rust race group. *Science* 341, 783–786. DOI: 10.1126/science.1239022.

Saintenac, C., Lee, W.-S., Cambon, F., Rudd, J.J., King, R.C. *et al.* (2018) Wheat receptor-kinase-like protein Stb6 controls gene-for-gene resistance to fungal pathogen *Zymoseptoria tritici*. *Nature Genetics* 50, 368–374. DOI: 10.1038/s41588-018-0051-x.

Saintenac, C., Cambon, F., Aouini, L., Verstappen, E., Ghaffary, S.M.T. *et al.* (2021) A wheat cysteine-rich receptor-like kinase confers broad-spectrum resistance against *Septoria tritici* blotch. *Nature Communications* 12, 433. DOI: 10.1038/s41467-020-20685-0.

Sánchez-Martín, J., Steuernagel, B., Ghosh, S., Herren, G., Hurni, S. *et al.* (2016) Rapid gene isolation in barley and wheat by mutant chromosome sequencing. *Genome Biology* 17, 221. DOI: 10.1186/s13059-016-1082-1.

Sánchez-Martín, J., Widrig, V., Herren, G., Wicker, T., Zbinden, H. *et al.* (2021) Wheat Pm4 resistance to powdery mildew is controlled by alternative splice variants encoding chimeric proteins. *Nature Plants* 7, 327–341. DOI: 10.1038/s41477-021-00869-2.

Santantonio, N., Atanda, S.A., Beyene, Y., Varshney, R.K., Olsen, M. *et al.* (2020) Strategies for effective use of genomic information in crop breeding programs serving Africa and South Asia. *Frontiers in Plant Science* 11. DOI: 10.3389/fpls.2020.00353.

Schmidt, S.M., Belisle, M. and Frommer, W.B. (2020) The evolving landscape around genome editing in agriculture. *EMBO Reports* 21, e50680. DOI: 10.15252/embr.202050680.

Sharma, D., Avni, R., Gutierrez-Gonzalez, J., Kumar, R., Sela, H. *et al.* (2024) A single NLR gene confers resistance to leaf and stripe rust in wheat. *Nature Communications* 15, 9925. DOI: 10.1038/s41467-024-54068-6.

Shewry, P.R., Lovegrove, A., Wingen, L.U. and Griffiths, S. (2022) Opinion exploiting genomics to improve the benefits of wheat: Prospects and limitations. *Journal of Cereal Science* 105, 103444. DOI: 10.1016/j.jcs.2022.103444.

Shi, G., Zhang, Z., Friesen, T.L., Raats, D., Fahima, T. *et al.* (2016) The hijacking of a receptor kinase–driven pathway by a wheat fungal pathogen leads to disease. *Science Advances* 2, e1600822. DOI: 10.1126/sciadv.1600822.

Shuai, J., Tu, Q., Zhang, Y., Xia, X., Wang, Y. *et al.* (2024) Silence of five F. graminearum genes in wheat host confers resistance to fusarium head blight. *Journal of Integrative Agriculture* S2095311924001837. DOI: 10.1016/j.jia.2024.04.026.

Singh, S.P., Hurni, S., Ruinelli, M., Brunner, S., Sanchez-Martin, J. *et al.* (2018) Evolutionary divergence of the rye Pm17 and Pm8 resistance genes reveals ancient diversity. *Plant Molecular Biology* 98, 249–260. DOI: 10.1007/s11103-018-0780-3.

Singh, J., Chhabra, B., Raza, A., Yang, S.H. and Sandhu, K.S. (2023) Important wheat diseases in the US and their management in the 21st century. *Frontiers in Plant Science* 13, 1010191. DOI: 10.3389/fpls.2022.1010191.

Srichumpa, P., Brunner, S., Keller, B. and Yahiaoui, N. (2005) Allelic series of four powdery mildew resistance genes at the *Pm3* locus in hexaploid bread wheat. *Plant Physiology* 139, 885–895. DOI: 10.1104/pp.105.062406.

Steed, A., King, J., Grewal, S., Yang, C., Clarke, M. *et al.* (2022) Identification of Fusarium head blight resistance in *Triticum timopheevii* accessions and characterization of Wheat-*T. timopheevii* Introgression lines for enhanced resistance. *Frontiers in Plant Science* 13, 943211. DOI: 10.3389/fpls.2022.943211.

Steuernagel, B., Periyannan, S.K., Hernández-Pinzón, I., Witek, K., Rouse, M.N. *et al.* (2016) Rapid cloning of disease-resistance genes in plants using mutagenesis and sequence capture. *Nature Biotechnology* 34, 652–655. DOI: 10.1038/nbt.3543.

Su, Z., Bernardo, A., Tian, B., Chen, H., Wang, S. *et al.* (2019) A deletion mutation in TaHRC confers Fhb1 resistance to Fusarium head blight in wheat. *Nature Genetics* 51, 1099–1105. DOI: 10.1038/s41588-019-0425-8.

Sunilkumar, V.P., Krishna, H., Devate, N.B., Manjunath, K.K., Chauhan, D. *et al.* (2022) Marker assisted improvement for leaf rust and moisture deficit stress tolerance in wheat variety HD3086. *Frontiers in Plant Science* 13, 1035016. DOI: 10.3389/fpls.2022.1035016.

The International Wheat Genome Sequencing Consortium (IWGSC), Appels, R., Eversole, K., Stein, N., Feuillet, C. *et al.* (2018) Shifting the limits in wheat research and breeding using a fully annotated reference genome. *Science* 361, eaar7191. DOI: 10.1126/science.aar7191.

Thind, A.K., Wicker, T., Šimková, H., Fossati, D., Moullet, O. *et al.* (2017) Rapid cloning of genes in hexaploid wheat using cultivar-specific long-range chromosome assembly. *Nature Biotechnology* 35, 793–796. DOI: 10.1038/nbt.3877.

Tomar, V., Singh, D., Dhillon, G.S., Singh, R.P., Poland, J. *et al.* (2021) New QTLs for spot blotch disease resistance in wheat (*Triticum aestivum* L.) using genome-wide association mapping. *Frontiers in Genetics* 11, 613217. DOI: 10.3389/fgene.2020.613217.

Turnbull, C., Lillemo, M. and Hvoslef-Eide, T.A.K. (2021) Global regulation of genetically modified crops amid the gene edited crop boom – A review. *Frontiers in Plant Science* 12, 630396. DOI: 10.3389/fpls.2021.630396.

United Nations (2012) *Feeding the World Sustainably [WWW Document]*. UN (accessed 5.12.24).

Upadhyaya, N.M., Mago, R., Panwar, V., Hewitt, T., Luo, M. *et al.* (2021) Genomics accelerated isolation of a new stem rust avirulence gene–wheat resistance gene pair. *Nature Plants* 7, 1220–1228. DOI: 10.1038/s41477-021-00971-5.

USDA ERS (2024) *Wheat Sector at a Glance* [WWW Document]. Available at: https://www.ers.usda.gov/topics/crops/wheat/wheat-sector-at-a-glance/ (accessed 5 August 24).

Várallyay, É., Giczey, G. and Burgyán, J. (2012) Virus-induced gene silencing of mlo genes induces powdery mildew resistance in *Triticum aestivum*. *Archives of Virology* 157, 1345–1350. DOI: 10.1007/s00705-012-1286-y.

Waldron, B.L., Moreno-Sevilla, B., Anderson, J.A., Stack, R.W. and Frohberg, R.C. (1999) RFLP mapping of QTL for Fusarium head blight resistance in wheat. *Crop Science* 39, 805–811. DOI: 10.2135/cropsci1999.0011183X003900030032x.

Walkowiak, S., Gao, L., Monat, C., Haberer, G., Kassa, M.T. *et al.* (2020) Multiple wheat genomes reveal global variation in modern breeding. *Nature* 588, 277–283. DOI: 10.1038/s41586-020-2961-x.

Wang, Y., Cheng, X., Shan, Q., Zhang, Y., Liu, J. *et al.* (2014) Simultaneous editing of three homoeoalleles in hexaploid bread wheat confers heritable resistance to powdery mildew. *Nature Biotechnology* 32, 947–951. DOI: 10.1038/nbt.2969.

Wang, H., Sun, S., Ge, W., Zhao, L., Hou, B. *et al.* (2020a) Horizontal gene transfer of *Fhb7* from fungus underlies Fusarium head blight resistance in wheat. *Science* 368, eaba5435. DOI: 10.1126/science.aba5435.

Wang, H., Zou, S., Li, Y., Lin, F. and Tang, D. (2020b) An ankyrin-repeat and WRKY-domain-containing immune receptor confers stripe rust resistance in wheat. *Nature Communications* 11, 1353. DOI: 10.1038/s41467-020-15139-6.

Wang, Y., Abrouk, M., Gourdoupis, S., Koo, D.-H., Karafiátová, M. *et al.* (2023) An unusual tandem kinase fusion protein confers leaf rust resistance in wheat. *Nature Genetics* 55, 914–920. DOI: 10.1038/s41588-023-01401-2.

Wulff, B.B.H. and Moscou, M.J. (2014) Strategies for transferring resistance into wheat: From wide crosses to GM cassettes. *Frontiers in Plant Science* 5. DOI: 10.3389/fpls.2014.00692.

Xie, J., Guo, G., Wang, Y., Hu, T., Wang, L. *et al.* (2020) A rare single nucleotide variant in *Pm5e* confers powdery mildew resistance in common wheat. *New Phytologist* 228, 1011–1026. DOI: 10.1111/nph.16762.

Xing, L., Hu, P., Liu, J., Witek, K., Zhou, S. *et al.* (2018) Pm21 from *Haynaldia villosa* encodes a CC-NBS-LRR protein conferring powdery mildew resistance in wheat. *Molecular Plant* 11, 874–878. DOI: 10.1016/j.molp.2018.02.013.

Yahiaoui, N., Srichumpa, P., Dudler, R. and Keller, B. (2004) Genome analysis at different ploidy levels allows cloning of the powdery mildew resistance gene Pm3b from hexaploid wheat. *The Plant Journal* 37, 528–538. DOI: 10.1046/j.1365-313X.2003.01977.x.

Yu, G., Matny, O., Champouret, N., Steuernagel, B., Moscou, M.J. *et al.* (2022) *Aegilops sharonensis* genome-assisted identification of stem rust resistance gene Sr62. *Nature Communications* 13, 1607. DOI: 10.1038/s41467-022-29132-8.

Yu, G., Matny, O., Gourdoupis, S., Rayapuram, N., Aljedaani, F.R. *et al.* (2023) The wheat stem rust resistance gene Sr43 encodes an unusual protein kinase. *Nature Genetics* 55, 921–926. DOI: 10.1038/s41588-023-01402-1.

Zhang, W., Chen, S., Abate, Z., Nirmala, J., Rouse, M. *et al.* (2017) Identification and characterization of Sr13, a tetraploid wheat gene that confers resistance to the Ug99 stem rust race group. Proceedings of National Academy of Sciences 114(45), E9483–E9492. DOI: 10.1073/pnas.1706277114.

Zhang, C., Huang, L., Zhang, H., Hao, Q., Lyu, B. *et al.* (2019) An ancestral NB-LRR with duplicated 3′UTRs confers stripe rust resistance in wheat and barley. *Nature Communications* 10, 4023. DOI: 10.1038/s41467-019-11872-9.

Zhang, J., Hewitt, T.C., Boshoff, W.H.P., Dundas, I., Upadhyaya, N. *et al.* (2021a) A recombined Sr26 and Sr61 disease resistance gene stack in wheat encodes unrelated NLR genes. *Nature Communications* 12, 3378. DOI: 10.1038/s41467-021-23738-0.

Zhang, Q., Li, Y., Li, Y., Fahima, T., Shen, Q. *et al.* (2021b). Introgression of the Powdery Mildew Resistance Genes Pm60 and Pm60b from Triticum urartu to Common Wheat Using Durum as a 'Bridge'. *Pathogens* 11, 25. DOI: 10.3390/pathogens11010025.

Zhang, Y., Yang, Z., Ma, H., Huang, L., Ding, F. *et al.* (2021c) Pyramiding of fusarium head blight resistance quantitative trait loci, Fhb1, Fhb4, and Fhb5, in modern Chinese wheat cultivars. *Frontiers in Plant Science* 12, 694023. DOI: 10.3389/fpls.2021.694023.

Zhang, Z., Running, K.L.D., Seneviratne, S., Peters Haugrud, A.R., Szabo-Hever, A. *et al.* (2021d) A protein kinase–major sperm protein gene hijacked by a necrotrophic fungal pathogen triggers disease susceptibility in wheat. *The Plant Journal* 106, 720–732. DOI: 10.1111/tpj.15194.

Zhang, L., Liu, Y., Wang, Q., Wang, C., Lv, S. *et al.* (2022) An alternative splicing isoform of wheat TaYRG1 resistance protein activates immunity by interacting with dynamin-related proteins. *Journal of Experimental Botany* 73, 5474–5489. DOI: 10.1093/jxb/erac245.

Zhang, M., Chen, Z., Wu, H., Kong, F., Guo, B. *et al.* (2024) Identification of candidate genes for English grain aphid resistance from QTLs using a RIL population in wheat. *Agronomy* 14, 637. DOI: 10.3390/agronomy14030637.

Zhao, Y., Dong, Z., Miao, J., Liu, Q., Ma, C. *et al.* (2024) Pm57 from *Aegilops searsii* encodes a tandem kinase protein and confers wheat powdery mildew resistance. *Nature Communications* 15, 4796. DOI: 10.1038/s41467-024-49257-2.

Zou, S., Wang, H., Li, Y., Kong, Z. and Tang, D. (2018) The NB - LRR gene *Pm60* confers powdery mildew resistance in wheat. *New Phytologist* 218, 298–309. DOI: 10.1111/nph.14964.

7 Optimizing Nitrogen Use Efficiency: Genomic Insights for Reducing Carbon Footprint in a Changing Climate

Nitika Sandhu*, Aman Kumar, Jasneet Singh, Gaurav Augustine, Jaismeen Kaur, Muskan Gupta, Harsh Shukla and Ekta Kharche
School of Agricultural Biotechnology, Punjab Agricultural University, Ludhiana-141004, India

Abstract

Nitrogen (N) is a critical macronutrient for crop productivity, yet its inefficient utilization in agricultural systems contributes to environmental degradation and increased carbon emissions. Enhancing Nitrogen Use Efficiency (NUE) in crops is essential for sustainable food production under changing climatic conditions. This chapter explores genomic and molecular approaches to optimize NUE, integrating advances in genomic technologies, marker-assisted selection and gene-editing technologies. Key regulatory networks governing nitrogen uptake, assimilation and remobilization are examined, with a focus on genes and quantitative trait loci (QTLs) associated with improved NUE. The role of root architecture, nitrate transporters and nitrogen-responsive transcription factors in enhancing nutrient acquisition is highlighted. Additionally, the chapter discusses the interplay between nitrogen metabolism and carbon sequestration, emphasizing how optimizing NUE can mitigate greenhouse gas emissions. The chapter also addresses the challenges associated with enhancing NUE through genomic approaches, emphasizing the need for integrating these strategies with sustainable agronomic practices and policies to achieve long-term food security and environmental sustainability. By leveraging genomic innovations, this study provides insights into breeding climate-resilient varieties with enhanced NUE, reducing the carbon footprint while ensuring food security in a resource-constrained world.

7.1 Introduction

Nitrogen (N) is an essential nutrient required by all plants for growth, development, and productivity. It plays a critical role in key biological processes such as photosynthesis, protein synthesis and DNA replication. In modern agriculture, nitrogen fertilizers have been instrumental in achieving high crop yields, significantly contributing to the success of the Green Revolution and feeding an ever-growing global population. However, the widespread and often excessive use of nitrogen fertilizers has given rise to several environmental challenges, including increased greenhouse gas emissions, soil degradation and water pollution.

Nitrogen fertilizers account for the largest share of all fertilizers applied globally, with usage exceeding 110 million metric tons annually which may increase up to 240 million tons by 2050 (FAO, 2024) (Figure 7.1A). The demand for nitrogen fertilizers continues to grow due to increasing

*Corresponding author. Email: nitikasandhu@pau.edu

 Genomic Solutions for Breeding Climate-resilient Crops (R.K. Varshney, R. Barmukh and R.J. Harper)
DOI: 10.1079/9781800627185.0007

food production needs. China and India are the largest consumers of nitrogen fertilizers globally, driven by their massive agricultural sectors (Figure 7.1B). China alone consumes around 30% of the world's nitrogen fertilizer (Sapkota and Takele, 2023). In India, nitrogen fertilizers are heavily subsidized, particularly urea, to ensure affordable prices for farmers. The USA is one of the top consumers, applying large amounts of nitrogen fertilizers for corn, wheat and soybean production (Figure 7.1B). Fertilizer consumption in Africa is relatively low compared with other regions (Figure 7.1B). However, countries such as Egypt and South Africa are notable consumers. Wheat, rice and maize are the largest global consumers of nitrogen fertilizers, collectively accounting for approximately 50% of total nitrogen fertilizer usage worldwide (Figure 7.1C). These staple crops are grown extensively across the world, and their high nitrogen requirements are crucial for achieving the high yields necessary to feed a growing population. Together, these three crops make up about half of the world's nitrogen fertilizer consumption, emphasizing their central role in global agriculture and the need for effective nitrogen management to minimize environmental impacts.

The nine planetary boundaries are a scientific model that outlines the limits of human impact on Earth's systems (Figure 7.1D). It was developed to define a "safe operating space" for humanity by identifying nine key Earth system processes crucial for maintaining the planet's stability and resilience. Crossing these boundaries increases the risk of destabilizing Earth's systems, potentially causing harm to human societies and ecosystems. The current rate of application of industrially fixed N fertilizer is 112 Tg of N per year, which exceeds the planetary boundary for nitrogen (62 Tg) (Richardson *et al.*, 2023). Excessive use of nitrogen fertilizers contributes to about 70% of total greenhouse gas emissions. The extensive use of commercial fertilizers has degraded air, soil and water quality. When nitrogen levels exceed crop needs, plants become more susceptible to diseases and pests. As a result, managing nitrogen effectively has become a priority for sustainable agriculture.

The efficiency with which crops use nitrogen, known as nitrogen use efficiency (NUE), is a critical factor in determining both agricultural productivity and environmental impact. NUE refers to the amount of biomass or yield produced per unit of nitrogen available in the soil, either from fertilizers or natural sources. Despite the economic and environmental importance of NUE, it remains relatively low in most cropping systems. Studies indicate that plants typically utilize only about 30–50% of the nitrogen applied through fertilizers, with the remainder being lost to the environment in the form of nitrogen gases, such as nitrous oxide (N_2O), or through nitrate leaching into water bodies (Ladha *et al.*, 2005). This inefficiency not only increases production costs for farmers but also exacerbates environmental problems, including climate change and water pollution.

One of the most pressing environmental concerns linked to inefficient nitrogen use is the emission of nitrous oxide, a potent greenhouse gas. N_2O has a global warming potential approximately 300 times greater than carbon dioxide (CO_2) and accounts for a significant portion of agriculture's contribution to climate change (Tian *et al.*, 2023). Reducing nitrogen-related emissions through improved NUE could, therefore, play a vital role in mitigating the agricultural sector's carbon footprint. Moreover, with the global population expected to reach nearly 10 billion by 2050, optimizing NUE becomes even more critical to ensuring food security while minimizing environmental degradation.

In the face of these challenges, there is a growing need for innovative strategies that enhance NUE in crops while maintaining or improving yields. Traditional methods, such as optimized fertilizer application and crop rotation, have been effective to some extent but are insufficient to meet the demands of sustainable agriculture in a changing climate. Advances in genomic technologies, however, offer new opportunities to address these challenges. By leveraging insights into the genetic and molecular mechanisms that control nitrogen uptake, assimilation, and metabolism, researchers can develop crop varieties with enhanced NUE. These advances could potentially reduce the reliance on synthetic fertilizers, lower nitrous oxide emissions and improve agricultural sustainability.

This chapter explores the genetic and molecular underpinnings of nitrogen use efficiency and discusses how genomic tools and breeding strategies can be harnessed to optimize NUE in crops. The chapter begins by outlining the key

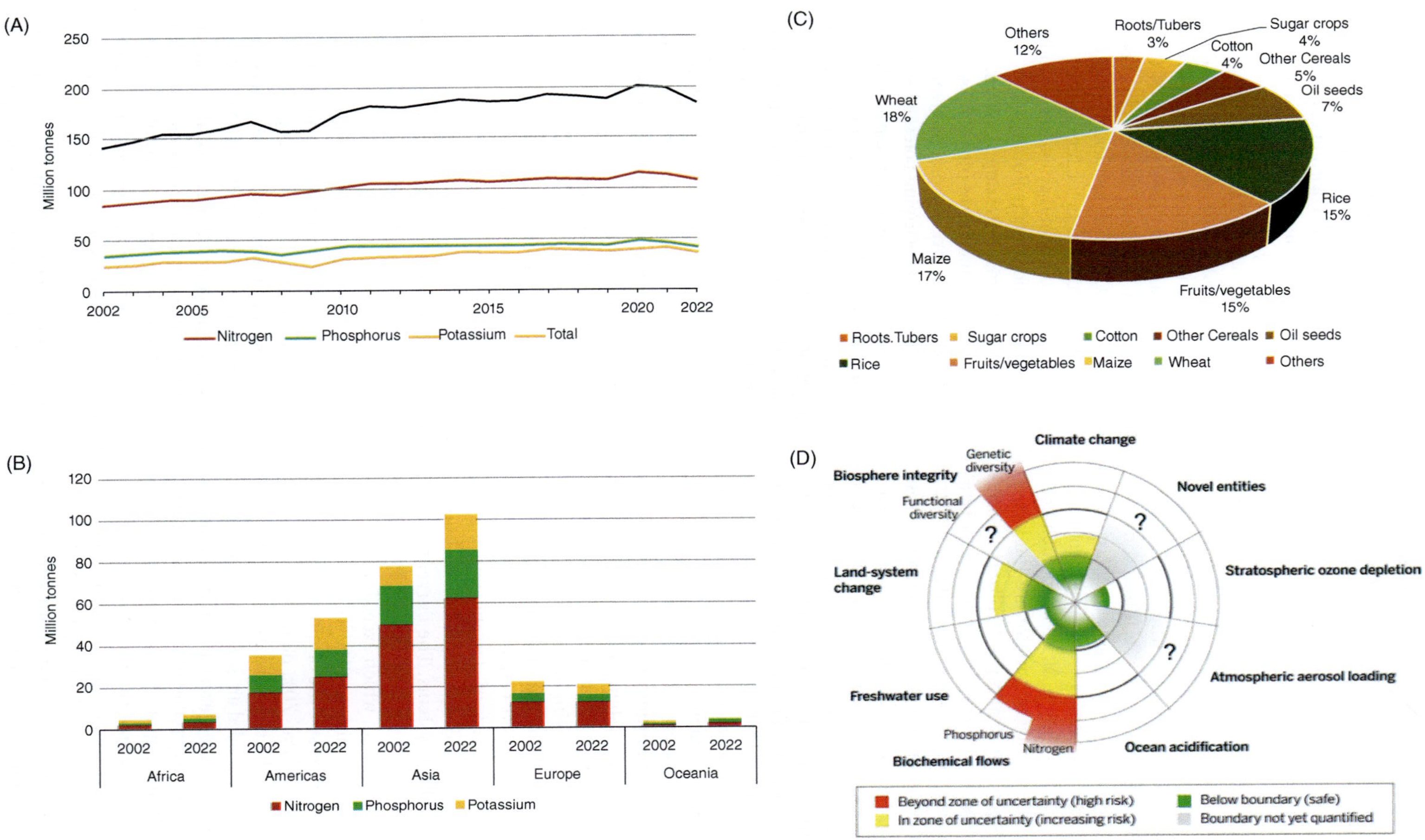

Figure 7.1. Consumption of nitrogen fertilizer across world. (A) Agriculture use of inorganic fertilizers across world. Source: FAO (2024) *FAOSTAT: Fertilizers by Nutrient* (accessed September 2024). (B) Inorganic fertilizers agricultural use by nutrient and region. Source: FAO (2024) *FAOSTAT: Fertilizers by Nutrient* (accessed September 2024). (C) Global share of nitrogen (%) fertilizer. Source: FAO (2024) *FAOSTAT: Fertilizers by Nutrient* (accessed September 2024). (D) Nine planetary boundaries and nitrogen use. Source: Richardson *et al.* (2023).

components of NUE – nitrogen uptake efficiency (NUpE) and nitrogen utilization efficiency (NUtE) – and their roles in plant nitrogen metabolism. It then delves into recent genomic insights, highlighting key genes involved in nitrogen uptake, assimilation and signaling pathways. The chapter also discusses how cutting-edge breeding technologies, such as genome-wide association studies (GWAS), marker-assisted selection (MAS) and CRISPR-Cas9 gene editing, are being used to develop nitrogen-efficient crop varieties.

In addition, the chapter addresses the environmental benefits of optimizing NUE, particularly in the context of reducing the carbon footprint of agricultural systems. By enhancing NUE, farmers can reduce the amount of nitrogen fertilizer required, thus lowering greenhouse gas emissions, and minimizing the impact on soil and water ecosystems. These benefits are especially important in the context of climate change, where increasing temperatures, altered precipitation patterns and extreme weather events are expected to further strain global food production systems.

The chapter concludes by discussing the future directions and challenges associated with improving NUE through genomic approaches. While significant progress has been made, the complex nature of nitrogen metabolism and its interaction with environmental factors poses several challenges for researchers. Furthermore, the success of genomic strategies in improving NUE will depend on their integration with agronomic practices and policies that promote sustainable fertilizer use and soil management.

Ultimately, optimizing nitrogen use efficiency through genomic insights represents a critical step toward achieving sustainable agriculture in the 21st century. It offers the potential to reduce the environmental impact of food production, mitigate climate change and ensure food security for future generations.

7.2 The Importance of Nitrogen in Agriculture and Its Environmental Impacts

Nitrogen (N) is a key macronutrient for plant growth, playing an indispensable role in the formation of vital biomolecules, including amino acids, proteins, chlorophyll and nucleic acids. Without adequate nitrogen, crops cannot reach their full yield potential, which directly impacts food security, especially in a world with an ever-growing population. Due to its central role in plant health, nitrogen deficiency can result in stunted growth, yellowing leaves (chlorosis), reduced biomass and lower crop yields. To avoid these issues, farmers often apply nitrogen fertilizers, which supplement the nitrogen available in the soil. The development of synthetic nitrogen fertilizers, particularly after the invention of the Haber–Bosch process in the early 20th century, enabled the mass production of ammonia (NH_3) from atmospheric nitrogen (N_2). This breakthrough allowed for the widespread application of nitrogen fertilizers, supporting the exponential increase in agricultural productivity and feeding billions of people.

Nitrogen fertilizers are essential for plant growth and development, with common types including ammonium-based, nitrate-based and urea fertilizers. These can be applied using methods such as broadcasting, banding, fertigation, foliar spraying and side-dressing. Plants absorb nitrogen mainly in the forms of ammonium (NH_4^+) and nitrate (NO_3^-) (Figure 7.2A). Factors such as soil pH, moisture, temperature and texture significantly affect nitrogen uptake. Proper management of nitrogen fertilizer is crucial to maximize plant absorption while minimizing environmental issues such as leaching, volatilization and denitrification (Figure 7.2A). Over the past century, the introduction of synthetic nitrogen fertilizers has revolutionized agriculture, increasing crop yields significantly. However, this increase in productivity has come with considerable environmental costs. The use and management of nitrogen in agriculture are now seen as central to discussions around sustainability and climate change.

7.2.1 Importance of Nitrogen in Plant Growth and Agricultural Productivity

Nitrogen is critical for a variety of physiological functions in plants:

Protein synthesis: Nitrogen is a key component of amino acids, which are the building blocks of proteins. Proteins are required for nearly all

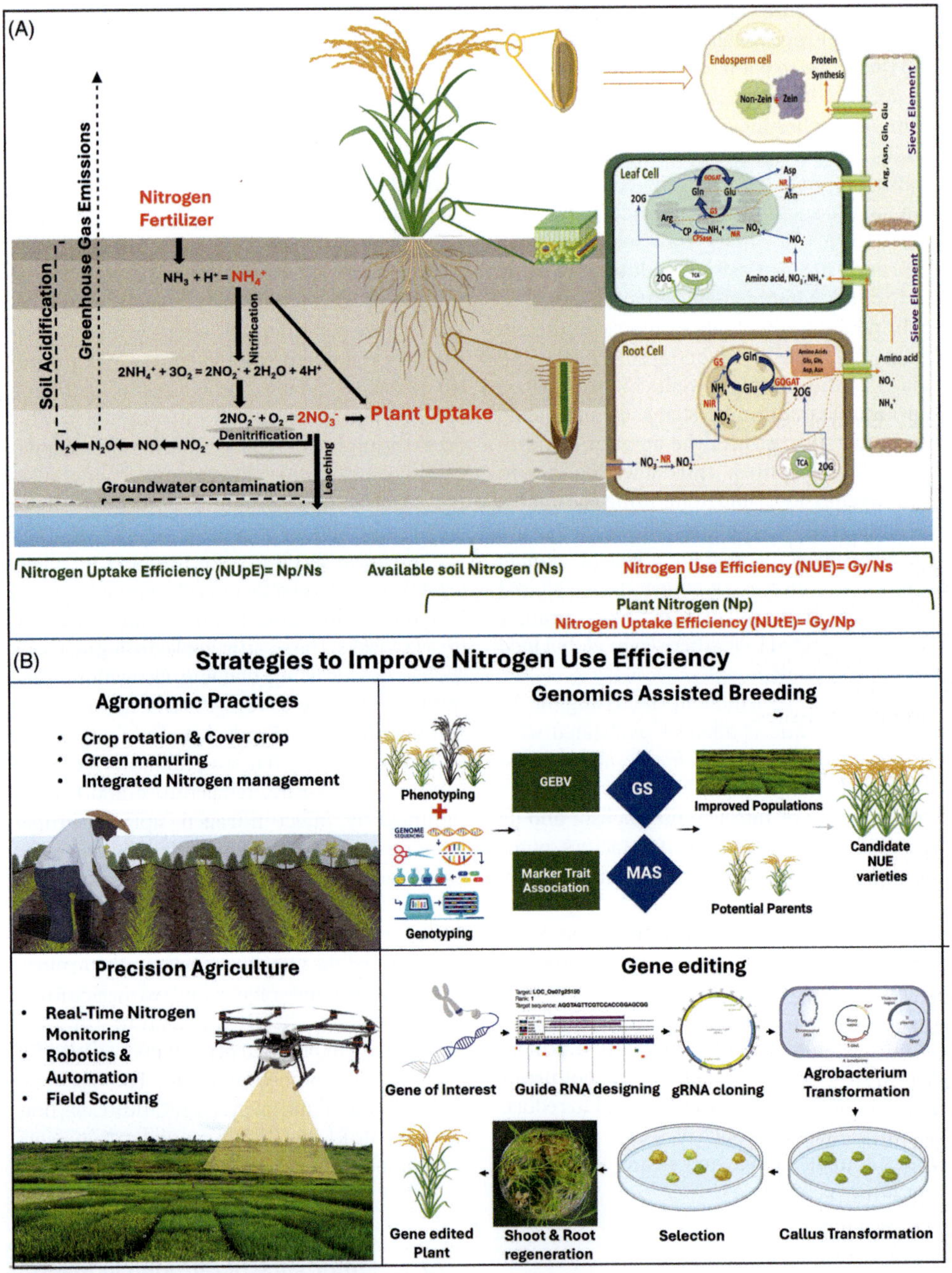

Figure 7.2. (A) A schematic representation of the relationship between the key physiological processes involved in nitrogen uptake, assimilation, translocation, remobilization, utilization, and its conversion into grain yield. (B) Strategies to improve NUE.

physiological functions in plants, including growth and metabolism.

Chlorophyll production: Chlorophyll, the pigment responsible for photosynthesis, contains nitrogen. Photosynthesis enables plants to convert sunlight into chemical energy, which fuels growth and productivity.

Nucleic acids: Nitrogen is an essential element in nucleotides, which make up DNA and RNA. These nucleic acids are critical for cell division, growth, and the transmission of genetic information.

Energy transfer: Nitrogen is involved in the formation of adenosine triphosphate (ATP), the molecule that stores and transfers energy within cells.

However, while nitrogen fertilizers are crucial for achieving high crop yields, their inefficient use can lead to serious environmental consequences.

7.2.2 Environmental Impacts of Nitrogen Use in Agriculture

Despite its importance, nitrogen is often poorly managed in agricultural systems, leading to significant environmental problems. Nitrogen use efficiency in most cropping systems is relatively low, with plants typically absorbing only 30–50% of the nitrogen applied as fertilizer. The remainder is lost to the environment, causing multiple environmental issues.

7.2.2.1 Nitrous Oxide (N_2O) Emissions and Climate Change

Nitrogen fertilizers significantly impact the environment by contributing to nitrous oxide (N_2O) emissions, a potent greenhouse gas. N_2O is produced through soil microbial processes such as nitrification and denitrification, which are intensified by excessive nitrogen application. N_2O has a global warming potential about 300 times greater than CO_2 over 100 years, making it a substantial contributor to climate change despite its lower atmospheric concentration. Agriculture accounts for over 50% of global anthropogenic N_2O emissions, primarily due to fertilizer use (Petersen *et al.*, 2023). Additionally, N_2O contributes to ozone layer depletion, making emission reductions crucial for mitigating climate change and environmental protection.

7.2.2.2 Nitrate Leaching and Water Pollution

Excessive nitrogen use leads to nitrate (NO_3^-) leaching, where soluble nitrate is washed into groundwater and surface water systems, contributing to water pollution. This reduces nitrogen availability for crops and poses health risks, particularly to infants, who may develop "blue baby syndrome" (methemoglobinemia) from nitrate-contaminated drinking water (Ward *et al.*, 2018). Nitrate runoff also fuels eutrophication, causing algal blooms that deplete oxygen in aquatic systems, leading to hypoxia, fish kills and biodiversity loss. This has resulted in "dead zones," such as in the Gulf of Mexico, caused by agricultural nutrient runoff (Rabalais *et al.*, 2010).

7.2.2.3 Soil Acidification and Degradation

Over-application of nitrogen fertilizers can cause soil acidification. Ammonium-based fertilizers are converted to nitrate by soil microbes, releasing hydrogen ions (H^+), which lower soil pH and degrade soil quality (Wang *et al.*, 2020). Acidic soils lead to nutrient imbalances, such as the leaching of calcium, magnesium and potassium, reducing fertility. Additionally, aluminum becomes more soluble in acidic conditions, posing toxicity risks to plant roots. Soil acidification also disrupts beneficial microbial communities, negatively impacting nitrogen fixation and organic matter decomposition (Li *et al.*, 2023). This degradation threatens long-term agricultural sustainability in heavily fertilized regions.

7.2.2.4 Biodiversity Loss

Excessive nitrogen inputs disrupt terrestrial ecosystems by favoring fast-growing, nitrogen-loving species over slower-growing ones, reducing plant biodiversity (Botez and Postolache, 2013). This imbalance affects species dependent on diverse plant communities for food and habitat. Grasslands, forests and wetlands are particularly sensitive, with nitrogen deposition linked to biodiversity loss in these ecosystems (Sala *et al.*, 2000). The shift in plant communities caused by nitrogen deposition can have cascading ecological impacts, threatening ecosystem health.

7.3 Nitrogen Use Efficiency: Definition and Components

Nitrogen use efficiency measures how effectively plants absorb and convert nitrogen into biomass or yield, reflecting the efficiency of nitrogen utilization in agriculture (Moll *et al.*, 1982).

Improving NUE is essential due to the high cost of nitrogen fertilizers and the environmental risks of nitrogen losses, such as water pollution and greenhouse gas emissions (Raun and Johnson, 1999). Enhancing NUE through better management practices is critical for achieving both economic and environmental sustainability in agriculture.

7.3.1 Definition of Nitrogen Use Efficiency

Nitrogen use efficiency is the ratio of crop yield to the nitrogen available in the soil from both natural and applied sources, reflecting how effectively plants convert nitrogen into harvestable products (Fageria and Baligar, 2005) (Figure 7.2A). Higher NUE indicates efficient nitrogen use, leading to better yields with less input, while low NUE results in significant nitrogen loss through volatilization, leaching or denitrification (Ladha *et al.*, 2005).

Nitrogen use efficiency is typically expressed as:

> NUE = Crop Yield (kg)/Total Nitrogen of both Soil and Input (kg)
>
> This equation highlights the importance of both plant genetic factors and agronomic practices in improving nitrogen use in agriculture.

Improving NUE is critical for:

> Maximizing crop yields: Efficient nitrogen use directly affects plant growth, development, and final yield.
>
> Reducing fertilizer costs: Enhanced NUE allows farmers to achieve higher yields with less fertilizer input, lowering production costs.
>
> Minimizing environmental impact: By reducing nitrogen loss, high NUE systems help mitigate negative environmental effects such as greenhouse gas emissions, water pollution, and soil degradation.

7.3.2 Components of Nitrogen Use Efficiency

Nitrogen use efficiency can be broken down into several interrelated components, each of which contributes to the overall efficiency of nitrogen utilization in plants. The two primary components are:

1. Nitrogen uptake efficiency (NUpE).
2. Nitrogen utilization efficiency (NUtE).

Each of these components reflects different stages of the nitrogen cycle within the plant and agricultural system.

7.3.2.1 Nitrogen Uptake Efficiency (NUpE)

Nitrogen uptake efficiency (NUpE) refers to the ability of a plant to absorb nitrogen from the soil. It is the proportion of available nitrogen in the soil that is taken up by the plant roots. NUpE depends on several factors, including the plant's root architecture, the presence of nitrogen transporters and soil conditions (Masclaux-Daubresse *et al.*, 2010).

> NUpE = Nitrogen absorbed by plant/Total nitrogen available in soil.

Key aspects of NUpE include:

> Root system development: A well-developed and extensive root system enhances a plant's ability to explore a larger soil volume for nitrogen uptake. Deep, branching roots are more effective at intercepting nitrate (NO_3^-) and ammonium (NH_4^+) ions in the soil (Bloom *et al.*, 2002).
>
> Nitrogen transporters: Specific genes encode nitrogen transporters that move nitrate and ammonium from the soil into plant roots. These transporters play a crucial role in determining how efficiently a plant can absorb nitrogen. For example, nitrate transporters such as *NRT1.1* and ammonium transporters such as *AMT1* are critical for nitrogen uptake in many crops (Muratore *et al.*, 2021).
>
> Mycorrhizal associations: Symbiotic associations between plant roots and mycorrhizal fungi can enhance nitrogen uptake, particularly in soils where nitrogen is in limited supply (Begum *et al.*, 2019). Mycorrhizal fungi extend the root system's reach and help capture nitrogen that would otherwise be inaccessible.

Plants with higher NUpE are more effective at extracting nitrogen from the soil, reducing the amount of nitrogen lost through leaching and volatilization. This efficiency is particularly important in nutrient-poor soils or when nitrogen fertilizer inputs are minimized for sustainability reasons.

7.3.2.2 Nitrogen Utilization Efficiency (NUtE)

Nitrogen utilization efficiency (NUtE) measures how effectively a plant converts absorbed nitrogen into biomass or yield. It reflects the plant's ability to use nitrogen for producing proteins, enzymes and other growth-related structures. High NUtE allows plants to generate more biomass or grain per unit of nitrogen absorbed, making them more efficient in utilizing available nitrogen for growth (Hirel *et al.*, 2007).

NUtE = Crop Yield (kg)/Nitrogen absorbed by plant (kg).

Key factors influencing NUtE include:
Nitrogen metabolism: After uptake, nitrogen is metabolized into amino acids, proteins and chlorophyll, crucial for plant growth and function. Efficient metabolism requires coordinated enzyme activity (Masclaux-Daubresse *et al.*, 2010).

Nitrogen remobilization: Plants redistribute nitrogen from older tissues to younger, actively growing tissues. This is vital during grain-filling in cereals, where nitrogen is transferred from leaves to developing grains, influencing yield (Masclaux-Daubresse *et al.*, 2008).

Genetic regulation: Genes for enzymes such as glutamine synthetase (GS) and glutamate dehydrogenase (GDH) regulate nitrogen metabolism and remobilization, converting inorganic nitrogen into usable forms (Fortunato *et al.*, 2023).

Environmental conditions: Nitrogen utilization is affected by factors such as light, temperature and water availability, as these influence photosynthesis, which depends on nitrogen-containing chlorophyll (Cossani and Sadras, 2021).

Plants with higher NUtE produce more output (such as grain or biomass) with the same or lower amounts of nitrogen compared with plants with lower NUtE. This makes NUtE a key focus for breeding programs aiming to reduce nitrogen inputs without sacrificing yield.

7.3.3 Interplay Between NUpE and NutE

Interdependence of NUpE and NUtE: NUpE (nitrogen uptake efficiency) and NUtE (nitrogen utilization efficiency) are interconnected components that together determine overall nitrogen use efficiency (NUE) (Fageria and Press, 2009).

High NUpE but Low NUtE: A plant with high NUpE but low NUtE absorbs nitrogen efficiently but fails to convert it into yield, leading to nitrogen accumulation in tissues rather than productive use, resulting in low overall NUE (Raun and Johnson, 1999).

Low NUpE but High NUtE: Conversely, a plant with low NUpE may take up less nitrogen but utilizes it effectively for biomass production, potentially exhibiting high NUE, especially in low nitrogen environments (Ladha *et al.*, 2005).

Optimal Balance: Enhancing NUE requires improving both NUpE and NUtE, such as increasing root nitrogen uptake efficiency for nitrogen capture (higher NUpE) and enhancing nitrogen remobilization and metabolism (higher NUtE) (Zhang *et al.*, 2015).

7.4 Factors Affecting NUE

Several factors influence NUE in crops, including:

Soil management: Soil characteristics, including texture, organic matter content and microbial activity, impact nitrogen availability and uptake. Agronomic practices that enhance soil fertility, such as using organic fertilizers and cover crops, can improve NUE (Zhang *et al.*, 2015).

Fertilizer management: The timing, type and amount of nitrogen fertilizer applied are critical for determining NUE. Precision farming techniques that align nitrogen application with crop needs can enhance both NUpE and NUtE (Raun and Johnson, 1999).

Environmental conditions: Factors such as temperature, water availability and light influence nitrogen uptake and metabolism. In the face of climate change, developing crop varieties that exhibit resilience in nitrogen efficiency may become increasingly important (Cossani and Sadras, 2021).

Irrigation: Well-irrigated conditions improve NUE by maintaining soil nitrogen availability (Ding *et al.*, 2021a). Drought stress reduces nitrogen uptake by limiting root activity and transporter function (Giller *et al.*, 2004).

Temperature: High temperatures increase nitrification and denitrification rates, leading to nitrogen losses.Low temperatures slow microbial activity, reducing nitrogen mineralization and uptake (Fageria *et al.*, 2014).

Rate and timing of nitrogen application: Excess application leads to losses via leaching, runoff and volatilization (Zhang *et al.*, 2021). Split nitrogen applications throughout the growing season improve NUE by matching plant demand. Foliar nitrogen application can enhance NUE under certain conditions (Blandino *et al.*, 2015).

Crop variety: The efficiency of nitrogen uptake per unit of biomass production varies significantly among different crops and crop varieties. Under well-irrigated conditions, crops tend to exhibit greater agronomic nitrogen use efficiency (NUE) compared with those grown under rainfed or limited irrigation conditions. Optimal soil moisture levels play a crucial role in enhancing crop growth rate, yield and nitrogen uptake (Giller *et al.*, 2004; Ding *et al.*, 2021b). Annual crops generally exhibit higher agronomic NUE than perennial crops due to their greater nitrogen uptake efficiency and concentration (Weih *et al.*, 2011). However, in terms of yield-specific nitrogen efficiency, perennials outperform wheat (Weih *et al.*, 2011). Additionally, fodder crops tend to have superior agronomic NUE compared with food crops, primarily due to their higher biomass production per unit area and time.

7.5 Strategies for Improving NUE

Inefficient nitrogen use in agriculture underscores the need for sustainable nitrogen management to mitigate environmental harm. Reducing reliance on synthetic nitrogen fertilizers, while maintaining crop yields, is crucial for limiting nitrogen's negative impacts. Improving NUE is key, as optimizing nitrogen uptake and utilization in crops can reduce fertilizer application and nitrogen losses to the environment (Zhang *et al.*, 2023). Several agronomic and genomic strategies offer pathways to balance agricultural productivity with environmental protection (Lassaletta *et al.*, 2014). A schematic representation of the agronomic and genomic strategies for improving NUE is shown in Figure 7.2(B).

7.5.1 Agronomic Practices for Improving NUE

7.5.1.1 Cover Crops and Crop Rotation

Cover crops, particularly legumes, enhance soil health and decrease nitrogen losses by capturing residual nitrogen in the soil post-harvest. They fix atmospheric nitrogen, improving soil fertility and reducing reliance on synthetic fertilizers in future planting cycles, thus aiding in the reduction of nitrous oxide emissions and enhancing nitrogen cycling efficiency. Additionally, implementing crop rotation, where different crops are grown in succession, can enhance nitrogen use efficiency by alternating crops with varying nitrogen needs. This practice reduces the continuous requirement for high fertilizer application, leading to lower nitrogen losses and emissions (Dabney *et al.*, 2010).

7.5.1.2 Integrated Nutrient Management

Integrated nutrient management (INM) enhances NUE by combining organic and inorganic nutrient sources, optimizing their availability to crops. It involves precise application of chemical fertilizers based on soil testing, alongside organic inputs such as manure, compost and biofertilizers that improve soil health and nutrient cycling. INM reduces nutrient losses through leaching or volatilization, promotes better root growth and enhances microbial activity. By balancing nutrient supply and demand, INM ensures that crops utilize nutrients more efficiently, leading to higher productivity and reduced environmental impact, while maintaining long-term soil fertility and sustainability.

7.5.1.3 Sustainable Fertilizer Innovations

The development of efficient, eco-friendly fertilizers is vital for enhancing NUE. Enhanced efficiency fertilizers (EEF) are formulated to slow the nitrogen release into the soil or inhibit processes that lead to nitrogen loss (Asadu *et al.*, 2024). Notable examples include:

Nitrification inhibitors: These chemicals slow the conversion of ammonium to nitrate, thereby reducing nitrogen loss through leaching or denitrification.

Slow-release and controlled-release fertilizers: Designed to release nitrogen gradually, these fertilizers align with the crop's uptake pattern, minimizing losses. The use of EEFs can significantly lower the amount of nitrogen lost to the environment, effectively decreasing the carbon footprint associated with nitrogen fertilization.

Biological nitrogen fixation: Engineering non-leguminous crops to form symbiotic relationships with nitrogen-fixing bacteria can reduce synthetic fertilizer reliance.

Microbial inoculants: Utilizing beneficial microorganisms as biofertilizers can enhance nitrogen uptake.

7.5.1.4 Precision Agriculture and Digital Farming

Precision agriculture employs techniques such as variable rate fertilization, enabling farmers to apply nitrogen fertilizers with high specificity (Anas *et al.*, 2020). By aligning fertilizer application with the unique needs of crops based on factors such as soil type and growth stage, and through real-time monitoring of nitrogen status, this approach optimizes nitrogen use. Consequently, it minimizes over-application and reduces nitrogen loss, leading to decreased nitrous oxide emissions and other environmental impacts.

The integration of digital agricultural technologies (DATs) has become a key element of this transformative shift, providing an innovative outlook in the agricultural sector (Papadopoulos *et al.*, 2024).

Precision agriculture technologies will be critical in enhancing NUE:

Real-time nitrogen monitoring: Advanced sensors and satellite tools will allow farmers to apply nitrogen precisely, minimizing losses and increasing NUE.

Machine learning and AI in nitrogen management: AI tools will analyze farm data to create predictive models, optimizing nitrogen application strategies.

Robotics and automation: Automated systems can perform tasks such as soil sampling and nitrogen application, improving efficiency and reducing labor costs.

7.5.2 Genomic Insights into NUE

Advances in genomic technologies have significantly enhanced the potential for improving NUE in crops. By understanding the genetic basis of NUE, researchers can identify key genes, pathways, and regulatory networks involved in nitrogen uptake, assimilation and utilization (Hou *et al.*, 2021). This knowledge facilitates the development of crop varieties that utilize nitrogen more efficiently, thereby reducing the reliance on excessive fertilizer application and mitigating environmental impacts (Hirel *et al.*, 2011). Breeding programs that focus on traits such as root architecture, nitrogen transporters and efficient nitrogen metabolism have led to varieties with improved NUE (Ladha *et al.*, 2005). Genomic approaches, including genome-wide association studies (GWAS), marker-assisted selection (MAS) and gene editing, are increasingly central to breeding nitrogen-efficient crops, paving the way for sustainable agricultural practices (Bohra *et al.*, 2020).

The genetic makeup of a crop significantly influences its nitrogen uptake and utilization efficiency. NUE is a complex trait regulated by multiple genes, each contributing to various aspects of nitrogen uptake, assimilation, transport and utilization. These genes are involved in processes such as nitrogen sensing, root development, nitrogen transport, metabolism and remobilization (Fageria and Press, 2009). While traditional breeding has made progress in enhancing NUE through phenotypic trait selection, genomic insights allow for a more precise identification of specific genes and molecular mechanisms that govern NUE (Ladha *et al.*, 2005).

7.5.3 Genomic Approaches for Enhancing NUE

Several genomic approaches are being employed to identify genes associated with NUE and to develop nitrogen-efficient crop varieties.

7.5.3.1 Quantitative Trait Loci (QTLs) Mapping

Quantitative trait loci (QTLs) mapping is a genomic technique that identifies specific regions of the genome associated with NUE. This approach involves crossing two plant varieties with differing nitrogen efficiencies and analyzing the genetic makeup of their offspring to pinpoint chromosomal regions linked to NUE traits. These

QTLs often encompass multiple genes contributing to the overall NUE phenotype.

In wheat, QTL mapping has identified several loci related to nitrogen uptake, metabolism, and grain yield under low nitrogen conditions, providing critical targets for breeding programs aimed at enhancing NUE. For example, Sandhu *et al.* (2021a) performed a meta-QTL analysis in rice, uncovering significant QTLs associated with NUE and exploring cross-genome relationships among cereal crops. Their analysis revealed essential genomic regions controlling NUE and highlighted phylogenetic connections among these regions across different cereal species. This integrated approach enhances understanding of the genetic basis of NUE and informs breeding programs focused on improving NUE in various cereal crops.

Numerous studies have identified quantitative trait loci (QTL) associated with NUE across various crops. For instance, Jewel *et al.* (2019a) focused on rice, uncovering key QTLs linked to nitrogen uptake and content in grains and leaves. In barley, Chen *et al.* (2024) mapped QTLs revealing associations with grain yield and nitrogen concentration. Zhang *et al.* (2010) identified significant QTLs in maize that influence ear length diameter under varying nitrogen levels. Similarly, Wang *et al.* (2023) utilized genome-wide association studies to find QTLs related to nitrogen fixation and biomass in soybean, highlighting genetic regions linked to nitrogen assimilation. Collectively, these studies demonstrate the potential of QTL mapping to improve NUE in crops, contributing to more sustainable agricultural practices. A summary of information on the QTLs associated with nitrogen use efficiency identified in maize, wheat and rice is presented in Table 7.1.

7.5.3.2 Marker-Assisted Selection (MAS)

Marker-assisted selection (MAS) is an effective breeding strategy that utilizes molecular markers associated with genes linked to desirable traits, such as NUE. By identifying key genetic markers through genome-wide association studies (GWAS) or quantitative trait loci (QTLs) mapping, breeders can screen large plant populations for these markers, enabling the selection of plants with improved NUE traits while minimizing reliance on phenotypic selection, which can be affected by environmental conditions.

MAS has been successfully implemented in various crops to enhance nitrogen uptake and utilization. In rice, for instance, the marker *NRT1.1B* is associated with improved nitrogen uptake efficiency, and MAS has facilitated its introduction into high-yielding rice varieties (Hu *et al.*, 2015). In maize, the marker for the *ZmNRT2.1* gene has been utilized to enhance nitrogen uptake efficiency. Similarly, in wheat, the *NUE1* QTL has been targeted for MAS to improve nitrogen utilization (Pandurangan *et al.*, 2021).

7.5.3.3 Genome-wide Association Studies (GWAS)

GWAS are essential for identifying genes linked to complex traits such as NUE. This method scans the genomes of plant populations to find genetic variations, such as single nucleotide polymorphisms (SNPs), that correlate with differences in nitrogen efficiency. By analyzing the relationship between genetic markers and NUE traits, researchers can locate genomic regions that influence nitrogen uptake and utilization.

GWAS has successfully identified key genomic regions associated with nitrogen efficiency in major crops such as rice, maize, and wheat. For instance, in rice, specific loci regulating nitrogen metabolism and root architecture have been pinpointed, which are vital for nitrogen uptake efficiency (NUpE) (Li *et al.*, 2018a). These insights provide valuable targets for breeding programs aimed at enhancing NUE in staple crops. Moreover, GWAS has facilitated the identification of QTLs related to NUE in crops, aiding the development of molecular markers for breeding nitrogen-efficient varieties. For example, Wang *et al.* (2022) identified QTLs in maize related to nitrogen uptake and utilization, while Yu *et al.* (2021) discovered significant loci in rice associated with nitrogen assimilation and grain yield. Shi *et al.* (2024) found QTLs in wheat linked to nitrogen uptake. In barley, Karunarathne *et al.* (2020) identified genetic loci relevant for NUE, and Rani *et al.* (2023) focused on soybean, revealing significant loci for nitrogen fixation. These studies provide valuable insights for breeding nitrogen-efficient crop varieties to promote sustainable agriculture.

Table 7.1. A summary of the QTLs associated with nitrogen use efficiency (NUE) and related traits identified in maize, wheat, and rice crops.

Sr No.	Crop	Number of QTLs	Chr	Trait	Phenotypic variance (%R^2)	Reference
1	Maize	1	1	Grain Yield under low nitrogen	15.2	Agrama *et al.*, 1999
2	Maize	5	1	Grain Yield under low nitrogen	0.5-8.0	Ribaut *et al.*, 2007
3	Maize	3	1,2, 9	NUE at abundant nitrogen	9-21.9	Coque *et al.*, 2006
4	Maize	3	1,9	NUE	8.4-12.3	Mandolino *et al.*, 2018
5	Maize	15	1,2,4,7,8	NUE under Low nitrogen	5.8-19.0	Li *et al.*, 2015
6	Maize	14	1,2,4,6	N remobilization	9.2-16.4	Gallais and Hirel, 2004
7	Maize	2	7,9	Grain Yield under low nitrogen	9.2-13.4	Agrama *et al.*, 1999
8	Maize	42	1,2,3,4,5,6,8,10	NUE and physiological traits under Normal and Low nitrogen	–	Hirel *et al.*, 2001
9	Maize	72	All	NUE and physiological traits under Normal and Low nitrogen	–	Coque *et al.*, 2008
10	Maize	17	1,2,3,5,6,7,8,10	Root system architecture under N stress	11-43.7	Liu *et al.*, 2012a, 2012b
11	Maize	30	2,3,4,5,6,8,9,10	Root system architecture under low and Normal nitrogen	17-45	Pestsova *et al.*, 2016
12	Wheat	34	1B,2B, 2D, 3B, 4A, 4B, 5A, 5B, 6A, 6B, 7B, 7D	nitrogen uptake	5.2-21.9	An *et al.*, 2006
13	Wheat	185	All	NUE related and physiological traits under normal nitrogen and nitrogen stress	–	Cormier *et al.*, 2016
14	Wheat	47	1A,1B, 2B2D, 3A,3B, 3D,4A,4B, 4D, 5B, 6A, 6D, 7A, 7D	NUE related and physiological traits	5.27-21.03	Deng *et al.*, 2017
15	Wheat	14	1D, 2B, 3B, 4B, 6B, 6D, 7A, 7D	Root system architecture under nitrogen stress	5.66-18.56	Fan *et al.*, 2018
16	Wheat	103	All	NUE related and physiological traits		Habash *et al.*, 2007
17	Wheat	57	1A, 2A, 2D, 3A, 3B, 3D,4B, 5A, 5B, 6A, 7A, 7B	Yield related traits in response to nitrogen	3.29-33	Laperche *et al.*, 2006, 2007
18	Wheat	5	4B, 4D, 6A, 2B, 5A	Root system architecture under nitrogen stress	6.08-21.6	Ren *et al.*, 2017
19	Wheat	66	1A, 1D, 2B, 2D, 4A, 4B, 5B, 5D, 6B, 7A, 7B	NUE related and physiological traits	8.2-47.9	Sun *et al.*, 2013
20	Wheat	18	2D, 3B, 4B, 4D,5A,5B,6A,6D,7A	NUE related traits	–	Xu *et al.*, 2014

Continued

Table 7.1. Continued.

Sr No.	Crop	Number of QTLs	Chr	Trait	Phenotypic variance (%R^2)	Reference
21	Wheat	5	1B, 2B, 2D, 4B	Yield related traits in response to nitrogen	5.8-26.8	Zheng *et al.*, 2010
22	Rice	57	All	Yield related traits in response to nitrogen	–	Tong *et al.*, 2011
23	Rice	15	1,2,6,7,11	NUE and grain Yield related traits	–	Wei *et al.*, 2011
24	Rice	13	2,3,4,5,8,10,11	NUE and nitrogen uptake	3.6-13.99	Zhou *et al.*, 2017
25	Rice	77	All	NUE and physiological traits	–	Nguyen *et al.*, 2016
26	Rice	14	1,2,3,4,6,7,9,10	NUE and nitrogen deficit tolerance	–	Wei *et al.*, 2012
27	Rice	16	1,3,5,7,8,9,11	NUE and physiological traits	15.7-50.30	Senthilvel *et al.*, 2008
28	Rice	49	1,2,3,4,5,6,8,9,10	NUE related traits	20.25-34.68	Jewel *et al.*, 2019a
29	Rice	7	1,2,3,8	NUE at seedling stage	9.07-14.45	Feng *et al.*, 2010
30	Rice	13	2,3,4,5,8,9,10,12	NUE at different nitrogen levels	–	Dai *et al.*, 2015

Sandhu *et al.* (2021b) identified genome-wide marker-trait associations for plant growth, yield and nitrogen utilization efficiency in synthetic wheat introgression libraries under varying nitrogen levels. Their findings highlight QTLs important for breeding nitrogen-efficient wheat varieties. In a subsequent study (Sandhu *et al.*, 2023), they leveraged natural genetic variations and identified novel genetic regions linked to enhanced NUE by incorporating genetic diversity from wild wheat relatives. This research offers valuable insights for developing wheat varieties that utilize nitrogen more efficiently, thereby reducing reliance on synthetic fertilizers and promoting sustainable agricultural practices.

7.5.3.4 Genes Influencing NUE

Nitrogen transporters: Genes encoding nitrate (NRT) and ammonium transporters (AMT) are crucial for nitrogen uptake. For instance, NRT1.1 not only facilitates nitrate uptake but also acts as a sensor, regulating absorption based on nitrogen availability (Krouk *et al.*, 2010).

Nitrogen metabolism enzymes: Enzymes such as nitrate reductase (NR), nitrite reductase (NiR), and glutamine synthetase (GS) play vital roles in converting absorbed nitrogen into usable forms. NR and NiR reduce nitrate to ammonium, while GS incorporates ammonium into amino acids (Stitt *et al.*, 2010).

Transcription factors: Proteins such as the NLP (NIN-Like Protein) family regulate the expression of nitrogen-related genes, playing a significant role in nitrogen sensing and responsiveness. Modifying these transcription factors can enhance nitrogen uptake and utilization (Maurya *et al.*, 2020).

Nitrogen remobilization genes: Genes involved in nitrogen remobilization from older tissues to developing organs are critical for improving nitrogen utilization efficiency (NUtE). For example, genes related to senescence-associated proteolysis and amino acid transport affect nitrogen recycling within the plant (Saidi *et al.*, 2021).

Improving nitrogen use efficiency (NUE) in staple crops such as rice, maize, and wheat is critical for sustainable agriculture and food production. In rice (*Oryza sativa*), key genes such as *NRT1.1B* and *OsGS1;1* have been identified. *NRT1.1B* encodes a nitrate transporter, with variations enhancing NUE and yield under low nitrogen conditions (Hu *et al.*, 2015). The overexpression of *OsGS1;1*, a cytosolic glutamine synthetase, increases growth by improving nitrogen assimilation into amino acids and proteins (Fuentes *et al.*, 2001). Root development in rice regulated by nitrogen accumulation involves the *OsMADS* 25 transcription factor (Yu *et al.*, 2015). In maize (*Zea mays*), the gene *ZmNLP5* regulates nitrogen responsiveness and has been linked to enhanced NUE through improved nitrogen assimilation and remobilization (Ge *et al.*, 2020). Additionally, *ZmAMT1*, an ammonium transporter gene, plays a crucial role in nitrogen uptake, with modifications increasing efficiency under low nitrogen conditions (Hao *et al.*, 2020). In wheat (*Triticum aestivum*), the transcription factor *TaNAC2-5A* is associated with nitrogen remobilization, and its overexpression contributes to higher NUE and grain yield (He *et al.*, 2015). Over-expression of *TaGS2-2Abpro* and *TaGS2-2Ab* increases nitrogen uptake and remobilization in wheat (Hu *et al.*, 2018). Zhou *et al.* (2017) using comparative transcriptomic and metabolomic analysis, found that OsDREB1C enhances photosynthesis and NUE in maize and rice (Wei *et al.*, 2022). Gao *et al.* demonstrated that *CsNRT2.1* functions as a high-affinity nitrate transporter based on its expression profile analysis (Li *et al.*, 2018b). Ming *et al.* examined the GS1 protein's expression and subcellular localization under varying nitrogen conditions, confirming its role in improving low-nitrogen tolerance in cucumber (Ming *et al.*, 2021). Sandhu *et al.* (2021c) review strategies to enhance NUE in cereal crops, focusing on optimizing nitrogen metabolism, utilizing genetic diversity and employing molecular markers and genetic engineering. They advocate integrating these approaches to reduce fertilizer reliance, minimize environmental impacts and promote sustainable agriculture.

A detailed description of network genes associated with NUE in rice and wheat crops is presented in Sandhu *et al.* (2021c).

7.5.4 Gene Editing and CRISPR-Cas9

The emergence of gene-editing technologies such as CRISPR-Cas9 has significantly transformed plant breeding, particularly in enhancing NUE. CRISPR-Cas9 enables precise modifications of genes related to nitrogen uptake, metabolism and remobilization, creating crop varieties with improved NUE. For instance, CRISPR has been employed to modify nitrogen transporters such as NRT and AMT in rice, leading to increased nitrogen uptake and higher yields in low nitrogen environments (Hu *et al.*, 2015). Additionally, gene editing can boost the activity of transcription factors that regulate nitrogen-responsive genes, further enhancing NUE (Xin *et al.*, 2021). The advantages of gene editing over traditional breeding include faster development and targeted modifications without introducing foreign DNA, making CRISPR-Cas9 a powerful tool for rapidly addressing challenges posed by climate change and optimizing NUE in crops.

In addition to rice, gene-editing technologies such as CRISPR-Cas9 have been successfully applied to other staple crops to enhance NUE. In barley, CRISPR-Cas9 has been utilized to enhance NUE by mutating barley *abnormal cytokinin response1 repressor 1* (*HvARE1*) gene (Karunarathne *et al.*, 2022). In wheat, gene editing has targeted *TaNAC* transcription factors associated with nitrogen remobilization, resulting in increased grain yield and enhanced NUE under low nitrogen conditions (Zhang *et al.*, 2016). These advancements highlight the potential of CRISPR-Cas9 in optimizing nitrogen efficiency across various cereal crops.

7.5.5 Multi-omics Approaches

Multi-omics strategies integrate genomics, transcriptomics, proteomics, and metabolomics to provide a comprehensive understanding of NUE:

Systems biology and network modeling: Analyzing multi-omics data can identify bottlenecks in nitrogen use and optimize pathways to enhance NUE.

Environment-specific omics: Research will focus on how changing climate conditions affect NUE, helping develop crops adaptable to diverse environments.

7.6 Challenges in Optimizing NUE and Reducing Carbon Footprint

While genomic insights have greatly advanced our understanding of NUE, several challenges remain. NUE is a complex trait influenced by both genetic and environmental factors, and achieving consistent improvements across different environments can be difficult. Additionally, NUE is often linked to other agronomic traits, such as drought tolerance or disease resistance, and improving NUE without negatively impacting these traits requires careful breeding strategies.

7.6.1 Complexity of Nitrogen Metabolism

The intricate nature of nitrogen metabolism complicates NUE trait manipulation. Increasing nitrogen uptake efficiency without corresponding utilization efficiency can lead to wasteful nitrogen consumption, necessitating a nuanced understanding of plant physiology.

7.6.2 Environmental Variability and NUE

Environmental factors significantly influence nitrogen availability and crop NUE. Variability in soil type, climate, and agronomic practices complicates the development of universally efficient crops, requiring extensive field trials to account for genotype-environment interactions.

7.6.3 Trade-offs with Other Traits

Optimizing NUE may lead to trade-offs with yield and other agronomic traits, such as pest resistance. Breeding strategies must carefully balance improving NUE while maintaining or enhancing overall productivity.

7.6.4 Imbalance in Nitrogen Application

Overuse of nitrogen fertilizers leads to low NUE, with much of the nitrogen lost through leaching, volatilization, and runoff. Under-application results in nutrient deficiency, limiting crop growth and yield potential.

7.6.5 Soil Variability

Soil variability, for example differences in organic composition, pH and microbial diversity influences nitrogen mineralization and availability. Microbial activity plays a crucial role in nitrogen cycling. Certain bacteria and fungi improve nitrogen availability, but their abundance and efficiency vary across soil type.

7.6.6 Adoption Barriers for New Technologies

Barriers to adopting NUE-enhancing technologies include:

Cost and accessibility: High costs can restrict access to advanced technologies, particularly for smallholder farmers. Making these innovations affordable is crucial.

Knowledge and training: Successful technology implementation requires farmer education and training programs to ensure effective use of precision agriculture tools and practices.

7.6.7 Policy and Regulatory Hurdles

Regulatory frameworks can hinder the adoption of NUE technologies, particularly gene-edited crops (Mbaya *et al.*, 2022). Policymakers must create supportive policies that incentivize sustainable practices, such as subsidies for precision agriculture technologies and tax breaks for biofertilizer adoption.

7.7 Future Directions

7.7.1 Reducing the Carbon Footprint Through Optimized NUE

Improving NUE in agriculture significantly enhances crop productivity while mitigating the environmental impacts linked to nitrogen fertilizer application. Optimizing NUE can lead to a substantial reduction in the carbon footprint of agricultural systems (Zhou *et al.*, 2023), which encompasses the total greenhouse gas (GHG) emissions associated with farming practices, expressed as CO_2 equivalents (CO_2e). Nitrogen fertilizers are a significant source of GHG emissions, primarily due to processes such as nitrous oxide (N_2O) emissions from soil and fertilizer application, which is a potent greenhouse gas with a much higher global warming potential than CO_2. By enhancing NUE through improved practices and technologies, farmers can decrease fertilizer inputs, thereby reducing nitrogen losses to the environment and lowering overall GHG emissions. This optimization is crucial for fostering sustainable and climate-resilient agricultural practices.

7.7.2 The Link Between Nitrogen Fertilizers and GHG Emissions

Nitrogen fertilizers contribute to both direct and indirect greenhouse gas emissions throughout their life cycle, including production, transportation, application, and biological processes in the soil. The energy-intensive production of synthetic nitrogen fertilizers, primarily through the Haber–Bosch process, significantly contributes to CO_2 emissions. The emission factors for synthetic N production range from 1–10 kg CO_2-eq./kg of nitrogen depending on type of fertilizer and source of hydrogen used by the steam reforming process (Walling and Vaneeckhaute, 2020). Once applied, a portion of nitrogen is lost via nitrification and denitrification, resulting in N_2O emissions, a potent greenhouse gas with a global warming potential around 300 times greater than CO_2 over a century (IPCC, 2013). Furthermore, excess nitrogen not taken up by plants can leach into water bodies, where it may convert into N_2O, leading to indirect emissions. Enhancing nitrogen use efficiency (NUE) helps minimize these losses, reducing both N_2O emissions and the overall carbon footprint of agriculture by decreasing the reliance on nitrogen fertilizers.

Optimizing NUE addresses two key issues in reducing the carbon footprint of agriculture: the overuse of nitrogen fertilizers and the inefficiency of nitrogen uptake and utilization by crops. The direct impact of improved NUE on GHG emissions is twofold.

7.7.3 Case Studies: Impact of Optimized NUE on Carbon Footprint Reduction

7.7.3.1 Wheat Production in Europe

In Europe, optimizing NUE through genetic improvements, precision fertilization, and sustainable soil management has reduced N_2O emissions by up to 50%, significantly lowering the carbon footprint of wheat farming (Berry *et al.*, 2008).

7.7.3.2 Maize in the USA

In the US Corn Belt, precision agriculture and enhanced efficiency fertilizers have allowed maize farmers to reduce nitrogen application rates while maintaining yields, leading to decreased CO_2 emissions from fertilizer production and reduced N_2O emissions from soils (Cook *et al.*, 2015; Graham *et al.*, 2018).

7.7.3.3 Rice Systems in Asia

In several Asian countries, adopting high-NUE rice varieties (Jewel *et al.*, 2019b) and precision fertilizer management has lowered nitrogen fertilizer use, resulting in reduced N_2O and methane emissions, enhancing the productivity of rice paddies.

7.7.4 Reduced Nitrogen Fertilizer Application

Improving NUE significantly reduces the need for nitrogen fertilizers, enabling higher crop yields with less input. Enhanced fertilizer use efficiency allows plants to absorb more nitrogen, which decreases the energy-intensive production of fertilizers, subsequently lowering CO_2 emissions (Lam *et al.*, 2022). Moreover, reduced fertilizer application leads to diminished nitrogen losses through volatilization, leaching and runoff, resulting in lower N_2O emissions, a potent greenhouse gas. Thus, optimizing NUE not only boosts productivity but also mitigates environmental impacts associated with nitrogen fertilizer use.

7.7.5 Improved Nitrogen Uptake and Utilization by Plants

Enhancing nitrogen uptake efficiency (NUpE) and nitrogen utilization efficiency (NUtE) is crucial for minimizing nitrogen losses in agriculture, which in turn reduces greenhouse gas emissions. When crops effectively absorb nitrogen, less nitrogen remains in the soil, mitigating losses through volatilization and leaching. Key strategies for improving NUpE include breeding for efficient nitrogen transporters, such as *NRT1* and *AMT* genes, which enhance nitrogen capture by plant roots. Additionally, optimizing nitrogen-sensing mechanisms in plants can improve NUtE by maximizing nitrogen metabolism and remobilization. These strategies collectively ensure that nitrogen is fully utilized, reducing environmental emissions associated with excess nitrogen.

7.7.6 Co-Benefits of Optimizing NUE for Climate Change Mitigation

Optimizing nitrogen use efficiency (NUE) not only helps reduce the carbon footprint of agriculture but also offers several co-benefits that support climate change mitigation and adaptation:

> Improved soil health: Efficient nitrogen use decreases the risk of soil degradation from excessive fertilizer application. Healthier soils can sequester more carbon, aiding long-term carbon storage and enhancing farming systems' resilience to climate change.
>
> Water quality protection: By minimizing nitrogen losses through leaching and runoff, optimized NUE protects water bodies from nitrate pollution and eutrophication. This reduces the release of greenhouse gases such as methane and N_2O from aquatic systems.
>
> Increased yield stability: Crops with optimized NUE tend to better tolerate fluctuating nitrogen availability, increasing their resilience to climate change impacts such as altered precipitation patterns and temperature extremes.

7.8 Conclusions

Optimizing nitrogen use efficiency is crucial for reducing the carbon footprint of agriculture in the face of climate change. Advances in genomic technologies, multi-omics approaches, precision agriculture and sustainable fertilizer innovations offer promising pathways to enhance NUE while maintaining crop productivity. However, challenges such as the complexity of nitrogen metabolism, environmental variability, trade-offs with other agronomic traits and barriers to technology adoption must be addressed to achieve meaningful progress. By fostering collaboration between researchers, farmers, and policymakers, we can develop resilient agricultural systems that not only optimize nitrogen use but also promote sustainability and environmental health. The future of agriculture hinges on our ability to integrate scientific advancements with practical strategies to create a more efficient and climate-smart food production system.

References

Agrama, H.A.S., Zakaria, A.G., Said, F.B. and Tuinstra, M. (1999) Identification of quantitative trait loci for nitrogen use efficiency in maize. *Molecular Breeding* 5, 187–195. https://doi.org/10.1023/A:1009669507144.

An, D., Su, J., Liu, Q., Zhu, Y., Tong, Y. *et al.* (2006) Mapping QTLs for nitrogen uptake in relation to the early growth of wheat (*Triticum aestivum* L.). *Plant and Soil* 284, 73–84. https://doi.org/10.1007/s11104-006-0030-3.

Anas, M., Liao, F., Verma, K.K., Sarwar, M.A., Mahmood, A. *et al.* (2020) Fate of nitrogen in agriculture and environment: Agronomic, eco-physiological and molecular approaches to improve nitrogen use efficiency. *Biological Research* 53, 1–20. https://doi.org/10.1186/s40659-020-00312-4.

Asadu, C.O., Ezema, C.A., Ekwueme, B.N., Onu, C.E., Onoh, I.M. *et al.* (2024) Enhanced efficiency fertilizers: Overview of production methods, materials used, nutrients release mechanisms, benefits and considerations. *Environmental Pollution and Management* 1, 32–48. https://doi.org/10.1016/j.epm.2024.07.002.

Begum, N., Qin, C., Ahanger, M.A., Raza, S., Khan, M.I. *et al.* (2019) Role of arbuscular mycorrhizal fungi in plant growth regulation: Implications in abiotic stress tolerance. *Frontiers in Plant Science* 10, 1068. https://doi.org/10.3389/fpls.2019.01068.

Berry, P.M., Kindred, D.R. and Paveley, N.D. (2008) Quantifying the effects of fungicides and disease resistance on greenhouse gas emissions associated with wheat production. *Plant Pathology* 57, 1000–1008. https://doi.org/10.1111/j.1365-3059.2008.01899.x.

Blandino, M., Vaccino, P. and Reyneri, A. (2015) Late-season nitrogen increases improver common and durum wheat quality. *Agronomy Journal* 107(2), 680–690. https://doi.org/10.2134/agronj14.0405.

Bloom, A.J., Meyerhoff, P.A., Taylor, A.R. and Rost, T.L. (2002) Root development and absorption of ammonium and nitrate from the rhizosphere. *Journal of Plant Growth Regulation* 21, 416–431. https://doi.org/10.1007/s00344-003-0009-8.

Bohra, A., Chand Jha, U., Godwin, I.D. and Kumar Varshney, R. (2020) Genomic interventions for sustainable agriculture. *Plant Biotechnology Journal* 18(12), 2388–2405. https://doi.org/10.1111/pbi.13472.

Botez, F. and Postolache, C. (2013) Nitrogen deposition impact on terrestrial ecosystems. *Romanian Biotechnological Letters* 18(6), 8723–8742.

Chen, B., Hou, Y., Huo, Y., Zeng, Z., Hu, D. *et al.* (2024) QTL mapping of yield, agronomic, and nitrogen-related traits in barley (*Hordeum vulgare* L.) under low nitrogen and normal nitrogen treatments. *Plants* 13(15), 2137. https://doi.org/10.3390/plants13152137.

Cook, R., Nail, A., Vigardt, A., Trlica, A., Hagarty, B., Williams, T. and Wolt, J. (2015) Meta-Analysis of Enhanced Efficiency Fertilizers in Corn Systems in the Midwest. International Plant Nutrition Institute Report, Canadian Center of Science and Education, Canada. ISSN: 1916-9752, 11(4), 1–580. http://research.ipni.net/project/IPNI-2014-USA-4RM06.

Coque, M., Bertin, P., Hirel, B. and Gallais, A. (2006) Genetic variation and QTLs for 15N natural abundance in a set of maize recombinant inbred lines. *Field Crops Research* 97(2–3), 310–321. https://doi.org/10.1016/j.fcr.2005.11.002.

Coque, M., Martin, A., Veyrieras, J.B., Hirel, B. and Gallais, A. (2008) Genetic variation for N-remobilization and postsilking N-uptake in a set of maize recombinant inbred lines. 3. QTL detection and coincidences. *Theoretical and Applied Genetics* 117(5), 729–747. https://doi.org/10.1007/s00122-008-0815-2.

Cormier, F., Foulkes, J., Hirel, B., Gouache, D., Moënne-Loccoz, Y. and Gouis, J. Le (2016) Breeding for increased nitrogen-use efficiency: A review for wheat (T. Aestivum L.). *Plant Breeding* 135(3), 255–278. https://doi.org/10.1111/pbr.12371.

Cossani, C.M. and Sadras, V.O. (2021) Nitrogen and water supply modulate the effect of elevated temperature on wheat yield. *European Journal of Agronomy* 124, 126227. https://doi.org/10.1016/j.eja.2020.126227.

Dabney, S.M., Delgado, J.A., Meisinger, J.J., Schomberg, H.H., Liebig, M.A. *et al.* (2010) Using cover crops and cropping systems for nitrogen management. *Advances in Nitrogen Management for Water Quality* 66, 231–282.

Dai, G.J., Cheng, S.H., Hua, Z.T., Zhang, M.L., Jiang, H.B. *et al.* (2015) Mapping quantitative trait loci for nitrogen uptake and utilization efficiency in rice (*Oryza sativa* L.) at different nitrogen fertilizer levels. *Genetics and Molecular Research* 14(3), 10404–10414. http://dx.doi.org/10.4238/2015.

Deng, Z., Cui, Y., Han, Q., Fang, W., Li, J. and Tian, J. (2017) Discovery of consistent QTLs of wheat spike-related traits under nitrogen treatment at different development stages. *Frontiers in Plant Science* 8, 2120. https://doi.org/10.3389/fpls.2017.02120.

Ding, J., Li, F., Le, T., Xu, D., Zhu, M. *et al.* (2021a) Tillage and seeding strategies for wheat optimizing production in harvested rice fields with high soil moisture. *Scientific Reports* 11(1), 119. https://doi.org/10.1038/s41598-020-80256-7.

Ding, W., Xu, X., Zhang, J., Huang, S., He, P. and Zhou, W. (2021b) Nitrogen balance acts an indicator for estimating thresholds of nitrogen input in rice paddies of China. *Environmental Pollution* 290, 118091. https://doi.org/10.1016/j.envpol.2021.118091.

Fageria, N.K. and Baligar, V.C. (2005) Enhancing nitrogen use efficiency in crop plants. *Advances in Agronomy* 88, 97–185. https://doi.org/10.1016/S0065-2113(05)88004-6.

Fageria, N.K. and Press, C. (2009) The use of nutrients in crop plants. *Cereal Research Communications* 37, 149–150. https://doi.org/10.1556/crc.37.2009.1.18.

Fageria, N.K., Baligar, V.C. and Li, Y.C. (2014) Nutrient uptake and use efficiency by tropical legume cover crops at varying pH of an oxisol. *Journal of Plant Nutrition* 37(2), 294–311. https://doi.org/10.1080/01904167.2013.851695.

Fan, X., Zhang, W., Zhang, N., Chen, M., Zheng, S. *et al.* (2018) Identification of QTL regions for seedling root traits and their effect on nitrogen use efficiency in wheat (*Triticum aestivum* L.). *Theoretical and Applied Genetics* 131, 2677–2698. https://doi.org/10.1007/s00122-018-3183-6.

FAO (2024) Inorganic fertilizers – 2002–2022. FAOSTAT Analytical Briefs, No. 90. Rome. https://doi.org/10.4060/cd1485en.

Feng, Y., Cao, L.Y., Wu, W.M., Shen, X.H., Zhan, X.D. *et al.* (2010) Mapping QTLs for nitrogen-deficiency tolerance at seedling stage in rice (*Oryza sativa* L.). *Plant Breeding* 129(6), 652–656. https://doi.org/10.1111/j.1439-0523.2009.01728.x.

Fortunato, S., Nigro, D., Lasorella, C., Marcotuli, I., Gadaleta, A. and Pinto, M.C. de (2023) The role of glutamine synthetase (GS) and glutamate synthase (GOGAT) in the improvement of nitrogen use efficiency in cereals. *Biomolecules* 13(12), 1771. https://doi.org/10.3390/biom13121771.

Fuentes, S.I., Allen, D.J., Ortiz-Lopez, A. and Hernández, G. (2001) Over-expression of cytosolic glutamine synthetase increases photosynthesis and growth at low nitrogen concentrations. *Journal of Experimental Botany* 52(358), 1071–1081. DOI: 10.1093/jexbot/52.358.1071.

Gallais, A. and Hirel, B. (2004) An approach to the genetics of nitrogen use efficiency in maize. *Journal of Experimental Botany* 55(396), 295–306. https://doi.org/10.1093/jxb/erh006.

Ge, M., Wang, Y., Liu, Y., Jiang, L.U., He, B. *et al.* (2020) The NIN-like protein 5 (ZmNLP5) transcription factor is involved in modulating the nitrogen response in maize. *The Plant Journal* 102(2), 353–368. https://doi.org/10.1111/tpj.14628.

Giller, K.E., Chalk, P., Dobermann, A., Hammond, L., Heffer, P. *et al.* (2004) Emerging technologies to increase the efficiency of use of fertilizer nitrogen. *Agriculture and the Nitrogen Cycle: Assessing the Impacts of Fertilizer Use on Food Production and the Environment* 65, 35–51.

Graham, R.F., Greer, K.D., Villamil, M.B., Nafziger, E.D. and Pittelkow, C.M. (2018) Enhanced-efficiency fertilizer impacts on yield-scaled nitrous oxide emissions in maize. *Soil Science Society of America Journal* 82(6), 1469–1481. https://doi.org/10.2136/sssaj2018.05.0196.

Habash, D.Z., Bernard, S., Schondelmaier, J., Weyen, J. and Quarrie, S.A. (2007) The genetics of nitrogen use in hexaploid wheat: N utilisation, development and yield. *Theoretical and Applied Genetics* 114, 403–419. https://doi.org/10.1007/s00122-006-0429-5.

Hao, D.L., Zhou, J.Y., Yang, S.Y., Huang, Y.N. and Su, Y.H. (2020) Functional and regulatory characterization of three AMTs in maize roots. *Frontiers in Plant Science* 11, 884. https://doi.org/10.3389/fpls.2020.00884.

He, X., Qu, B., Li, W., Zhao, X., Teng, W. *et al.* (2015) The nitrate-inducible NAC transcription factor TaNAC2-5A controls nitrate response and increases wheat yield. *Plant Physiology* 169(3), 1991–2005. https://doi.org/10.1104/pp.15.00568.

Hirel, B., Bertin, P., Quillere, I., Bourdoncle, W., Attagnant, C. *et al.* (2001) Towards a better understanding of the genetic and physiological basis for nitrogen use efficiency in maize. *Plant Physiology* 125(3), 1258–1270. https://doi.org/10.1104/pp.125.3.1258.

Hirel, B., Gouis, J. Le, Ney, B. and Gallais, A. (2007) The challenge of improving nitrogen use efficiency in crop plants: Towards a more central role for genetic variability and quantitative genetics within integrated approaches. *Journal of Experimental Botany* 58(9), 2369–2387. https://doi.org/10.1093/jxb/erm097.

Hirel, B., Tétu, T., Lea, P.J. and Dubois, F. (2011) Improving nitrogen use efficiency in crops for sustainable agriculture. *Sustainability* 3(9), 1452–1485. https://doi.org/10.3390/su3091452.

Hou, M., Yu, M., Li, Z., Ai, Z. and Chen, J. (2021) Molecular regulatory networks for improving nitrogen use efficiency in rice. *International Journal of Molecular Sciences* 22(16), 9040. https://doi.org/10.3390/ijms22169040.

Hu, B., Wang, W., Ou, S., Tang, J., Li, H. *et al.* (2015) Variation in NRT1. 1B contributes to nitrate-use divergence between rice subspecies. *Nature Genetics* 47(7), 834–838. https://doi.org/10.1038/ng.3337.

Hu, M., Zhao, X., Liu, Q., Hong, X., Zhang, W. *et al.* (2018) Transgenic expression of plastidic glutamine synthetase increases nitrogen uptake and yield in wheat. *Plant Biotechnology Journal* 16(11), 1858–1867. https://doi.org/10.1111/pbi.12921.

IPCC (2013) Climate change 2013: The physical science basis. In: *Contribution of Working Group I to the Fifth Assessment Report of the Intergovernmental Panel on Climate Change*. Cambridge University Press. Cambridge, United Kingdom and New York, NY, USA, 1535 pp.

Jewel, Z.A., Ali, J., Mahender, A., Hernandez, J., Pang, Y. and Li, Z. (2019a) Identification of quantitative trait loci associated with nutrient use efficiency traits, using SNP markers in an early backcross population of rice (*Oryza sativa* L.). *International Journal of Molecular Sciences* 20(4), 900. https://doi.org/10.3390/ijms20040900.

Jewel, Z.A., Ali, J., Pang, Y., Mahender, A., Acero, B. *et al.* (2019b) Developing green super rice varieties with high nutrient use efficiency by phenotypic selection under varied nutrient conditions. *The Crop Journal* 7(3), 368–377. https://doi.org/10.1016/j.cj.2019.01.002.

Karunarathne, S.D., Han, Y., Zhang, X.Q., Zhou, G., Hill, C.B. *et al.* (2020) Genome-wide association study and identification of candidate genes for nitrogen use efficiency in barley (*Hordeum vulgare* L.). *Frontiers in Plant Science* 11, 571912. https://doi.org/10.3389/fpls.2020.571912.

Karunarathne, S.D., Han, Y., Zhang, X.Q. and Li, C. (2022) CRISPR/Cas9 gene editing and natural variation analysis demonstrate the potential for HvARE1 in improvement of nitrogen use efficiency in barley. *Journal of Integrative Plant Biology* 64(3), 756–770. https://doi.org/10.1111/jipb.13214.

Krouk, G., Lacombe, B., Bielach, A., Perrine-Walker, F., Malinska, K. *et al.* (2010) Nitrate-regulated auxin transport by NRT1. 1 defines a mechanism for nutrient sensing in plants. *Developmental Cell* 18(6), 927–937. DOI: 10.1016/j.devcel.2010.05.008.

Ladha, J.K., Pathak, H., Krupnik, T.J., Six, J. and Kessel, C. van (2005) Efficiency of fertilizer nitrogen in cereal production: Retrospects and prospects. *Advances in Agronomy* 87, 85–156. https://doi.org/10.1016/S0065-2113(05)87003-8.

Lam, S.K., Wille, U., Hu, H.W., Caruso, F., Mumford, K. *et al.* (2022) Next-generation enhanced-efficiency fertilizers for sustained food security. *Nature Food* 3(8), 575–580. https://doi.org/10.1038/s43016-022-00542-7.

Laperche, A., Brancourt-Hulmel, M., Heumez, E., Gardet, O. and Gouis, J.L. (2006) Estimation of genetic parameters of a DH wheat population grown at different N stress levels characterized by probe genotypes. *Theoretical and Applied Genetics* 112(5), 797–807. https://doi.org/10.1007/s00122-005-0176-z.

Laperche, A., Brancourt-Hulmel, M., Heumez, E., Gardet, O., Hanocq, E., Devienne-Barret, F. and Gouis, J. Le (2007) Using genotypex nitrogen interaction variables to evaluate the QTL involved in wheat tolerance to nitrogen constraints. *Theoretical and Applied Genetics* 115(3), 399–415. https://doi.org/10.1007/s00122-007-0575-4.

Lassaletta, L., Billen, G., Grizzetti, B., Anglade, J. and Garnier, J. (2014) 50 year trends in nitrogen use efficiency of world cropping systems: The relationship between yield and nitrogen input to cropland. *Environmental Research Letters* 9(10), 105011. DOI: 10.1088/1748-9326/9/10/105011.

Li, P., Chen, F., Cai, H., Liu, J., Pan, Q. *et al.* (2015) A genetic relationship between nitrogen use efficiency and seedling root traits in maize as revealed by QTL analysis. *Journal of Experimental Botany* 66(11), 3175–3188. https://doi.org/10.1093/jxb/erv127.

Li, S., Tian, Y., Wu, K., Ye, Y., Yu, J. *et al.* (2018a) Modulating plant growth–metabolism coordination for sustainable agriculture. *Nature* 560(7720), 595–600. https://doi.org/10.1038/s41586-018-0415-5.

Li, Y., Li, J., Yan, Y., Liu, W., Zhang, W., Gao, L. and Tian, Y. (2018b) Knock-down of CsNRT2. 1, a cucumber nitrate transporter, reduces nitrate uptake, root length, and lateral root number at low external nitrate concentration. *Frontiers in Plant Science* 9, 722. https://doi.org/10.3389/fpls.2018.00722.

Li, X., Chen, D., Carrión, V.J., Revillini, D., Yin, S. *et al.* (2023) Acidification suppresses the natural capacity of soil microbiome to fight pathogenic Fusarium infections. *Nature Communications* 14(1), 5090. https://doi.org/10.1038/s41467-023-40810-z.

Liu, J., Jung, C., Xu, J., Wang, H., Deng, S. *et al.* (2012a) Genome-wide analysis uncovers regulation of long intergenic noncoding RNAs in Arabidopsis. *The Plant Cell* 24(11), 4333–4345. https://doi.org/10.1105/tpc.112.102855.

Liu, R., Zhang, H., Zhao, P., Zhang, Z., Liang, W., Tian, Z. and Zheng, Y. (2012b) Mining of candidate maize genes for nitrogen use efficiency by integrating gene expression and QTL data. *Plant Molecular Biology Reporter* 30, 297–308. https://doi.org/10.1007/s11105-011-0346-x.

Mandolino, C.I., D'andrea, K.E., Olmos, S.E., Otegui, M.E. and Eyhérabide, G.H. (2018) Maize nitrogen use efficiency: QTL mapping in a US Dent x Argentine-Caribbean flint RILs population, Maydica 63: 17. https://api.semanticscholar.org/CorpusID:89988023.

Masclaux-Daubresse, C., Reisdorf-Cren, M. and Orsel, M. (2008) Leaf nitrogen remobilisation for plant development and grain filling. *Plant Biology* 10, 23–36. https://doi.org/10.1111/j.1438-8677.2008.00097.x.

Masclaux-Daubresse, C., Daniel-Vedele, F., Dechorgnat, J., Chardon, F., Gaufichon, L. and Suzuki, A. (2010) Nitrogen uptake, assimilation and remobilization in plants: Challenges for sustainable and productive agriculture. *Annals of Botany* 105(7), 1141–1157. https://doi.org/10.1093/aob/mcq028.

Maurya, J., Bandyopadhyay, T. and Prasad, M. (2020) Transcriptional regulators of nitrate metabolism: Key players in improving nitrogen use in crops. *Journal of Biotechnology* 324, 121–133. https://doi.org/10.1016/j.jbiotec.2020.10.001.

Mbaya, H., Lillico, S., Kemp, S., Simm, G. and Raybould, A. (2022) Regulatory frameworks can facilitate or hinder the potential for genome editing to contribute to sustainable agricultural development. *Frontiers in Bioengineering and Biotechnology* 10, 959236. https://doi.org/10.3389/fbioe.2022.959236.

Ming, X.I.N., Qin, Z.W., Jing, Y.A.N.G., Zhou, X.Y. and Lei, W.A.N.G. (2021) Functional analysis of the nitrogen metabolism-related gene CsGS1 in cucumber. *Journal of Integrative Agriculture* 20(6), 1515–1524. https://doi.org/10.1016/S2095-3119(20)63305-6.

Moll, R.H., Kamprath, E.J. and Jackson, W.A. (1982) Analysis and interpretation of factors which contribute to efficiency of nitrogen utilization 1. *Agronomy Journal* 74(3), 562–564. https://doi.org/10.2134/agronj1982.00021962007400030037x.

Muratore, C., Espen, L. and Prinsi, B. (2021) Nitrogen uptake in plants: The plasma membrane root transport systems from a physiological and proteomic perspective. *Plants* 10(4), 681. https://doi.org/10.3390/plants10040681.

Nguyen, H.T.T., Dang, D.T., Pham, C. Van and Bertin, P. (2016) QTL mapping for nitrogen use efficiency and related physiological and agronomical traits during the vegetative phase in rice under hydroponics. *Euphytica* 212, 473–500. https://doi.org/10.1007/s10681-016-1778-z.

Pandurangan, S., Workman, C., Nilsen, K. and Kumar, S., 2021. Introduction to marker-assisted selection in wheat breeding. In *Accelerated Breeding of Cereal Crops*. Springer US, pp. 77–117. https://doi.org/10.1007/978-1-0716-1526-3_3.

Papadopoulos, G., Arduini, S., Uyar, H., Psiroukis, V., Kasimati, A. and Fountas, S. (2024) Economic and environmental benefits of digital agricultural technologies in crop production: A review. Smart Agricultural Technology, 8: 100441. https://doi.org/10.1016/j.atech.2024.100441.

Pestsova, E., Lichtblau, D., Wever, C., Presterl, T., Bolduan, T., Ouzunova, M. and Westhoff, P. (2016) QTL mapping of seedling root traits associated with nitrogen and water use efficiency in maize. *Euphytica* 209, 585–602. https://doi.org/10.1007/s10681-015-1625-7.

Petersen, S.O., Peixoto, L.E., Sørensen, H., Tariq, A., Brændholt, A. *et al.* (2023) Higher N2O emissions from organic compared to synthetic N fertilisers on sandy soils in a cool temperate climate. *Agriculture, Ecosystems & Environment* 358, 108718. https://doi.org/10.1016/j.agee.2023.108718.

Rabalais, N.N., Díaz, R.J., Levin, L.A., Turner, R.E., Gilbert, D. and Zhang, J. (2010) Dynamics and distribution of natural and human-caused coastal hypoxia. *Biogeosciences* 7, 585–619. DOI: 10.5194/bg-7-585-2010.

Rani, R., Raza, G., Ashfaq, H., Rizwan, M., Razzaq, M.K. *et al.* (2023) Genome-wide association study of soybean (*Glycine max* [L.] Merr.) germplasm for dissecting the quantitative trait nucleotides and candidate genes underlying yield-related traits. *Frontiers. Plant Science* 14, 1229495. https://doi.org/10.3389/fpls.2023.1229495.

Raun, W.R. and Johnson, G.V. (1999) Improving nitrogen use efficiency for cereal production. *Agronomy Journal* 91(3), 357–363. https://doi.org/10.2134/agronj1999.00021962009100030001x.

Ren, Y., Qian, Y., Xu, Y., Zou, C., Liu, D. *et al.* (2017) Characterization of QTLs for root traits of wheat grown under different nitrogen and phosphorus supply levels. *Frontiers in Plant Science* 8, 2096. https://doi.org/10.3389/fpls.2017.02096.

Ribaut, J.M., Fracheboud, Y., Monneveux, P., Banziger, M., Vargas, M. and Jiang, C. (2007) Quantitative trait loci for yield and correlated traits under high and low soil nitrogen conditions in tropical maize. *Molecular Breeding* 20, 15–29. https://doi.org/10.1007/s11032-006-9041-2.

Richardson, K., Steffen, W., Lucht, W., Bendtsen, J., Cornell, S.E. *et al.* (2023) Earth beyond six of nine planetary boundaries. *Science Advances* 9(37), p.eadh2458. DOI: 10.1126/sciadv.adh2458.

Saidi, A., Hajibarat, Z. and Ghaffari, M.R. (2021) The role of effective factors in cell senescence and material remobilization in cereals. *Genetic Engineering and Biosafety Journal* 10(1), 108–120. https://doi.org/10.3389/fpls.2025.1721421.

Sala, O.E., Stuart Chapin, F.I.I.I., Armesto, J.J., Berlow, E., Bloomfield, J. *et al.* (2000) Global biodiversity scenarios for the year 2100. *Science* 287(5459), 1770–1774. DOI: 10.1126/science.287.5459.1770.

Sandhu, N. *et al.* (2021a) Meta-QTL analysis in rice and cross-genome talk of the genomic regions controlling nitrogen use efficiency in cereal crops revealing phylogenetic relationship. *Frontiers in Genetics* 12, 807210. DOI: 10.3389/fgene.2021.807210.

Sandhu, N., Kaur, A., Sethi, M., Kaur, S., Singh, V. *et al.* (2021b) Genetic dissection uncovers genome wide marker-trait associations for plant growth, yield and yield related traits under varying nitrogen levels in nested synthetic wheat introgression libraries. *Frontiers in Plant Science* 12, 738710. DOI: 10.3389/fpls.2021.738710.

Sandhu, N., Sethi, M., Kumar, A., Dang, D., Singh, J. and Chhuneja, P. (2021c) Biochemical and genetic approaches improving nitrogen use efficiency in cereal crops: A review. *Frontiers in Plant Science* 12, 757. https://doi.org/10.3389/fpls.2021.657629.

Sapkota, T.B. and Takele, R. (2023) Improving nitrogen use efficiency and reducing nitrogen surplus through best fertilizer nitrogen management in cereal production: The case of India and China. *Advances in Agronomy* 178, 233–294. https://doi.org/10.1016/bs.agron.2022.11.006.

Senthilvel, S., Vinod, K.K., Malarvizhi, P. and Maheswaran, M. (2008) QTL and QTL× environment effects on agronomic and nitrogen acquisition traits in rice. *Journal of Integrative Plant Biology* 50(9), 1108–1117. https://doi.org/10.1111/j.1744-7909.2008.00713.x.

Shi, H., Wang, W., Gao, L., Wu, J., Hu, C. *et al.* (2024) Genome-wide association study of seedling nitrogen-use efficiency-associated traits in common wheat (*Triticum aestivum* L.). *The Crop Journal* 12(1), 222–231. https://doi.org/10.1016/j.cj.2023.10.014.

Stitt, M., Lunn, J. and Usadel, B. (2010) Arabidopsis and primary photosynthetic metabolism–more than the icing on the cake. *The Plant Journal* 61(6), 1067–1091. https://doi.org/10.1111/j.1365-313X.2010.04142.x.

Sun, J.J., Guo, Y., Zhang, G.Z., Gao, M.G., Zhang, G.H. *et al.* (2013) QTL mapping for seedling traits under different nitrogen forms in wheat. *Euphytica* 191, 317–331. https://doi.org/10.1007/s10681-012-0834-6.

Tian, H., Pan, N., Thompson, R.L., Canadell, J.G., Suntharalingam, P. *et al.* (2023) Global nitrous oxide budget 1980–2020. *Earth System Science Data Discussions* 2023, 1–98. https://doi.org/10.5194/essd-16-2543-2024.

Tong, H.H., Chen, L., Li, W.P., Mei, H.W., Xing, Y.Z. *et al.* (2011) Identification and characterization of quantitative trait loci for grain yield and its components under different nitrogen fertilization levels in rice (*Oryza sativa* L.). *Molecular Breeding* 28, 495–509. https://doi.org/10.1007/s11032-010-9499-9.

Walling, E. and Vaneeckhaute, C. (2020) Greenhouse gas emissions from inorganic and organic fertilizer production and use: A review of emission factors and their variability. *Journal of Environmental Management* 276, 111211p. https://doi.org/10.1016/j.jenvman.2020.111211.

Wang, J., Tu, X., Zhang, H., Cui, J., Ni, K. *et al.* (2020) Effects of ammonium-based nitrogen addition on soil nitrification and nitrogen gas emissions depend on fertilizer-induced changes in pH in a tea plantation soil. *Science of the Total Environment* 747, 141340. https://doi.org/10.1016/j.scitotenv.2020.141340.

Wang, Y., Zhu, T., Yang, J., Wang, H., Ji, W. *et al.* (2022) GWAS and transcriptome analysis reveal key genes affecting root growth under low nitrogen supply in maize. *Genes* 13(9), 1632. https://doi.org/10.3390/genes13091632.

Wang, X., Zhou, S., Wang, J., Lin, W., Yao, X. *et al.* (2023) Genome-wide association study for biomass accumulation traits in soybean. *Molecular Breeding* 43(5), 33. https://doi.org/10.1007/s11032-023-01380-6.

Ward, M.H., Jones, R.R., Brender, J.D., Kok, T.M. De, Weyer, P.J. *et al.* (2018) Drinking water nitrate and human health: An updated review. *International Journal of Environmental Research and Public Health* 15(7), 1557. https://doi.org/10.3390/ijerph15071557.

Wei, D., Cui, K., Pan, J., Ye, G., Xiang, J., Nie, L. and Huang, J. (2011) Genetic dissection of grain nitrogen use efficiency and grain yield and their relationship in rice. *Field Crops Research* 124(3), 340–346. https://doi.org/10.1016/j.fcr.2011.07.003.

Wei, D., Cui, K., Ye, G., Pan, J., Xiang, J., Huang, J. and Nie, L. (2012) QTL mapping for nitrogen-use efficiency and nitrogen-deficiency tolerance traits in rice. *Plant and Soil* 359, 281–295. https://doi.org/10.1007/s11104-012-1142-6.

Wei, S., Li, X., Lu, Z., Zhang, H., Ye, X. *et al.* (2022) A transcriptional regulator that boosts grain yields and shortens the growth duration of rice. *Science* 377(6604), p.eabi8455. DOI: 10.1126/science.abi8455.

Weih, M., Asplund, L. and Bergkvist, G. (2011) Assessment of nutrient use in annual and perennial crops: A functional concept for analyzing nitrogen use efficiency. *Plant and Soil* 339, 513–520. https://doi.org/10.1007/s11104-010-0599-4.

Xin, W., Wang, J., Li, J., Zhao, H., Liu, H. *et al.* (2021) Candidate gene analysis for nitrogen absorption and utilization in japonica rice at the seedling stage based on a genome-wide association study. *Frontiers in Plant Science* 12, 670861. https://doi.org/10.3389/fpls.2021.670861.

Xu, Y., Wang, R., Tong, Y., Zhao, H., Xie, Q. *et al.* (2014) Mapping QTLs for yield and nitrogen-related traits in wheat: Influence of nitrogen and phosphorus fertilization on QTL expression. *Theoretical and Applied Genetics* 127, 59–72.

Yu, C., Liu, Y., Zhang, A., Su, S., Yan, A., Huang, L., Ali, I., Liu, Y., Forde, B.G. and Gan, Y. (2015) MADS-box transcription factor OsMADS25 regulates root development through affection of nitrate accumulation in rice. *PLoS One* 10(8), p.e0135196. https://doi.org/10.1371/journal.pone.0135196.

Yu, J., Xuan, W., Tian, Y., Fan, L., Sun, J. *et al.* (2021) Enhanced OsNLP4-OsNiR cascade confers nitrogen use efficiency by promoting tiller number in rice. *Plant Biotechnology Journal* 19(1), 167–176.

Zhang, H., Zheng, Z., Liu, X., Li, Z., He, C. *et al.* (2010) QTL mapping for ear length and ear diameter under different nitrogen regimes in maize. *African Journal of Agricultural Research* 5(8), 626–630. http://www.academicjournals.org/AJAR.

Zhang, L., Garneau, M.G., Majumdar, R., Grant, J. and Tegeder, M. (2015) Improvement of pea biomass and seed productivity by simultaneous increase of phloem and embryo loading with amino acids. *The Plant Journal* 81(1), 134–146. https://doi.org/10.1111/tpj.12716.

Zhang, Y., Liang, Z., Zong, Y., Wang, Y., Liu, J. *et al.* (2016) Efficient and transgene-free genome editing in wheat through transient expression of CRISPR/Cas9 DNA or RNA. *Nature Communications* 7(1), 12617. https://doi.org/10.1038/ncomms12617.

Zhang, S., Zhang, G., Wang, D., Liu, Q. and Xu, M. (2021) Investigation into runoff nitrogen loss variations due to different crop residue retention modes and nitrogen fertilizer rates in rice-wheat cropping systems. *Agricultural Water Management* 247, 106729. https://doi.org/10.1016/j.agwat.2020.106729.

Zhang, H., Zhang, J. and Yang, J. (2023) Improving nitrogen use efficiency of rice crop through an optimized root system and agronomic practices. *Crop and Environment* 2(4), 192–201. https://doi.org/10.1016/j.crope.2023.10.001.

Zheng, B.S., Gouis, J. Le, Leflon, M., Rong, W.Y., Laperche, A. and Brancourt-Hulmel, M. (2010) Using probe genotypes to dissect QTL× environment interactions for grain yield components in winter wheat. *Theoretical and Applied Genetics* 121(8), 1501–1517. https://doi.org/10.1007/s00122-010-1406-6.

Zhou, Y., Tao, Y., Tang, D., Wang, J., Zhong, J. *et al.* (2017) Identification of QTL associated with nitrogen uptake and nitrogen use efficiency using high throughput genotyped CSSLs in rice (*Oryza sativa* L.). *Frontiers in Plant Science* 8, 1166. https://doi.org/10.3389/fpls.2017.01166.

Zhou, W., Long, W., Wang, H., Long, P., Xu, Y. *et al.* (2023) Reducing carbon footprints and increasing net ecosystem economic benefits through dense planting with less nitrogen in double-cropping rice systems. *Science of the Total Environment* 891, 164756. https://doi.org/10.1016/j.scitotenv.2023.164756.

8 Genebank Genomics for Adapting Crops to Climate Change

Hyeonah Shim[1], Martin Mascher[1,2], and Murukarthick Jayakodi[3,*]
[1]Leibniz Institute of Plant Genetics and Crop Plant Research (IPK) Gatersleben, Seeland, Germany; [2]German Centre for Integrative Biodiversity Research (iDiv) Halle-Jena-Leipzig, Leipzig, Germany; [3]Department of Soil and Crop Sciences, Texas A&M AgriLife Research, Dallas, Texas, USA

Abstract

Plant diversity is preserved in genebanks around the world, usually *ex situ*, creating a large reservoir of plant genetic resources. As we face global challenges such as climate change, plant productivity is threatened by various abiotic and biotic stresses. Genebank collections could afford solutions to global challenges such as climate change and biotic stresses, and enhance nutritional value in crops. Rapid advancements in sequencing technologies and remarkable improvements in generating high-quality reference genomes and cost-effective high-throughput genotyping platforms paved the way to convert the genebanks into digital form or so-called "bio-digital resource centers." These digital platforms further enable efficient conservation, characterization and utilization of plant genetic resources. This chapter discusses the genomics-driven systematic approaches for the exploitation of genetic resources of genebanks to mobilize valuable diversity and promising genotypes for modern breeding in the face of changing circumstances.

8.1 Introduction

Though about 7000 species have been cultivated or collected over time, only about 30 crops are responsible for 95% of human food energy needs, of which wheat, rice, and maize provide for more than half of global plant-derived energy intake. Domestication of major crop plants including cereals and grain legumes dates to about 10,000 years ago (Weiss *et al.*, 2006). Since then, cultivated plants have been intentionally and non-intentionally selected to meet the needs of humans and livestock. As selection intensifies in crops, we observe a greater loss of diversity.

Due to changes in climate patterns and influence of different biotic and abiotic factors to the agriculture sector, we face challenges that could threaten global food security. In the last century, plant genetic resources, including landraces, old varieties and wild relatives, have been consciously collected and preserved in genebanks to prevent the loss of biodiversity. It is now time to tap into the diversity held in the genebanks to cope with the changing environment. Genetic diversity stored in genebanks represents resources for developing crop varieties that can withstand relatively less favorable conditions.

Recent advancements in genomic tools and technologies have increased our capacity

*Corresponding author. Email: muru.jayakodi@ag.tamu.edu

 Genomic Solutions for Breeding Climate-resilient Crops (R.K. Varshney, R. Barmukh and R.J. Harper)
DOI: 10.1079/9781800627185.0008

to analyze the vast reservoirs of genetic resources. High-throughput sequencing, genotyping, phenotyping and bioinformatics analyses have enabled more precise identification of valuable traits in genebank collections. The genetic profiling of these traits offers clues for strategies that can help improve resilience, yield and nutritional quality. Genomic data with historical documents can facilitate the discovery of novel traits and genotypes that may be crucial for environmental adaptation (Figure 8.1).

In this chapter, we discuss the systematic approaches to search and exploit the genetic resources in genebanks for cataloging diversity, trait discovery and identifying promising donors for prebreeding. In this vein, first, we summarize the trait-specific promising genotypes identified in genebank collections. Then, we describe the potential path to characterize the genebank collection of a species.

8.2 Genebank, a Repository of Germplasm Collections

Crop plants are deposited in about 1750 genebanks around the world, having an accumulative number of 7.4 million germplasm accessions (Consultative Group on International Agriculture Research (CGIAR), https://www.cgiar.org/), and requests for deposition are increasing year on year. A huge number of accessions were collected

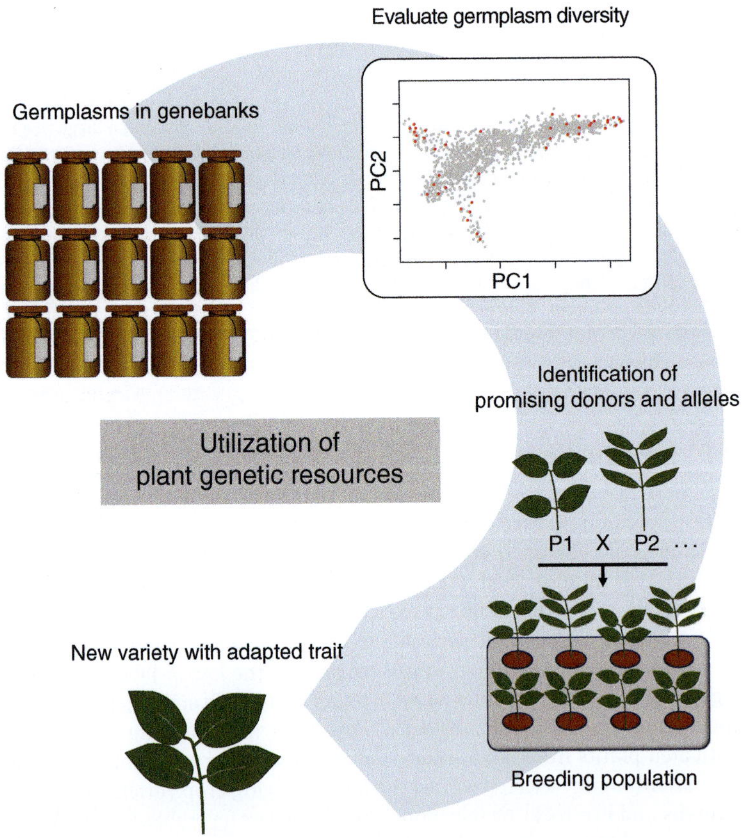

Figure 8.1. Schematic representation of the utilization of plant genetic resources stored in genebanks for crop improvement. The genetic and phenotypic diversity of germplasms is assessed through various methods, such as principal component analysis, allowing for a comprehensive evaluation of variation. This analysis facilitates the identification of promising donors harboring beneficial alleles, which are critical for addressing specific breeding targets. The selected genetic resources can be incorporated into breeding programs, where they are used to develop improved cultivars with enhanced adaptability and traits tailored to specific environmental or agronomic challenges. (Authors' own image.)

for some of the major crops including wheat (*Triticum aestivum*), rice (*Oryza sativa*), barley (*Hordeum vulgare*), maize (*Zea mays*) and soybean (*Glycine max*) (Wambugu *et al.*, 2018). These collections have been stored in major international genebanks including the International Maize and Wheat Improvement Center (CIMMYT), International Center for Agricultural Research in the Dry Areas (ICARDA), International Rice Research Institute (IRRI), International Center for Tropical Agriculture (CIAT), International Institute of Tropical Agriculture (IITA), International Crops Research Institute for the Semi-Arid Tropics (ICRISAT), under the Consortium of International Agricultural Research Centers (CGIAR) based on the International Treaty on Plant Genetic Resources for Food and Agriculture (ITPGRFA), and large national or regional genebanks such as the National Genebank in Norway, Federal *Ex situ* Genebank in the Leibniz Institute of Plant Genetics and Crop Plant Research (IPK), Germany, and The National Bureau of Plant Genetic Resources (NBPGR) in New Delhi, India.

Germplasm collections in these genebanks are most commonly *ex situ*, away from their natural habitat, in forms such as seeds, grains and/or parts of plants. These germplasms hold passport data containing basic information such as taxonomy, collection sites, accession number and/or phenotypic data, when available. Currently, the loss of biodiversity is a pronounced global challenge, diminishing the resilience of agricultural and food systems, menacing nutritional security, and putting the foundation of crop improvement at risk. Further, the current and future breeding seeks novel valuable alleles or genes, which are absent in the breeding populations, to confer resistance or tolerance to climate change, different biotic and abiotic stresses, and desirable nutritional profiles.

8.3 Genomic Characterization of Plant Genetic Resources

8.3.1 Advanced Genomic Tools and Sequencing Technologies

The initial step in utilizing plant genetic resources is to characterize the vast number of genetic resources stored in genebanks. Previously, gel-based methodologies such as simple sequence repeats (SSR) (Huang *et al.*, 2002; Khlestkina *et al.*, 2007), restriction fragment length polymorphism (RFLP) (Siedler *et al.*, 1994), and inter simple sequence repeats (ISSR) (Sofalian *et al.*, 2008) were utilized to evaluate the genetic diversity of germplasm collections. These methods have largely been superseded by more advanced genomic technologies. Next-generation sequencing (NGS) revolutionized the way in which population-scale genome characterization is conducted in major plant species. Especially, cost-effective NGS genotyping platforms such as DArT-seq (Diversity Array Technology) (Wenzl *et al.*, 2004) and genotyping-by-sequencing (GBS) (Elshire *et al.*, 2011), use restriction enzymes (REs) to reduce genome complexity. They are more practical for the simultaneous discovery of SNPs and short insertions or deletions (INDELs) and high-resolution genotyping of a large number of accessions, including those with extensive genome sizes, or even the ones that lack reference genomes.

Genotyping-by-sequencing, in particular, is being actively utilized in characterizing thousands of genebank accessions. It has facilitated the genotyping of diverse collections such as selected sets of the Swiss National Genebank wheat and spelt accessions (Muller *et al.*, 2018), and reference core sets of wheat accessions at the Wheat Genetics Resource Center (WGRC) (Yadav *et al.*, 2022). Extensive studies have categorized other wheat genebank accessions (Sehgal *et al.*, 2015; Kobayashi *et al.*, 2016; Pascual *et al.*, 2020; Lehnert *et al.*, 2022; Schulthess *et al.*, 2022), barley accessions (Milner *et al.*, 2019; Guo *et al.*, 2022) and pea collections (Gali *et al.*, 2019; Bari *et al.*, 2021), to name just a few. Therefore, the first major step of genomic characterization or genotyping of a species' entire collections can be done by a cost-effective genotyping platform such as GBS without ascertainment biases. Nevertheless, optimization of appropriate REs for GBS assay is crucial to circumvent potential biases in genome reduction and marker discovery.

8.3.2 High-quality Reference Genomes

High-quality reference genomes for many plant species are more readily available thanks to advancements in sequencing technologies and lower costs. These reference genomes can serve as backbones for comparing the genotypes and identifying variants. Previously, draft genome sequences were constructed based on short read sequencing technologies. This limited the quality of the genome by having too many contigs or scaffolds and missing information which made it difficult to order them into chromosome-scale.

Due to the robust advancement of sequencing technologies in the past decade or so in both accuracy and length, along with refined assembly pipelines, chromosome-scale reference genomes are available for not only most major crops such as wheat (Chapman *et al.*, 2015; Clavijo *et al.*, 2017; Zimin *et al.*, 2017; Appels *et al.*, 2018), barley (Mascher *et al.*, 2017), rice (Goff *et al.*, 2002; Yu *et al.*, 2002; Du *et al.*, 2017), soybean (Xie *et al.*, 2019; Yi *et al.*, 2022), but also for many underutilized crops such as cassava (*Manihot esculentum*) (Xu *et al.*, 2023) and foxtail millet (*Setaria italica*) (Zhang *et al.*, 2012). We are now at a phase in which genome-wide genotyping is available for entire genebank collections (Mascher *et al.*, 2019); coupled with high-quality reference genomes, efficient discovery of genome-wide variations within and among species and the utilization of genomic data of genebank germplasms is possible.

The passport record alone is inadequate to grasp the characteristics of each accession in the genebanks. The genomic information obtained by GBS can complement the original passport data along with available phenotypic data. Such a generation of molecular passport data for the accessions enable efficient genebank management and conservation by tracking potential duplicated accession and spotting collection gaps (Mascher *et al.*, 2019). Furthermore, the SNP matrices obtained from GBS would permit to perform genome-wide association studies (GWAS) and genomic selection in plant genetic resources.

8.4 Screen Genetic Resources for Agronomic, Climate-resilient and Quality Traits

8.4.1 Biotic Stress

Plants are sessile and thus always vulnerable to various environmental factors. Among these, biotic stresses from pests and microbial pathogens significantly impact plant survival and crop yields. Integrating genetic resistance into crop varieties serves as a sustainable alternative to the extensive use of pesticides and fungicides. Breeding for disease resistance is one of the major breeding goals for many crops. The identification of putative genes linked to desired traits is crucial for their incorporation into effective breeding programs. In some cases, crop wild relatives (CWRs) harbor unique traits beneficial for enhancing cultivated varieties. For instance, resistance to powdery mildew (PMR) caused by *Erysiphe pisi* was discovered in a CWR line of pea (*Pisum fulvum*) (Sharma and Yadav, 2003), and molecular markers linked to this gene have been developed to facilitate the identification of resistant individuals (Fondevilla *et al.*, 2008). Similarly, in chickling pea (*Lathyrus cicera* L.), germplasms from the Iberian Peninsula have been screened for various degrees of resistance to powdery mildew (Patto *et al.*, 2007) and rust fungi (Patto *et al.*, 2009). The extensive collections maintained in genebanks provide a valuable repository of such alleles, which can be utilized for targeted and efficient breeding programs.

Ongoing efforts include the development of a resistant gene atlas for major crops such as wheat, utilizing genetic resources to enhance disease resistance (Hafeez *et al.*, 2021; Kale *et al.*, 2022). In barley, loci for barley yellow mosaic virus (*BaYMV*) resistance were detected on chromosome 3H, through fingerprinting 22,626 barley collections from the IPK genebank (Milner *et al.*, 2019). Additionally, Arora *et al.* (2019) cloned candidate genes for stem rust resistance from the pangenome variation observed in wild diploid wheat. A geographically diverse set of wheat accessions was also characterized and screened to uncover novel alleles associated with the powdery mildew resistance gene (*Pm3*) (Bhullar *et al.*, 2010). Iranian common wheat landraces from the Watkins collection have been newly characterized for the *Yr47* gene related to stripe rust (Bansal *et al.*, 2011), and the accession displaying *Lr67/Yr46*, resistance to stem rust and powdery mildew (Herrera-Foessel *et al.*, 2014). Moreover, the IRRI genebank has been delivering promising genetic resources for disease-resistant breeding in rice over several years (Jamora and Ramaiah, 2022).

8.4.2 Abiotic Stress

Abiotic stress can arise from a variety of environmental factors including extreme temperatures, varying water and salt levels, presence of heavy metals. Each crop has its optimal growth conditions, and deviations from these conditions can threaten plant survival. Screening a genetically diverse pool from genebanks is crucial to identify candidates that exhibit resilience to these stresses, which helps in the development of breeding materials with enhanced abiotic stress tolerance. For example, Mwando *et al.* (2020) analyzed 350 barley accessions from 32 countries using GWAS to identify loci related to salinity tolerance. Similarly, an association panel of sorghum was evaluated under thermal stress conditions (Chopra *et al.*, 2017).

Drought and heat also pose significant challenges to crop production, particularly affecting cool-season crops such as wheat, barley, lentil (*Lens culinaris*) and faba bean (*Vicia faba*). Singh *et al.* (2018) identified rare haplotypes that enhance biomass and grain yield under drought and heat stress by utilizing exotic wheat resources from the genebank crossed with elite breeding lines of CIMMYT. Additionally, Vikram *et al.* (2016) analyzed Creole wheats from various climatic regions of Mexico and identified rare alleles in the core reference set that are associated with different climatic conditions. Extensive GBS data analysis of 1423 spring bread wheat accessions from diverse environments were evaluated to discover single-nucleotide polymorphism (SNP) variations specific to landraces adapted to drought and heat conditions (Sehgal *et al.*, 2015). Likewise, Sukumaran *et al.* (2015) identified loci associated with grain yield in a wheat association mapping initiative (WAMI) population derived from CIMMYT resources. Research on sorghum collections from Uganda, which included 3000 accessions, identified potential cold-tolerant genes that could be crucial under changing climate conditions (Chakrabarty *et al.*, 2022). Grando *et al.* (2001) identified lines of barley germplasms that had high adaptability to drought conditions, which are extensively used in breeding programs today. Furthermore, Torres *et al.* (2013) successfully selected candidate genotypes from the IRRI genebank rice accessions that showed drought resistance, demonstrating them as viable candidates for breeding materials. The identification of rare alleles, combined with geographic data, enables researchers to select candidate genes or germplasms that are particularly suited for abiotic stress tolerance. This strategic use of genebank resources not only preserves genetic diversity but also provides practical solutions to the challenges posed by global climate variability, thereby enhancing resilience and sustainability of agricultural systems.

8.4.3 Nutritional Quality

Genebanks harbor valuable genetic variability not only to increase crop performance in varying stress conditions but also to improve food grains quality. Faba beans are one of the highest yielding legumes with high protein content of about 30%. However, faba bean also contains two pyrimidine glycosides, vicine and convicine (VC), that cause favism, an acute hemolytic anemia, in people who have glucose-6-phosphate dehydrogenase (G6PD) deficiency. They also cause a significant reduction in the efficiency of production systems for broiler chickens, laying hens, and pigs (Khazaei *et al.*, 2019). In 1989, the first, as yet only source of low VC genotype [genebank accession: 1268(4)(1) from Radzikov (Poland)] was identified after screening a germplasm collection of genebank accessions (Duc *et al.*, 1989). Identification of this promising genotype opened a realistic possibility for mapping the causal genes and breeding for low VC faba beans (Bjornsdotter *et al.*, 2021).

Wheat resources have also been exhaustively characterized for the enhancement of nutritional qualities utilizing the genetic diversity in genebanks. Sansaloni *et al.* (2020) evaluated 80,000 wheat accessions that include domesticated and CWRs from CIMMYT and ICARDA germplasm banks and found quantitative trait loci (QTLs) and candidate genes related to SDS sedimentation, values that are associated with gluten quality. Also, genomic regions exhibiting stable expression and their epistatic interactions affecting grain yield and yield stability were detected, achieved by genome-wide association mapping (GWAM) on a panel of 720 elite wheat varieties (Sehgal *et al.*, 2017). An evaluation of a mini-core collection of chickpeas comprising 600 accessions identified QTLs for raffinose family oligosaccharides (RFO) and high sucrose content and provided promising accessions with low RFOs (Elango *et al.*, 2022). Similarly, from these accessions, 10 nutrient-specific and 15 multinutrient rich accessions were identified for the development of biofortified varieties in superior agronomic background (Susmitha *et al.*, 2022).

8.5 Mobilization of Genebank Biodiversity: From a Cold Room to Field Plots

8.5.1 Catalog Variants Beyond SNPs with Pangenomics

Now as high-quality reference genomes are readily available for major crops, it is more feasible to identify variants based on mapping to the correct positions. However, relying on a single reference genome does not sufficiently capture the genetic diversity of a specific crop. Thanks to advancements in sequencing technologies, which have become more affordable, accessions from genebanks can be evaluated, selected and sequenced to assemble high-quality genomes, enabling the construction of a pangenome (Bayer *et al.*, 2020; Jayakodi *et al.*, 2021). This approach allows for the observation of diversity on a larger scale, including larger structural variants, which provides a richer landscape than examining one genome. A pangenome encompasses high-quality assemblies of representative genotypes. Pangenomes have been constructed for many crop species such as wheat (Montenegro *et al.*, 2017), barley (Jayakodi *et al.*, 2020; Jayakodi *et al.*, 2021), rice (Sun *et al.*, 2017; Zhao *et al.*, 2018; Qin *et al.*, 2021), and soybean (Li *et al.*, 2014b; Liu *et al.*, 2020; Bayer *et al.*, 2022). Single nucleotide polymorphisms can be extracted in larger-scale, potentially covering the whole genome, which can provide insight for genetic diversity that better represents the population.

Growing evidence clearly shows that genomic structural variants (approximately >50 nt) including presence/absence, inversions, and translocations play a key role in trait innovations and adaptations in plants (Saxena *et al.*, 2014; Gabur *et al.*, 2019) and thus capturing sequence variants beyond SNPs is important for better understanding of genome to phenome. Pangenome analysis with multiple genome assemblies provides a unique opportunity to catalog structural variants (SVs) with precise break-point resolution (Bayer *et al.*, 2020; Jayakodi *et al.*, 2021). Since genebanks could capture a global diversity of a particular species, a narrowed-down discovery panel comprising a few tens to hundreds of representative accessions (covering wide geography or gene pool) can be developed from genebank accessions. For example, selection of 76 representative barley genotypes from the IPK genebank detected numerous presence/absence variants and large polymorphic inversions with functional implications (Himmelbach *et al.*, 2018; Jayakodi *et al.*, 2024). Similarly, a rice pangenome selected representative accessions from the IRRI genebank (Qin *et al.*, 2021).

8.5.2 Core Set or Association Panel for Phenotyping

Phenotyping a large collection of genebank accessions is unrealistic for many species. Selection of a subset or a so-called core set comprising a few hundreds to thousands of accessions based on either neutral diversity or traits (also referred to as trait-customized selection) could create possibilities for conducting phenotype experiments and trait discovery and mapping via association scans. The core set can be sequenced with a whole genome shotgun approach to get short variants across chromosomes. Moreover, the SVs detected from the pangenome analysis can be genotyped to perform a SV-based association scan (Yuan *et al.*, 2021). Furthermore, resequencing of such an association panel with wild populations sheds light on domestication, diversification, migration, and demographic histories. The analysis of 80,000 wheat accessions including domesticated and wild accession with different ploidy level from CIMMYT and ICARDA revealed a wide range of genetic diversity aside from what can be seen in the elite cultivars, exposing a vast proportion of untapped alleles (Sansaloni *et al.*, 2020). The resequencing of 3000 rice genomes selected from different genebanks generated high dense SNPs to serve as a foundation for rice improvement (Rellosa *et al.*, 2014; Li *et al.*, 2014a). In addition, these results discovered unreported subpopulations correlating with geographic locations.

8.5.3 Identifying Novel Alleles and Promising Donors

Significant variants associated with phenotypic traits can be discovered with an association scan of a core set. However, resequencing of elite germplasm or breeding material is requisite to check whether the alleles are novel or already present in breeding population. This can systematically be done when comparing a variant matrix containing plant genetic resources and elite germplasm. Based on genotype and phenotype evaluation and comparison with modern types, promising genotypes with highly desirable trait expression (ex. resistance to specific disease) can be identified (if not for all traits) for pre-breeding. Further, crossing the donor genotype with a modern elite line would permit mapping QTLs and subsequently, develop markers for marker-assisted selection (MAS) in the pre-breeding programs.

For example, as discussed by Tanksley and McCouch, wild relatives of crops such as wheat have been extensively utilized to introduce new genetic variability into domesticated varieties (Tanksley and McCouch, 1997). By crossing wild genotypes with modern elite lines, researchers can map QTLs and develop markers for MAS in pre-breeding programs. This allows the incorporation of traits such as disease resistance and drought tolerance into commercially viable crops. Genomic selection uses all marker data for a population to predict the breeding value by integrating genotype and phenotype information. Genomic estimated breeding values (GEBVs) are calculated for all genotypes in the population, which predicts the performance of the genotypes as parents for crossing, a concept first proposed by Meuwissen *et al.* (2001). The integration of novel alleles from genebanks not only revitalizes the genetic base of crops, but also facilitates the development of varieties adapted to changing environmental conditions and agricultural demands.

8.6 Database and Management

As the plant genetic resources in genebanks are fingerprinted and evaluated for various traits, it is important to keep track of the data for the use of researchers and breeders around the world without restrictions. For this purpose, CGIAR (https://www.cgiar.org/) connects large genebanks worldwide, facilitating access and exchange of information and resources. Other sources of germplasm data can be easily accessible by various web-based platforms. Genesys portal contains information on genebank crop diversity of more than half of the collections in the world's genebanks (https://www.genesys-pgr.org/). Through this portal, one can obtain information on the collection and request for germplasms. Another one that serves a similar purpose is the European search catalog for plant genetic resources (EURISCO, https://eurisco.ipk-gatersleben.de/), and Genebank Information System (GBIS) (Oppermann *et al.*, 2015). Some are more tailored towards molecular and phenotypic information related to a specific crop, such as the BRIDGE web portal that stores information on the barley accessions (Konig *et al.*, 2020) or the SNP-Seek database of the SNP genotype information from the 3000 rice genomes project that contains passport data as well as phenotypic data (Alexandrov *et al.*, 2015). Currently, genebank genomics studies often focus on characterizing "precision collection" comprised of a single genotype for each accession, for which sequence and phenotype data are generated. Genebanks should also create systems to access such precision collection seeds for community utilization. Such systems could readily be used to select a genotype of interest for future breeding and research.

8.7 Conclusion and Future Outlook

Genebanks house thousands of accessions as treasure sources that should be accessible to mitigate various threats such as climate crisis, genetic erosions, diseases and pests, and lack of crop/diet diversity. More and more germplasm will continuously be stored in the genebanks and thus an efficient management system is essential. Against unexpected loss, a common backup vault, the Svalbard Global Seed Vault has also been constructed. Modern genomics tools have the capacity to transform the genebank into digital form for efficient management and utilization in breeding and research.

Currently, efforts are being made to fingerprint the collections from different genebanks ("pan-genebank collection"). Nevertheless, to date, the extensive genomic characterization of genebank collections has mostly been done for major crops such as wheat, rice and barley. Characterization and mobilization of minor and potential under-utilized crops are important to establish a sustainable and resilient agri-food system. The intra-accessions diversity or heterogeneity has largely been unexplored even for the major crop species. Genotyping all the genotypes in all genebank accessions with a genome-wide method such as GBS is cost-prohibitive. However, such evaluation might become feasible in the future with the development of targeted low-cost genotyping methods.

Biotic and abiotic stress tolerant alleles are rare and thus a core set generated from neutral genetic diversity might not have sufficient power to identify rare, beneficial variants via association scan. Trait-customized selection is preferable to increase the frequencies of alleles associated with desirable traits in the panel as well as detection power in GWAS. Currently, phenotyping is a major bottleneck for plant genetic resources. Genomic selection or prediction is an efficient tool to predict the phenotype in the plant genetic resources. Mexican and Iranian landraces from CIMMYT were evaluated for testing the genomic prediction accuracy of reference core sets with genotype and phenotype information (Crossa *et al.*, 2016). Fingerprinting of winter wheat collections hosted at the IPK genebank compared the different genotyping methods to optimize a suitable approach for genome-wide prediction (Chu *et al.*, 2020). Similarly, Afghan wheat landraces characterized by SNP markers (Manickavelu *et al.*, 2014) and the Watkins collection from Asia, Europe and some parts of Africa were genotyped using microsatellite markers (Wingen *et al.*, 2014) to cover a wide range of diversity within bread wheat. In general, plant genetic resources are non-adaptive material and thus, the selected promising donors can be crossed with a widely adapted high-yielding elite line for simultaneous evaluation of the population in multiple locations, incorporating the exotic alleles from plant genetic resource to elite line. Additionally, is the development of a multi-parent population such as nested association mapping (NAM) with an elite line as a common parent and crossed with different trait-specific promising genotypes. This will serve as a permanent community resource for trait mapping and improvement.

References

Alexandrov, N., Tai, S.S., Wang, W.S., Mansueto, L., Palis, K. *et al.* (2015) SNP-seek database of SNPs derived from 3000 rice genomes. *Nucleic Acids Research* 43, D1023–D1027. DOI: 10.1093/nar/gku1039.

Appels, R., Eversole, K., Feuillet, C., Keller, B., Rogers, J. *et al.* (2018) Shifting the limits in wheat research and breeding using a fully annotated reference genome. *Science* 361, eaar7191. DOI: 10.1126/science.aar7191.

Arora, S., Steuernagel, B., Gaurav, K., Chandramohan, S., Long, Y.M. *et al.* (2019) Resistance gene cloning from a wild crop relative by sequence capture and association genetics. *Nature Biotechnology* 37, 139–143. DOI: 10.1038/s41587-018-0007-9.

Bansal, U.K., Forrest, K.L., Hayden, M.J., Miah, H., Singh, D. *et al.* (2011) Characterisation of a new stripe rust resistance gene Yr47 and its genetic association with the leaf rust resistance gene Lr52. *Theoretical and Applied Genetics* 122, 1461–1466. DOI: 10.1007/s00122-011-1545-4.

Bari, M.a.A., Zheng, P., Viera, I., Worral, H., Szwiec, S. *et al.* (2021) Harnessing genetic diversity in the USDA pea germplasm collection through genomic prediction. *Frontiers in Genetics* 12, 707754. DOI: 10.3389/fgene.2021.707754.

Bayer, P.E., Golicz, A.A., Scheben, A., Batley, J. and Edwards, D. (2020) Plant pan-genomes are the new reference. *Nature Plants* 6, 914–920. DOI: 10.1038/s41477-020-0733-0.

Bayer, P.E., Valliyodan, B., Hu, H.F., Marsh, J.I., Yuan, Y.X. *et al.* (2022) Sequencing the USDA core soybean collection reveals gene loss during domestication and breeding. *Plant Genome* 15, e20109. DOI: 10.1002/tpg2.20109.

Bhullar, N.K., Zhang, Z.Q., Wicker, T. and Keller, B. (2010) Wheat gene bank accessions as a source of new alleles of the powdery mildew resistance gene Pm3: A large scale allele mining project. *BMC Plant Biology* 10, 88. DOI: 10.1186/1471-2229-10-88.

Bjornsdotter, E., Nadzieja, M., Chang, W., Escobar-Herrera, L. *et al.* (2021) VC1 catalyses a key step in the biosynthesis of vicine in faba bean. *Nature Plants* 7, 923–931. DOI: 10.1038/s41477-021-00950-w.

Chakrabarty, S., Mufumbo, R., Windpassinger, S., Jordan, D., Mace, E. *et al.* (2022) Genetic and genomic diversity in the sorghum gene bank collection of Uganda. *BMC Plant Biology* 22, 378. DOI: 10.1186/s12870-022-03770-y.

Chapman, J.A., Mascher, M., Buluc, A., Barry, K., Georganas, E. *et al.* (2015) A whole-genome shotgun approach for assembling and anchoring the hexaploid bread wheat genome. *Genome Biology* 16, 26. DOI: 10.1186/s13059-015-0582-8.

Chopra, R., Burow, G., Burke, J.J., Gladman, N. and Xin, Z.G. (2017) Genome-wide association analysis of seedling traits in diverse sorghum germplasm under thermal stress. *BMC Plant Biology* 17, 12. DOI: 10.1186/s12870-016-0966-2.

Chu, J.T., Zhao, Y.S., Beier, S., Schulthess, A.W., Stein, N. *et al.* (2020) Suitability of single-nucleotide polymorphism arrays versus genotyping-by-sequencing for genebank genomics in wheat. *Frontiers in Plant Science* 11, 42. DOI: 10.3389/fpls.2020.00042.

Clavijo, B.J., Venturini, L., Schudoma, C., Accinelli, G.G., Kaithakottil, G. *et al.* (2017) An improved assembly and annotation of the allohexaploid wheat genome identifies complete families of agronomic genes and provides genomic evidence for chromosomal translocations. *Genome Research* 27, 885–896. DOI: 10.1101/gr.217117.116.

Crossa, J., Jarquin, D., Franco, J., Perez-Rodriguez, P., Burgueno, J. *et al.* (2016) Genomic prediction of gene bank wheat landraces. *G3-Genes Genomes Genetics* 6, 1819–1834. DOI: 10.1534/g3.116.029637.

Du, H.L., Yu, Y., Ma, Y.F., Gao, Q., Cao, Y.H. *et al.* (2017) Sequencing and de novo assembly of a near complete indica rice genome. *Nature Communications* 8, 15324. DOI: 10.1038/ncomms15324.

Duc, G., Sixdenier, G., Lila, M. and Furstoss, V. (1989) *Search of Genetic Variability for Vicine and Convicine Content in Vicia Faba L. A First Report of A Gene Which Codes for Nearly Zero-Vicine and Zero-Convicine Contents*. In Recent advances of Research in antinutritional factors in legume seeds (pp. 305–313). Wageningen Academic.

Elango, D., Wang, W., Thudi, M., Sebastiar, S., Ramadoss, B.R. *et al.* (2022) Genome-wide association mapping of seed oligosaccharides in chickpea. *Frontiers in Plant Science* 13, 1024543. DOI: 10.3389/fpls.2022.1024543.

Elshire, R.J., Glaubitz, J.C., Sun, Q., Poland, J.A., Kawamoto, K. *et al.* (2011) A robust, simple genotyping-by-sequencing (GBS) approach for high diversity species. *PLoS ONE* 6, e19379. DOI: 10.1371/journal.pone.0019379.

Fondevilla, S., Rubiales, D., Moreno, M.T. and Torres, A.M. (2008) Identification and validation of RAPD and SCAR markers linked to the gene Er3 conferring resistance to *Erysiphe pisi* DC in pea. *Molecular Breeding* 22, 193–200. DOI: 10.1007/s11032-008-9166-6.

Gabur, I., Chawla, H.S., Snowdon, R.J. and Parkin, I.a.P. (2019) Connecting genome structural variation with complex traits in crop plants. *Theoretical and Applied Genetics* 132, 733–750. DOI: 10.1007/s00122-018-3233-0.

Gali, K.K., Sackville, A., Tafesse, E.G., Lachagari, V.B.R., Mcphee, K. *et al.* (2019) Genome-wide association mapping for agronomic and seed quality traits of fieldpea (*Pisum sativum* L.). *Frontiers in Plant Science* 1–19. DOI: 10.3389/fpls.2019.01538.

Goff, S.A., Ricke, D., Lan, T.H., Presting, G., Wang, R.L. *et al.* (2002) A draft sequence of the rice genome (*Oryza sativa* L. ssp. *japonica*). *Science* 296, 92–100. DOI: 10.1126/science.1068275.

Grando, S., von Bothmer, R. and Ceccarelli, S. (2001) Genetic diversity of barley: Use of locally adapted germplasm to enhance yield and yield stability of barley in dry areas. In Cooper, H.D., Spillane, C. and Hodgkin, T. (eds) *Broadening the Genetic Base of Crop Production*. CABI Publishing, pp. 351–371. https://doi.org/10.1079/9780851994116.0351.

Guo, Y., Himmelbach, A., Weiss, E., Stein, N. and Mascher, M. (2022) Six-rowed wild-growing barleys are hybrids of diverse origins. *Plant Journal* 111, 849–858. DOI: 10.1111/tpj.15861.

Hafeez, A.N., Arora, S., Ghosh, S., Gilbert, D., Bowden, R.L. *et al.* (2021) Creation and judicious application of a wheat resistance gene atlas. *Molecular Plant* 14, 1053–1070. DOI: 10.1016/j.molp.2021.05.014.

Herrera-Foessel, S.A., Singh, R.P., Lillemo, M., Huerta-Espino, J., Bhavani, S. *et al.* (2014) Lr67/Yr46 confers adult plant resistance to stem rust and powdery mildew in wheat. *Theoretical and Applied Genetics* 127, 781–789. DOI: 10.1007/s00122-013-2256-9.

Himmelbach, A., Ruban, A., Walde, I., Simková, H., Dolezel, J. *et al.* (2018) Discovery of multi-megabase polymorphic inversions by chromosome conformation capture sequencing in large-genome plant species. *Plant Journal* 96, 1309–1316. DOI: 10.1111/tpj.14109.

Huang, X.Q., Borner, A., Roder, M.S. and Ganal, M.W. (2002) Assessing genetic diversity of wheat (*Triticum aestivum* L.) germplasm using microsatellite markers. *Theoretical and Applied Genetics* 105, 699–707. DOI: 10.1007/s00122-002-0959-4.

Jamora, N. and Ramaiah, V. (2022) Global demand for rice genetic resources. *CABI Agriculture and Biosciences* 3, 26. DOI: 10.1186/s43170-022-00095-6.

Jayakodi, M., Lu, Q., Pidon, H., Rabanus-Wallace, M.T., Bayer, M., Lux, T. and Stein, N. (2024) Structural variation in the pangenome of wild and domesticated barley. *Nature* 636(8043), 654–662. DOI: 10.1038/s41586-024-08187-1.

Jayakodi, M., Padmarasu, S., Haberer, G., Bonthala, V.S., Gundlach, H. *et al.* (2020) The barley pan-genome reveals the hidden legacy of mutation breeding. *Nature* 588, 284–289. DOI: 10.1038/s41586-020-2947-8.

Jayakodi, M., Schreiber, M., Stein, N. and Mascher, M. (2021) Building pan-genome infrastructures for crop plants and their use in association genetics. *DNA Research* 28, dsaa030. DOI: 10.1093/dnares/dsaa030.

Kale, S.M., Schulthess, A.W., Padmarasu, S., Boeven, P.H.G., Schacht, J. *et al.* (2022) A catalogue of resistance gene homologs and a chromosome-scale reference sequence support resistance gene mapping in winter wheat. *Plant Biotechnology Journal* 20, 1730–1742. DOI: 10.1111/pbi.13843.

Khazaei, H., Purves, R.W., Hughes, J., Link, W., O'sullivan, D.M. *et al.* (2019) Eliminating vicine and convicine, the main anti-nutritional factors restricting faba bean usage. *Trends in Food Science & Technology* 91, 549–556. DOI: 10.1016/j.tifs.2019.07.051.

Khlestkina, E., Varshney, R., Röder, M., Graner, A. and Börner, A. (2007) A comparative assessment of genetic diversity in cultivated barley collected in different decades of the last century in Austria, Albania and India by using genomic and genic simple sequence repeat (SSR) markers. *Plant Genetic Resources* 4, 125–133. DOI: 10.1079/PGR2006109.

Kobayashi, F., Tanaka, T., Kanamori, H., Wu, J., Katayose, Y. *et al.* (2016) Characterization of a mini core collection of Japanese wheat varieties using single-nucleotide polymorphisms generated by genotyping-by-sequencing. *Breeding Science* 66, 213–225. DOI: 10.1270/jsbbs.66.213.

Konig, P., Beier, S., Basterrechea, M., Schuler, D., Arend, D. *et al.* (2020) BRIDGE - A visual analytics web tool for barley genebank genomics. *Frontiers in Plant Science* 11, 701. DOI: 10.3389/fpls.2020.00701.

Lehnert, H., Berner, T., Lang, D., Beier, S., Stein, N. *et al.* (2022) Insights into breeding history, hotspot regions of selection, and untapped allelic diversity for bread wheat breeding. *Plant Journal* 112, 897–918. DOI: 10.1111/tpj.15952.

Li, J.Y., Wang, J. and Zeigler, R.S. (2014a) The 3,000 rice genomes project: New opportunities and challenges for future rice research. *GigaScience* 3, 8. DOI: 10.1186/2047-217X-3-8.

Li, Y.H., Zhou, G.Y., Ma, J.X., Jiang, W.K., Jin, L.G. *et al.* (2014b) De novo assembly of soybean wild relatives for pan-genome analysis of diversity and agronomic traits. *Nature Biotechnology* 32, 1045–1052. DOI: 10.1038/nbt.2979.

Liu, Y.C., Du, H.L., Li, P.C., Shen, Y.T., Peng, H. *et al.* (2020) Pan-genome of wild and cultivated soybeans. *Cell* 182, 162–176. DOI: 10.1016/j.cell.2020.05.023.

Manickavelu, A., Jighly, A. and Ban, T. (2014) Molecular evaluation of orphan Afghan common wheat (*Triticum aestivum* L.) landraces collected by Dr. Kihara using single nucleotide polymorphic markers. *BMC Plant Biology* 14, 320. DOI: 10.1186/s12870-014-0320-5.

Mascher, M., Gundlach, H., Himmelbach, A., Beier, S., Twardziok, S.O. *et al.* (2017) A chromosome conformation capture ordered sequence of the barley genome. *Nature* 544, 427–433. DOI: 10.1038/nature22043.

Mascher, M., Schreiber, M., Scholz, U., Graner, A., Reif, J.C. *et al.* (2019) Genebank genomics bridges the gap between the conservation of crop diversity and plant breeding. *Nature Genetics* 51, 1076–1081. DOI: 10.1038/s41588-019-0443-6.

Meuwissen, T.H.E., Hayes, B.J. and Goddard, M.E. (2001) Prediction of total genetic value using genome-wide dense marker maps. *Genetics* 157, 1819–1829. DOI: 10.1093/genetics/157.4.1819.

Milner, S.G., Jost, M., Taketa, S., Mazon, E.R., Himmelbach, A. *et al.* (2019) Genebank genomics highlights the diversity of a global barley collection. *Nature Genetics* 51, 319–326. DOI: 10.1038/s41588-018-0266-x.

Montenegro, J.D., Golicz, A.A., Bayer, P.E., Hurgobin, B., Lee, H. *et al.* (2017) The pangenome of hexaploid bread wheat. *Plant Journal* 90, 1007–1013. DOI: 10.1111/tpj.13515.

Muller, T., Schierscher-Viret, B., Fossati, D., Brabant, C., Schori, A. *et al.* (2018) Unlocking the diversity of genebanks: Whole-genome marker analysis of Swiss bread wheat and spelt. *Theoretical and Applied Genetics* 131, 407–416. DOI: 10.1007/s00122-017-3010-5.

Mwando, E., Han, Y., Angessa, T.T., Zhou, G.F., Hill, C.B. *et al.* (2020) Genome-wide association study of salinity tolerance during germination in barley (*Hordeum vulgare* L.). *Frontiers in Plant Science* 11, 118. DOI: 10.3389/fpls.2020.00118.

Oppermann, M., Weise, S., Dittmann, C. and Knupffer, H. (2015) GBIS: The information system of the German genebank. *Database – The Journal of Biological Databases and Curation* bav021, bav021. DOI: 10.1093/database/bav021.

Pascual, L., Ruiz, M., Lopez-Fernandez, M., Perez-Pena, H., Benavente, E. *et al.* (2020) Genomic analysis of Spanish wheat landraces reveals their variability and potential for breeding. *BMC Genomics* 21, 122. DOI: 10.1186/s12864-020-6536-x.

Patto, M.C.V., Fernández-Aparicio, M., Moral, A. and Rubiales, D. (2007) Resistance reaction to powdery mildew (*Erysiphe pisi*) in a germplasm collection of *Lathyrus cicera* from Iberian origin. *Genetic Resources and Crop Evolution* 54, 1517–1521. DOI: 10.1007/s10722-006-9142-0.

Patto, M.C.V., Fernández-Aparicio, M., Moral, A. and Rubiales, D. (2009) Pre and posthaustorial resistance to rusts in *Lathyrus cicera* L. *Euphytica* 165, 27–34. DOI: 10.1007/s10681-008-9737-y.

Qin, P., Lu, H.W., Du, H.L., Wang, H., Chen, W.L. *et al.* (2021) Pan-genome analysis of 33 genetically diverse rice accessions reveals hidden genomic variations. *Cell* 184, 3542–3558. DOI: 10.1016/j.cell.2021.04.046.

Rellosa, M.C., Reano, R.A., Capilit, G.L.S., De Guzman, F.C., Ali, J. *et al.* (2014) The 3,000 rice genomes project. *GigaScience* 3, 7. DOI: 10.1186/2047-217x-3-7.

Sansaloni, C., Franco, J., Santos, B., Percival-Alwyn, L., Singh, S. *et al.* (2020) Diversity analysis of 80,000 wheat accessions reveals consequences and opportunities of selection footprints. *Nature Communications* 11, 4572. DOI: 10.1038/s41467-020-18404-w.

Saxena, R.K., Edwards, D. and Varshney, R.K. (2014) Structural variations in plant genomes. *Briefings in Functional Genomics* 13, 296–307. DOI: 10.1093/bfgp/elu016.

Schulthess, A.W., Kale, S.M., Liu, F., Zhao, Y., Philipp, N. *et al.* (2022) Genomics-informed prebreeding unlocks the diversity in genebanks for wheat improvement. *Nature Genetics* 54, 1544–1552. DOI: 10.1038/s41588-022-01189-7.

Sehgal, D., Autrique, E., Singh, R., Ellis, M., Singh, S. *et al.* (2017) Identification of genomic regions for grain yield and yield stability and their epistatic interactions. *Scientific Reports* 7, 41578. DOI: 10.1038/srep41578.

Sehgal, D., Vikram, P., Sansaloni, C.P., Ortiz, C., St Pierre, C. *et al.* (2015) Exploring and mobilizing the gene bank biodiversity for wheat improvement. *PLoS ONE* 10, e0132112. DOI: 10.1371/journal.pone.0132112.

Sharma, B. and Yadav, Y. (2003) *Pisum fulvum* carries a recessive gene for powdery mildew resistance. *Pisum Genetics* 35. pp. 31 [Brief Communications], https://pisum.icgbio.ru/pg/35/31.htm.

Siedler, H., Messmer, M.M., Schachermayr, G.M., Winzeler, H., Winzeler, M. *et al.* (1994) Genetic diversity in European wheat and spelt breeding material based on RFLP data. *Theoretical and Applied Genetics* 88, 994–1003. DOI: 10.1007/BF00220807.

Singh, S., Vikram, P., Sehgal, D., Burgueno, J., Sharma, A. *et al.* (2018) Harnessing genetic potential of wheat germplasm banks through impact-oriented-prebreeding for future food and nutritional security. *Scientific Reports* 8, 12527. DOI: 10.1038/s41598-018-30667-4.

Sofalian, O., Chaparzadeh, N., Javanmard, A. and Hejazi, M.S. (2008) Study of the genetic diversity of wheat landraces from northwest of Iran based on ISSR molecular markers. *International Journal of Agriculture & Biology* 10, 465–468.

Sukumaran, S., Dreisigacker, S., Lopes, M., Chavez, P. and Reynolds, M.P. (2015) Genome-wide association study for grain yield and related traits in an elite spring wheat population grown in temperate irrigated environments. *Theoretical and Applied Genetics* 128, 353–363. DOI: 10.1007/s00122-014-2435-3.

Sun, C., Hu, Z.Q., Zheng, T.Q., Lu, K.C., Zhao, Y. *et al.* (2017) RPAN: Rice pan-genome browser for similar to 3000 rice genomes. *Nucleic Acids Research* 45, 597–605. DOI: 10.1093/nar/gkw958.

Susmitha, D., Kalaimagal, T., Senthil, R., Vetriventhan, M., Manonmani, S. *et al.* (2022) Grain nutrients variability in pigeonpea genebank collection and its potential for promoting nutritional security in dryland ecologies. *Frontiers in Plant Science* 13, 934296. DOI: 10.3389/fpls.2022.934296.

Tanksley, S.D. and Mccouch, S.R. (1997) Seed banks and molecular maps: Unlocking genetic potential from the wild. *Science* 277, 1063–1066. DOI: 10.1126/science.277.5329.1063.

Torres, R.O., Mcnally, K.L., Cruz, C.V., Serraj, R. and Henry, A. (2013) Screening of rice genebank germplasm for yield and selection of new drought tolerance donors. *Field Crops Research* 147, 12–22. DOI: 10.1016/j.fcr.2013.03.016.

Vikram, P., Franco, J., Burgueno-Ferreira, J., Li, H.H., Sehgal, D. *et al.* (2016) Unlocking the genetic diversity of Creole wheats. *Scientific Reports* 6, 23092. DOI: 10.1038/srep23092.

Wambugu, P.W., Ndjiondjop, M.N. and Henry, R.J. (2018) Role of genomics in promoting the utilization of plant genetic resources in genebanks. *Briefings in Functional Genomics* 17, 198–206. DOI: 10.1093/bfgp/ely014.

Weiss, E., Kislev, M.E. and Hartmann, A. (2006) Autonomous cultivation before domestication. *Science* 312, 1608–1610. DOI: 10.1126/science.1127235.

Wenzl, P., Carling, J., Kudrna, D., Jaccoud, D., Huttner, E. *et al.* (2004) Diversity arrays technology (DArT) for whole-genome profiling of barley. *Proceedings of the National Academy of Sciences USA* 101, 9915–9920. DOI: 10.1073/pnas.0401076101.

Wingen, L.U., Orford, S., Goram, R., Leverington-Waite, M., Bilham, L. *et al.* (2014) Establishing the A. E. Watkins landrace cultivar collection as a resource for systematic gene discovery in bread wheat. *Theoretical and Applied Genetics* 127, 1831–1842. DOI: 10.1007/s00122-014-2344-5.

Xie, M., Chung, C.Y.L., Li, M.W., Wong, F.L., Wang, X. *et al.* (2019) A reference-grade wild soybean genome. *Nature Communications* 10, 1216. DOI: 10.1038/s41467-019-09142-9.

Xu, X.D., Zhao, R.P., Xiao, L., Lu, L.Y., Gao, M. *et al.* (2023) Telomere-to-telomere assembly of cassava genome reveals the evolution of cassava and divergence of allelic expression. *Horticulture Research* 10, uhad200. DOI: 10.1093/hr/uhad200.

Yadav, I.S., Singh, N., Wu, S., Raupp, J., Wilson, D.L. *et al.* (2022) Exploring genetic diversity of wild and related tetraploid wheat species *Triticum turgidum* and *Triticum timopheevii*. *Journal of Advanced Research* 48, 47–60. DOI: 10.1016/j.jare.2022.08.020.

Yi, X.X., Liu, J., Chen, S.C., Wu, H., Liu, M. *et al.* (2022) Genome assembly of the JD17 soybean provides a new reference genome for comparative genomics. *G3 – Genes Genomes Genetics* 12, jkac017. DOI: 10.1093/g3journal/jkac017.

Yu, J., Hu, S.N., Wang, J., Wong, G.K.S., Li, S.G. *et al.* (2002) A draft sequence of the rice genome (*Oryza sativa* L. ssp. *indica*). *Science* 296, 79–92. DOI: 10.1126/science.1068037.

Yuan, Y.X., Bayer, P.E., Batley, J. and Edwards, D. (2021) Current status of structural variation studies in plants. *Plant Biotechnology Journal* 19, 2153–2163. DOI: 10.1111/pbi.13646.

Zhang, G.Y., Liu, X., Quan, Z.W., Cheng, S.F., Xu, X. *et al.* (2012) Genome sequence of foxtail millet (*Setaria italica*) provides insights into grass evolution and biofuel potential. *Nature Biotechnology* 30, 549–554. DOI: 10.1038/nbt.2195.

Zhao, Q., Feng, Q., Lu, H.Y., Li, Y., Wang, A. *et al.* (2018) Pan-genome analysis highlights the extent of genomic variation in cultivated and wild rice. *Nature Genetics* 50, 278–284. DOI: 10.1038/s41588-018-0041-z.

Zimin, A.V., Puiu, D., Hall, R., Kingan, S., Clavijo, B.J. *et al.* (2017) The first near-complete assembly of the hexaploid bread wheat genome, *Triticum aestivum*. *GigaScience* 6, gix097. DOI: 10.1093/gigascience/gix097.

9 Genomic Prediction for Climate Resilience Breeding

Oluwaseun J. Akinlade[1], Rutwik Barmukh[1], Babatunde Olasege[2], Bhargava Kotte[1], Ajaya K. Karkee[1], and Rajeev K. Varshney[1]*

[1]WA State Agricultural Biotechnology Centre, Centre for Crop and Food Innovation, Food Futures Institute, Murdoch University, Murdoch 6150, Western Australia, Australia; [2]The Institute for Molecular Bioscience, The University of Queensland, Brisbane, Queensland, Australia

Abstract

Climate change imposes significant challenges on global agriculture through extreme temperatures, erratic precipitation, drought and salinity, threatening food security. Conventional breeding methods, while effective, are too slow to meet the urgent need for developing climate-resilient crops. Genomic prediction (GP) offers a transformative solution by enabling early and accurate selection of superior genotypes through genome-wide markers and machine learning (ML) algorithms. This chapter outlines the application of GP for breeding abiotic stress-tolerant crops, highlights the role of multi-environment trials (MET) and emphasizes the integration of environmental data (enviromics) to improve predictive accuracy. Recent advances in computational approaches, including deep learning and transformer-based models, are discussed for their potential to enhance GP efficiency under complex genotype-by-environment interactions. Despite challenges, such as limited adoption in low-resource breeding programs, prospects of integrating multi-omics, pan-genomics and gene editing promise to further optimize GP. These innovations position GP as a critical tool for developing climate-resilient crops and ensuring future food security.

9.1 Introduction

Global agriculture is facing the challenges of climate change, which is being manifested as extreme temperatures, erratic rainfall, extended drought periods and higher occurrence of pests and diseases. Ensuring global food and nutritional security in the face of these increasing environmental fluctuations necessitates the development of climate-smart crops with improved stress resilience, higher yields and enhanced adaptability to changing climates (Varshney *et al.*, 2021; Raza *et al.*, 2024). Conventional breeding approaches have been effective in delivering climate-resilient crops in the past decades (Kumar *et al.*, 2014). However, these approaches are often time-consuming, typically involving initial evaluation of parental phenotypes and genotypes, through crossing and selection, to the development of commercially viable varieties. As a result, these approaches alone may not be sufficient to meet the growing demand for stress-resilient crops within the required timeframes, requiring the integration of modern technologies such as genomic prediction to accelerate crop improvement.

*Corresponding author: Rajeev.Varshney@murdoch.edu.au

 Genomic Solutions for Breeding Climate-resilient Crops (R.K. Varshney, R. Barmukh and R.J. Harper)
DOI: 10.1079/9781800627185.0009

Genomic prediction (GP) is a cutting-edge statistical framework that harnesses genome-wide marker data to generate predictive values – typically genetic merits or breeding values – that represent the genetic potential of genotypes for important agronomic traits (Crossa *et al.*, 2025). In contrast to conventional marker-assisted selection (MAS) methods, which rely on a limited number of significant loci, genomic selection (GS) captures the effects of thousands or millions of genome-wide markers simultaneously, making it effective for complex, polygenic traits such as abiotic stress tolerance and disease resistance (Heffner *et al.*, 2009). By leveraging genome-wide ML algorithms, GP estimates the breeding values of individuals within a population, guiding informed selection decisions. Beyond selection, GP is also employed to optimize breeding strategies, predict genotype-by-environment interactions, and enhance the design of multi-environment trials. One prominent application of GP, known as genomic selection (GS), has revolutionized plant breeding by enabling earlier and more precise identification of superior genotypes, thereby reducing reliance on extensive and time-consuming phenotypic evaluations (Ballen-Taborda *et al.*, 2022; Dreisigacker *et al.*, 2023). GS is a powerful tool to accelerate genetic gain across various crop breeding programs through refining predictive models and optimizing training population design to enhance accuracy – areas that hold considerable promise, although current performance remains suboptimal and requires further innovation (Roorkiwal *et al.*, 2018; Rembe *et al.*, 2022).

With an increase in global population and food demand, breeding crops that can withstand abiotic/biotic stresses are critical for sustainable agricultural production. Stresses induced by variations in climatic patterns often lead to grain yield losses, impacting the livelihoods of millions of farmers, particularly in vulnerable regions. Developing (a)biotic stress-resilient cultivars through advanced breeding strategies that incorporate genomic prediction is a promising avenue to mitigate these impacts. Genomic selection (GS), a key application of genomic prediction, has transformed plant breeding by combining high-throughput genotyping, advanced statistical methods and ML algorithms to increase the accuracy and efficiency of selection (Alemu *et al.*, 2024). By leveraging historical and training population datasets, GP models can estimate the breeding values of genotypes with high precision, enabling early selection of superior lines. This approach has been particularly beneficial in crop breeding programs focused on improving climate resilience, as it accelerates the development of cultivars with increased tolerance to (a)biotic stresses while maintaining high yield potential (He *et al.*, 2016; Arojju *et al.*, 2018). Notably, GS provides a transformative approach to expedite climate resilience crop breeding, helping to ensure global food security and agricultural sustainability. By integrating this approach with emerging tools and technologies such as artificial intelligence (AI) and high-throughput phenotyping, plant breeders can drive innovation and design climate-resilient crops that can thrive under untested environmental conditions.

This chapter explores the growing impact of climate change on crop productivity and highlights the potential of GP as a strategic breeding technique. It explores the application of GP for selecting abiotic stress-tolerant genotypes, the importance of multi-environment trials (MET) and the integration of environmental data to increase prediction accuracy. Advances in computational methods tailored for climate-smart breeding are also discussed. Finally, the article outlines current challenges and future directions, providing insights into how GP can be further optimized to design crops adapted to a changing climate.

9.2 Climate Change and its Impact on Crop Production

Climate change significantly impacts global weather patterns, leading to an increase in temperature and a change in rainfall patterns (Habib-Ur-Rahman *et al.*, 2022). These shifting climatic conditions directly affect plant growth and development (Kumari *et al.*, 2020). Essential physiological processes such as respiration, photosynthesis and stomatal movement are highly dependent on stable environmental conditions that favor plant health (Oshunsanya *et al.*, 2019) and deviations from these conditions create stress in plants. The escalating global warming and anticipated alterations in climate are inducing multiple stresses – including drought, heat, salinity and erratic precipitation – in crop plants, resulting in severe losses to global agriculture.

9.2.1 Drought Stress

Drought, characterized by the unavailability or deficiency of soil moisture, is a critical factor impacting agricultural productivity (Isfahani *et al.*, 2022). The effect of drought on crops depends on the severity, duration and growth stage at which it occurs (Martignago *et al.*, 2020). Severe drought conditions lead to cellular dehydration, destabilization of proteins and membranes, and ultimately irreversible cell damage and death (Sato *et al.*, 2024).

Major food crops, including cereals and legumes, are highly susceptible to drought stress, contributing to global food insecurity. Rice, in particular, exhibits high sensitivity to soil water deficits, especially at the flowering stage, resulting in reduced photosynthesis, impaired grain filling and yield losses of between 23% and 21% compared with traditional irrigation (Zhang *et al.*, 2009; Yang *et al.*, 2019). Drought stress also affects legumes by significantly lowering germination rates and consequently reducing pod numbers by 80–90%, leading to a 73–82% decline in soybean yield (Wei *et al.*, 2018). Drought often occurs concurrently with heat stress under field conditions, further compounding its effects.

9.2.2 Heat Stress

Temperature and water status in plants are intricately linked, making it difficult to separate the impacts of heat and drought stress. Both stresses disrupt crop physiological processes, biochemical pathways and cellular structures (Ru *et al.*, 2023; Awasthi *et al.*, 2024). Since 1880, global temperatures have increased by 0.85°C, with a projected rise of 0.2°C per decade (Fahad *et al.*, 2017). Heat stress shortens the growing period and alters the distribution of agricultural species (Porter, 2005).

Even a 1°C rise in temperature can decrease wheat yields by 6% (Akter and Rafiqul Islam, 2017). Heat stress damages plant proteins, enzymes and chlorophyll, adversely affecting photosynthesis and yield (Fahad *et al.*, 2017). The reproductive phase, particularly pollination, is highly sensitive to elevated temperatures, resulting in up to 80–90% yield loss in maize (Hatfield and Prueger, 2015) and poor seed set and reduced seed size in soybean (Wei *et al.*, 2018). Moreover, each 1°C increase above optimum conditions may decrease rice yields by 10% and wheat yields by 3–4% (Wang *et al.*, 2020; Xu *et al.*, 2020), while prolonged exposure to temperatures above 30°C can lead to a 1% daily reduction in barley and maize yields (Rezaei *et al.*, 2015).

9.2.3 Erratic Precipitation

Climate change intensifies the water cycle, causing more frequent floods and droughts. The altered rainfall patterns associated with these through extreme weather events disrupt crop growth (Habib-Ur-Rahman *et al.*, 2022) and lead to losses during planting and harvesting. In addition, waterlogging hampers photosynthesis and nutrient uptake, inducing anaerobic respiration (Manghwar *et al.*, 2024). For instance, in legumes, waterlogging restricts nodulation and reduces yields (Iizumi *et al.*, 2024).

9.2.4 Salinity Stress

Climate change also exacerbates soil salinization through increased evapotranspiration rates, adversely affecting crop production (Khamidov *et al.*, 2022). Salinity stress currently affects approximately 20% of global arable land (Hasanuzzaman and Fujita, 2022). Saline conditions impair plant growth by reducing nutrient uptake and causing osmotic stress, ion toxicity, oxidative damage and nutrient deficiencies (Shrivastava and Kumar, 2015; Atta *et al.*, 2023).

In saline soils, the accumulation of ions such as Na^+, Ca^{2+}, K^+ and Mg^{2+} around the root zone leads to growth inhibition (Atta *et al.*, 2022). Salinity stress damages the photosynthetic apparatus, suppresses chlorophyll synthesis, alters enzymatic activity and affects gene expression, ultimately impairing photosynthesis

(Zahra *et al.*, 2022). In wheat, salinity stress has caused yield losses of between 13.4% and 39.1% depending on the growth stage (El Sabagh *et al.*, 2021).

The cumulative effects of drought, heat, erratic precipitation and salinity pose a serious threat to plant growth and agricultural productivity. Addressing these challenges through the development of climate-resilient cultivars is vital to ensuring global food security. Understanding genotype-by-environment interactions will be crucial for breeding crops that can withstand these evolving environmental stresses.

9.2.5 Genotype-by-environment (G×E) Interactions and their Significance

Genotype-by-environment (G×E) interactions describe variations in the relative performance of genotypes in diverse environments. The G×E interaction consists of two categories: (a) crossover type and (b) non-crossover type. The crossover G×E interaction clarifies the varying responses of cultivars to different environments, suggesting that cultivar rankings can change under diverse environmental conditions. Non-crossover interactions suggest differences in the extent of genotype performance (quantitative), while the rank order of genotypes across various environments remains stable; specifically, genotypes that perform well in one environment consistently perform well in others (Kang, 2002; De Leon *et al.*, 2016). The intricate nature of grain yield as a quantitative trait, along with substantial environmental interactions, makes the selection of genotypes based solely on performance in a single environment inadequate for varietal improvement. Selection must emphasize the evaluation of yield stability rather than focusing solely on average performance across diverse environmental conditions (Dia *et al.*, 2016; Islam *et al.*, 2015).

G×E interaction demonstrates the varying behaviors of genotypes in different environments, affecting the effectiveness of selection in breeding programs (Begna, 2020). Breeders have the ability to rank varieties across different environments (De Leon *et al.*, 2016). G×E interactions influence the selection of environments and helps the breeder to aim for broad or specific adaptation, as well as genetic uniformity versus genetic diversity or heterogeneity (Liang *et al.*, 2015).

9.3 Fundamentals of Genomic Prediction

The primary goal of GS is to enhance breeding efficiency by enabling early selection, thereby reducing cycle times and accelerating genetic gains. This approach functions by developing statistical prediction models trained on a reference population, which is genotyped and phenotyped extensively. These models estimate the genomic estimated breeding values (GEBVs) of new individuals based solely on genotypic data. GS has successfully demonstrated significant advantages over traditional selection methods, particularly for traits with low heritability or challenging phenotypic assessment, such as abiotic stress resilience (Crossa *et al.*, 2017).

Several statistical models are widely employed in genomic prediction, including best linear unbiased prediction (BLUP), genomic BLUP (G-BLUP), Bayesian methods (BayesA, BaycsB, BayesC, BayesR), ridge regression BLUP (rrBLUP), reproducing kernel Hilbert spaces (RKHS) and emerging ML models such as random forests, support vector machines and deep learning algorithms. A summary of the prediction accuracies achieved by these models across different crops and traits is presented in Table 9.1.

One of the primary statistical models employed in GP is G-BLUP. It is an extension of the traditional BLUP model that incorporates genomic relationships matrices among individuals focusing on additive genetic effects (VanRaden, 2008) rather than pedigree-based relationships. This method assumes that all genetic markers contribute equally to trait variation, allowing breeders to make more informed selection decisions based on GEBVs (Ferrao *et al.*, 2021). It is the most frequently used method. Another widely used approach is the Bayesian framework, which includes methods such as BayesA, BayesB, BayesC and BayesR. These Bayesian models provide more flexibility in accounting for varying genetic effects and sparseness (Gianola, 2021) and are particularly useful for traits controlled

Table 9.1. Prediction accuracy of various genomic prediction models across different crops and traits. (Authors' own table.)

Model	Crop	Trait	Accuracy	Reference
Best linear unbiased prediction (BLUP)	Maize	Striga resistant	0.58	(Badu-Apraku *et al.*, 2019)
		Drought resistant	0.42–0.65	
		Grain yield (GY)	0.71–0.75	(Vivek *et al.*, 2017)
		Anthesis silking interval (ASI)	0.35–0.43	
		10 agro-morphological traits	0.63–0.96	(Guo *et al.*, 2013)
	Wheat	GY, thousand grain weight (TGW), days to heading	0.3–0.5	(Poland *et al.*, 2012)
	Rice	Agro-morphological and quality traits	0.35–0.92	(Cui *et al.*, 2020)
		100 kernel weight	F2:3 (0.77) > RIL&DH (0.65) > NP (0.48)	(Liu *et al.*, 2018)
	Wheat	Leaf rust resistance (LRR)	0.35	(Daetwyler *et al.*, 2014)
		Yellow rust resistance (YRR)	0.44	
		Stem rust resistance (SRR)	0.27	
	Rice	GY	0.5	(Xu *et al.*, 2018)
		TGW	0.88	(Wang *et al.*, 2018)
		GY	0.362	(Tiede and Smith, 2018)
Ridge regression BLUP model (rrBLUP)	wheat	Powdery mildew resistance	0.6	(Sarinelli *et al.*, 2019)
	Rice	Blast resistance	0.17–0.73	(Huang *et al.*, 2019)
	Barley	GY	0.57	(Sallam *et al.*, 2015)
	Chickpea	Drought tolerance	0.56–0.61	(Li *et al.*, 2018)
	Soybean	GY	0.68	(Beche *et al.*, 2021)
		Oil content	0.76	
RKHS	Maize	GY	0.78	(Rio *et al.*, 2019)
	Wheat	GY	0.43–0.79	(Crossa *et al.*, 2011)
		Deoxynivalenol (DON) resistance	0.118–0.575	(Rutkoski *et al.*, 2012)
	Rice	Panicle weight	0.3	(Ben Hassen *et al.*, 2018)
	Lentil	Maturity duration	0.58–0.84	(Haile *et al.*, 2020)
Bayes A	Maize	100 kernel weight	0.92	(Shikha *et al.*, 2017)
	Wheat	GY	0.48	(Heffner *et al.*, 2011)
	Common bean	Cooking time	0.22–0.55	(Diaz *et al.*, 2020)

Continued

Table 9.1. Continued.

Model	Crop	Trait	Accuracy	Reference
Bayes LASSO	Soybean	Protein content	0.76	(Beche *et al.*, 2021)
	Brassica napus	GY	0.27–0.55	(Xiong *et al.*, 2020)
	Chickpea	Drought tolerance	0.56–0.61	(Xu *et al.*, 2018)
	Lentil	Maturity duration	0.58–0.84	(Haile *et al.*, 2020)
	Rice	GY	0.309	(Grenier *et al.*, 2015)
Dual CNN	Soybean	GY	0.452	(Montesinos-Lopez *et al.*, 2021)
		Protein	0.619	
		Oil	0.668	
		Moisture	0.463	
		Plant height	0.615	
DeepGS	Soybean	GY	0.391	
		Protein	0.506	
		Oil	0.531	
		Moisture	0.31	
		Plant height	0.452	
Single CNN	Soybean	GY	0.463	
		Protein	0.573	
		Oil	0.627	
		Moisture	0.449	
		Plant height	0.565	
RF	Tomato	Soluble solids content and total fruit weight per plant	0.22–0.70	(Su *et al.*, 2012)
	Cassava	Dry matter, harvest index, root and shoot weight, root number, mosaic disease severity, and early plant vigor	–0.03–0.68	(Wolfe *et al.*, 2017)
	Chickpea	Seed yield, 100 seed weight, days to 50% flowering, and days to maturity	0.13–0.91	(Roorkiwal *et al.*, 2016)
	Wheat	13 yield and yield related traits	0.52–0.81	(Farooq *et al.*, 2022)
		Grain protein	0.37	(Sandhu *et al.*, 2021)
XGBoost	Wheat	13 yield and yield-related traits	0.53–0.8	(Farooq *et al.*, 2022)
SVR	Wheat	Thousand kernel weight (TKW)	0.489	(Wang *et al.*, 2023a)
DNNGP	Wheat	TKW	0.634	(Wang *et al.*, 2023a)

by a few large-effect loci alongside many small-effect loci. Bayesian models consider marker-specific effects with variable shrinkage to capture complex genetic architectures, including loci with varying effect sizes. This approach has been successfully implemented in breeding programs for traits with complex genetic architecture, enhancing predictive accuracy significantly compared with traditional models. Among them, BayesR has gained prominence due to its flexibility in fitting marker effects with different distributions, improving accuracy for traits with complex genetic bases (Moser *et al.*, 2015). BayesR categorizes markers into classes based on their effect size, allowing for a flexible modeling framework suitable for complex traits influenced by both major and minor genes (Moser *et al.*, 2015).

RKHS is used when a non-linear relationship is perceived between genetic markers with phenotype and/or environmental data in a population that is intended to be used for a GP analysis (Gianola and van Kaam, 2008). In short, GP model of choice is critical to implement GS in breeding program and the understanding of the genetic architecture of the focused traits is needed to select the best model(s). Hence, understanding of the factors that affect the efficiency of the prediction model should be considered when planning a genomic selection project.

Machine learning (ML) techniques have also emerged as powerful tools in GP. Algorithms such as random forests, gradient boosting and deep learning offer improved predictive capabilities by capturing complex, non-linear interactions between genetic markers and environmental factors. Unlike traditional parametric models, ML approaches do not rely on explicit assumptions about the genetic architecture of traits, making them particularly suitable for analysing high-dimensional genomic datasets. The integration of artificial intelligence (AI) in GP continues to enhance its predictive accuracy, particularly for traits influenced by G×E interactions.

The effectiveness of GP is further enhanced by the incorporation of high-throughput phenotyping and enviromics. High-throughput phenotyping platforms, such as LeasyScan, facilitate rapid, non-invasive measurement of traits under controlled and field conditions, generating precise phenotypic data essential for robust genomic prediction models (Ferrao *et al.*, 2021). These data, obtained through drones, hyperspectral imaging and automated field sensors, provide detailed insights into plant traits under diverse environmental conditions. Enviromics, the systematic collection of environmental data through advanced sensing technologies and remote sensing, further refines GP by integrating environmental variables with genomic data. Incorporating enviromics enables models to explicitly account for environmental variability and G×E interactions, thus enhancing predictive accuracy across diverse and changing climates (He *et al.*, 2021). This combined approach ensures that breeding programs can develop crops that not only exhibit superior genetic potential but also perform optimally across varied climatic conditions.

Despite its advantages, GP presents several challenges. The computational complexity associated with processing vast genomic datasets requires significant infrastructure and expertise. Additionally, the cost of genotyping and high-throughput phenotyping can be prohibitive for resource-limited breeding programs. Efforts to address these challenges include the development of cost-effective genotyping methods such as genotyping-by-sequencing (GBS) and the use of imputation techniques to reduce genotyping expenses.

As GP continues to evolve, future advancements will likely focus on integrating multi-omics data, including transcriptomics, proteomics and metabolomics, to refine predictive models. The incorporation of functional genomics and gene editing technologies, such as CRISPR-Cas9, holds promise for accelerating the identification and incorporation of beneficial alleles into breeding programs. The increasing accessibility of cloud-based computing and AI-driven analytical tools will further enhance the scalability and accuracy of GP, ensuring its widespread adoption in modern plant breeding.

9.4 Genomic Prediction for Climate Resilience Breeding

Climate change has introduced new challenges, such as increased drought and heat stress, erratic precipitation, and shifting disease patterns, necessitating the development of crop varieties that can thrive under such harsh conditions. Sometimes, two or more of these conditions could occur simultaneously thereby worsening the situation and could lead to complete crop yield loss (Akinlade *et al.*, 2022). By integrating high-throughput phenotyping, enviromics and ML, GS provides a more precise and efficient breeding methodology for varieties that are resilient to these conditions. A general workflow for applying GP to develop climate-resilient cultivars, integrating phenotypic and sequencing data, model training, GEBV estimation and field validation, is depicted in Figure 9.1. This section explores key aspects of GP in climate resilience breeding, including drought and heat tolerance, multi-environment trials (MET) and the integration of environmental data into GP models.

9.4.1 Selection for Drought and Heat Tolerance

Accurate prediction and selection for drought and heat-related traits can substantially improve crop performance under climatic stress (Bonnett *et al.*, 2021). Crops must develop physiological and molecular mechanisms to survive under these conditions, such as deep root systems, improved water-use efficiency, and heat shock protein expression. Traditional breeding methods for drought and heat tolerance rely on field evaluations under controlled stress conditions, which are time-consuming and often fail to capture the complexity of G×E interactions. GP and GS provide breeders with advanced tools to develop cultivars resilient to these stresses by identifying key genomic regions linked to traits such as water use efficiency, stomatal regulation and temperature stress response.

By leveraging large datasets from phenotypic and genotypic evaluations, breeders can identify candidate genes associated with stress tolerance. Key traits for selection include:

> Root system architecture: Deep and dense root systems enhance water uptake and drought tolerance.
>
> Transpiration efficiency: Optimized stomatal conductance ensures water conservation without compromising growth.
>
> Stay-green phenotype: Delayed leaf senescence improves photosynthetic efficiency and prolongs productivity under stress.
>
> Heat shock protein expression: Molecular mechanisms protect cells from heat-induced damage.

Recent studies have utilized GP models that effectively incorporate genomic and environmental data to enhance predictive accuracy for drought and heat-related traits. For instance, the use of genome-wide markers in maize breeding programs has significantly improved selection efficiency for grain yield under drought stress conditions, showcasing genomic selection's practical benefits (Vivek *et al.*, 2017). Similarly, wheat breeding programs at CIMMYT have leveraged GP and GS to achieve significant genetic gains in heat tolerance by integrating comprehensive phenotypic and genomic data across diverse climatic environments (Tomar *et al.*, 2021). Moreover, genomic-enabled models employing sparse phenotyping strategies have effectively predicted genotypic performance under limited water availability, significantly enhancing selection efficiency in early generations (He *et al.*, 2021).

9.4.2 Leveraging MET and Enviromics for Climate-Resilient Breeding

Multi-environment trials (METs) are essential for breeding climate-resilient crops by evaluating genotype performance across diverse agro-climatic conditions. METs data improve GP models by capturing G×E interactions, leading to greater reliability of the model and predictive accuracy. Integrating METs into GP has proven effective for predicting complex traits such as grain yield and adaptability, particularly when environmental covariates are included (Crespo-Herrera *et al.*, 2021; Tomar *et al.*, 2021). Studies in wheat, maize and chickpea demonstrate that models incorporating G×E interactions, such as reaction norm models and gradient boosting methods with detailed environmental data, significantly enhance prediction accuracy and cultivar selections (Tomar *et al.*, 2021; Westhues *et al.*, 2021). Empirical validations across independent datasets consistently show that GS surpasses conventional breeding, delivering higher genetic gains and better management of environmental stress (Ferrao *et al.*, 2021).

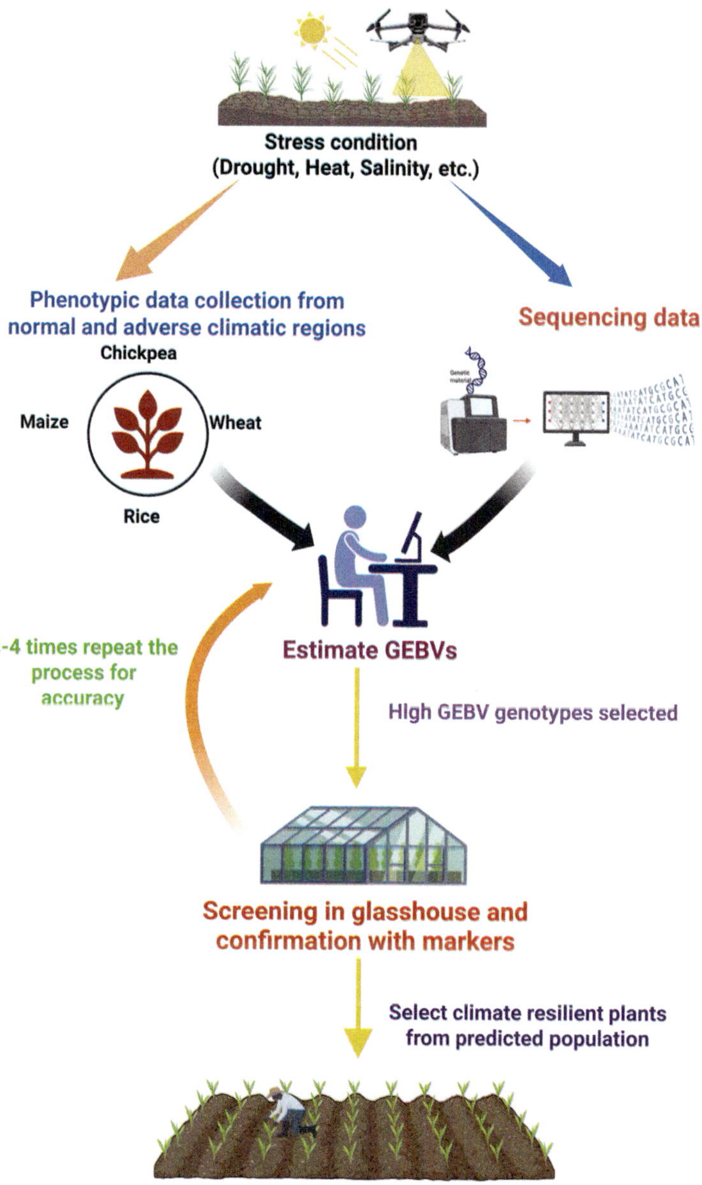

Figure 9.1. Workflow for developing climate-resilient cultivars using GP. Phenotypic data are collected from normal and stress-affected environments (drought, heat, salinity, etc.) alongside high-throughput sequencing to generate sequencing data. These datasets are integrated to subsequently drive marker effects and thereby estimate genomic estimated breeding values (GEBVs) for target crops such as maize, wheat, rice and chickpea. The process is repeated multiple times to enhance prediction accuracy. High-GEBV genotypes are selected, validated through greenhouse/field screening and marker confirmation, and ultimately used to identify and deploy climate-resilient plants. (Authors' own image.)

In METs, genotypes are tested across locations and seasons, enriching GP training datasets, improving prediction accuracy and enabling the identification of genetically stable, high-performing cultivars across environments. Additionally, integrating environmental data – termed enviromics –

into GP models has revolutionized climate resilience breeding. Environmental variables such as temperature, soil moisture and CO_2 levels are now routinely captured through remote sensing and crop growth models such as APSIM and DSSAT, enhancing predictions by modeling physiological responses under variable conditions. Combined genomic and environmental modeling has shown promise, particularly in maize, for predicting performance under future climate scenarios (Westhues *et al.*, 2021).

The growing availability of high-resolution environmental data and ML tools further supports the integration of enviromics and genomics. Together, MET and enviromics approaches provide a holistic framework for developing climate-resilient crops, empowering breeders to make data-driven decisions and contributing to global food security in the face of climate change.

9.5 Advances in Computational Approaches for Climate-Resilient Genomic Prediction

Traditional statistical and computational techniques in GP face significant limitations in delivering accurate predictions required for modern agricultural demands. The high dimensionality of genomic data increases the complexity of analysis, and the amount of data necessary for accurate generalization grows exponentially, challenging classical statistical theories (Bellman and Kalaba, 1959; Fan and Li, 2006). Moreover, recent technological advancements have enabled the large-scale acquisition of diverse omics data (e.g. phenomics, enviromics, transcriptomics, metabolomics, epigenomics and metagenomics), which offer deeper insights into trait variation beyond genomics alone (Azodi *et al.*, 2020). These emerging data types, often sparse, noisy and collected irregularly under varying conditions, pose substantial challenges to traditional prediction methodologies, resulting in poor predictive accuracy and reliability (Xu *et al.*, 2022). To overcome these limitations, researchers have increasingly turned to artificial intelligence (AI). AI has demonstrated particular strengths in domains such as robotics, medical diagnostics, natural language processing and autonomous systems, where it excels in managing high-dimensional, heterogeneous and redundant datasets – qualities that characterize many of the emerging omics data types (Wang *et al.*, 2022; Chafai *et al.*, 2023; Farooq *et al.*, 2024). Consequently, several start-of-the art ML have been developed to tackle the complexity of modern breeding data. This model excels at managing multi-dimension and emerging data types at massive scales with high prediction accuracy and reliability, serving as an accelerator for developing novel varieties to meet the growing demands for higher yield, improved quality and enhanced resilience in crops (Wang *et al.*, 2023b). The following section presents a brief overview of some of the most advanced computational tools and algorithms that have been designed to enhance the accuracy and efficiency of genomic prediction, particularly in the face of climate-induced stress conditions.

9.5.1 Deep Neural Network Genomic Prediction (DNNGP)

Deep Neural Network Genomic Prediction (DNNGP) is a recently developed ML technique based on a deep multi-layered hidden neural network architecture that successfully integrates multi-omics data in a genomic prediction model (Wang *et al.*, 2023b). DNNGP advances GP uniquely in five ways. Firstly, DNNGP effectively predicts phenotypes using diverse omics datasets, including traits associated with climate resilience. Secondly, it adaptively learns features from raw data, while avoiding over-fitting. This is accomplished using batch normalization (standardized layer outputs for smoother optimization), rectified linear unit (ReLU) activation functions (set negative activations to zero, a feature that relaxes the strict assumption of linearity while keeping the computation simple) and an early-stopping rule that halts training once validation error no longer improves. Thirdly, DNNGP provides robust estimates even with small sample sizes. Fourthly, it significantly reduces computational time. Lastly, it allows convenient hyperparameter tuning on local machines without extensive computational resources.

The DNNGP model receives omics data as input with a convolutional neural network serving as a key component for computation (Liu *et al.*, 2024). In brief, the model architecture comprises an input layer, multiple convolutional layers, batch normalization (BN) layers, dropout layers, a flattening layer, a fully connected dense layer and an output layer for phenotype prediction (Wang *et al.*, 2023b). The convolutional layer, central to DNNGP, performs feature extraction by convolving input data with kernel functions:

$$s(t)=(f*k)(t)=\sum_{x}(t-x)f(x)$$

where $f(x)$ represents the input data and k is the kernel used for convolution.

The ReLU activation function introduces non-linearity into the model and is mathematically defined as:

$$h(x)=\begin{cases} x, & if\ x>0 \\ 0, & otherwise \end{cases}$$

Regularization is a common solution to over-fitting in ML and can reduce model complexity. The DNNGP model integrates L2 regularization (often called weight decay) to reduce model complexity (i.e. generalization error):

$$J=J_0+\alpha\sum_{\omega}\omega^2$$

where J is the regularized loss function, J_0 represents the initial loss, α is the regularization coefficient, and ω are model weights. Early stopping, a callback function (Khan *et al.*, 2020), further enhances predictive accuracy and computational efficiency.

DNNGP has shown to provide robust phenotype prediction across climate-resilient crops such as maize and wheat (Wang *et al.*, 2023b), outperforming traditional and other start-of-the-art methods.

9.5.2 GPFormer

GPFormer is a framework built upon a transformer-based deep learning algorithm (Wu *et al.*, 2024). Unlike convolutional-based methods such as DNNGP, GPFormer offers three core advantages. First, it captures global interactions between SNPs at varying genomic distances, offering improved predictive accuracy and interpretability. Second, it can integrate prior information from GWAS into the model to further improve predictive accuracy. Finally, it is robust and stable to hyper-parameters and generalizes well to datasets with varying sample sizes. The GPFormer architecture consists of three key components: an embedding block, an encoder block, and a regression block. The embedding block encodes raw SNP input data into meaningful numerical representations via two mechanisms: SNP feature embedding and positional embedding. The former extracts informative SNP characteristics using convolutional operations while the latter integrates the physical genomic positions of SNPs, capturing positional relationships as:

$$pe_t^{(i)}=\begin{cases} \sin(\varphi_k t), & if\ i=2k \\ \cos(\varphi_k t), & if\ i=2k+1 \end{cases}$$

Where $\varphi_k=\frac{1}{10000^{2k/d_{model}}}$, d_{model} represents the dimension of embedding vectors, and t is the physical position index of SNPs. These embeddings enable GPFormer to distinguish SNPs effectively based on genomic locations.

The encoder block is the core component of GPFormer, designed to capture complex patterns across SNPs by leveraging an auto-correlation attention layer that models long-range dependencies through time-delay correlations computed via Fast Fourier Transform:

$$AutoCorrelation(Q,K,V)=\sum_{j=1}^{k}Delay(V,\tau_j)\widehat{AR_{Q,K}}(\tau_j)$$

where Q, K, V are query, key, and value matrices respectively, and delay τ_j captures relevant time-step shifts.

The encoder block further incorporates a series decomposition layer to remove trend noise, a convolutional layer for local feature extraction, Gaussian Error Linear Unit activation for smooth non-linearity, and both layer normalization and dropout to enhance training stability and prevent overfitting.

The final block, known as the regression block, transforms the extracted features into final phenotype predictions. It consists of a reshape operation followed by a linear layer, mapping learned representations to the predicted phenotypic values. To improve prediction accuracy, GPFormer incorporates a Knowledge-Guided Module, which integrates prior biological knowledge from GWAS into the model by assigning weights to SNPs based on their statistical significance. These weighted SNP representations are combined with the input embeddings, guiding the model to focus on biologically relevant markers. The comprehensive architecture of GPFormer and knowledge-guided integration of biological prior knowledge make the ML model distinctively powerful compared with other methods.

9.5.3 GEFormer

More recently, Yao *et al.* (2025) developed GEFormer, an extension of GPFormer transformer-based design that adds several pivotal improvements. Firstly, GEformer employs a gating multilayer mechanism to extract the hidden complex, non-linear genotype–environment interactions. Secondly, it replaces conventional self-attention with a linear attention mechanism to better capture both local and global features among SNPs. Finally, GEFormer gating mechanism dynamically fuses genomics and environmental features, remarkably improving prediction accuracy even in unknown environments. GEFormer integrates two modalities of data (i.e. genomic features [G*] and environmental features [E*]) into a single model through separate embedding blocks:

$$\boldsymbol{H}^{(1)} = E_1(\boldsymbol{G}),\ \boldsymbol{H}^{(2)} = E_2(\boldsymbol{E}),$$

This step maps both modalities into a common latent space, preparing them for joint processing. GEformer then employs a cross-attention mechanism to capture interdependencies between the two modalities by querying Q from one modality and K and V from the other:

$$Attention(Q,K,V) = softmax\left(\frac{QK^T}{\sqrt{d}}\right)V$$

where d is the dimension of Q or K.

This fuses both local and global interactions, allowing the model to integrate complementary phenotype signals effectively, an approach that contrasts with GPFormer's focus on genomic data. A gating mechanism dynamically balances the contributions from each modality. The gate G is computed as:

$$\mathbf{G} = \sigma\left(\pi_g\left[\boldsymbol{H}^{(1)} : \boldsymbol{H}^{(2)}\right] + b\right)$$

Where σ is the sigmoid activation π_g and b are learnable parameters. The final integrated feature is then obtained by:

$$\mathbf{Z} = G \odot \boldsymbol{H}^{(1)} + (1-G) \odot \boldsymbol{H}^{(2)}$$

Here, the gating mechanism modulates the influence of each modality to produce a balanced, fused representation (**Z**). **Z** is then processed by transformer encoder layer and a regression block to map the predicted phenotype using linear and non-linear transformations. The Integration of environmental data, as seen in GEFormer, could help to accurately predict crop phenotypes under complex and ever-changing climatic conditions (Yao *et al.*, 2025).

9.5.4 K_pRR

K_pRR, unlike the others, is a novel ML technique introduced to capture non-additive genetic effects that conventional genomic prediction models often miss (Li *et al.*, 2025). K_pRR offers two main advantages. Firstly, by integrating a polynomial kernel into ridge regression, it captures non-additive genetic effects (i.e. dominance, epistasis and other higher-order interactions) providing a more accurate prediction of crop phenotypes. Secondly, it retains the ridge regression's closed-form solution, so model parameters can directly obtain optimal solution without costly iterations.

In K_pRR, genomic data **X** are transformed into a higher dimensional feature space via a polynomial kernel k:

$$k(\boldsymbol{x}_i, \boldsymbol{x}_j) = (\gamma \boldsymbol{x}_i^T \boldsymbol{x}_j + c)^d$$

where $\boldsymbol{x}_i$ and $\boldsymbol{x}_j$ represent genotype vectors for individual i and j, γ is a positive scalar (i.e. $\gamma > 0$) that serves as a scaling factor on the dot product of $\boldsymbol{x}_i^T \boldsymbol{x}_j$, c is a constant term and d is the polynomial degree. This transformation allows linear methods to capture non-linear (non-additive) effects in the original space. Once the kernel matrix **K** is computed from the transformed data, K_pRR fits a ridge regression model:

$$\min_{\beta} \|y - \mathbf{K}\beta\|^2 + \lambda \beta^T \beta$$

Here, y is the vector of observed phenotypes, β are the regression coefficients in the kernel space and λ is the regularization parameter controlling model complexity. By virtue of the polynomial kernel, K_pRR incorporates cross-terms among SNPs, enabling it to model dominance and epistasis. Despite the potentially large dimensionality of **X**, the kernel-based formulation and efficient linear solvers allow KpRR to remain computationally tractable.

The rise of AI, particularly DL, has shown great promise in advancing GP for developing climate-resilient crops (Wu *et al.*, 2025). Recent computational tools such as DNNGP (Wang *et al.*, 2023b), GPFormer (Wu *et al.*, 2024), GEFormer (Yao *et al.*, 2025) and KpRR (Li *et al.*, 2025) have demonstrated improved predictive performance over traditional methods across key crops including wheat, rice, maize and soybean. With continued innovation, the next decade is expected to see the development of increasingly sophisticated ML algorithms and emerging AI-driven methodologies (Saad *et al.*, 2022). These advancements hold the potential to transform agriculture and make sustainable food security an attainable goal. The integration of such tools may enable the development of climate-resilient crop varieties capable of meeting the demands of a growing global population.

One of the major challenges in the decade ahead will be balancing predictive performance with computational efficiency. While ML and DL techniques can significantly improve prediction accuracy, they often come with high computational costs, including increased resource usage, training time and energy consumption (Nissar *et al.*, 2024). This is largely due to the vast number of parameters that must be optimized during model training (Stergiou *et al.*, 2023). In the era of smart breeding, where high-throughput, multi-modal data can now be collected at scales reaching millions or even tens of millions of observations, these computational demands are expected to rise exponentially. Thus, we need not only more efficient algorithms and high-performance computing solutions (Xu *et al.*, 2022; Lourenço *et al.*, 2024), but also turnkey, user-friendly pipelines that bundle data cleaning, model training and reporting into a single, easy-to-use interface. Such turnkey tools will let breeding teams harness large-scale omics data without needing advanced programming skills.

Another critical challenge for the research community is promoting open data sharing (Harfouche *et al.*, 2019). While individual-level data remains restricted due to commercial interests, contractual obligations and intellectual property concerns, the sharing of summary-level data presents a practical and impactful alternative. In human genetics, summary statistics have enabled major breakthroughs through large-scale meta-analyses and cross-study comparisons. Similarly, in plant genomics, broader access to summary data can facilitate the refinement of prediction algorithms, enhance reproducibility, and ultimately improve the accuracy and efficiency of genomic prediction models.

9.6 Challenges and Future Directions

Despite the potential of GS, several challenges limit its widespread adoption, specifically in (but not limited to) low-input crop breeding programs. Furthermore, improving the accuracy of GP models under diverse environments remains a major concern. Overcoming these challenges needs the integration of emerging technologies such as genome-enviromics, pan-genomics and gene editing, which hold potential to enhance the predictive power and efficiency of GS. This section discusses some key limitations of GS in climate-resilient crop breeding and describes future research directions to overcome these barriers.

9.6.1 Low-input Crop Breeding Programs

One of the major restrictions in implementing GS is its inadequate application in low-input crop breeding programs, particularly in developing countries where resources for large-scale genotyping and phenotyping are limited (Burrow *et al.*, 2021). GS mainly depends on the availability of extensive genomic and phenotypic datasets to precisely train predictive models. However, in resource-limited environments, the costs associated with high-throughput sequencing/genotyping can be high. The lack of well-characterized reference populations and historical phenotypic data further restrains the efficiency of GS in such programs.

In addition to these financial and data-related limitations, is the availability of computational resources and expertise required for implementation of GS (Crossa *et al.*, 2025). Several breeding programs lack access to modern bioinformatics tools essential to process and interpret large-scale genomic data. Furthermore, the implementation of GS – even in larger and better-resourced breeding programs – is often hindered by logistical complexities, lack of user-friendly tools and the need for centralized expertise, which sometimes necessitates outsourcing computational tasks. There is also a degree of scepticism among some breeders regarding the return on investment and operational feasibility of GS, particularly in the absence of clear integration strategies into existing workflows. Additionally, legal and regulatory constraints such as biosecurity and plant protection laws may restrict the international exchange of genetic material, delaying collaborative research and deployment of improved lines. These technical, infrastructural and institutional challenges limit the broader adoption of GS and contribute to slower breeding progress, especially in low-input systems.

9.6.2 Enhancing the Accuracy of Genomic Prediction Models under Diverse Environments

Variations in environment significantly impact the predictive accuracy of GP models, as G×E interactions play an important role in determining crop performance. Enhancing GP accuracy under diverse conditions can be achieved through several strategies:

Integrating environmental covariates: Combining environmental data such as temperature, rainfall and soil properties, among others, into GP models can improve their predictive ability (Heslot *et al.*, 2014). This approach aims to capture G×E interactions more accurately, resulting in more robust selection decisions.

Multi-environment trials: Expanding GP training populations by incorporating data from multiple locations and seasons can in turn enhance model accuracy. Conducting METs across different environments can help GP models to consider climatic variations and provide more reliable predictions (Crespo-Herrera *et al.*, 2021).

Advanced machine learning approaches: The application of DL and ML techniques can improve the performance of GP models by capturing complex interactions between genomic and environmental variables (Montesinos-Lopez *et al.*, 2022).

9.6.3 Future Prospects

To further improve the power of GS, future research needs to focus on integrating multiple emerging technologies:

Genome-enviromics: Incorporating environmental data into genomic prediction models is crucial for enhancing its accuracy under diverse climatic scenarios. Recent progress in remote sensing, climate modeling and precision agriculture offers new opportunities to combine environmental data with GP, resulting in informed breeding decisions (Wang *et al.*, 2023c).

Integration of multi-omics data: Integrating genomics with transcriptomics, proteomics, metabolomics and epigenomics can offer deeper insights into the molecular mechanisms underlying climate resilience traits (Hu *et al.*, 2021; Wang *et al.*, 2023c). Combining multi-omics data can improve the accuracy of GS by identifying biomarkers and pathways associated with stress resistance, leading to more precise selection strategies.

Pangenomics: Traditional GP models mostly rely on single reference genomes, which may not capture the entire genetic diversity of a species. Pan-genomics, which involves analyzing the complete set of genes within a species as opposed to a single reference genome, provides a comprehensive understanding of genetic variation (Jayakodi *et al.*, 2024). Integrating pan-genomic data into GP models can help to enhance predictive accuracy by identifying novel alleles linked to climate resilience traits.

Gene editing: Recent innovations in CRISPR-Cas and related gene editing technologies offer exciting opportunities to complement GS. By detecting and modifying favorable alleles in elite germplasm lines, gene editing can accelerate the development of stress-tolerant genotypes (Chen *et al.*, 2024). Importantly, integrating GS with gene editing enables precise selection of genomic variants that can improve drought tolerance, disease resistance and other adaptive traits.

Robust predictive modes: Recent advances in AI and machine learning can be leveraged to develop dynamic GP models that can adapt to fluctuating environmental conditions.

By including real-time environmental and phenotyping data, these GP models can enhance the accuracy of trait prediction and improve breeding efficiency.

Taken together, while GP holds immense potential for expediting climate resilience crop breeding, several challenges still need to be addressed to maximize its impact. Improving GP accuracy under diverse environmental conditions, decreasing the reliance on extensive training populations, and integrating advanced technologies will be key to overcoming these limitations. Future research should focus on developing more robust GP models and fostering international collaboration for ensuring that GP-based methods are accessible to all breeding programs, regardless of resource constraints.

Acknowledgements

RKV thanks Grains Research & Development Corporation (GRDC), Australia, and Hort Innovation Australia for supporting research projects on genomics and pre-breeding in wheat (UMU2404-003RTX, WSU2303-001RTX), legumes (UMU2403-009RTX, UMU2303-003RTX, UOQ2403-012RTX, and UOQ2402-010RTX) and horticultural crops (AS21006, AS23003).

References

Akinlade, O.J., Voss-Fels, K., Costilla, R., Kholova, J., Choudhary, S. *et al.* (2022) Designing chickpea for a hotter drier world. *Euphytica* 218, 100. https://doi.org/10.1007/s10681-022-03048-2.

Akter, N. and Rafiqul Islam, M. (2017) Heat stress effects and management in wheat. A review. *Agronomy for Sustainable Development* 37, 37. https://doi.org/10.1007/s13593-017-0443-9.

Alemu, A., Astrand, J., Montesinos-Lopez, O.A., Isidro, Y.S.J., Fernandez-Gonzalez, J. *et al.* (2024) Genomic selection in plant breeding: Key factors shaping two decades of progress. *Molecular Plant* 17, 552–578. DOI: 10.1016/j.molp.2024.03.007.

Arojju, S.K., Conaghan, P., Barth, S., Milbourne, D., Casler, M.D. *et al.* (2018) Genomic prediction of crown rust resistance in Lolium perenne. *BMC Genetics* 19, 35. https://doi.org/10.1186/s12863-018-0613-z.

Atta, K., Adhikary, S., Mondal, S., Mukherjee, S., Pal, A. *et al.* (2022) A review on stress physiology and breeding potential of an underutilized, multipurpose legume: Rice bean (Vigna umbellata). In: Jha, U.C., Nayyar, H., Agrawal, S.K., Siddique, K.H.M. (eds) *Developing Climate Resilient Grain and Forage Legumes*. Springer, Singapore, pp. 235–253. https://doi.org/10.1007/978-981-16-9848-4_11.

Atta, K., Mondal, S., Gorai, S., Singh, A.P., Kumari, A. *et al.* (2023) Impacts of salinity stress on crop plants: Improving salt tolerance through genetic and molecular dissection. *Frontiers in Plant Science* 14. https://doi.org/10.3389/fpls.2023.1241736.

Awasthi, R., Devi, P., Jha, U.C., Sharma, K.D., Roorkiwal, M. *et al.* (2024) Exploring the synergistic effects of drought and heat stress on chickpea seed development: Insights into nutritional quality and seed yield. Transcriptome-based prediction of complex traits in maize. *The Plant Cell* 32, 139–151. https://doi.org/10.1016/j.stress.2024.100635.

Azodi, C.B., Pardo, J., VanBuren, R., de Los Campos, G. *et al.* (2020) Transcriptome-Based Prediction of Complex Traits in Maize. *Plant Cell* 32, 139–151. DOI: 10.1105/tpc.19.00332.

Badu-Apraku, B., Talabi, A.O., Fakorede, M.A.B., Fasanmade, Y., Gedil, M. *et al.* (2019) Yield gains and associated changes in an early yellow bi-parental maize population following genomic selection for Striga resistance and drought tolerance. *BMC Plant Biology* 19, 129. DOI: 10.1186/s12870-019-1740-z.

Ballen-Taborda, C., Lyerly, J., Smith, J., Howell, K., Brown-Guedira, G. *et al.* (2022) Utilizing genomics and historical data to optimize gene pools for new breeding programs: A case study in winter wheat. *Frontiers in Genetics* 13, 964684. https://doi.org/10.3389/fgene.2022.964684.

Beche, E., Gillman, J.D., Song, Q., Nelson, R., Beissinger, T. *et al.* (2021) Genomic prediction using training population design in interspecific soybean populations. *Molecular Breeding* 41, 15. https://doi.org/10.1007/s11032-021-01203-6.

Begna, T. (2020) The role of genotype by environmental interaction in plant breeding. *International Journal of Agriculture and Biosciences* 9, 209–215. DOI: 10.7176/JNSR/11-20-02.

Bellman, R. and Kalaba, R. (1959) A mathematical theory of adaptive control processes. *Proceedings of the National Academy of Sciences* 45, 1288–1290. https://www.jstor.org/stable/90152.

Ben Hassen, M., Bartholome, J., Vale, G., Cao, T.V. and Ahmadi, N. (2018) Genomic prediction accounting for genotype by environment interaction offers an effective framework for breeding simultaneously for adaptation to an abiotic stress and performance under normal cropping conditions in rice. *G3 (Bethesda)* 8, 2319–2332. DOI: 10.1534/g3.118.200098.

Bonnett, D., Li, Y., Crossa, J., Dreisigacker, S., Basnet, B. *et al.* (2021) Response to early generation genomic selection for yield in wheat. *Frontiers in Plant Science* 12, 718611. https://doi.org/10.3389/fpls.2021.718611.

Burrow, H.M., Mrode, R., Mwai, A.O., Coffey, M.P. and Hayes, B.J. (2021) Challenges and opportunities in applying genomic selection to ruminants owned by smallholder farmers. *Agriculture* 11, 1172. https://doi.org/10.3390/agriculture11111172.

Chafai, N., Hayah, I., Houaga, I. and Badaoui, B. (2023) A review of machine learning models applied to genomic prediction in animal breeding. *Frontiers in Genetics* 14, 1150596. https://doi.org/10.3389/fgene.2023.1150596.

Chen, F., Chen, L., Yan, Z., Xu, J., Feng, L. *et al.* (2024) Recent advances of CRISPR-based genome editing for enhancing staple crops. *Frontiers in Plant Science* 15, 1478398. https://doi.org/10.3389/fpls.2024.1478398.

Crespo-Herrera, L., Howard, R., Piepho, H.P., Perez-Rodriguez, P., Montesinos-Lopez, O. *et al.* (2021) Genome-enabled prediction for sparse testing in multi-environmental wheat trials. *Plant Genome* 14, e20151. DOI: 10.1002/tpg2.20151.

Crossa, J., Martini, J.W.R., Vitale, P., Perez-Rodriguez, P., Costa-Neto, G. *et al.* (2025) Expanding genomic prediction in plant breeding: Harnessing big data, machine learning, and advanced software. *Trends in Plant Science* 7, 756–774. https://doi.org/10.1016/j.tplants.2024.12.009.

Crossa, J., Pérez, P., de los Campos, G., Mahuku, G., Dreisigacker, S. *et al.* (2011) Genomic selection and prediction in plant breeding. *Journal of Crop Improvement* 25, 239–261. https://doi.org/10.1080/15427528.2011.558767.

Crossa, J., Perez-Rodriguez, P., Cuevas, J., Montesinos-Lopez, O., Jarquin, D. *et al.* (2017) Genomic selection in plant breeding: Methods, models, and perspectives. *Trends in Plant Science* 22, 961–975. https://doi.org/10.1016/j.tplants.2017.08.011.

Cui, Y., Li, R., Li, G., Zhang, F., Zhu, T. *et al.* (2020) Hybrid breeding of rice via genomic selection. *Plant Biotechnology Journal* 18, 57–67. DOI: 10.1111/pbi.13170.

Daetwyler, H.D., Bansal, U.K., Bariana, H.S., Hayden, M.J. and Hayes, B.J. (2014) Genomic prediction for rust resistance in diverse wheat landraces. *Theoretical and Applied Genetics* 127, 1795–1803. DOI: 10.1007/s00122-014-2341-8.

De Leon, N., Jannink, J.L., Edwards, J.W. and Kaeppler, S.M. (2016) Introduction to a special issue on genotype by environment interaction. *Crop Science* 56, 2081–2089. https://doi.org/10.2135/cropsci2016.07.0002in.

Dia, M., Wehner, T.C., Hassell, R., Price, D.S., Boyhan, G.E., Olson, S., King, S., Davis, A.R. and Tolla, G.E. (2016) Genotype × Environment Interaction and Stability Analysis for Watermelon Fruit Yield in the United States. *Crop Science* 56, 1645–1661. https://doi.org/10.2135/cropsci2015.10.0625.

Diaz, S., Ariza-Suarez, D., Ramdeen, R., Aparicio, J., Arunachalam, N. *et al.* (2020) Genetic architecture and genomic prediction of cooking time in common bean (Phaseolus vulgaris L.). *Frontiers in Plant Science* 11, 622213. https://doi.org/10.3389/fpls.2020.622213.

Dreisigacker, S., Perez-Rodriguez, P., Crespo-Herrera, L., Bentley, A.R. and Crossa, J. (2023) Results from rapid-cycle recurrent genomic selection in spring bread wheat. *G3 (Bethesda)* 13. DOI: 10.1093/g3journal/jkad025.

El Sabagh, A., Islam, M.S., Skalicky, M., Ali Raza, M., Singh, K. *et al.* (2021) Salinity stress in wheat (Triticum aestivum L.) in the changing climate: Adaptation and management strategies. *Frontiers in Agronomy* 3, 661932. DOI: 10.3389/fagro.2021.661932.

Fahad, S., Bajwa, A.A., Nazir, U., Anjum, S.A., Farooq, A. *et al.* (2017) Crop production under drought and heat stress: Plant responses and management options. *Frontiers in Plant Science* 8, 1147. DOI: 10.3389/fpls.2017.01147.

Fan, J. and Li, R. (2006) Statistical challenges with high dimensionality: Feature selection in knowledge discovery. In: *Proceedings of the International Congress of Mathematicians*. European Mathematical Society, Zurich, pp. 595–622. https://doi.org/10.48550/arXiv.math/0602133.

Farooq, M.A., Gao, S., Hassan, M.A., Huang, Z., Rasheed, A. *et al.* (2024) Artificial intelligence in plant breeding. *Trends in Genetics* 40, 891–9525. https://doi.org/10.1016/j.tig.2024.07.001.

Farooq, M., van Dijk, A.D.J., Nijveen, H., Mansoor, S. and de Ridder, D. (2022) Genomic prediction in plants: Opportunities for ensemble machine learning based approaches. *F1000Res* 11, 802. DOI: 10.12688/f1000research.122437.2.

Ferrao, L.F.V., Amadeu, R.R., Benevenuto, J., de Bem Oliveira, I. and Munoz, P.R. (2021) Genomic selection in an outcrossing autotetraploid fruit crop: Lessons from blueberry breeding. *Frontiers in Plant Science* 12, 676326. https://doi.org/10.3389/fpls.2021.676326.

Gianola, D. (2021) Opinionated views on genome-assisted inference and prediction during a pandemic. *Frontiers in Plant Science* 12, 717284. https://doi.org/10.3389/fpls.2021.717284.

Gianola, D. and van Kaam, J.B. (2008) Reproducing kernel Hilbert spaces regression methods for genomic assisted prediction of quantitative traits. *Genetics* 178, 2289–2303. https://doi.org/10.1534/genetics.107.084285.

Grenier, C., Cao, T.V., Ospina, Y., Quintero, C., Chatel, M.H. *et al.* (2015) Accuracy of genomic selection in a rice synthetic population developed for recurrent selection breeding. *PLoS One* 10, e0136594. https://doi.org/10.1371/journal.pone.0136594.

Guo, T., Li, H., Yan, J., Tang, J., Li, J. *et al.* (2013) Performance prediction of F1 hybrids between recombinant inbred lines derived from two elite maize inbred lines. *Theoretical and Applied Genetics* 126, 189–201. DOI: 10.1007/s00122-012-1973-9.

Habib-Ur-Rahman, M., Ahmad, A., Raza, A., Hasnain, M.U., Alharby, H.F., Alzahrani, Y.M. *et al.* (2022) Impact of climate change on agricultural production; Issues, challenges, and opportunities in Asia. *Frontiers in Plant Science* 13, 925548. DOI: 10.3389/fpls.2022.925548.

Haile, T.A., Heidecker, T., Wright, D., Neupane, S., Ramsay, L. *et al.* (2020) Genomic selection for lentil breeding: Empirical evidence. *Plant Genome* 13, e20002. DOI: 10.1002/tpg2.20002.

Harfouche, A.L., Jacobson, D.A., Kainer, D., Romero, J.C., Harfouche, A.H. *et al.* (2019) Accelerating climate resilient plant breeding by applying next-generation artificial intelligence. *Trends in Biotechnology* 37, 1217–1235. https://doi.org/10.1016/j.tibtech.2019.05.007.

Hasanuzzaman, M. and Fujita, M. (2022) Plant responses and tolerance to salt stress: Physiological and molecular interventions. *International Journal of Molecular Sciences* 23, 4810. DOI: 10.3390/ijms23094810.

Hatfield, J.L. and Prueger, J.H. (2015) Temperature extremes: Effect on plant growth and development. *Weather and Climate Extremes* 10, 4–10. https://doi.org/10.1016/j.wace.2015.08.001.

He, S., Jiang, Y., Thistlethwaite, R., Hayden, M.J., Trethowan, R. *et al.* (2021) Improving selection efficiency of crop breeding with genomic prediction aided sparse phenotyping. *Frontiers in Plant Science* 12, 735285. DOI: 10.3389/fpls.2021.735285.

He, S., Schulthess, A.W., Mirdita, V., Zhao, Y., Korzun, V. *et al.* (2016) Genomic selection in a commercial winter wheat population. *Theoretical and Applied Genetics* 129, 641–651. DOI: 10.1007/s00122-015-2655-1.

Heffner, E.L., Jannink, J.L. and Sorrells, M.E. (2011) Genomic selection accuracy using multifamily prediction models in a wheat breeding program. *The Plant Genome* 4. https://doi.org/10.3835/plantgenome2010.12.0029.

Heffner, E.L., Sorrells, M.E. and Jannink, J.-L. (2009) Genomic selection for crop improvement. *Crop Science* 49, 1–12. https://doi.org/10.2135/cropsci2008.08.0512.

Heslot, N., Akdemir, D., Sorrells, M.E. and Jannink, J.L. (2014) Integrating environmental covariates and crop modeling into the genomic selection framework to predict genotype by environment interactions. *Theoretical and Applied Genetics.* 127, 463–480. https://doi.org/10.1007/s00122-013-2231-5.

Hu, H., Campbell, M.T., Yeats, T.H., Zheng, X., Runcie, D.E. *et al.* (2021) Multi-omics prediction of oat agronomic and seed nutritional traits across environments and in distantly related populations. *Theoretical and Applied Genetics* 134, 4043–4054. DOI: 10.1007/s00122-021-03946-4.

Huang, M., Balimponya, E.G., Mgonja, E.M., Mchale, L.K., Luzi-Kihupi, A. *et al.* (2019) Use of genomic selection in breeding rice (Oryza sativa L.) for resistance to rice blast (Magnaporthe oryzae). *Molecular Breeding* 39, 114. https://doi.org/10.1007/s11032-019-1023-2.

Iizumi, T., Iseki, K., Ikazaki, K., Sakai, T., Shiogama, H. *et al.* (2024) Increasing heavy rainfall events and associated excessive soil water threaten a protein-source legume in dry environments of West Africa. *Agricultural and Forest Meteorology* 344, 109783. https://doi.org/10.1016/j.agrformet.2023.109783.

Isfahani, P.M., Soltani, S. and Modarres, R. (2022) Assessing agrometeorological drought trends in Iran during 1985–2018. *Theoretical and Applied Climatology* 150, 251–262. https://doi.org/10.1007/s00704-022-04159-5.

Islam, M.S., Halder, T., Hossain, J., Mahmud, F. and Rahman, J. (2015) Genotype-environment interaction in spring wheat (*Triticum aestivum* L.) of Bangladesh. *Bangladesh Journal of Plant Breeding and Genetics* 2, 17–24. https://doi.org/10.3329/bjpbg.v28i2.29957.

Jayakodi, M., Shim, H. and Mascher, M. (2024) What are we learning from plant pangenomes? *Annual Review of Plant Biology* 76, 663–686. DOI: 10.1146/annurev-arplant-090823-015358.

Kang, M.S. (2002) Genotype-environment interaction: Progress and prospects. In: KANG, M.S. (ed) *Quantitative Genetics, Genomics and Plant Breeding*. CABI Publishing, New York, NY 10016, USA, 221–243. DOI: 10.1079/9780851996011.0221.

Khamidov, M., Ishchanov, J., Hamidov, A., Donmez, C. and Djumaboev, K. (2022) Assessment of soil salinity changes under the climate change in the Khorezm region, Uzbekistan. *International Journal of Environmental Research and Public Health* 19, 8784. https://doi.org/10.3390/ijerph19148794.

Khan, A., Sohail, A., Zahoora, U. and Qureshi, A.S. (2020) A survey of the recent architectures of deep convolutional neural networks. *Artificial Intelligence Review* 53, 5455–5516. https://doi.org/10.1007/s10462-020-09825-6.

Kumar, A., Dixit, S., Ram, T., Yadaw, R.B., Mishra, K.K. *et al.* (2014) Breeding high-yielding drought-tolerant rice: Genetic variations and conventional and molecular approaches. *Journal of Experimental Botany* 65, 6265–6278. DOI: 10.1093/jxb/eru363.

Kumari, P., Rastogi, A. and Yadav, S. (2020) Effects of heat stress and molecular mitigation approaches in orphan legume, chickpea. *Molecular Biology Reports* 47, 4659–4670. DOI: 10.1007/s11033-020-05358-x.

Li, M., Hall, T., Machugh, D.E., Chen, L., Garrick, D. *et al.* (2025) KPRR: A novel machine learning approach for effectively capturing nonadditive effects in genomic prediction. *Briefings in Bioinformatics* 26, bbae683. https://doi.org/10.1093/bib/bbae683.

Li, Y., Ruperao, P., Batley, J., Edwards, D., Khan, T. *et al.* (2018) Investigating drought tolerance in chickpea using genome-wide association mapping and genomic selection based on whole-genome resequencing data. *Frontiers in Plant Science* 9, 190. https://doi.org/10.3389/fpls.2018.00190.

Liang, S., Ren, G., Liu, J., Zhao, X., Zhou, M. *et al.* (2015) Genotype-by-environment interaction is important for grain yield in irrigated lowland rice. *Field Crops Research* 180, 90–99. https://doi.org/10.1016/j.fcr.2015.05.014.

Liu, Y., Lu, S., Liu, F., Shao, C., Zhou, Q. *et al.* (2018) Genomic selection using BayesCpi and GBLUP for resistance against Edwardsiella tarda in Japanese flounder (Paralichthys olivaceus). *Marine Biotechnology (New York, N.Y.)* 20, 559–565. DOI: 10.1007/s10126-018-9839-z.

Liu, Q., Zuo, S.-M., Peng, S., Zhang, H., Peng, Y. *et al.* (2024) Development of machine learning methods for accurate prediction of plant disease resistance. *Engineering* 40, 100–110. https://doi.org/10.1016/j.eng.2024.03.014.

Lourenço, V.M., Ogutu, J.O., Rodrigues, R.A., Posekany, A. and Piepho, H.-P. (2024) Genomic prediction using machine learning: A comparison of the performance of regularized regression, ensemble, instance-based and deep learning methods on synthetic and empirical data. *BMC Genomics* 25, 152. https://doi.org/10.1186/s12864-023-09933-x.

Manghwar, H., Hussain, A., Alam, I., Khoso, M.A., Ali, Q. *et al.* (2024) Waterlogging stress in plants: Unraveling the mechanisms and impacts on growth, development, and productivity. *Environmental and Experimental Botany* 224. https://doi.org/10.1016/j.envexpbot.2024.105824.

Martignago, D., Rico-Medina, A., Blasco-Escámez, D., Fontanet-Manzaneque, J.B. and Caño-Delgado, A.I. (2020) Drought resistance by engineering plant tissue-specific responses. *Frontiers in Plant Science* 10, 1676. DOI: 10.3389/fpls.2019.01676.

Montesinos-Lopez, O.A., Montesinos-Lopez, A., Kismiantini, Roman-Gallardo, A., Gardner, K. *et al.* (2022) Partial least squares enhances genomic prediction of new environments. *Frontiers in Genetics* 13, 920689. https://doi.org/10.3389/fgene.2022.920689.

Montesinos-Lopez, O.A., Montesinos-Lopez, A., Perez-Rodriguez, P., Barron-Lopez, J.A., Martini, J.W.R. *et al.* (2021) A review of deep learning applications for genomic selection. *BMC Genomics* 22, 19. https://doi.org/10.1186/s12864-020-07319-x.

Moser, G., Lee, S.H., Hayes, B.J., Goddard, M.E., Wray, N.R. *et al.* (2015) Simultaneous discovery, estimation and prediction analysis of complex traits using a Bayesian mixture model. *PLoS Genetics* 11, e1004969. https://doi.org/10.1371/journal.pgen.1004969.

Nissar, I., Alam, S. and Masood, S. (2024) Computationally efficient LC-SCS deep learning model for breast cancer classification using thermal imaging. *Neural Computing and Applications* 36, 16233–16250. https://doi.org/10.1007/s00521-024-09968-5.

Oshunsanya, S.O., Li, Y. and Yu, H. (2019) Vetiver grass hedgerows significantly reduce nitrogen and phosphorus losses from fertilized sloping lands. *Science of the Total Environment* 661, 86–94. https://doi.org/10.1016/j.scitotenv.2019.01.129.

Poland, J., Endelman, J., Dawson, J., Rutkoski, J., Wu, S. *et al.* (2012) Genomic selection in wheat breeding using genotyping-by-sequencing. *The Plant Genome* 5. https://doi.org/10.3835/plantgenome2012.06.0006.

Porter, J.R. (2005) Rising temperatures are likely to reduce crop yields. *Nature* 436, 174. DOI: 10.1038/436174b. PMID: 16015304.

Raza, A., Bashir, S., Khare, T., Karikari, B., Copeland, R.G.R. *et al.* (2024) Temperature-smart plants: A new horizon with omics-driven plant breeding. *Physiologia Plantarum* 176, e14188. https://doi.org/10.1111/ppl.14188.

Rembe, M., Zhao, Y., Wendler, N., Oldach, K., Korzun, V. *et al.* (2022) The potential of genome-wide prediction to support parental selection, evaluated with data from a commercial barley breeding program. *Plants (Basel)* 11, 2564. DOI: 10.3390/plants11192564.

Rezaei, E.E., Webber, H., Gaiser, T., Naab, J. and Ewert, F. (2015) HS in cereals: Mechanisms and modelling. *European Journal of Agronomy* 64, 98–113. https://doi.org/10.1016/j.eja.2014.10.003.

Rio, S., Mary-Huard, T., Moreau, L. and Charcosset, A. (2019) Genomic selection efficiency and a priori estimation of accuracy in a structured dent maize panel. *Theoretical and Applied Genetics* 132, 81–96. DOI: 10.1007/s00122-018-3196-1.

Roorkiwal, M., Jarquin, D., Singh, M.K., Gaur, P.M., Bharadwaj, C. *et al.* (2018) Genomic-enabled prediction models using multi-environment trials to estimate the effect of genotype x environment interaction on prediction accuracy in chickpea. *Scientific Reports* 8, 11701. https://doi.org/10.1038/s41598-018-30027-2.

Roorkiwal, M., Rathore, A., Das, R.R., Singh, M.K., Jain, A. *et al.* (2016) Genome-enabled prediction models for yield related traits in chickpea. *Frontiers in Plant Science* 7, 1666. https://doi.org/10.3389/fpls.2016.01666.

Ru, C., Hu, X., Chen, D., Wang, W., Zhen, J. *et al.* (2023) Individual and combined effects of heat and drought and subsequent recovery on winter wheat (Triticum aestivum L.) photosynthesis, nitrogen metabolism, cell osmoregulation, and yield formation. *Plant Physiology and Biochemistry* 196, 222–235. https://doi.org/10.1016/j.plaphy.2023.01.038.

Rutkoski, J., Benson, J., Jia, Y., Brown-Guedira, G., Jannink, J.-L. *et al.* (2012) Evaluation of genomic prediction methods for fusarium head blight resistance in wheat. *The Plant Genome* 5, 51–61. https://doi.org/10.3835/plantgenome2012.02.0001.

Saad, N.S.M., Neik, T.X., Thomas, W.J., Amas, J.C., Cantila, A.Y. *et al.* (2022) Advancing designer crops for climate resilience through an integrated genomics approach. *Current Opinion in Plant Biology* 67, 102220. https://doi.org/10.1016/j.pbi.2022.102220.

Sallam, A.H., Endelman, J.B., Jannink, J.L. and Smith, K.P. (2015) Assessing genomic selection prediction accuracy in a dynamic barley breeding population. *Plant Genome* 8, eplantgenome2014 05 0020. https://doi.org/10.3835/plantgenome2014.05.0020.

Sandhu, K.S., Mihalyov, P.D., Lewien, M.J., Pumphrey, M.O. and Carter, A.H. (2021) Combining genomic and phenomic information for predicting grain protein content and grain yield in spring wheat. *Frontiers in Plant Science* 12, 613300. https://doi.org/10.3389/fpls.2021.613300.

Sarinelli, J.M., Murphy, J.P., Tyagi, P., Holland, J.B., Johnson, J.W. *et al.* (2019) Training population selection and use of fixed effects to optimize genomic predictions in a historical USA winter wheat panel. *Theoretical and Applied Genetics* 132, 1247–1261. DOI: 10.1007/s00122-019-03276-6.

Sato, H., Mizoi, J., Shinozaki, K. and Yamaguchi-Shinozaki, K. (2024) Complex plant responses to drought and heat stress under climate change. *The Plant Journal* 117, 1873–1892. https://doi.org/10.1111/tpj.16612.

Shikha, M., Kanika, A., Rao, A.R., Mallikarjuna, M.G., Gupta, H.S. *et al.* (2017) Genomic selection for drought tolerance using genome-wide SNPs in maize. *Frontiers in Plant Science* 8, 550. https://doi.org/10.3389/fpls.2017.00550.

Shrivastava, P. and Kumar, R. (2015) Soil salinity: A serious environmental issue and plant growth promoting bacteria as one of the tools for its alleviation. *Saudi Journal of Biological Sciences* 22, 123–131. DOI: 10.1016/j.sjbs.2014.12.001.

Stergiou, K., Ntakolia, C., Varytis, P., Koumoulos, E., Karlsson, P. *et al.* (2023) Enhancing property prediction and process optimization in building materials through machine learning: A review. *Computational Materials Science* 220, 112031. https://doi.org/10.1016/j.commatsci.2023.112031.

Su, G., Madsen, P., Nielsen, U.S., Mantysaari, E.A., Aamand, G.P. *et al.* (2012) Genomic prediction for nordic red cattle using one-step and selection index blending. *Journal of Dairy Science* 95, 909–917. https://doi.org/10.3168/jds.2011-4804.

Tiede, T. and Smith, K.P. (2018) Evaluation and retrospective optimization of genomic selection for yield and disease resistance in spring barley. *Molecular Breeding* 38, 55–61. https://doi.org/10.1007/s11032-018-0820-3.

Tomar, V., Singh, D., Dhillon, G.S., Chung, Y.S., Poland, J. *et al.* (2021) Increased predictive accuracy of multi-environment genomic prediction model for yield and related traits in spring wheat (Triticum aestivum L.). *Frontiers in Plant Science* 12, 720123. https://doi.org/10.3389/fpls.2021.720123.

VanRaden, P.M. (2008) Efficient methods to compute genomic predictions. *Journal of Dairy Science* 91, 4414–4423. https://doi.org/10.3168/jds.2007-0980.

Varshney, R.K., Barmukh, R., Roorkiwal, M., Qi, Y., Kholova, J. *et al.* (2021) Breeding custom-designed crops for improved drought adaptation. *Advanced Genetics (Hoboken)* 2, e202100017. DOI: 10.1002/ggn2.202100017.

Vivek, B.S., Krishna, G.K., Vengadessan, V., Babu, R., Zaidi, P.H. *et al.* (2017) Use of genomic estimated breeding values results in rapid genetic gains for drought tolerance in maize. *Plant Genome* 10, plantgenome2016.07.0070. https://doi.org/10.3835/plantgenome2016.07.0070.

Wang, K., Abid, M.A., Rasheed, A., Crossa, J., Hearne, S. *et al.* (2023a) DNNGP, a deep neural network-based method for genomic prediction using multi-omics data in plants. *Molecular Plant* 16, 279–293. https://doi.org/10.1016/j.molp.2022.11.004.

Wang, W., Cai, C., He, J., Gu, J., Zhu, G. *et al.* (2020) Yield, dry matter distribution and photosynthetic characteristics of rice under elevated CO2 and increased temperature conditions. *Field Crops Research* 248. https://doi.org/10.1016/j.fcr.2019.107605.

Wang, W., Guo, W., Le, L., Yu, J., Wu, Y. *et al.* (2023c) Integration of high-throughput phenotyping, GWAS, and predictive models reveals the genetic architecture of plant height in maize. *Molecular Plant* 16, 354–373. DOI: 10.1016/j.molp.2022.11.016.

Wang, X., Shi, S., Wang, G., Luo, W., Wei, X. *et al.* (2022) Using machine learning to improve the accuracy of genomic prediction of reproduction traits in pigs. *Journal of Animal Science and Biotechnology* 13, 60. DOI: 10.1186/s40104-022-00708-0.

Wang, X., Xu, Y., Hu, Z. and Xu, C. (2018) Genomic selection methods for crop improvement: Current status and prospects. *The Crop Journal* 6, 330–340. https://doi.org/10.1016/j.cj.2018.03.001.

Wei, Y., Jin, J., Jiang, S., Ning, S. and Liu, L. (2018) Quantitative response of soybean development and yield to drought stress during different growth stages in the Huaibei plain, China. *Agronomy* 8, 97. https://doi.org/10.3390/agronomy8070097.

Westhues, C.C., Mahone, G.S., da Silva, S., Thorwarth, P., Schmidt, M. *et al.* (2021) Prediction of maize phenotypic traits with genomic and environmental predictors using gradient boosting frameworks. *Frontiers in Plant Science* 12, 699589. https://doi.org/10.3389/fpls.2021.699589.

Wolfe, M.D., del Carpio, D.P., Alabi, O., Ezenwaka, L.C., Ikeogu, U.N. *et al.* (2017) Prospects for genomic selection in cassava breeding. *The Plant Genome* 10, plantgenome2017.03.0015. https://doi.org/10.3835/plantgenome2017.03.0015.

Wu, H., Han, R., Zhao, L., Liu, M., Chen, H. *et al.* (2025) AutoGP: An intelligent breeding platform for enhancing maize genomic selection. *Plant Communications* 6, 101240. https://doi.org/10.1016/j.xplc.2025.101240.

Wu, C., Zhang, Y., Ying, Z., Li, L., Wang, J. *et al.* (2024) A transformer-based genomic prediction method fused with knowledge-guided module. *Briefings in Bioinformatics* 25, bbad438. https://doi.org/10.1093/bib/bbad438.

Xiong, S., Wang, M., Zou, J., Meng, J. and Liu, Y. (2020) A two-stage method for improving the prediction accuracy of complex traits by incorporating genotype by environment interactions in Brassica napus. *Discrete Dynamics in Nature and Society* 2020, 1–12. https://doi.org/10.1155/2020/7959508.

Xu, Y., Wang, X., Ding, X., Zheng, X., Yang, Z. *et al.* (2018) Genomic selection of agronomic traits in hybrid rice using an NCII population. *Rice (N Y)* 11, 32. https://doi.org/10.1186/s12284-018-0223-4.

Xu, Y., Zhang, X., Li, H., Zheng, H., Zhang, J. *et al.* (2022) Smart breeding driven by big data, artificial intelligence, and integrated genomic-enviromic prediction. *Molecular Plant* 15, 1664–1695. https://doi.org/10.1016/j.molp.2022.09.001.

Xu, U., Zhang, L., Ou, S., Wang, R., Wang, Y. *et al.* (2020) Natural variations of SLG1 confer high-temperature tolerance in indica rice. *Nature Communications* 11, 5441. https://doi.org/10.1038/s41467-020-19320-9.

Yang, X., Wang, B., Chen, L., Li, P. and Cao, C. (2019) The different influences of drought stress at the flowering stage on rice physiological traits, grain yield, and quality. *Scientific Reports* 9, 1–12. https://doi.org/10.1038/s41598-019-40161-0.

Yao, Z., Yao, M., Wang, C., Li, K., Guo, J. *et al.* (2025) GEFormer: A genomic prediction method of genotype-environment interaction in maize by integrating gating mechanism MLP and linear attention mechanism. *Molecular Plant* 18, 527–549. https://doi.org/10.1016/j.molp.2025.01.020.

Zahra, N., Al Hinai, M.S., Hafeez, M.B., Rehman, A., Wahid, A. *et al.* (2022) Regulation of photosynthesis under salt stress and associated tolerance mechanisms. *Plant Physiology and Biochemistry* 178, 55–69. https://doi.org/10.1016/j.plaphy.2022.03.003.

Zhang, S.W., Li, C.H., Cao, J., Zhang, Y.C., Zhang, S.Q. *et al.* (2009) Altered architecture and enhanced drought tolerance in rice via the down-regulation of indole-3-acetic acid by TLD1/OsGH3.13 activation. *Plant Physiology* 151, 1889–1901. DOI: 10.1104/pp.109.146803.

10 Predictive Breeding for Crop Genetic Improvement

Abdulqader Jighly and Reem Joukhadar
AgriSapiens PTY LTD, Melbourne, 3108 VIC, Australia

Abstract

Genomic prediction has revolutionized breeding by accelerating genetic gain, yet its accuracy can be significantly enhanced by effectively integrating environmental data, particularly to account for complex genotype-by-environment (G×E) interactions. This chapter provides a comprehensive overview of methodologies for leveraging environmental information in genomic prediction models. It explores various approaches, categorized into linear, non-linear and hybrid models. Linear models, including the direct inclusion of environmental covariates, environmental clustering and G×E interaction models (such as reaction norms and multi-trait models utilizing environmental relationship matrices), offer interpretable frameworks for capturing G×E. Non-linear approaches, including kernel-based methods and deep learning models, are highlighted for their ability to capture complex non-additive genetic effects and complex G×E patterns, with examples demonstrating the fusion of diverse data types. The chapter also explores hybrid models, specifically the integration of deep learning with linear mixed models, and ensemble methods, which combine the strengths of different modeling paradigms for improved accuracy and robustness. Finally, future directions emphasizing the integration of diverse omics and phenomics data, advanced AI approaches and the development of user-friendly tools to bridge the gap between research and practical application in breeding are discussed. Such strategies are crucial for developing climate-resilient crops and advancing sustainable agriculture.

10.1 Introduction

Global food security faces critical challenges with the growing human population, coupled with the impacts of climate change, including more frequent extreme weather events, shifting rainfall patterns, and novel pest and disease pressures (Vanaja *et al.*, 2024; Islam, 2025). These challenges require applying transformative strategies to accelerate the rate of genetic gain for different crops. Traditional breeding programs, while successful, often struggle to consistently deliver superior genotypes across the wide range of diverse environments and require a large number of huge field trials beyond financial feasibility (Jauhar, 2006). This inconsistency is largely attributable to genotype-by-environment interaction (G×E), a phenomenon where the relative performance of genotypes changes depending on the local environmental conditions in which they are grown (Kang, 1997). Overcoming G×E is utmost for developing resilient and high-yielding varieties that can succeed in variable conditions, thereby ensuring stable food production (De Leon *et al.*, 2016).

*Emails: a.jighly@agrisapiens.com.au; r.joukhadar@agrisapiens.com.au

 Genomic Solutions for Breeding Climate-resilient Crops (R.K. Varshney, R. Barmukh and R.J. Harper)
DOI: 10.1079/9781800627185.0010

The introduction of genomic selection (GS) has revolutionized breeding by enabling the prediction of an individual's genetic merit using genome-wide molecular markers (Meuwissen *et al.*, 2001). GS has been demonstrated to significantly shorten breeding cycles and increase selection efficiency for different crops (Heffner *et al.*, 2010). However, conventional GS models assume that genetic effects are stable across environments and they either average the trials or analyze them independently. While powerful, these models often fall short in predicting performance accurately in novel or highly variable environments, because they do not account for the complex interaction between genotype and environment (Burgueño *et al.*, 2012). Therefore, they usually ignore the wealth of information that can be leveraged by modeling the relations across different trials (Lopez-Cruz *et al.*, 2015).

To address this gap, there is a critical need to synergize environmental and genomic data to shift GS from genotype-centric predictions to a more holistic understanding of performance (Heslot *et al.*, 2014). Such advance seeks not only achieving greater precision in predicting breeding values but also establishing predictions that are highly robust and reliable even when meeting the variability of different environments (Jarquín *et al.*, 2014). This involves moving beyond the "genotype-only" paradigm to systematically incorporate detailed environmental information, allowing breeders to better understand and exploit G×E. Integrating environmental characterization (envirotyping) with genomic data could make it possible to predict how different genotypes will perform under specific environmental conditions, facilitate targeted breeding for specific environments, and identify genotypes with broad adaptation (Prado *et al.*, 2024).

This book chapter aims to provide a comprehensive overview of the methodologies employed to integrate environmental and genomic data, ultimately to improve genetic gain in diverse environments. We will first explore the critical aspects of environmental data acquisition and processing, a process termed "envirotyping", highlighting how raw environmental observations are transformed into meaningful covariates. Subsequently, we will categorize various modeling approaches, from established linear mixed models that explicitly account for G×E, to cutting-edge non-linear and hybrid methods, including deep learning, that unravel the most complex genotype–environment interactions. Finally, we will discuss future directions in this rapidly evolving field, emphasizing its transformative potential for sustainable agriculture.

10.2 Environmental Data: Acquisition and Processing

The effective integration of environmental data into genomic prediction models depends on envirotyping, a sophisticated process dedicated to the comprehensive characterization and quantification of environmental factors that influence phenotypic expression (Xu *et al.*, 2012; Resende *et al.*, 2021). At its core, envirotyping aims to precisely define the "environment" encountered by an organism, moving beyond simple location identifiers to capture the dynamic interplay of various climate variables (Xu, 2015). This detailed characterization is crucial for understanding the complexities of genotype-by-environment interaction (G×E) and, ultimately, for optimizing breeding programs by enabling targeted selection and variety deployment (Elmerich *et al.*, 2023).

The initial phase of envirotyping involves the systematic collection of diverse environmental data. This incorporates a broad spectrum of information, including macro-environmental factors such as temperature, precipitation and solar radiation, typically sourced from weather stations and advanced remote sensing technologies such as satellites and drones (Xu, 2016; Olson and Anderson, 2021). It also extends to micro-environmental details, often captured by on-site sensors, and includes critical edaphic factors derived from soil analyses (e.g., soil type, pH, nutrient content, water holding capacity) (Whelan and Taylor, 2013). Furthermore, management practices, such as irrigation, fertilization, and pest control, are components of the environmental context and should be recorded (Yue *et al.*, 2022). A significant portion of this data is collected within multi-environment trials (METs), which serve as invaluable platforms for observing genotypic performance across a range of real-world conditions.

Once collected, raw environmental data requires rigorous processing and feature engineering. This crucial step transforms environmental data into biologically meaningful metrics suitable for statistical modeling. A key output of this phase is the derivation of environmental covariates (ECs), quantitative descriptors that capture specific environmental stresses (Bicard *et al.*, 2025). These ECs are then used to construct environmental relatedness matrices (ERM), which quantify the similarities or differences between environments, similar to how genomic relatedness matrices (GRM) describe genetic relationships (Jarquín *et al.*, 2014; Saint Pierre *et al.*, 2016). ECs can be calculated by evaluating environmental variable values for each field plot at specific developmental stages, such as heading, flowering and maturity, to capture their stage-specific impacts on plant performance at crucial developmental phases. This approach aligns the EC data with the scale of field phenotypic data, i.e. equivalent matrix size, thereby generating equivalent plot-level environmental descriptors (He *et al.*, 2019). Furthermore, environmental data can be directly utilized, often on a daily base, within biophysical crop growth models to simulate day-to-day plant development and biomass accumulation under different environmental scenarios, providing a deeper mechanistic understanding of the effect of environmental data on crop performance (Bellocchi *et al.*, 2010).

Beyond the initial data acquisition and detailed processing, envirotyping often extends to more detailed characterization of grouping the environments. Using the carefully derived ECs, advanced statistical methods are applied to detect patterns and identify distinct "envirotypes." These envirotypes represent groups of environments that share similar environmental profiles, and consequently, are expected to lead to comparable genotypic responses (Costa-Neto *et al.*, 2024). This process often involves multivariate statistical techniques such as clustering algorithms (e.g. K-means, hierarchical clustering or factor analytic) to group environments into "mega-environments" or smaller, more homogeneous clusters (Heslot *et al.*, 2013; Winans *et al.*, 2024). A mega-environment is typically defined as a large geographic region comprising a set of environments that consistently share similar environmental characteristics and to which genotypes respond in a similar manner (Yan *et al.*, 2000). Such characterization provides a powerful framework for breeders to understand environmental similarities and differences, optimize the design of future METs, and make informed decisions regarding the selection and deployment of genotypes best suited for specific environmental conditions (Yan *et al.*, 2023).

10.3 Integrating Environmental and Genomic Data in Prediction

The effective utilization of envirotyping data in genomic prediction models requires a diverse toolkit of statistical and machine learning methodologies. These methods aim to capture the complex relationships between genotypes, environments and phenotypes, thereby improving prediction accuracy and enabling more informed selection decisions (Jubair and Domaratzki, 2023). Broadly, these approaches can be categorized into linear, non-linear and hybrid models (Figure 10.1), each offering distinct advantages in modeling the complex interaction of genetic and environmental effects. This section will detail the various methodologies employed to integrate environmental and genomic information, highlighting their underlying principles and applications in enhancing genetic gain across diverse environments.

10.3.1 Linear Models

Linear models form the foundational method of genomic prediction, largely due to their interpretability, computational efficiency and well-established statistical properties. These models typically assume additive genetic effects and linear relationships between genetic markers, environmental covariates and phenotypic outcomes (Crossa *et al.*, 2017). Despite the complexity of biological systems, linear models have proven remarkably effective in capturing a significant portion of the genetic variation and G×E interactions. Their strength lies in their ability to incorporate environmental information as fixed or random effects, or through interaction terms, providing a direct framework for understanding how environmental factors modulate genotypic performance (Desta and Ortiz, 2014). We will explore various linear modeling

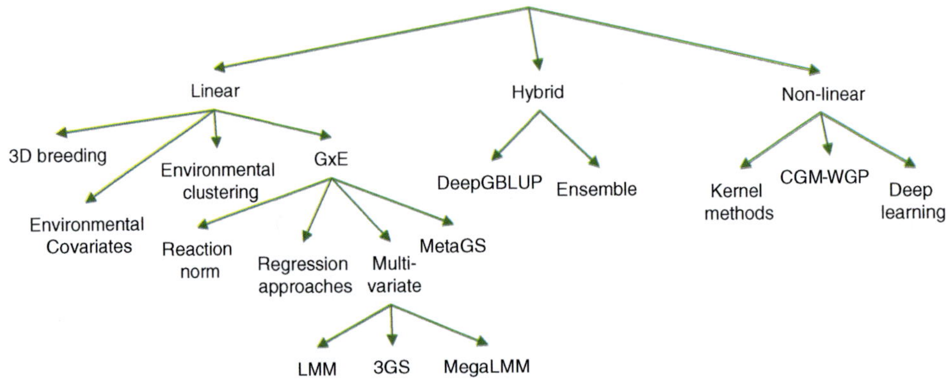

Figure 10.1. The three major classifications of genomic prediction models (linear, non-linear and hybrid) that integrate environmental data. (Authors' own image.)

approaches that integrate environmental data to enhance genomic prediction accuracy.

10.3.1.1 Inclusion of Environmental Covariates

One of the most straightforward approaches to integrate environmental data into genomic prediction is the direct inclusion of ECs into linear mixed models. In this method, relevant environmental factors (e.g. average temperature during flowering, cumulative rainfall during grain filling, soil pH at planting) are recorded for each field plot, often considering their specific developmental phases such as heading or maturity. These plot-level, phase-specific ECs are then incorporated in the prediction model alongside genomic information. For instance, a model might include terms for the overall genomic effects, the main effects of specific environments and the direct effects of these detailed environmental covariates. Montesinos-López *et al.* (2024b) followed this approach to demonstrate a baseline for their investigations into more advanced feature engineering. Environmental factors serve as direct descriptors or predictors alongside genomic information, influencing prediction accuracy based on their main effects rather than through a statistical interaction term that specifically quantifies how different genotypes respond uniquely to varying environmental conditions.

10.3.1.2 Environmental Clusters

Another strategy within the linear modeling framework involves leveraging environmental clusters. After environments are grouped into "mega-environments" based on shared environmental characteristics, genomic prediction models can be developed either *within* each cluster or by explicitly accounting for these clusters in a multi-environment model. For within-cluster prediction, separate genomic prediction models are trained and validated for each identified environmental cluster. This approach assumes that genotypes behave more consistently within a homogeneous environmental cluster, leading to higher prediction accuracies (Lin *et al.*, 2021). Such clustering can be achieved through factor analytic linear mixed models (FALMM). This approach systematically defines "interaction classes" (iClasses) based on the estimated parameters of the FALMM (Smith *et al.*, 2021). Specifically, iClasses are formed by categorizing environments according to the consistent signs (positive or negative) of their estimated factor loadings across multiple factors. This method effectively groups environments where crossover variety by environment interaction is minimized internally but can be substantial between groups, thereby simplifying the analysis of complex G×E patterns by creating more homogeneous environmental contexts.

10.3.1.3 G×E Interaction Models

Genotype-by-environment (G×E) interaction occurs when the relative performance of genotypes changes across different environments (Kang, 1997). Effectively modeling G×E is crucial for accurate genomic prediction, especially in diverse agricultural settings where environmental conditions vary significantly. Linear models provide several powerful frameworks for explicitly accounting for G×E (Malosetti *et al.*, 2013), allowing breeders to identify genotypes that are either broadly adapted or specifically suited to specific environmental conditions. This section will describe major strategies to model G×E for genomic prediction.

10.3.1.3.1 REACTION NORM MODELS. Reaction norm models offer a robust approach to explicitly model G×E when environmental variation can be described by continuous variables, such as temperature, humidity or nutrient levels (Hayes *et al.*, 2016). These models describe the phenotypic response of each genotype as a function of the environmental covariate, effectively modeling a "reaction norm" for each genotype across an environmental gradient (Jarquín *et al.*, 2014). EC are integrated in these models by estimated environmental relatedness matrices (ERM) which represent a covariance structure among environments. An ERM quantifies the similarities or differences between environments, analogous to Genomic Relationship Matrices (GRMs) that describe genetic relationships among individuals (He *et al.*, 2019). A reaction norm model incorporates high-dimensional genomic and environmental data by modeling interactions between markers and ECs using covariance functions, ultimately increasing prediction accuracy (Jarquín *et al.*, 2014, He *et al.*, 2019). Reaction norm models estimate a single GEBV for each reference environment. This framework enables the identification of "robust" genotypes that exhibit stable performance across a range of environments or "sensitive" genotypes that show a strong response to specific environmental changes.

10.3.1.3.2 REGRESSION-BASED APPROACHES. Regression-based approaches offer a powerful framework for integrating ECs into genomic prediction models, thereby enhancing the capture and prediction of G×E (Rebollo *et al.*, 2023). Heslot *et al.* (2014) propose extending factorial regression models for genomic selection, where they derive "stress covariates" from daily weather data using a crop model, allowing them to model genotype-specific sensitivities to these environmental stresses and improve prediction in unobserved environments. Tolhurst *et al.* (2019) develop an integrated FALMM that combines both known environmental covariates and latent covariates, leveraging random regressions to make the modeled G×E predictable and enable accurate predictions into current and future environments. Collectively, these papers demonstrated the evolution from simply including ECs to developing complex models that leverage them to understand and predict specific genotype responses to varied environmental conditions.

10.3.1.3.3 MULTI-VARIATE MODELS. Multi-trait models or multi-variate linear mixed models (MvLMM) provide another flexible framework for handling G×E. In this approach, the performance of a genotype for a given trait in each different environment is treated as a separate but genetically correlated trait (Hayes *et al.*, 2016). For example, grain yield production in Environment A and grain yield production in Environment B are considered two distinct, yet related, traits. The genetic correlation between these "traits" (performance in different environments) quantifies the degree of re-ranking of genotypes across environments. A genetic correlation significantly less than one indicates substantial G×E, suggesting that separate breeding programs or selection strategies might be beneficial for those environments. In genomic prediction, MvLMMs can be used, where the genomic relationship matrix is incorporated to model the genetic covariance structure among the performances in different environments. This allows for the estimation of GEBVs for each genotype in each specific environment, even if a genotype has not been directly evaluated in all environments, by leveraging genetic relationships and shared environmental information.

10.3.1.3.4 ADVANCED MULTI-VARIATE MODELS FOR HIGH-DIMENSIONAL DATA. As high-throughput phenotyping and multi-environment trials generate increasingly vast and complex datasets, computational efficiency becomes a critical factor in genomic prediction. Traditional MvLMMs, while statistically powerful, can become computationally prohibitive when applied to hundreds or thousands of traits due to the quadratic increase in the number of parameters and the need for repeated large matrix inversions. To address these challenges, innovative methods have been developed to enhance the scalability and robustness of multi-trait genomic prediction, often by leveraging dimensionality reduction or re-parameterization techniques.

One such method is 3GS (Genotype plus Genotype × Environment Genomic Selection) (Jighly *et al.*, 2021). 3GS offers a computationally efficient framework for modeling G×E by integrating genomic selection with Genotype plus G×E (GGE) biplot analysis. The core innovation of 3GS lies in its ability to calculate the principal components (PCs) of the environmental phenotypes using GGE analysis. Instead of directly predicting the complex multi-environment phenotypes, 3GS estimates SNP effects for each of these GGE PCs. This effectively transforms a high-dimensional multi-environment prediction problem into a series of more manageable, independent univariate genomic prediction problems (one for each PC). The GEBVs of these PCs (pcGEBVs) are then converted back to the original phenotypic scale. This approach significantly reduces computational complexity, allowing the analysis to increase linearly with the number of environments, making it hundreds to thousands of times faster than standard models for large datasets. Furthermore, 3GS demonstrated higher prediction accuracy, particularly in environments with low or negative correlations to other environments and showed robustness in predicting new genotypes in unobserved environments.

Another powerful advancement in handling high-dimensional multi-trait data is MegaLMM (Mega-scale Linear Mixed Models) (Runcie *et al.*, 2021). MegaLMM is a novel statistical framework designed for fitting massive-scale MvLMMs to datasets with thousands of traits, overcoming the computational burden of naive MvLMM implementations. It re-parameterizes the MvLMM as a mixed-effect factor model and introduces a set of unobserved (latent) factors that account for all sources of correlation among the observed traits. Modeling these latent factors with independent linear mixed models avoids computationally demanding large matrix inversions, dramatically improving computational performance. MegaLMM can leverage thousands of traits simultaneously to significantly improve genetic value prediction accuracy, demonstrating scalability that extends the feasible range of input data for MvLMMs by at least two orders of magnitude over standard methods. Recently Hu *et al.* (2025) extended the application of MegaLMM to incorporate ECs, which allowed them to predict new environments.

10.3.1.3.5 USING SUMMARY STATISTICS (META-ANALYSIS). Meta-genomic analysis refers to the integration of genomic information from multiple studies or populations, often leveraging summary statistics rather than raw individual-level data (Speed and Balding, 2019). This approach is particularly valuable when raw data sharing is restricted due to privacy concerns or logistical challenges, yet combining information across diverse populations can significantly boost statistical power and prediction accuracy (Chen *et al.*, 2021). A key method in this domain for genomic prediction is MetaGS, which stands out for its ability to mimic the performance of MvLMM, but by using only summary statistics (Jighly *et al.*, 2022).

MetaGS operates by exploiting the genetic correlations among different populations or studies to generate more accurate, population-specific SNP effects. Similar to MvLMM that treat performance in different environments as correlated traits, MetaGS views SNP effects across different populations as genetically correlated, allowing information to be borrowed from related populations. This is achieved by requiring participating data holders to share summary-level information, such as linkage disequilibrium (LD) matrices, marker allele frequencies, SNP effect estimates and variance components. When shared, this type of information cannot be used to infer the original data and can only serve the meta-analysis run. A notable advantage of MetaGS is its flexibility in handling situations where different populations are genotyped with varying sets of SNPs. MetaGS can impute missing summary statistics without requiring access to raw individual data, which overcome a significant practical limitation of many meta-analysis applications. MetaGS facilitates more flexible collaborations among breeding programs and research groups, leading to improved genomic predictions. This enhances the reliability of GEBVs across a wider range of genetic backgrounds and environments given that it increases the reference population size.

10.3.1.4 The 3D Breeding Method

The 3D breeding method, or data-driven decentralized breeding, represents a paradigm shift in crop improvement, with major application so far for smallholder agricultural systems operating in challenging and diverse environments (de Sousa *et al.*, 2021). This approach addresses the limitations of conventional centralized breeding programs, which often struggle to develop varieties optimally adapted to the highly variable conditions encountered on actual farms (van Etten *et al.*, 2023). Unlike centralized breeding that relies on data collected from limited research stations, 3D-breeding decentralizes variety evaluation by engaging numerous farmers. Farmers test candidate genotypes in small, blinded, and randomized plots directly on their fields. This generates a large volume of on-farm phenotypic and environmental data, which complements the "big data" dimension of genomics. The method leverages this rich, real-world data to capture complex Genotype-by-Environment-by-Management (G×E×M) interactions that are often missed in controlled research settings. 3D breeding aims to enhance local adaptation by combining genomic prediction with farmer knowledge and detailed environmental characterization (Gesesse *et al.*, 2023). 3D breeding could significantly increase prediction accuracy for traits such as grain yield and overall farmer appreciation, outperforming centralized genomic prediction. This method offers a promising pathway to develop and deploy resilient crop varieties tailored to the specific needs and diverse conditions of smallholder farmers, ultimately contributing to global food security in the face of climate change.

10.3.2 Non-linear Models

While linear models provide a powerful and interpretable framework for genomic prediction, biological systems often exhibit complex, non-linear relationships that may not be adequately captured by linear assumptions (Montesinos-López *et al.*, 2024a). Non-linear models offer the flexibility to model these complex patterns, potentially leading to improved prediction accuracy, especially when dealing with highly complex traits or environments where genotypic responses are not simply proportional to environmental changes. These models can uncover hidden interactions and dependencies that linear approaches might overlook, providing a more comprehensive understanding of the genetic and environmental architecture of traits (Montesinos-López *et al.*, 2021). This subsection will explore various non-linear modeling approaches that integrate environmental and genomic data for enhanced prediction.

10.3.2.1 Kernel-based Methods

Kernel-based methods represent a sophisticated class of statistical approaches that have significantly advanced genomic prediction, offering a robust framework for unraveling the complex relationships between genotypes, phenotypes and environments (Morota and Gianola, 2014). These methods utilize kernel functions to map genomic data from its original space into a high-dimensional feature space, where complex, non-linear relationships can be more easily identified. This inherent capability allows kernel-based models, such as those employing reproducing kernel Hilbert spaces (RKHS), to effectively capture and account for non-additive genetic effects, including both dominance and various forms of epistasis, which are often overlooked by simpler linear genomic prediction models (Morota and Gianola, 2014; Li *et al.*, 2025).

Beyond capturing complex genetic architectures, kernel-based methods are particularly powerful in modeling G×E. Traditional models often struggle to adequately capture these dynamic interactions, leading to reduced prediction accuracy when genotypes are moved to new or different environments. Kernel methods address this by constructing G×E interaction kernels, which can integrate environmental information (Cuevas *et al.*, 2016). This integration typically involves creating an "environmental kernel" or an environmental similarity matrix based on relevant environmental covariates (e.g., temperature, rainfall, soil type). This environmental similarity matrix is then combined with the genomic similarity matrix (often via the Hadamard product) to form a comprehensive G×E kernel (Jarquín *et al.*, 2014). This approach effectively models the varying performance of genotypes across different environments (Cuevas *et al.*, 2017).

Furthermore, kernel-based methods are powerful for modeling non-linear G×E interactions.

This can be achieved by constructing genomic and environmental kernels derived using non-linear kernel functions (e.g. a Gaussian kernel based on environmental distances). The environmental kernel itself can be constructed directly from ECs, as demonstrated by Gevartosky *et al.* (2023) with their genotype × environmental covariates × trait kernel, which creates a relationship matrix between environments based on ECs. This allows the model to learn complex patterns where a genotype's response to an environmental factor is not a simple linear function, but rather a more complex relationship. These methods provide a flexible framework to account for complex biological processes and have shown promise in enhancing prediction accuracy in various breeding contexts.

10.3.2.2 CGM-WGP Model

The Crop Growth Model-Whole Genome Prediction (CGM-WGP) model represents a significant advancement in genomic prediction by directly integrating mechanistic crop growth models (CGMs) with genomic prediction within a single Bayesian framework. This approach, initially proposed by Technow *et al.* (2015) and further developed by Messina *et al.* (2018), aims to overcome the limitations of purely statistical genomic prediction models in capturing complex genotype-by-environment-by-management (G×E×M) and the inability of traditional CGMs to predict novel genotypes. A key improvement of CGM-WGP is transferring the standard CGM models from single cultivar level simulation to a population level simulation, allowing for more accurate predictions.

CGM-WGP estimates the effects of genomic variants not on the observed phenotypes directly, but on genotype-specific parameters (GSPs), also known as latent traits. These GSPs describe essential physiological processes within the deterministic equations of a CGM (e.g. phenology, biomass accumulation, nutrient uptake). These GSPs are supposed to be highly heritable with very limited interaction with the environment (Heslot *et al.*, 2014). The model can then simulate the final observed phenotypes (such as grain yield, heading date or protein content) under various environmental conditions using these predicted underlying physiological parameters. This hierarchical structure allows CGM-WGP to account for non-linear relationships and G×E×M interactions that emerge from the biophysical processes modeled by the CGM (Technow *et al.*, 2015). CGM-WGP sampled more accurate and more heritable GSPs, compared with the standard CGM, and allowed for predicting new traits that have no phenotypic records in any reference environments if they can be modeled through the CGM (Jighly *et al.*, 2023a).

A key advantage of CGM-WGP is its ability to predict the performance of new, non-phenotyped genotypes in untested environments, including future climates, by leveraging the genomic information to infer their GSPs (Jighly *et al.*, 2023b). This is a substantial improvement over traditional CGMs and genomic prediction models, which are typically calibrated for specific cultivars and cannot be readily extended to new genetic material, and deviated environments, respectively. Furthermore, another recent study (Jighly *et al.*, 2024) has shown that CGM-WGP can effectively address the challenge of missing environmental records in historical reference populations which is a major limitation of standard CGMs (Hammer *et al.*, 2014). By statistically sampling and optimizing crucial environmental variables (e.g. initial soil water content) within the CGM-WGP algorithm, prediction accuracy for traits such as grain yield can be significantly improved, enabling the utilization of valuable historical data that would otherwise be unusable. This integration of biological knowledge through CGMs offers a more comprehensive and robust approach to genomic prediction, particularly for complex traits influenced by G×E dynamics.

10.3.2.3 Deep Learning

Deep learning (DL), a subset of machine learning characterized by multi-layered neural networks, has emerged as a powerful tool for modeling complex, non-linear relationships in high-dimensional data, making it increasingly relevant for genomic prediction (Azodi *et al.*, 2019; Montesinos-López *et al.*, 2021). Unlike traditional linear models or even some kernel methods, DL models can automatically learn complex patterns, hierarchical features and non-additive interactions (e.g. dominant and epistasis) directly from raw genomic and environmental data without requiring explicit feature engineering. This capacity is advantageous for capturing the highly complex and often non-linear nature of G×E, resulting in an improved prediction accuracy (Feng *et al.*, 2024).

A significant strength of deep learning lies in its ability to integrate diverse data modalities, including environmental covariates. Multi-modal neural networks are specifically designed for this purpose, enabling the fusion of heterogeneous data types such as genomic sequences, high-throughput phenotypic data and environmental information. These networks typically process each modality through specialized sub-networks (e.g. Convolutional Neural Networks for images, Fully Connected Networks for genomic data) and then combine their learned representations at various layers within the overall architecture. In this context, Togninalli *et al.* (2023) introduced PheGeMIL (Phenotype-Genotype Multiple Instance Learning), a deep learning framework that integrates multispectral and thermal imagery, digital elevation models and genomic variants to predict grain yield in wheat. Their model effectively leverages a multiple instance learning approach with attention mechanisms to weigh the contributions of different input channels and temporal observations, demonstrating superior accuracy in predicting yield in both similar and unseen environments. This multi-modal integration allows deep learning models to capture the dynamics and interactions between genetic potential and environmental influences, leading to more accurate predictions.

While deep learning offers high potential for uncovering hidden genetic and G×E architectures as well as improving prediction accuracy, challenges remain in terms of model interpretability (understanding why a prediction is made), computational demands and the need for very large training datasets to prevent overfitting (Qazi *et al.*, 2022; Bhat *et al.*, 2023; Novielli *et al.*, 2024). Nevertheless, the ongoing development of more efficient architectures and training strategies continues to position deep learning as a frontier in non-linear genomic prediction.

10.3.3 Hybrid/Integrated Models (Combining Linear and Non-Linear Approaches)

The increasing complexity of genetic architecture and G×E often requires models that can capture both linear (e.g. additive genetic effects) and non-linear (e.g. dominant, epistatic interactions and complex G×E patterns) relationships. Hybrid or integrated models could represent a powerful class of approaches that combine the strengths of different statistical and machine learning paradigms, aiming to achieve higher prediction accuracy and better interpretability than either linear or non-linear models alone. These models often leverage the robustness and interpretability of linear mixed models while incorporating the flexibility of non-linear methods to capture more complex biological signals. This section will explore various strategies for integrating linear and non-linear components to enhance genomic prediction.

10.3.3.1 Deep Learning Integrated with Linear Mixed Models

A promising direction in genomic prediction involves the integration of deep learning with the widely demonstrated LMM, such as the genomic best linear unbiased prediction (GBLUP) framework, named deepGBLUP (Lee *et al.*, 2023). This hybrid approach aims to harness the LMM's ability to model additive genetic effects efficiently, while leveraging deep learning's capacity to capture complex non-linear relationships and interactions. Such hybrid approach can serve multiple functions. Firstly, it can be utilized to extract sophisticated, non-linear features directly from raw genomic marker data. These learned features are then be passed to the LMM, either to construct a non-linear genomic relationship matrix or to directly estimate genetic effects that extend beyond additive assumptions. Secondly, the deep learning part is adept at modeling the complex, non-linear components of G×E. Its output can be integrated into the LMM, allowing the model to capture both the additive genetic and environmental effects, while the deep learning segment focuses on capturing the complex non-linear G×E patterns. Thirdly, deep learning can act as a powerful pre-processor and integrator for diverse data types. It can process various inputs, such as environmental covariates, high throughput phenomics data, and other multi-omics data, transforming them into a unified, information-rich representation. This integrated representation can then be used in the LMM, significantly enhancing its predictive power by providing a more comprehensive view of the underlying biological and environmental effects.

The primary advantage of these hybrid models lies in their ability to potentially achieve higher prediction accuracies than purely linear or non-linear models. Looking ahead, this method holds significant promise for future applications, particularly in its capacity to fully leverage the increasing volume and diversity of environmental data. A key strategy for future development involves designing deep learning architectures that are specifically optimized to learn and represent environmental dynamics and their interactions with genomic features. These rich environmental representations can be subsequently fed into LMMs to enhance G×E modeling and prediction across a wide range of real-world scenarios. Ultimately, this will enable plant breeders to make more accurate, data-driven decisions across a wide array of real-world scenarios such as predicting crop yields under shifting climates or assessing livestock performance in diverse production systems.

10.3.3.2 Ensemble Methods

Ensemble methods represent a robust class of hybrid approaches in genomic prediction that combine the predictions from multiple individual models to achieve higher accuracy and improved robustness compared with any single model (Cooper *et al.*, 2025). This strategy is motivated by fundamental machine learning principles, the "No Free Lunch Theorem" (Wolpert and Macready, 2002), which suggests that no single prediction model is universally superior across all prediction scenarios, and the "Diversity Prediction Theorem" (Page, 2008), which suggests that combining diverse predictions from multiple models can reduce overall prediction error. Ensemble methods can capture a more comprehensive view of the complex genetic architecture and G×E by leveraging the strengths of different algorithms and mitigating their individual weaknesses.

Various ensemble strategies are employed in genomic prediction. One prominent approach is stacking ensemble learning, as demonstrated by Liang *et al.* (2021) with their stacking ensemble learning framework (SELF). In this framework, multiple "base learners" are trained on the original genomic data. Their predictions then serve as meta-data which are fed into a meta-learner to generate the final prediction. This hierarchical structure allows the meta-learner to learn how to optimally combine the strengths of the diverse base models. Another common strategy is the naïve ensemble-average model, (Tomura *et al.*, 2025). This simpler approach combines predictions from various individual genomic prediction models by simply averaging their outputs. Tomura *et al.* (2025) showed that this ensemble approach significantly increased prediction accuracies and reduced prediction errors for traits in maize, attributing the advantage to the diversity of predictions among the individual models. While these studies primarily focus on combining different genomic prediction algorithms, the principles of ensemble learning are highly applicable to integrating environmental data. Future advancements could involve developing ensemble frameworks where individual models are specifically designed to incorporate different types of environmental data and then combining their environmentally informed predictions to further enhance the accuracy and robustness of genomic prediction across diverse environments.

10.4 Conclusion and Future Directions

The field of genomic prediction is continuously evolving, driven by advancements in data generation technologies and computational methodologies. Future directions should focus on further enhancing prediction accuracy, improving model utility to integrate more data types, and bridging the gap between cutting-edge research and practical breeding applications.

One key area for future development is the deeper integration of diverse data types. Beyond genomics and traditional environmental covariates, the incorporation of high-resolution phenomics data (e.g. from drones, sensors and imaging platforms), epigenomics (e.g. DNA methylation, histone modifications), transcriptomics, proteomics, soil microbial and metabolomics will provide a more holistic understanding of the genotype-to-phenotype pathway. Advanced multi-modal deep learning architectures will be crucial for effectively integrating these disparate data sources, enabling models to capture more complex biological processes and their interactions with the environment.

Further advancements in advanced machine learning and AI approaches will continue to boost genomic prediction accuracy. This includes the exploration of novel deep learning architectures tailored for specific biological problems, the development of explainable AI (XAI) techniques to enhance model interpretability, and the application of emerging deep learning structures for optimizing breeding strategies over multiple generations. The development of more sophisticated ensemble methods that can intelligently weight and combine predictions from various models, including those incorporating different environmental data types, will also be a critical area of focus.

The development of user-friendly tools and software is paramount for translating research innovations into practical impact. Complex models require user-friendly interfaces and robust computational pipelines that can be easily adopted by breeders and geneticists without extensive statistical and programming knowledge. This includes tools for efficient data preprocessing, model training, prediction, visualization of results, and chatbots all designed to seamlessly integrate into existing breeding program workflows.

Ultimately, bridging the gap between research and practical application remains a central objective. This involves not only developing powerful models but also validating their effectiveness in diverse real-world breeding scenarios, demonstrating clear benefits in terms of genetic gain and resource efficiency. Collaborative efforts between researchers, breeders and technology developers will be essential to ensure that latest methodologies are translated into practical improvements in crop and livestock breeding, contributing to global food security and sustainable agriculture.

References

Azodi, C.B., Bolger, E., McCarren, A., Roantree, M., de Los Campos, G. *et al.* (2019) Benchmarking parametric and machine learning models for genomic prediction of complex traits. *G3: Genes, Genomes, Genetics* 9(11), 3691–3702. https://doi.org/10.1534/g3.119.400498.

Bellocchi, G., Rivington, M., Donatelli, M. and Matthews, K. (2010) Validation of biophysical models: issues and methodologies. A review. *Agronomy for Sustainable Development* 30(1), 109–130.

Bhat, J.A., Feng, X., Mir, Z.A., Raina, A. and Siddique, K.H. (2023) Recent advances in artificial intelligence, mechanistic models, and speed breeding offer exciting opportunities for precise and accelerated genomics-assisted breeding. *Physiologia Plantarum* 175(4), e13969.

Bicard, M., Faucon, M.P., Dockter, C., Vequaud, D., Pin, P.A. *et al.* (2025) Envirotyping to drive spring barley adaptation in Northwestern Europe. *Field Crops Research* 326, 109793.

Burgueño, J., de los Campos, G., Weigel, K. and Crossa, J. (2012) Genomic prediction of breeding values when modeling genotype× environment interaction using pedigree and dense molecular markers. *Crop Science* 52(2), 707–719.

Chen, W., Wu, Y., Zheng, Z., Qi, T., Visscher, P.M. *et al.* (2021) Improved analyses of GWAS summary statistics by reducing data heterogeneity and errors. *Nature Communications* 12(1), 7117.

Cooper, M., Tomura, S., Wilkinson, M.J., Powell, O. and Messina, C.D. (2025) Breeding perspectives on tackling trait genome-to-phenome (G2P) dimensionality using ensemble-based genomic prediction. *Theoretical and Applied Genetics* 138(7), 172.

Costa-Neto, G., Matta, D.H.D., Fernandes, I.K., Stone, L.F. and Heinemann, A.B. (2024) Environmental clusters defining breeding zones for tropical irrigated rice in Brazil. *Agronomy Journal* 116(3), 931–955.

Crossa, J., Pérez-Rodríguez, P., Cuevas, J., Montesinos-López, O., Jarquín, D. *et al.* (2017) Genomic selection in plant breeding: Methods, models, and perspectives. *Trends in Plant Science* 22(11), 961–975.

Cuevas, J., Crossa, J., Montesinos-López, O.A., Burgueño, J., Pérez-Rodríguez, P. *et al.* (2017) Bayesian genomic prediction with genotype× environment interaction kernel models. *G3: Genes, Genomes, Genetics* 7(1), 41–53.

Cuevas, J., Crossa, J., Soberanis, V., Pérez-Elizalde, S., Pérez-Rodríguez, P. *et al.* (2016) Genomic prediction of genotype× environment interaction kernel regression models. *The Plant Genome* 9(3), plantgenome2016-03.

De Leon, N., Jannink, J.L., Edwards, J.W. and Kaeppler, S.M. (2016) Introduction to a special issue on genotype by environment interaction. *Crop Science* 56(5), 2081–2089.

de Sousa, K., van Etten, J., Poland, J., Fadda, C., Jannink, J.L. *et al.* (2021) Data-driven decentralized breeding increases prediction accuracy in a challenging crop production environment. *Communications Biology* 4(1), 944.

Desta, Z.A. and Ortiz, R. (2014) Genomic selection: Genome-wide prediction in plant improvement. *Trends in Plant Science* 19(9), 592–601.

Elmerich, C., Faucon, M.P., Garcia, M., Jeanson, P., Boulch, G. *et al.* (2023) Envirotyping to control genotype x environment interactions for efficient soybean breeding. *Field Crops Research* 303, 109113.

Feng, W., Gao, P. and Wang, X. (2024) AI breeder: Genomic predictions for crop breeding. *New Crops* 1, 100010.

Gesesse, C.A., Nigir, B., de Sousa, K., Gianfranceschi, L., Gallo, G.R. *et al.* (2023) Genomics-driven breeding for local adaptation of durum wheat is enhanced by farmers' traditional knowledge. *Proceedings of the National Academy of Sciences* 120(14), e2205774119.

Gevartosky, R., Carvalho, H.F., Costa-Neto, G., Montesinos-López, O.A., Crossa, J. *et al.* (2023) Enviromic-based kernels may optimize resource allocation with multi-trait multi-environment genomic prediction for tropical maize. *BMC Plant Biology* 23(1), 10.

Hammer, G.L., McLean, G., Chapman, S., Zheng, B., Doherty, A. *et al.* (2014) Crop design for specific adaptation in variable dryland production environments. *Crop and Pasture Science* 65(7), 614–626.

Hayes, B.J., Daetwyler, H.D. and Goddard, M.E. (2016) Models for genome× environment interaction: Examples in livestock. *Crop Science* 56(5), 2251–2259.

He, S., Thistlethwaite, R., Forrest, K., Shi, F., Hayden, M.J. *et al.* (2019) Extension of a haplotype-based genomic prediction model to manage multi-environment wheat data using environmental covariates. *Theoretical and Applied Genetics* 132, 3143–3154.

Heffner, E.L., Lorenz, A.J., Jannink, J.L. and Sorrells, M.E. (2010) Plant breeding with genomic selection: Gain per unit time and cost. *Crop Science* 50(5), 1681–1690.

Heslot, N., Akdemir, D., Sorrells, M.E. and Jannink, J.L. (2014) Integrating environmental covariates and crop modeling into the genomic selection framework to predict genotype by environment interactions. *Theoretical and Applied Genetics* 127, 463–480.

Heslot, N., Jannink, J.L. and Sorrells, M.E. (2013) Using genomic prediction to characterize environments and optimize prediction accuracy in applied breeding data. *Crop Science* 53(3), 921–933.

Hu, H., Rincent, R. and Runcie, D.E. (2025) MegaLMM improves genomic predictions in new environments using environmental covariates. *Genetics* 229(1), iyae171.

Islam, S. (2025) Agriculture, food security, and sustainability: A review. *Exploration of Foods and Foodomics* 3, 101082.

Jarquín, D., Crossa, J., Lacaze, X., Du Cheyron, P., Daucourt, J. *et al.* (2014) A reaction norm model for genomic selection using high-dimensional genomic and environmental data. *Theoretical and Applied Genetics* 127, 595–607.

Jauhar, P.P. (2006) Modern biotechnology as an integral supplement to conventional plant breeding: The prospects and challenges. *Crop Science* 46(5), 1841–1859.

Jighly, A., Benhajali, H., Liu, Z. and Goddard, M.E. (2022) MetaGS: An accurate method to impute and combine SNP effects across populations using summary statistics. *Genetics Selection Evolution* 54(1), 37.

Jighly, A., Hayden, M. and Daetwyler, H. (2021) Integrating genomic selection with a genotype plus genotype x environment (GGE) model improves prediction accuracy and computational efficiency. *Plant, Cell & Environment* 44(10), 3459–3470.

Jighly, A., Thayalakumaran, T., Kant, S., Panozzo, J., Aggarwal, R. *et al.* (2024) Statistical sampling of missing environmental variables improves biophysical genomic prediction in wheat. *Theoretical and Applied Genetics* 137(5), 108.

Jighly, A., Thayalakumaran, T., O'Leary, G.J., Kant, S., Panozzo, J. *et al.* (2023a) Using genomic prediction with crop growth models enables the prediction of associated traits in wheat. *Journal of Experimental Botany* 74(5), 1389–1402.

Jighly, A., Weeks, A., Christy, B., O'Leary, G.J., Kant, S. *et al.* (2023b) Integrating biophysical crop growth models and whole genome prediction for their mutual benefit: A case study in wheat phenology. *Journal of Experimental Botany* 74(15), 4415–4426.

Jubair, S. and Domaratzki, M. (2023) Crop genomic selection with deep learning and environmental data: A survey. *Frontiers in Artificial Intelligence* 5, 1040295.

Kang, M.S. (1997) Using genotype-by-environment interaction for crop cultivar development. *Advances in Agronomy* 62, 199–252.

Lee, H.J., Lee, J.H., Gondro, C., Koh, Y.J. and Lee, S.H. (2023) deepGBLUP: Joint deep learning networks and GBLUP framework for accurate genomic prediction of complex traits in Korean native cattle. *Genetics Selection Evolution* 55(1), 56.

Li, M., Hall, T., MacHugh, D.E., Chen, L., Garrick, D. *et al.* (2025) KPRR: A novel machine learning approach for effectively capturing nonadditive effects in genomic prediction. *Briefings in Bioinformatics* 26(1), bbae683.

Liang, M., Chang, T., An, B., Duan, X., Du, L. *et al.* (2021) A stacking ensemble learning framework for genomic prediction. *Frontiers in Genetics* 12, 600040.

Lin, Z., Robinson, H., Godoy, J., Rattey, A., Moody, D. *et al.* (2021) Genomic prediction for grain yield in a barley breeding program using genotype× environment interaction clusters. *Crop Science* 61(4), 2323–2335.

Lopez-Cruz, M., Crossa, J., Bonnett, D., Dreisigacker, S., Poland, J. *et al.* (2015) Increased prediction accuracy in wheat breeding trials using a marker× environment interaction genomic selection model. *G3: Genes, Genomes Genetics* 5(4), 569–582.

Malosetti, M., Ribaut, J.M. and van Eeuwijk, F.A. (2013) The statistical analysis of multi-environment data: Modeling genotype-by-environment interaction and its genetic basis. *Frontiers in Physiology* 4, 44.

Messina, C.D., Technow, F., Tang, T., Totir, R., Gho, C. *et al.* (2018) Leveraging biological insight and environmental variation to improve phenotypic prediction: Integrating crop growth models (CGM) with whole genome prediction (WGP). *European Journal of Agronomy* 100, 151–162.

Meuwissen, T.H., Hayes, B.J. and Goddard, M. (2001) Prediction of total genetic value using genome-wide dense marker maps. *Genetics* 157(4), 1819–1829.

Montesinos-López, O.A., Chavira-Flores, M., Kismiantini, Crespo-Herrera, L., Saint Piere, C. *et al.* (2024a) A review of multimodal deep learning methods for genomic-enabled prediction in plant breeding. *Genetics* 228(4), iyae161.

Montesinos-López, O.A., Crespo-Herrera, L., Pierre, C.S., Cano Paez, B., Huerta Prado, G.I. *et al.* (2024b) Feature engineering of environmental covariates improves plant genomic-enabled prediction. *Frontiers in Plant Science* 15, 1349569.

Montesinos-López, O.A., Montesinos-López, A., Pérez-Rodríguez, P., Barrón-López, J.A., Martini, J.W. *et al.* (2021) A review of deep learning applications for genomic selection. *BMC Genomics* 22, 19.

Morota, G. and Gianola, D. (2014) Kernel-based whole-genome prediction of complex traits: A review. *Frontiers in Genetics* 5, 363.

Novielli, P., Romano, D., Pavan, S., Losciale, P., Stellacci, A.M. *et al.* (2024) Explainable artificial intelligence for genotype-to-phenotype prediction in plant breeding: A case study with a dataset from an almond germplasm collection. *Frontiers in Plant Science* 15, 1434229.

Olson, D. and Anderson, J. (2021) Review on unmanned aerial vehicles, remote sensors, imagery processing, and their applications in agriculture. *Agronomy Journal* 113(2), 971–992.

Page, S. (2008) *The Difference: How the Power of Diversity Creates Better Groups, Firms, Schools, and Societies-New Edition*. Princeton University Press, Princeton, NJ (USA).

Prado, M., Famoso, A., Guidry, K. and Fritsche-Neto, R. (2024) Optimizing multi-environment trials in the southern US rice belt via smart-climate-soil prediction-based models and economic importance. *Frontiers in Plant Science* 15, 1458701.

Qazi, S., Khawaja, B.A. and Farooq, Q.U. (2022) IoT-equipped and AI-enabled next generation smart agriculture: A critical review, current challenges and future trends. *IEEE Access* 10, 21219–21235.

Rebollo, I., Aguilar, I., Pérez de Vida, F., Molina, F., Gutiérrez, L. *et al.* (2023) Genotype by environment interaction characterization and its modeling with random regression to climatic variables in two rice breeding populations. *Crop Science* 63(4), 2220–2240.

Resende, R.T., Piepho, H.P., Rosa, G.J., Silva-Junior, O.B., e Silva, F.F. *et al.* (2021) Enviromics in breeding: Applications and perspectives on envirotypic-assisted selection. *Theoretical and Applied Genetics* 134, 95–112.

Runcie, D.E., Qu, J., Cheng, H. and Crawford, L. (2021) MegaLMM: Mega-scale linear mixed models for genomic predictions with thousands of traits. *Genome Biology* 22, 213.

Saint Pierre, C., Burgueño, J., Crossa, J., Fuentes Dávila, G., Figueroa López, P. *et al.* (2016) Genomic prediction models for grain yield of spring bread wheat in diverse agro-ecological zones. *Scientific Reports* 6(1), 27312.

Smith, A., Norman, A., Kuchel, H. and Cullis, B. (2021) Plant variety selection using interaction classes derived from factor analytic linear mixed models: Models with independent variety effects. *Frontiers in Plant Science* 12, 737462.

Speed, D. and Balding, D.J. (2019) SumHer better estimates the SNP heritability of complex traits from summary statistics. *Nature Genetics* 51(2), 277–284.

Technow, F., Messina, C.D., Totir, L.R. and Cooper, M. (2015) Integrating crop growth models with whole genome prediction through approximate Bayesian computation. *PLoS One* 10(6), e0130855.

Togninalli, M., Wang, X., Kucera, T., Shrestha, S., Juliana, P. *et al.* (2023) Multi-modal deep learning improves grain yield prediction in wheat breeding by fusing genomics and phenomics. *Bioinformatics* 39(6), btad336.

Tolhurst, D.J., Mathews, K.L., Smith, A.B. and Cullis, B.R. (2019) Genomic selection in multi-environment plant breeding trials using a factor analytic linear mixed model. *Journal of Animal Breeding and Genetics* 136(4), 279–300.

Tomura, S., Wilkinson, M.J., Cooper, M. and Powell, O. (2025) Improved genomic prediction performance with ensembles of diverse models. *G3: Genes, Genomes, Genetics* 15(5), jkaf048.

van Etten, J., de Sousa, K., Cairns, J.E., Dell'Acqua, M., Fadda, C. *et al.* (2023) Data-driven approaches can harness crop diversity to address heterogeneous needs for breeding products. *Proceedings of the National Academy of Sciences* 120(14), e2205771120.

Vanaja, M., Sarkar, B., Sathish, P., Jyothi Lakshmi, N., Yadav, S.K. *et al.* (2024) Elevated CO2 ameliorates the high temperature stress effects on physio-biochemical, growth, yield traits of maize hybrids. *Scientific Reports* 14(1), 2928.

Whelan, B. and Taylor, J. (2013) *Precision Agriculture for Grain Production Systems*. CSIRO Publishing, Collingwood.

Winans, N.D., Fonseca, J.M., Perumal, R., Klein, P.E., Klein, R.R. *et al.* (2024) Envirotyping can increase genomic prediction accuracy of new environments in grain sorghum trials depending on mega-environment. *Crop Science* 64(5), 2519–2533.

Wolpert, D.H. and Macready, W.G. (2002) No free lunch theorems for optimization. *IEEE transactions on evolutionary computation* 1(1), 67–82.

Xu, Y. (2015) Envirotyping and its applications in crop science. *Scientia Agricultura Sinica* 48(17), 3354–3371.

Xu, Y. (2016) Envirotyping for deciphering environmental impacts on crop plants. *Theoretical and Applied Genetics* 129, 653–673.

Xu, Y., Lu, Y., Xie, C., Gao, S., Wan, J. *et al.* (2012) Whole-genome strategies for marker-assisted plant breeding. *Molecular Breeding* 29, 833–854.

Yan, W., Hunt, L.A., Sheng, Q. and Szlavnics, Z. (2000) Cultivar evaluation and mega-environment investigation based on the GGE biplot. *Crop Science* 40(3), 597–605.

Yan, W., Nilsen, K.T. and Beattie, A. (2023) Mega-environment analysis and breeding for specific adaptation. *Crop Science* 63(2), 480–494.

Yue, H., Olivoto, T., Bu, J., Li, J., Wei, J. *et al.* (2022) Multi-trait selection for mean performance and stability of maize hybrids in mega-environments delineated using envirotyping techniques. *Frontiers in Plant Science* 13, 1030521.

11 Integrated Multi-omics Approaches for Climate-Resilient Crop Improvement

Vasundhara Gaddam[1,2], Anusha Kutala[1,3], Mahesh Madineni[1,3], Sivasakthi Kaliamoorthy[1], Jagdish Jaba[1], Senthil Ramachandran[1], Someswar Rao Sagurthi[2], Yogendra Kalenahalli[1], Rakesh Srivastava[1], Mamta Sharma[1], Kuldeep Singh[1] and Himabindu Kudapa[1,*]

[1]*Global Research Program-Accelerated Crop Improvement, International Crops Research Institute for the Semi-Arid Tropics (ICRISAT), Patancheru 502324, India;* [2]*Department of Genetics and Biotechnology, Osmania University, Hyderabad 500007, India;* [3]*Department of Genetics and Plant Breeding, Acharya N. G. Ranga Agricultural University, Guntur 522034, India*

Abstract

Climate change poses significant challenges to global food security, necessitating the development of crops varieties with enhanced tolerance to a wide range of biotic and abiotic stresses and improved nutritional quality. Crop productivity and resilience are increasingly threatened by biotic stresses such as pests and diseases, as well as abiotic stresses including drought, salinity, extreme temperatures (both heat and cold stress), heavy metal toxicity and other environmental constraints. Multi-omics approaches – encompassing genomics, transcriptomics, proteomics, metabolomics, epigenomics and phenomics – offer a comprehensive framework for investigating the complex biological processes underlying plant stress responses, nutritional quality and resilience mechanisms. Integrating data from various -omics platforms enables the identification of key genes, proteins, metabolites and molecular pathways involved in stress adaptation and the enhancement of nutritional quality traits. Recent advances in high-throughput sequencing technologies, computational biology and systems-level analysis have further strengthened our understanding and ability to connect genotype to phenotype, thereby facilitating targeted breeding and genetic engineering strategies. This chapter highlights recent advancements in the integration and application of multi-omics approaches for developing climate-resilient crops, demonstrating how omics-driven research can accelerate the development of crop varieties capable of thriving under changing environmental conditions.

11.1 Introduction

Agriculture is facing an unprecedented crisis due to global climate change, which is threatening food security worldwide. Rising temperatures create favorable conditions for pest and disease incidence, while soil erosion, degradation and flooding increasingly challenge crop resilience and yield potential. By 2050, the global population is projected to reach 9.8 billion; the real challenge will lie in meeting the food demands of an exponentially growing population, while facing the constraints of limited and increasingly depleted arable land (http://unstats.un.org).

*Corresponding author. Email: himabindu.kudapa@icrisat.org

DOI: 10.1079/9781800627185.0011

Furthermore, the biotic and abiotic stressors together are rendering current crop varieties more vulnerable, highlighting the urgency in developing resilient crop varieties. Given these significant challenges, traditional agricultural practices are no longer sufficient to meet the demands of modern farming and are increasingly becoming unsustainable. Plants employ a range of adaptive strategies to cope with environmental changes and combat stress. For example, plants modify their morphology, such as altering leaf area, root architecture, stem elongation etc., to optimize resource acquisition and minimize stress impact. While others adjust their physiological process including photosynthetic efficiency, water-use efficiency etc. to maintain plant growth and development under adverse conditions (Leisner *et al.*, 2023).

Leveraging genetic and molecular tools to improve crop performance under harsh and unpredictable environmental conditions is therefore crucial. At the molecular level, changes in gene expression activate stress-responsive pathways, modulate key cellular mechanisms and trigger the production of protective proteins and metabolites, enabling plants to tolerate challenging environments (Raza *et al.*, 2021). Multi-omics aims to understand biological systems involving the analysis of various data layers, including genomics as it provides a comprehensive understanding of the structure and functions of whole genomes. Complementary -omics approaches such as transcriptomics, proteomics, metabolomics and epigenomics, offer deeper insights into gene expression dynamics, regulatory mechanisms, protein functions and alteration in metabolic pathways. Each layer captures a different aspect of cellular activity, from transcriptional regulation to post-translational modification and metabolite accumulation. Integration of these data layers enables a comprehensive understanding of how biological systems function and respond to internal and external cues (Pazhamala *et al.*, 2021). This systems level understanding is essential for identifying key molecular drivers of stress tolerance and crop improvement.

By modeling these layers, systems biology can explain existing patterns and predict the outcomes of specific gene or base pair modifications. These insights are particularly valuable for breeders, enabling the design of crop breeding programs aimed at improving traits with greater precision and efficiency. Combining -omics approaches provides a more complete and powerful way to understand complex biological systems (Pinu *et al.*, 2019). Conventional plant breeding involving selective breeding and cross-breeding for the transfer of desirable traits requires multiple generations and is often labor-intensive and time consuming. In contrast, -omics-based strategies employ cutting-edge molecular techniques to comprehensively examine the genetic composition, protein expression and metabolic pathways of organisms. This high-throughput, highly accurate multi-omics approach expedites the breeding cycle by facilitating the precise identification and modifications of target genes, leading to more rapid and targeted improvements.

For example, in rice (*Oryza sativa*), the use of multi-omics approach in a recombinant inbred population derived from Smaba Mahsuri and IR36, led to identification of key candidate genes (*OsSBEIIb*) and genomics regions associated with low glycemic index (GI) and high protein content (Badoni *et al.*, 2024). Similarly, in maize (*Zea mays*), a temperature-sensitive crop whose growth and maturation are heavily influenced by temperature, metabolite profiling has been conducted in both cold-tolerant variety B144 and susceptible variety Q319. Specifically, the glyoxylate and dicarboxylate metabolism pathways, along with associated proteins, lipids and metabolites, were analyzed to better understand molecular and metabolic adjustments underlying temperature stress tolerance (Yu *et al.*, 2024). In a model legume crop, *Medicago truncatula*, an integrated multi-omics approach was used to identify novel aphid effector candidates linked to virulence and non-virulence phenotypes. The study revealed key molecular factors involved in aphid-plant interactions providing insights for developing pest-resistant crop varieties (Thorpe *et al.*, 2024).

Developing successful crop improvement programs that enhance resilience to these challenges requires a comprehensive understanding of crop biology at the molecular level. The most effective strategy for achieving this is the application of integrated multi-omics approach in breeding, which can facilitate the development of climate-resilient crops capable of withstanding the extreme environmental conditions.

11.2 Effects of Changing Climate on crops

11.2.1 Overview of Cereals

Cereals are staple food crops that form the backbone of global food security, providing essential nutrients for human populations worldwide. Rice is one of the world's most important food crops, yet it is both a major cause and victim of climate change. Cultivated across roughly 165 million hectares, rice also serves as a model crop for research. However, rice yields are highly sensitive to climate fluctuations. Studies show a 3.85% decrease in yield with every 1°C rise in temperature (Li *et al.*, 2024). Rising temperatures also intensify the spread of pests and diseases, causing both direct and indirect yield losses. For example, a study on rice sheath blight shows that regions with higher-than-average temperatures experience 7.2–38.9% disease incidence, leading to 14.3–39.7% yield loss (Ahamad and Khan, 2023). This highlights the significant impact of climate change on rice production and the need for mitigative strategies. Maize, another globally vital food crop, ranks first among cereals in total production. Yet, like rice, it is highly vulnerable to the effects of climate change. Even a small increase in temperature can lead to delayed maturity, ranging from days to weeks. Crop simulation models predict that maize yields decline by 7.4%. for every 1°C rise in temperature (Abbas *et al.*, 2023). These projections highlight the potential threats of climate change to future maize production. Wheat (*Triticum aestivum*), the most widely cultivated crop worldwide, plays a crucial role in food security. However, wheat yields are expected to decline significantly under climate stress, with projections indicating a 15% and 16% reduction in Africa and South Asia, respectively, by 2050 (Pequeno *et al.*, 2021). Drought is one of the most limiting abiotic stresses for wheat because it induces the production of reactive oxygen species, leading to programmed cell death in the plant. To mitigate the effects of drought stress, it is important to develop wheat lines combining multiple stress-resilient traits and employ trait engineering approaches to enhance crops' ability to avoid or tolerate water deficit conditions (Nyaupane *et al.*, 2024).

Millets are widely recognized for their excellent nutritional profile with rich protein, dietary fiber, minerals and other essential nutrients. These crops can also thrive in challenging environments with poor soil fertility offering a sustainable solution to bridge the gap between resource extraction and food security. However, for every 1% increase in mean annual temperature, millets yields decrease by 3.06% (Jadaun and Khan, 2023). Researchers also found that millet releases fewer greenhouse gases than other crops, making it a promising option for reducing agriculture's carbon footprint. For instance, sorghum (*Sorghum bicolor*) a nutrient major millet, is valued for its drought and salinity tolerance, and is increasingly recognized as a climate-resilient crop. It is a dual-purpose crop and its ability to thrive in low fertility soils makes it essential for dryland agriculture. However, there are not enough measures to fully compensate for yield losses in areas where temperatures rise above 2°C. Even if precipitation could double, such adjustments are unlikely to prevent yield losses in most sorghum-growing areas (Talwar *et al.*, 2013). This highlights the serious challenges climate change poses to millet production and underscores the need for continued research and adaptation strategies to protect yields of these important crops. Encouragingly, studies show that drought- and heat-tolerant millet varieties can achieve yield increase of up to 14% compared with conventional cultivars (Wang *et al.*, 2018).

11.2.2 Overview of Legumes

Legumes are a critical source of global food and nutrition security, valued for their high protein content, essential minerals and dietary fiber. Soybean (*Glycine max*), a protein-rich annual legume is an important component of human and animal diets. However, soybean yields are highly sensitive to climate change, particularly shifts in rainfall patterns (de Souza Batista *et al.*, 2023). Future climate projections suggest a shrinking suitable cultivation area and declining yields for soybean, mainly due to increasing drought and heat stress (Timilsina *et al.*, 2023). Chickpea (*Cicer arietinum*), the third most important legume globally, is highly nutritious and tolerant to cooler temperatures. However, chickpeas are vulnerable to high-temperature stress, especially during germination and pod set, which can lead to significant yield losses (Gaur *et al.*, 2013; Jha *et al.*, 2025). This highlights the growing challenges that climate change poses to pulse production and food security. Pigeonpea (*Cajanus cajan*) is known for its resilience and deep root system, allowing it to withstand stress better than many other legumes (Saxena *et al.*, 2019). Although pigeonpea can thrive in marginal environments, recent trends in climate change have led to a decline in its production, highlighting the increasing challenges posed by climate change. Similarly, mung bean (*Vigna radiata*), a fast-growing, protein- and nutrient-rich legume experiences substantial yield reductions under high temperatures (Mahajan *et al.*, 2023).

Collectively these examples illustrate the vulnerability of cereal and legume crops to climate change, highlighting the urgent need for crop breeding strategies that enhance resilience, maintain yield stability and secure global nutrition.

11.3 The Multi-omics Revolution in Crop Science

Multi-omics enables the large-scale study of biological molecules such as DNA, RNA, protein and metabolites, providing detailed insights into their structure, function, and their role in cellular and physiological processes. By integrating genomics, transcriptomics, proteomics, metabolomics, epigenomics and other -omics approaches with advanced computation biology, a comprehensive understanding of plant responses to environmental stresses can be obtained, facilitating the development of nutrient-rich, climate-resilient crops. The schematic flowchart in Figure 11.1 represents the integrative multi-omics strategies for developing climate-resilient crops.

11.3.1 Genomics

Dissecting the genes associated with biotic and abiotic stress, and nutritional and agronomic traits, is possible through genomics, utilizing whole-genome sequencing via a high-throughput platform such as next-generation sequencing (NGS). This knowledge also facilitates the development of new/ improved "climate change ready" cultivars (Varshney et al., 2005, 2015). The evolution of genomic technologies from first-generation to the next- and third-generation sequencing has greatly enhanced the efficiency and scope of genome sequencing. The era of modern plant genomics began in December 2000 with the publication of the complete genome sequence of *Arabidopsis thaliana*, a model plant species with a compact genome (~135 Mb). This milestone was followed by the release of the high-quality rice genome in 2005 by IRGSP, followed by maize (2009), sorghum (2010), soybean (2012), pigeonpea (2012), chickpea (2013) and many other crop species, laying the foundation for advanced crop improvement programs. To date, genome assemblies of hundreds of crop species are available in public repositories (*see* Pazhamala et al., 2021; Maharajan *et al.*, 2024) and several genome databases have also been created (https://phytozome-next.jgi.doe.gov/).

Whole genome sequencing (WGS) and whole genome re-sequencing (WGRS) are powerful genomic tools for crop improvement. The WGS involves sequencing the entire genome of a species to generate a reference genome, whereas WGRS involves sequencing the genomes of multiple individuals of the same species. Pan-genome refers to the complete set of genes within a species, encompassing both the core and dispensable genomes, collectively representing the total genetic diversity across the species. Pan-genomes have grown in popularity over the past decade and are currently present for a large number of important crops.

For instance, the International Crops Research Institute for the Semi-Arid Tropics (ICRISAT) led international partnership succeeded in the development of a draft genome sequence of pearl millet (*Pennisetum glaucum*). Whole genome sequencing of the genotype, Tift 23D2B1-P1-P5 resulted in 1.79 Gb sequence data and 38,579 genes, with 50% of the scaffolds (N50) at 884.95 kb, which constitutes 90% of the genome. About 994 accessions in PMiGAP (Pearl Millet inbred Germplasm Association Panel) were re-sequenced and identified 29,542,173 SNPs; 3,844,446 indels and 423,118 SVs (Varshney *et al.*, 2017). Similarly, whole-genome

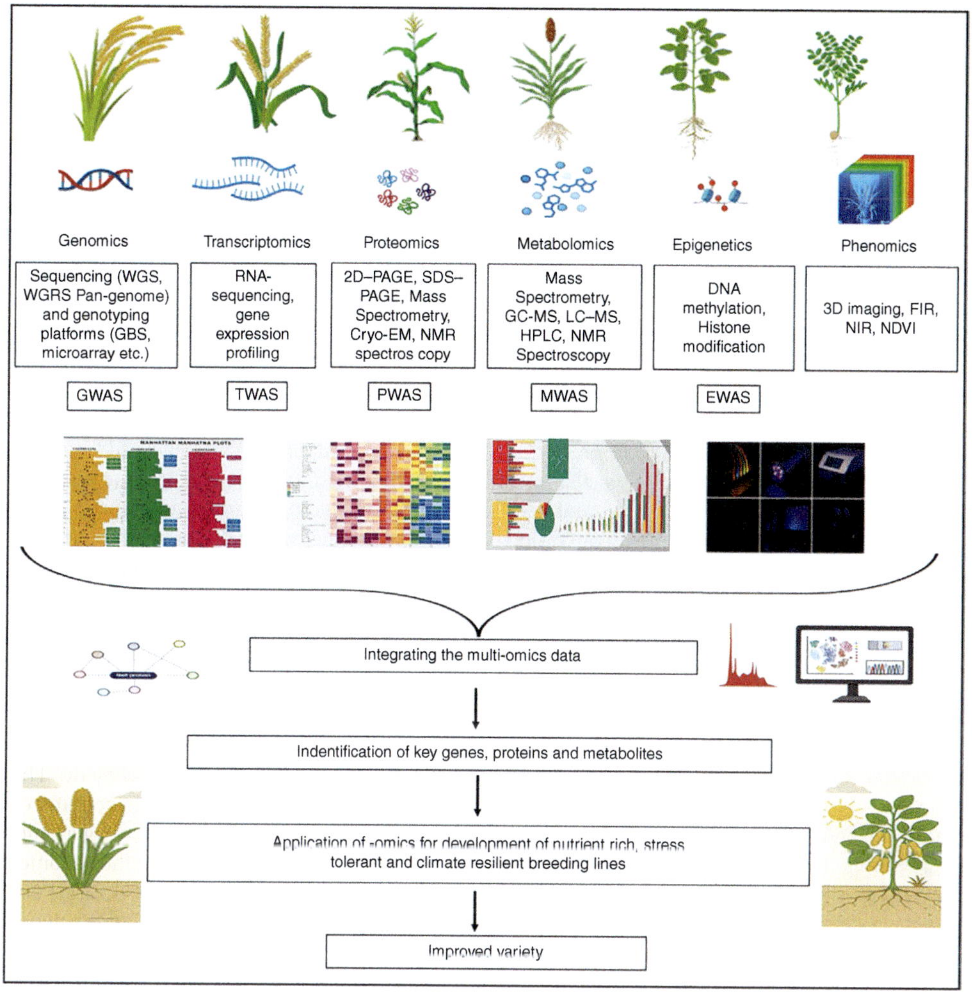

Figure 11.1. Integrative multi-omics strategies for developing climate-resilient crops. (Authors' own image.)

shotgun sequence of a kabuli chickpea variety 'CDC Frontier' was generated using the Illumina HiSeq 2000 platform and BAC-end sequencing (Varshney *et al.*, 2013). The draft assembly spanned ~532.29 Mb across 7163 scaffolds, representing around 73.8% of the estimated 738 MB genome and contained 28,269 genes.

Beyond the development of the reference genome in these crops, graph-based pangenomes were constructed. In pearl millet, a pan-genome was developed by assembling 10 high-quality *de novo* chromosome level genomes together with the earlier released genome assembly. Across the pangenome, 14,608 (46.6–52.08%) core gene families were identified accounting for more than half of the total gene family sets, while the dispensable gene families are 39.75–49.94%. A total of 744,364 SVs were identified when the assemblies were realigned against the PI537069 reference genome. The SVs from all accessions were merged to build a graph-based pangenome, resulting in 424,085 non-redundant structural variations (SVs). Comparative genomics revealed the involvement of the RWP-RK transcription factor family and genes related to the endoplasmic reticulum (ER) in heat tolerance adaptation in pearl millet (Yan *et al.*, 2023). Similarly, WGRS of 3366 chickpea accessions including 3171 cultivated and 195 wild accessions representing global diversity identified ~3.94 m SNPs in the cultivated set and ~19.577m SNPs in the wild accessions. This massive resequencing effort identified a total of 29,870 genes in the pangenome, including over 1000 novel genes, providing an essential resource for chickpea genomics-assisted breeding (Varshney *et al.*, 2021). Further, Khan *et al.* (2024) established a *Cicer* super-pangenome providing insights into species evolution and agronomic trait loci for chickpea improvement. The study sequenced and assembled chromosome-level genomes for eight wild *Cicer* species. These *Cicer* wild species genomes were integrated with existing cultivated chickpea genomes to build a super-pangenome. Further, gene clustering identified 24,827 gene families, including 14,748 core genes, 2958 softcore genes, 6212 dispensable and 909 species-specific gene families. The dispensable genome was found to be a rich source of genetic variation, containing genes associated with key agronomic traits, providing a roadmap on how to utilize wild species to broaden the genetic base of cultivated chickpea.

Pangenomes have revealed that, in addition to SNPs and simple InDels, significant structural rearrangements and sequence variations are typical among individuals of the same species. Genes that are not present in every member of the population are known as "dispensable" or "variable" genes, often contributing to important adaptive traits, including stress responses (Tettelin *et al.*, 2005). Dispensable genes are crucial for preserving crop variety and raising quality, according to different pangenomic studies in major crops such as rice (Zhao et al., 2018), maize (Hirsch et al., 2014), wheat (Montenegro et al., 2017) and soybean (Li et al., 2014). Importantly, within-species copy number variations (CNVs) and presence/absence variations (PAVs) are typically found in pangenomes, and these SVs have been shown to affect agronomically significant characteristics in crops. For example, in barley (*Hordeum vulgare*), variation in *HvCBF4* and *HvCBF2* copy number at the *FR-H2* gene was associated with frost tolerance (Francia *et al.*, 2016). Similarly, a receptor-like cytoplasmic kinase gene that provides tolerance to phosphorus deficiency has been linked to the Pup1 locus in rice (Gamuyao *et al.*, 2012). These investigations indicated that SVs are projected to have a major impact on crop improvement in the context of climate change. In general, the comprehensive genomes and pangenomes facilitate the investigation of functional alleles for phenotypic variations and increase genetic resources to enhance crop germplasm in changing climates.

11.3.2 Transcriptomics

Early transcriptomic studies employed traditional profiling techniques such as differential display-PCR (DD-PCR), cDNA-AFLP and suppression subtractive hybridization (SSH), but these methods were limited by low resolution (Anusha and Amir, 2024). To overcome this, advanced approaches such as serial analysis of gene expression (SAGE) and digital gene expression analysis were introduced. The extensively used technologies during the mid-1990s to 2000s in various transcriptomics studies were microarrays and RNA-sequencing, which became popular due to high-throughput and in-depth understanding of the ontology of genes (Lowe *et al.*, 2017).

To interpret a genome's biological functions, cellular components, molecular functions and the pathways that are involved in carrying out all these functions, it is important to have a comprehensive understanding of the transcriptome. The nature of a transcriptome being variable under different stages and conditions, it provides critical insights into how plants respond to specific stimuli at the molecular level (Zenda *et al.*, 2021).

A transcriptome study involving 10 rice accessions exposed to heat stress revealed the role of the UPR signaling cascade in the tolerant genotype TEJ-5, while the gene *OsbZIP58* was implicated in moderately tolerant genotype IND-2. These findings correspond to phenotypic differences during the early stages of grain development (Nalini *et al.*, 2024). In another study, several genes and complex mechanisms involved in drought tolerance, particularly jasmonic acid signaling and nitrogen metabolism, were identified (Park and Jeong, 2023), which can help develop drought-resistant rice varieties. A study in maize examined the response to heat stress, identifying genes associated with tolerance and sensitivity in two varieties, CML 25 (heat-tolerant) and LM 11 (heat-sensitive). The transcriptome analysis identified various genes involved in the generation of heat shock proteins, photosynthesis and antioxidant defense in heat stress. The pathways related to metabolism and secondary metabolite production activated by heat stress have the potential to improve heat-tolerant maize (Jagtap *et al.*, 2023).

A transcriptomic study in wheat investigated genes involved in cold stress tolerance by exposing seedlings to 4°C for 30 days. This led to the identification of numerous differentially expressed genes (DEGs) and high expression of transcription factors from the MYB, bHLH, WRKY, ERF and NAC families. Additionally, genes associated with key metabolic pathways such as hormone signal transduction, benzoxazinoid biosynthesis, phenylpropanoid biosynthesis, flavonoid biosynthesis and glutathione metabolism were significantly linked to cold stress tolerance. Eight DEGs were validated through expression profiling, providing potential targets for developing cold-tolerant wheat varieties (Li *et al.*, 2023).

Similar studies were also carried out in various other crops such as soybeans, peas, chickpeas and lentils (Garg *et al.*, 2023). In soybeans, comparing a drought-sensitive variety, Suinong 14 (SN14), and a tolerant variety, Heinong 44 (HN44), 8 key candidate genes, i.e. *LOC100808190*, *GmWRKY49*, *LOC100798375*, *LOC100804384*, *LOC100817546*, *GmbZIP4*, *LOC100819313* and GmbZIP1, which belong to various drought resistance transcription factors, were identified. These genes are related to the plant hormone signal transduction and MAPK signaling pathways, which are known to impart drought resistance mechanisms in plants. The identification of these genes provides a basis for molecular breeding and developing soybean varieties with high drought resistance and high water-use efficiency (Xu *et al.*, 2023). Further, the RNA seq-Atlas in soybean provides a comprehensive transcriptional profile across 14 diverse tissues and developmental stages, enabling deeper insights into gene function and regulatory networks (Severin *et al.*, 2010). Similarly, the chickpea gene expression atlas offers a comprehensive overview of transcript abundance across 27 tissues from different developmental stages, providing valuable insights into genes and regulatory networks associated with plant growth and development (Kudapa *et al.*, 2018).

Using comparative transcriptomic analysis in lentils, molecular responses for individual and combined drought and heat stresses, several genes were identified, including members from the CYP, MYB and CDPK families and mitochondrial Mn-SOD, that were specifically upregulated under combined heat and drought stress. Apart from these, various other lentil transcription factors (TFs), such as MtMYB, Mtjumonji and MtbZIP124 domain proteins, were also discovered to be viable candidates for breeding lentils with increased tolerance to drought stress and heat stress, as well as combined stress (Hosseini *et al.*, 2021).

11.3.3 Proteomics

Proteomics involves the study of entire proteins in an organism, collectively known as the proteome, to understand the structure, functions and interactions on a wider scale, such as how proteins control various developmental stages of a plant, how they are expressed and regulated with respect to stress conditions, and other critical processes like nutritional quality by studying proteins individually. Structural proteomics is associated with the examination of the structure of a protein and its analysis using various computational modeling and experimental methods such as X-ray crystallography (helps in determining 3D structures and arrangement of proteins with high resolution). For instance, recent advances in cryo-EM (cryo-electron microscopy) made it possible to observe and understand the flexible and dynamic nature of proteins. This gives us an understanding of how a protein's physical conformation is related to its functional performance. On the other hand, functional proteomics' primary focus is to understand the roles of proteins in the cells and tissues. Affinity chromatography and interaction studies are proved to be a great help in elucidating the protein functions. Expression levels of proteins can be studied and analyzed using expression proteomics with the help of various techniques such as two-dimensional gel electrophoresis (2-DE), mass spectrometry, MALDI-TOF, two-dimensional differential gel electrophoresis (2D-DIGE), SDS-PAGE, electrospray ionization (ESI) and collision-induced dissociation (CID) (Anusha and Amir, 2024).

Proteomic studies in wheat have highlighted key molecular responses to abiotic stresses. Under sodic alkaline stress (pH 9.7), 3104 proteins were detected, with 69 specifically upregulated, including superoxide dismutases, dehydrins, V-ATPase and malate dehydrogenases, which contribute to osmotic regulation and ion homeostasis.

Downregulation of malic enzymes and TCA cycle proteins suggested enhanced malate accumulation, an adaptive mechanism to alkaline stress (Han *et al.*, 2019). Similarly, heat stress during grain filling significantly altered 37 proteins, of which 25 were induced, affecting grain protein composition and thereby influencing dough properties and overall wheat quality (Majoul *et al.*, 2003). Proteomic studies conducted in two different genotypes of barley, with contrasting tolerance to drought stress, revealed proteomic changes in both roots and leaves. By combining three different techniques, i.e. two-dimensional electrophoresis (2D-PAGE), MALDI-TOF and MALDI-TOF/TOF mass spectrometry, 121 and 182 drought-responsive proteins were identified in leaves and roots, respectively. It was also found that many of these proteins were involved in processes such as photosynthesis and carbon metabolism, which are commonly affected by drought conditions (Chmielewska *et al.*, 2016).

Proteomic analysis of salinity-induced soybean revealed 440 proteins in roots and 278 proteins in leaves. Key proteins such as 14-3-3, MMK2, PP1 and TRX-h, were transcriptionally regulated under salt stress. These findings indicate that proteins involved in signaling, stress defense, metabolism, membrane function and transport play an important role in response to saline stress in soybean (Ji *et al.*, 2016).

Various proteomic studies were conducted in different crops, to understand various morphological and physiological responses and functional differences under a variety of stress conditions that are directly or indirectly related to climate changes (Raza *et al.*, 2021). All these studies shed light on various proteins, their structure and functions, and the mechanisms they are involved in bringing about the resistance or susceptibility in the particular crop, which can be helpful in future studies as it offers various advantages by enabling us to directly measure the proteins at a given time in a plant, their functional relevance, and dynamic changes in their levels of expression in response to stress, thereby providing insights into its genome's functional output.

11.3.4 Metabolomics

A metabolite is a small molecule crucial for processes such as maintaining growth, respiration, photosynthesis, defense against biotic and abiotic stresses, attraction of pollinators and seed dispersers, energy production and overall health of a plant. Plants produce two types of metabolites, i.e. primary and secondary metabolites. The primary metabolites are vital for primary processes such as the growth and reproduction of a plant, whereas the secondary metabolites are involved in mechanisms such as plant defense and also serve vital physiological functions (Guerriero *et al.*, 2018; Elshafie *et al.*, 2023). Studies conducted on rice cultivars with varying levels of tolerance to combined drought and heat stress under field conditions using gas chromatography and mass spectrometry techniques showed distinct metabolic responses in different tissues. Using correlation analysis, the potential metabolic markers that are linked to grain yield and quality under combined drought and heat stress, including stress-regulated metabolites and those with inherent differences between the cultivars, were identified (Lawas *et al.*, 2019). Metabolomic profiling in soybeans revealed key metabolites related to sugar, nitrogen and phytochemical metabolism, emphasizing their importance under the conditions of drought and heat stress (Das *et al.*, 2017).

In peanuts, research was conducted under water scarcity conditions at the seedling stage of two contrasting genotypes: JL 24, drought-sensitive and TAG 24, drought tolerant. The analysis of metabolites was done using high-performance liquid chromatography (HPLC) and gas chromatography-mass spectrometry (GC-MS) for the presence of polyamines and polyphenols. Specific polyamines such as agmatine and cadaverine, and polyphenols such as catechin, vanillic acid and syringic acid were found to be linked to drought tolerance in both genotypes. Seven metabolic pathways, including sucrose metabolism, mannose metabolism, fructose metabolism, starch metabolism, etc. were highly impacted by drought. This research provides important insights into how peanuts metabolically respond to drought (Gundaraniya *et al.*, 2020). Examples of multi-omics studies and their integration across different crops are summarized in Table 11.1.

11.3.5 Epigenomics

Epigenomics is the study of epigenetic changes in a cell that are heritable without changing the

Table 11.1. Examples of -omics studies in various crops. (Authors' own table.)

S. No	Crop	Tissue	Type of stress	Approaches used	Findings	References
1	Rice	–	Cold stress	Transcriptomics	Identified 14213 and 14730 DEGs in cold-tolerant genotypes CTWR1 and CTR and 720 and 9219 DEGs in cold-sensitive genotypes CTWR2 and SCR, respectively. Numerous common and special DEGs with upregulation and downregulation were observed in cold-tolerant and cold-sensitive genotypes involved in critical processes	(Guan *et al.*, 2019)
2	Rice	Leaf sheath	Salt stress	Proteomics	Proteins such as LSY081, LSY262, LSY363, PSII, SOD, OEE2, fructose bisphosphate aldolases etc. each expressed in specific region of plant and had a coordinated response to salt stress	(Abbasi and Komatsu, 2004)
3	Maize	Seeds	Drought stress	Comparative proteomics	721 differentially abundant proteins (DAPs)	(Zenda *et al.*, 2018)
4	Wheat	Flag leaves	Cold stress	Transcriptomics	Identified 876 unigenes (DEGs) that are involved in many critical processes and 28 transcription factors (TFs) belonging to 14 families resulted differentially expressed from which eight families comprised of only TFs induced by cold. 31 differentially expressed Long non-coding RNAs (lncRNAs) were also found, most of them being upregulated in plants treated with cold stress	(Díaz *et al.*, 2019)
5	Wheat	Seeds	Cold stress	Transcriptomics	Several key genes belonging to the bHLH, MYB, NAC, WRKY and ERF transcription factor families related to glutamate metabolism pathway were identified to be highly expressed	(Li *et al.*, 2023)
6	Barley	Root tissues	Salinity stress	Integrated transcriptomics, metabolomics and lipidomics	Revealed two distinctive salt tolerance mechanisms i.e. phenylpropanoid biosynthesis and an intense salt-induced lignin impregnation	(Ho *et al.*, 2020)
7	Sorghum	Root tissues	Salinity stress	Multi-omics	45 genes, 58 secondary metabolites and 246 proteins were recognized	(Ren *et al.*, 2022)
8	Sorghum	Seedling, stems, leaves, panicles	Variable field conditions	Metabolomics	Dhurrin, is identified as a biomarker of crop quality and development, p-glucosyloxy-2-hydroxyphenylacetic acid was also identified for the first time as a main metabolite after gradual neutralization of dhurrin	(Mandrone *et al.*, 2021)

Continued

Table 11.1. Continued.

S. No	Crop	Tissue	Type of stress	Approaches used	Findings	References
9	Finger millet	Leaf tissue	Heat stress	Comparative transcriptome analysis	The study revealed activation of genes related to heat stress, oxidation-reduction processes, water deprivation and key signaling pathways (calcium and kinase). Basal regulatory genes such as bZIP were upregulated, indicating a core stress response. A significant number of DEGs were uncharacterized proteins (PUFs). Overall, several candidate genes and pathways potentially contributing to heat stress tolerance in ragi were identified	(Goyal *et al.*, 2024)
10	Finger millet	Leaves	Salinity stress	Transcriptomics	Multiple genes associated with cell growth and differentiation were upregulated in both sensitive and tolerant genotypes, with stronger expression in the tolerant genotype. In contrast, genes linked to flavonoid biosynthesis were specifically downregulated in the salinity-tolerant Trichy 1	(Rahman *et al.*, 2014)
11	Soybean	Leaves and roots	Salt, saline-alkali and drought stresses	Transcriptomics	A total of 874, 1897 and 535 upregulated genes in leaves, and 1822, 1731 and 1690 upregulated genes in roots, respectively. GO analysis showed a similar distribution across stresses. 69 genes showed consistent expression patterns in both leaves and roots. Genes involved in calcium signaling and nucleic acid pathways were commonly upregulated under all stresses	(Fan *et al.*, 2013)
12	Chickpea	Root tissues	Drought stress	Comparative transcriptomics, proteomics and metabolomics	The integrated multi-omics analysis revealed co-expressed genes, proteins and metabolites involved in phosphatidylinositol signaling, glutathione metabolism and glycolysis/gluconeogenesis pathways, specifically in tolerant genotype under drought. These stress-responsive pathways were coordinately regulated by the differentially abundant transcripts, proteins and metabolites to circumvent the drought stress response/tolerance in tolerant genotype	(Singh *et al.*, 2023)
13	Chickpea	Root tissues	Drought stress	Integrated transcriptomics, proteomics and metabolomics	Integration of transcriptomics and proteomics data identified proteins encoding isoflavone 4′-O-methyltransferase, UDP-d-glucose/UDP-d-galactose 4-epimerase and delta-1-pyrroline-5-carboxylate synthetase. The integration of metabolomics data identified six metabolites (fructose, galactose, glucose, myoinositol, galactinol and raffinose). Integration of root-omics data also revealed some key candidate genes underlying the drought-responsive "QTL-hotspot" region	(Kudapa *et al.*, 2024)

underlying DNA sequence. Epigenetic mechanisms such as DNA methylation, histone modifications, histone variants, small RNAs, long noncoding RNAs and other undefined epigenetic mechanisms can regulate the expression of genes in response to abiotic stress to protect plants against damage caused by variable environments (Miryeganeh, 2021). For example, in rice under drought stress, alterations in DNA methylation in the *OsDREB1* gene can trigger drought-responsive pathways (Begcy and Dresselhaus, 2018). These epigenetic modifications activate drought-responsive pathways, enhancing the plant stress tolerance. In maize, histone marks like H3K4me3 have been found to facilitate the activation of stress genes under drought conditions (Li *et al.*, 2019). Small RNAs such as miR398 are also essential in post-transcriptional gene silencing during oxidative stress in *Arabidopsis* (Sunkar *et al.*, 2012). Overall, these epigenetic processes enable plants to modulate gene expression and increase their tolerance to environmental stresses. In a different study, overexpression of the histone acetyltransferase gene *HAC1* enhanced histone acetylation at stress-responsive loci and improved salinity and oxidative stress tolerance in *Arabidopsis* (Hirakawa *et al.*, 2023). These studies highlight the potential of epigenome editing and multi-layered data integration to create climate-resilient crops designed to resist intricately complex environmental stresses. Examples of stress responsive epigenetic modifications are summarized in Table 11.2. Despite the promising prospects of epigenetic breeding, some challenges persist. One of them is the tissue-specificity of epigenetic modifications, which may differ from one plant organ to another. Another concern is the stability and heritability of the modifications while certain epigenetic changes are reset across generations (Zheng *et al.*, 2017). Recent developments, such as epi-base editors, are now coming up with more precise and targeted solutions for editing such epigenetic marks.

In summary, epigenomics provides important clues regarding the basis for rapid and reversible stress adjustment in crops. Coupled with multi-omics platforms, it promises much towards precision breeding to achieve climate-tolerant varieties of crops. Future research should focus on strengthening the heritability and stability of favorable epigenetic markers, along with improving tools of genome and epigenome editing for practical application in agriculture.

11.3.6 Phenomics

Phenomics involves the detailed study of the traits that make up the phenotype of an organism. Usage of advanced computing and imaging

Table 11.2. Some examples of epigenomic mechanisms in response to climate change across species. (Authors' own table.)

Crop	Abiotic stress	Epigenetic mechanism	Target gene(s)/pathway	Reference
Rice	Drought	DNA methylation (transgenerational epimutations)	Various drought-responsive genes	(Zheng *et al.*, 2017)
Maize	Drought	DNA methylation (RdDM, siRNAs)	Drought-related genes; gene bodies (alternative splicing)	(Wang *et al.*, 2021)
Pearl millet	Salicylic acid (stress/hormone signal)	DNA methylation (global demethylation via ROS1)	Defense genes (R genes, GH3.5, hormone pathways) vs development genes	(Ngom *et al.*, 2017)
Soybean	Salinity	DNA methylation (global hypomethylation)	Genome-wide (hypomethylated regions in salt-responsive genes)	(Yung *et al.*, 2024)
Chickpea	Drought	DNA methylation (CHH demethylation)	Ribosomal protein genes (e.g. *RPS6*)	(Yadav *et al.*, 2024)

technologies in phenomics speeds up the process of phenotyping by analyzing the data from diverse platforms and provides meaningful insights for evaluating the diversity of the crops. Various tools such as magnetic resonance imaging (MRI), three-dimensional (3D) imaging, far-infrared (FIR) imaging systems and near-infrared (NIR) imaging systems are used for various purposes, such as capturing 3D images of roots, shoots and various parts of plants and also to study temperature differences, mechanisms such as photosynthesis efficiency and to measure water content, its movement patterns in leaves and soil, water usage, etc. (Sivasakthi *et al.*, 2017; Ndlovu, 2020). Image-based phenotyping is potentially used to measure and quantify plant traits in various environments and analyzes various traits such as color, texture, quantity and geometry.

A study in pea (*Pisum sativum*) under cold tolerance was conducted using a high-throughput phenotyping method, with automated analyses of photosystem II efficiency and shoot biomass. New software was also developed for chlorophyll fluorescence imaging and RGB image analysis. These developments proved to be effective in measuring the crop's response to cold stress (Humplík *et al.*, 2015). Another study conducted on common bean (*Phaseolus vulgaris*) highlights the importance of hyperspectral remote sensing as a powerful tool for phenotyping the physiological drought response to evaluate drought tolerance, which helps in screening genotypes and improving drought tolerance in breeding programs (Wong *et al.*, 2023).

11.4 Integrating Multi-omics Approaches

A key component of modern crop breeding is the genetic analysis of complex traits using multi-omics approaches. Understanding the molecular foundations of phenotypic features involves combining several biological data layers, including genomics, transcriptomics, proteomics, metabolomics and epigenomics. This is particularly useful for complex traits such as yield, drought tolerance and disease resistance that are controlled by multiple genes and environmental variables.

11.4.1 QTL-omics

The integration of -omics approaches enables identification of quantitative trait loci (QTLs) and the underlying genes associated with traits of interest. QTL-omics combines multi-omics data to dissect complex traits such as stress tolerance, yield and quality traits. By analyzing multi-omics datasets, QTL-omics not only pinpoints genomic regions linked to phenotypic variation but also reveals the molecular mechanisms and regulatory networks driving these complex traits. For example, in rice QTL-omics was used to identify loci associated with drought tolerance, integrating transcriptomics and genomics data to uncover candidate genes regulating stress response (Kumar *et al.*, 2023). In maize, combining QTL mapping and transcriptome revealed candidate genes linked to cold tolerance (He *et al.*, 2023). Similarly in chickpea, multi-omics analysis on drought-tolerant and sensitive genotypes revealed hub proteins and metabolites associated with drought stress response. Further, the study identified candidate genes within drought responsive '*QTL-hotspot*' region bridging genomic loci with molecular mechanisms of drought adaptation in chickpea (Kudapa *et al.*, 2024).

11.4.2 Genome-wide Association Studies (GWAS)

Genome-wide distributed high-density SNPs help in delineating the possible genome region associated with phenotypic variation in wide germplasm pools through GWAS. Through the identification of important genetic loci linked to temperature sensitivity, drought tolerance and water usage efficiency, GWAS has made it possible to improve genetic resilience to stress in a variety of crops. With significant contributions from a variety of rice germplasm panels and fine-mapped QTLs, rice GWAS have identified genomic areas associated with drought tolerance, cold stress resilience, root system architecture and grain quality attributes impacted by heat stress (Padmashree *et al.*, 2023; Ghazy *et al.*, 2024; Rastogi *et al.*, 2025). Breeding for climate adaptation has accelerated as a result of the discovery of genotypic variation in drought tolerance and water-use efficiency features in legume GWAS on crops such as common beans, cowpeas and chickpeas (Susmitha *et al.*, 2023). Through GWAS identifying loci associated with nitrogen responsiveness, drought tolerance and yield stability under stress, millets such as finger millet, sorghum, foxtail millet and pearl millet are becoming more recognized for their climate-resilient qualities (Mukherjee *et al.*, 2025).

11.4.3 Transcriptome-wide Association Studies (TWAS)

Compared with GWAS, transcriptome wide association studies (TWAS) is a powerful method for associating transcript abundance with phenotypic variation instead of genetic markers. TWAS uses RNA transcript levels as explanatory factors, offering quantitative metrics, in contrast to GWAS, which depends on genetic markers that show several functioning haplotypes. By facilitating actionable findings for functional validation, this direct gene-to-phenotype link in TWAS saves the time and effort needed for downstream approaches such as gene editing. Using this method, cuticular conductance, a trait associated with drought tolerance, was investigated in maize by combining the TWAS-GWAS ensemble approach and RNA expression measurements. About 310 genetically different maize lines were studied for *ISTL1*, a gene that is similar to an intracellular protein and membrane trafficking gene in *Arabidopsis*. This gene appears to be a promising candidate for controlling cuticular conductance (Kremling *et al.*, 2019). Even though transcriptomics has been widely utilized to study how crops respond to abiotic stress, the majority of research has either examined more specialized gene activities or concentrated on small sample sizes. Priority should be given to large-scale transcriptome investigations as they can provide a more thorough understanding of how plants react to abiotic stressors (Derbyshire *et al.*, 2022). Furthermore, multi-omics data modeling can explain how a final phenotype is controlled by the differential expression of mRNA, protein and metabolite in the central dogma. Combining the aforementioned studies can be used to predict the quality traits to aid crop breeding programs. As a result, networks with comparatively limited sets of interconnected -omics variables are more capable of forecasting the desired characteristic.

11.4.4 Epigenome-wide Association Studies (EWAS)

Epigenome-wide Association Studies (EWAS) is the method for identifying epigenetic variations associated with climate resilience in several crops. Epigenetic regulation of stress-responsive genes, such as DNA methylation, histone acetylation and methylation, helps plants to acclimate under temperature extremes. In maize, a study mapped genome-wide DNA methylation and histone methylation (H3K9me2) in seedlings, revealing enrichment of both marks within transposable elements but depletion near gene boundaries highlighting a complex interplay between transposons and genes that shape chromatin state (West *et al.*, 2014). Folsom *et al.* (2014) demonstrated that epigenetic modifications regulate the expression of the *OsFIE* gene in rice, resulting in smaller seeds. In maize, reduced methylation was seen in drought-tolerant conditions, while the upregulation of cold-responsive genes was caused by histone acetylation, including cold-regulated 413 protein (COR413) and drought-responsive element binding 1 (DREB1) (Hu *et al.*, 2011). Barley's epigenetic changes in response to drought stress were also investigated; methylation of the DEMETER (DME) homolog was shown to boost seed growth (Kapazoglou *et al.*, 2013). DNA methylation and acetylation were responsible for the stress tolerance in several legume crops. However, only a few applications were available with EWAS studies for several crops.

11.5 Computational Modeling for Multi-omics Data Analysis

Multi-omics approaches aim to gain a deeper understanding of how different biological processes play a role in the expression of desirable traits in various crops. An ample amount of data is generated during this process, which is vast and can be quite challenging to interpret. This is where computational modeling techniques become crucial tools by enabling integration and data analysis from various multi-omics approaches to gain meaningful insights. The integration of multi-omics data requires various software tools, models, methodologies and databases (Zhang *et al.*, 2022). Some examples of databases are mentioned in Table 11.3.

Apart from databases, there are also many software packages available online that are used for analysis of multi-omics data in crops. Some

Table 11.3. Databases for integration of muti-omics datasets. (Authors' own table.)

Name	Species/crops	Omics focus	Key features/applications	Reference
CARMO	Rice	Multi-omics	Co-regulator networks, heatmaps, functional clusters, visualization tools; supports transcriptome, open chromatin, epigenetic and resequencing data interpretation	(Wang *et al.*, 2015)
IOMA	General	Proteomics + metabolomics	Predicts metabolic flux by integrating proteomic and metabolomic data	(Yizhak *et al.*, 2010)
GpemDB	General	Genomics, phenomics	Two-layer structure for data integration, visualization, sharing; supports precision breeding	(Gong *et al.*, 2022)
Gramene	Multiple	Multi-omics	Comparative genomics and multi-omics data	(Yang *et al.*, 2021)
GabiPD	Multiple	Multi-omics	Plant functional genomics and systems biology	
KaPPA-View4 KEGG	Multiple	Multi-omics	Pathway-based integration and visualization	
KOMICS	Multiple	Metabolomics	Plant metabolome data integration	
KNApSAcK	Multiple	Metabolomics	Plant metabolite database	
PMND	Multiple	Metabolomics	Plant metabolic network database	
PlantTFDB	Multiple	Transcriptomics	Plant transcription factor database	
PPDB	Multiple	Proteomics	Plant proteome database	
GrainGenes	Wheat, Barley, Oat, Rye	Genomics	Genomic data integration for cereals	
plantGDB	Multiple	Comparative genomics	Plant comparative genomics database	
RiceXPro	Rice	Functional genomics, transcriptomics	Transcriptome expression profiling	
SoyBase	Soybean	Multi-omics	Comprehensive soybean genomics and genetics	
SFGD	Soybean	Multi-omics	Soybean functional genomics database	
MaizeGDB	Maize	Multi-omics	Integrated maize genetics and genomics	
MaizeDIG	Maize	Phenomics, genomics	Digital integration for maize phenomics/genomics	
CTDB	Chickpea	Functional genomics, transcriptomics	Chickpea genomics and transcriptomics resource	
BARLEX	Barley	Genomics	Barley genome exploration	
SorghumFDB	Sorghum	Functional genomics	Sorghum functional genomics database	

of them are EasyPCC (in multiple species, for phenomics), STRING (in multiple species, for proteomics), Gromacs (in multiple species, for genomics, proteomics and metabolomics), PTools (in multiple species, for multi-omics), GSDS and GAP4 (in multiple species, for structural genomics), VISTA (in multiple species, for comparative genomics), AMDIS (in multiple species, for metabolomics) and SIMCA-P14.0 (in multiple species, for integrated ionomics and metabolomics) (Yang *et al.*, 2021).

11.6 Application of Multi-omics Approaches for Improving Climate-resilient Crops

Advances in -omics technologies and the rapid accumulation of highly reliable -omics data offers significant opportunities for their application in crop improvement. By integrating the data produced from various multi-omics approaches such as genomics, transcriptomics, proteomics, metabolomics and phenomics, researchers can gain a better understanding of molecular mechanisms underlying complex traits. Examples of several -omics and integrated -omics studies in different crops are summarized in Table 11.1.

One such example of integrated multi-omics approach in developing climate-resilient crops can be seen in rice, where GWAS is combined with transcriptomic data which resulted in the identification of a gene named *DROUGHT1* (*DROT1*). *DROT1* encodes for a COBRA-like protein whose role is vital when it comes to improving drought resistance. This protein modifies the structure of the cell wall especially by increasing cellulose content and keeping of cellulose crystallinity (Sun *et al.*, 2022). In rice, using data generated by the integration of transcriptomic and proteomic profiling, 236 small secreted proteins (SSPs) and 52 unannotated SSPs were identified and a new gene immune response peptide (IRP) was discovered among these SSPs which seemed to encode for the precursor of an

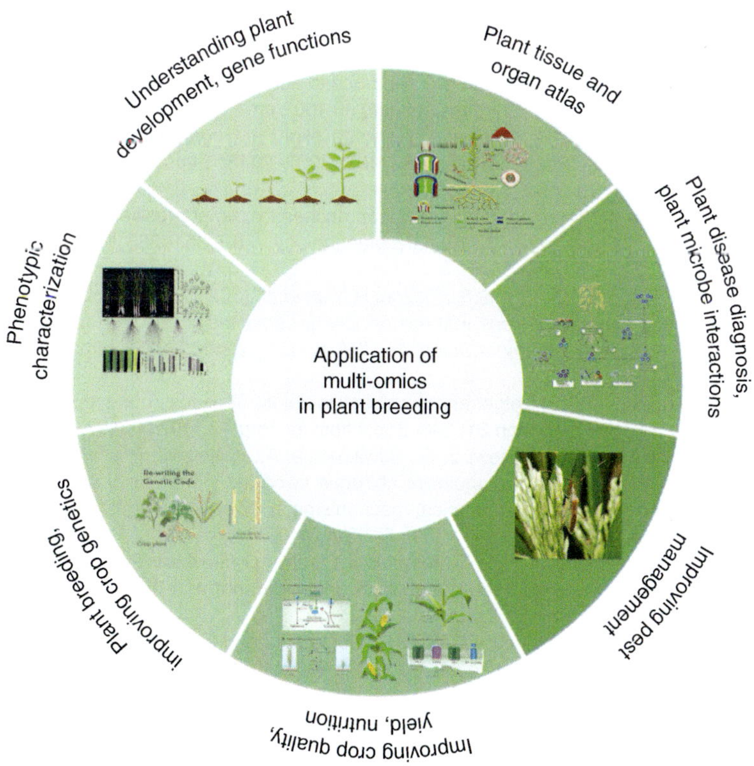

Figure 11.2. Applications of multi-omics approaches in plant breeding. (Authors' own image.)

SSP which also plays a key role in regulating immunity against rice blast caused by fungus *Magnaporthe grisea* (Wang *et al.*, 2020). Some of the applications of multi-omics in crop improvement are illustrated in Figure 11.2.

11.7 Conclusions

Climate change adversely impacts crop productivity, phenological development, growth cycles and food quality, posing significant threat to future food and nutritional security. To meet future food demands and reduce these risks, efforts are focused on developing novel crop varieties and stress resilient cultivars through genetic improvement and the integration of multi-omics technologies. Over the past two decades, multi-omics approaches have emerged as a powerful tool for creating climate-smart crops. Combining genomics with transcriptomics, proteomics, metabolomics and epigenomics enables a comprehensive understanding of the genotype-to-phenotype relationship under specific environmental stresses. The top-down (phenotype to genotype) and bottom-up (genotype to phenotype) strategies facilitates the identification of key candidate genes and pathways that regulate key traits. When coupled with contemporary plant breeding, advanced bioinformatics and computational modeling, these integrative strategies provide time- and cost-effective frameworks for developing climate-resilient, nutrient-rich and high-yielding crop varieties. Collectively, such integrative approaches offer a sustainable pathway towards safeguarding global food security in the face of climate change.

References

Abbas, G., Ahmed, M., Fatima, Z., Hussain, S., Kheir, A.M. *et al.* (2023) Modeling the potential impact of climate change on maize-maize cropping system in semi-arid environment and designing of adaptation options. *Agricultural and Forest Meteorology* 341, 109674. http://doi.org/10.1016/j.agrformet.2023.109674.

Abbasi, F.M. and Komatsu, S. (2004) A proteomic approach to analyze salt-responsive proteins in rice leaf sheath. *Proteomics* 4(7), 2072–2081. https://doi.org/10.1002/pmic.200300741.

Ahamad, F. and Khan, M.R. (2023) Incidence of sheath blight in irrigated rice and associated yield losses in northern India. *Plant Disease* 107(10), 2907–2915. https://doi.org/10.1094/PDIS-12-22-2905-RE.

Anusha, M.R. and Amir, M. (2024) Integrating omics approaches for climate-resilient crops: A comprehensive review. *Journal of Advances in Biology & Biotechnology* 27(6), 351–363. https://doi.org/10.9734/jabb/2024/v27i6895.

Badoni, S., Pasion-Uy, E.A., Kor, S., Kim, S.R., Tiozon, R.N.Jr, *et al.* (2024) Multiomics of a rice population identifies genes and genomic regions that bestow low glycemic index and high protein content. *Proceedings of the National Academy of Sciences USA* 121(36), e2410598121. https://doi.org/10.1073/pnas.2410598121.

Begcy, K. and Dresselhaus, T. (2018) Epigenetic responses to abiotic stresses during reproductive development in cereals. *Plant Reproduction* 31, 343–355. https://doi.org/10.1007/s00497-018-0343-4.

Chmielewska, K., Rodziewicz, P., Swarcewicz, B., Sawikowska, A., Krajewski, P. *et al.* (2016) Analysis of drought-induced proteomic and metabolomic changes in barley (*Hordeum vulgare* L.) leaves and roots unravels some aspects of biochemical mechanisms involved in drought tolerance. *Frontiers in Plant Science* 7, 1108. https://doi.org/10.3389/fpls.2016.01108.

Das, A., Rushton, P.J. and Rohila, J.S. (2017) Metabolomic profiling of soybeans (*Glycine max* L.) reveals the importance of sugar and nitrogen metabolism under drought and heat stress. *Plants* 6(2), 21. https://doi.org/10.3390/plants6020021.

de Souza Batista, F., Duku, C. and Hein, L. (2023) Deforestation-induced changes in rainfall decrease soybean-maize yields in Brazil. *Ecological Modelling* 486, 110533. https://doi.org/10.1016/j.ecolmodel.2023.110533.

Derbyshire, M.C., Batley, J. and Edwards, D. (2022) Use of multiple 'omics techniques to accelerate the breeding of abiotic stress tolerant crops. *Current Plant Biology* 32, p.100262. https://doi.org/10.1016/j.cpb.2022.100262.

Díaz, M.L., Soresi, D.S., Basualdo, J., Cuppari, S.J. and Carrera, A. (2019) Transcriptomic response of durum wheat to cold stress at reproductive stage. *Molecular Biology Reports* 46(2), 2427–2445. https://doi.org/10.1007/s11033-019-04704-y.

Elshafie, H.S., Camele, I. and Mohamed, A.A. (2023) A comprehensive review on the biological, agricultural and pharmaceutical properties of secondary metabolites based-plant origin. *International Journal of Molecular Sciences* 24(4), 3266. https://doi.org/10.3390/ijms24043266.

Fan, X.D., Wang, J.Q., Yang, N., Dong, Y.Y., Liu, L. *et al.* (2013) Gene expression profiling of soybean leaves and roots under salt, saline–alkali and drought stress by high-throughput Illumina sequencing. *Gene* 512(2), 392–402. https://doi.org/10.1016/j.gene.2012.09.100.

Folsom, J.J., Begcy, K., Hao, X., Wang, D. and Walia, H. (2014) Rice fertilization-independent endosperm1 regulates seed size under heat stress by controlling early endosperm development. *Plant Physiology* 165(1), 238–248. https://doi.org/10.1104/pp.113.232413.

Francia, E., Morcia, C., Pasquariello, M., Mazzamurro, V., Milc, J.A. *et al.* (2016) Copy number variation at the HvCBF4–HvCBF2 genomic segment is a major component of frost resistance in barley. *Plant Molecular Biology* 92, 161–175. https://doi.org/10.1007/s11103-016-0505-4.

Gamuyao, R., Chin, J.H., Pariasca-Tanaka, J., Pesaresi, P., Catausan, S. *et al.* (2012) The protein kinase Pstol1 from traditional rice confers tolerance of phosphorus deficiency. *Nature* 488(7412), 535–539. https://doi.org/10.1038/nature11346.

Garg, V., Chitikineni, A., Sharma, M., Ghosh, R., Samineni, S. *et al.* (2023) Transcriptome profiling reveals the expression and regulation of genes associated with Fusarium wilt resistance in chickpea (*Cicer arietinum* L.). *The Plant Genome* 16(3), e20340. https://doi.org/10.1002/tpg2.20340.

Gaur, P.M., Jukanti, A.K., Samineni, S., Chaturvedi, S.K., Basu, P.S. *et al.* (2013) Climate change and heat stress tolerance in chickpea. In Climate Change and Plant Abiotic Stress Tolerance. Tuteja, N., Gill, S.S., Eds.; Wiley: Weinheim, Germany, 2013; pp. 837–856. http://dx.doi.org/10.1002/9783527675265.ch31.

Ghazy, M.I., El-Naem, S.A., Hefeina, A.G., Sallam, A. and Eltaher, S. (2024) Genome-wide association study of rice diversity panel reveals new QTLs for tolerance to water deficit under the Egyptian conditions. *Rice* 17(1), 29. https://doi.org/10.1186/s12284-024-00703-1.

Gong, L., Lou, Q., Yu, C., Chen, Y., Hong, J. *et al.* (2022) GpemDB: A scalable database architecture with the multi-omics entity-relationship model to integrate heterogeneous big-data for precise crop breeding. *Frontiers in Bioscience-Landmark* 27(5), 159. https://doi.org/10.31083/j.fbl2705159.

Goyal, E., Singh, A.K., Mahajan, M.M. and Kanika, K. (.2024) Comparative transcriptome profiling of contrasting finger millet (*Eleusine coracana* (L.) Gaertn) genotypes under heat stress. *Molecular Biology Reports* 51(1), 283. https://doi.org/10.1007/s11033-024-09233-x.

Guan, S., Xu, Q., Ma, D., Zhang, W., Xu, Z. *et al.* (2019) Transcriptomics profiling in response to cold stress in cultivated rice and weedy rice. *Gene* 685, 96–105. https://doi.org/10.1016/j.gene.2018.10.066.

Guerriero, G., Berni, R., Muñoz-Sanchez, J.A., Apone, F., Abdel-Salam, E.M. *et al.* (2018) Production of plant secondary metabolites: Examples, tips and suggestions for biotechnologists. *Genes* 9(6), 309. https://doi.org/10.3390/genes9060309.

Gundaraniya, S.A., Ambalam, P.S. and Tomar, R.S. (2020) Metabolomic profiling of drought-tolerant and susceptible peanut (*Arachis hypogaea* L.) genotypes in response to drought stress. *ACS Omega* 5(48), 31209–31219. https://doi.org/10.1021/acsomega.0c04601.

Han, L., Xiao, C., Xiao, B., Wang, M., Liu, J. *et al.* (2019) Proteomic profiling sheds light on alkali tolerance of common wheat (*Triticum aestivum* L.). *Plant Physiology and Biochemistry* 138, 58–64. https://doi.org/10.1016/j.plaphy.2019.02.024.

He, R.Y., Yang, T., Zheng, J.J., Pan, Z.Y., Chen, Y. *et al.* (2023) QTL mapping and a transcriptome integrative analysis uncover the candidate genes that control the cold tolerance of maize introgression lines at the seedling stage. *International Journal of Molecular Sciences* 24(3), 2629. https://doi.org/10.3390/ijms24032629.

Hirakawa, T., Tanno, S. and Ohara, K. (2023) N-acetylglutamic acid alleviates oxidative stress based on histone acetylation in plants. *Frontiers in Plant Science* 14, 1165646. https://doi.org/10.3389/fpls.2023.1165646.

Hirsch, C.N., Foerster, J.M., Johnson, J.M., Sekhon, R.S., Muttoni, G. *et al.* (2014) Insights into the maize pan-genome and pan-transcriptome. *Plant Cell* 26, 121–135. https://doi.org/10.1105/tpc.113.119982.

Ho, W.W.H., Hill, C.B., Doblin, M.S., Shelden, M.C., van de Meene, A. *et al.* (2020) Integrative multi-omics analyses of barley rootzones under salinity stress reveal two distinctive salt tolerance mechanisms. *Plant Communications* 1(3), 1–17. https://doi.org/10.1016/j.xplc.2020.100031.

Hosseini, S.Z., Ismaili, A., Nazarian-Firouzabadi, F., Fallahi, H., Nejad, A.R. *et al.* (2021) Dissecting the molecular responses of lentil to individual and combined drought and heat stresses by comparative transcriptomic analysis. *Genomics* 113(2), 693–705. https://doi.org/10.1016/j.ygeno.2020.12.038.

Hu, Y., Zhang, L., Zhao, L., Li, J., He, S. *et al.* (2011) Trichostatin A selectively suppresses the cold-induced transcription of the ZmDREB1 gene in maize. *PLoS ONE* 6(7), e22132. https://doi.org/10.1371/journal.pone.0022132.

Humplík, J.F., Lazár, D., Fürst, T., Husičková, A., Hýbl, M. *et al.* (2015) Automated integrative high-throughput phenotyping of plant shoots: A case study of the cold-tolerance of pea (*Pisum sativum* L.). *Plant Methods* 11, 1–11. https://doi.org/10.1186/s13007-015-0063-9.

Jadaun, K.K. and Khan, A.A. (2023) Impact of climate change on millet yield in India since 1991: An econometric analysis. *The Journal of Research ANGRAU* 51(1), 129–139. https://doi.org/10.58537/jorangrau.2023.51.1.15.

Jagtap, A.B., Yadav, I.S., Vikal, Y., Praba, U.P., Kaur, N. *et al.* (2023) Transcriptional dynamics of maize leaves, pollens and ovules to gain insights into heat stress-related responses. *Frontiers in Plant Science* 14, 1117136. https://doi.org/10.3389/fpls.2023.1117136.

Jha, U.C., Naik, Y.D., Priya, M., Nayyar, H., Sofi, P.A. *et al.* (2025) Chickpea (*Cicer arietinum* L.) battling against heat stress: Plant breeding and genomics advances. *Plant Molecular Biology* 115(4), 101. https://doi.org/10.1007/s11103-025-01628-z.

Ji, W., Cong, R., Li, S., Li, R., Qin, Z. *et al.* (2016) Comparative proteomic analysis of soybean leaves and roots by iTRAQ provides insights into response mechanisms to short-term salt stress. *Frontiers in Plant Science* 7, 573. https://doi.org/10.3389/fpls.2016.00573

Kapazoglou, A., Drosou, V., Argiriou, A. and Tsaftaris, A.S. (2013) The study of a barley epigenetic regulator, HvDME, in seed development and under drought. *BMC Plant Biology* 13, 1–16. https://doi.org/10.1186/1471-2229-13-172.

Khan, A.W., Garg, V., Sun, S., Gupta, S., Dudchenko, O. *et al.* (2024) *Cicer* super-pangenome provides insights into species evolution and agronomic trait loci for crop improvement in chickpea. *Nature Genetics* 56(6), 1225–1234. https://doi.org/10.1038/s41588-024-01760-4.

Kremling, K.A., Diepenbrock, C.H., Gore, M.A., Buckler, E.S. and Bandillo, N.B. (2019) Transcriptome-wide association supplements genome-wide association in *Zea mays*. *G3: Genes, Genomes, Genetics*, 9(9), 3023–3033. https://doi.org/10.1534/g3.119.400549.

Kudapa, H., Garg, V., Chitikineni, A. and Varshney, R.K. (2018) The RNA-Seq-based high resolution gene expression atlas of chickpea (*Cicer arietinum* L.) reveals dynamic spatio-temporal changes associated with growth and development. *Plant, Cell & Environment* 41(9), 2209–2225. https://doi.org/10.1111/pce.13210.

Kudapa, H., Ghatak, A., Barmukh, R., Chaturvedi, P., Khan, A. *et al.* (2024) Integrated multi-omics analysis reveals drought stress response mechanism in chickpea (*Cicer arietinum* L.). *The Plant Genome* 17(1), e20337. https://doi.org/10.1002/tpg2.20337.

Kumar, D., Ramkumar, M.K., Dutta, B., Kumar, A., Pandey, R. *et al.* (2023) Integration of miRNA dynamics and drought tolerant QTLs in rice reveals the role of miR2919 in drought stress response. *BMC Genomics* 24(1), 526. https://doi.org/10.1186/s12864-023-09609-6.

Lawas, L.M.F., Li, X., Erban, A., Kopka, J., Jagadish, S.K. *et al.* (2019) Metabolic responses of rice cultivars with different tolerance to combined drought and heat stress under field conditions. *GigaScience* 8(5), giz050. https://doi.org/10.1093/gigascience/giz050.

Leisner, C.P., Potnis, N. and Sanz-Saez, A. (2023) Crosstalk and trade-offs: Plant responses to climate change-associated abiotic and biotic stresses. *Plant, Cell & Environment* 46(10), 2946–2963. https://doi.org/10.1111/pce.14532.

Li, E., Liu, H., Huang, L., Zhang, X. and Dong, X. (2019) Long-range interactions between proximal and distal regulatory regions in maize. *Nature Communications* 10(1), 2633. https://doi.org/10.1038/s41467-019-10603-4

Li, G.Y., Zhou, J.X., Ma, W.K., Jiang, L.G., Jin, Z.H. *et al.* (2014) De novo assembly of soybean wild relatives for pan-genome analysis of diversity and agronomic traits. *Nature Biotechnology* 32, 1045–1052. https://doi.org/10.1038/nbt.2979.

Li, L., Han, C., Yang, J., Tian, Z., Jiang, R. *et al.* (2023) Comprehensive transcriptome analysis of responses during cold stress in wheat (*Triticum aestivum* L.). *Genes* 14(4), 844. https://doi.org/10.3390/genes14040844.

Li, N., Zhao, Y., Han, J., Yang, Q., Liang, J. *et al.* (2024) Impacts of future climate change on rice yield based on crop model simulation – A meta-analysis. *Science of the Total Environment* 949, 175038. https://doi.org/10.1016/j.scitotenv.2024.175038.

Lowe, R., Shirley, N., Bleackley, M., Dolan, S. and Shafee, T. (2017) Transcriptomics technologies. *PLoS Computational Biology* 13(5), e1005457. https://doi.org/10.1371/journal.pcbi.1005457.

Mahajan, G., Wenham, K. and Chauhan, B.S. (2023) Mungbean (*Vigna radiata*) growth and yield response in relation to water stress and elevated day/night temperature conditions. *Agronomy* 13(10), 2546. https://doi.org/10.3390/agronomy13102546.

Maharajan, T., Krishna, T.P.A., Krishnakumar, N.M., Vetriventhan, M., Kudapa, H. *et al.* (2024) Role of genome sequences of major and minor millets in strengthening food and nutritional security for future generations. *Agriculture* 14(5), 670. https://doi.org/10.3390/agriculture14050670.

Majoul, T., Bancel, E., Triboï, E., Ben Hamida, J. and Branlard, G. (2003) Proteomic analysis of the effect of heat stress on hexaploid wheat grain: Characterization of heat-responsive proteins from total endosperm. *Proteomics* 3(2), 175–183. https://doi.org/10.1002/pmic.200390026.

Mandrone, M., Chiocchio, I., Barbanti, L., Tomasi, P., Tacchini, M. *et al.* (2021) Metabolomic study of sorghum (*Sorghum bicolor*) to interpret plant behavior under variable field conditions in view of smart agriculture applications. *Journal of Agricultural and Food Chemistry* 69(3), 1132–1145. https://doi.org/10.1021/acs.jafc.0c06533

Miryeganeh, M. (2021) Plants' epigenetic mechanisms and abiotic stress. *Genes* 12(8), 1106. https://doi.org/10.3390/genes12081106.

Montenegro, J.D., Golicz, A.A., Bayer, P.E., Hurgobin, B., Lee, H. *et al.* (2017) The pangenome of hexaploid bread wheat. *The Plant Journal* 90, 1007–1013. https://doi.org/10.1111/tpj.13515.

Mukherjee, B., Jha, R.K., Sattar, A., Dutta, S., Bhattacharya, U. *et al.* (2025) Harnessing the potential of millets for climate-resilient and sustainable agriculture. *Frontiers in Plant Science* 16, 1574699. https://doi.org/10.3389/fpls.2025.1574699.

Nalini Chandran, A.K., Paul, P., Dhatt, B.K., Sandhu, J., Irvin, L. *et al.* (2024) Phenotypic and transcriptomic responses of diverse rice accessions to transient heat stress during early grain development. *Frontiers in Plant Science* 15, 1429697. https://doi.org/10.3389/fpls.2024.1429697.

Ndlovu, N. (2020) Application of genomics and phenomics in plant breeding for climate resilience. *Asian Plant Research Journal* 6(4), 53–66. https://doi.org/10.9734/APRJ/2020/v6i430137.

Ngom, B., Sarr, I., Kimatu, J., Mamati, E. and Kane, N.A. (2017) Genome-wide analysis of cytosine DNA methylation revealed salicylic acid promotes defense pathways over seedling development in pearl millet. *Plant Signaling & Behavior* 12(9), e1356967. https://doi.org/10.1080/15592324.2017.1356967.

Nyaupane, S., Poudel, M.R., Panthi, B., Dhakal, A., Paudel, H. *et al.* (2024) Drought stress effect, tolerance, and management in wheat – A review. *Cogent Food & Agriculture* 10(1), 2296094. https://doi.org/10.1080/23311932.2023.2296094.

Padmashree, R., Barbadikar, K.M., Honnappa, Magar, N.D., Balakrishnan, D. *et al.* (2023) Genome-wide association studies in rice germplasm reveal significant genomic regions for root and yield-related traits under aerobic and irrigated conditions. *Frontiers in Plant Science* 14, 1143853. https://doi.org/10.3389/fpls.2023.1143853.

Park, S.Y. and Jeong, D.H. (2023) Comprehensive analysis of rice seedling transcriptome during dehydration and rehydration. *International Journal of Molecular Sciences* 24(9), 8439. https://doi.org/10.3390/ijms24098439.

Pazhamala, L.T., Kudapa, H., Weckwerth, W., Millar, A.H. and Varshney, R.K. (2021) Systems biology for crop improvement. *The Plant Genome* 14(2), e20098. https://doi.org/10.1002/tpg2.20098.

Pequeno, D.N., Hernandez-Ochoa, I.M., Reynolds, M., Sonder, K., MoleroMilan, A. *et al.* (2021) Climate impact and adaptation to heat and drought stress of regional and global wheat production. *Environmental Research Letters* 16(5), 054070. https://doi.org/10.1088/1748-9326/abd970.

Pinu, F.R., Beale, D.J., Paten, A.M., Kouremenos, K., Swarup, S. *et al.* (2019) Systems biology and multi-omics integration: Viewpoints from the metabolomics research community. *Metabolites* 9(4), 76. https://doi.org/10.3390/metabo9040076.

Rahman, H., Jagadeeshselvam, N., Valarmathi, R., Sachin, B., Sasikala, R. *et al.* (2014) Transcriptome analysis of salinity responsiveness in contrasting genotypes of finger millet (*Eleusine coracana* L.) through RNA-sequencing. *Plant Molecular Biology* 85(4), 485–503. https://doi.org/10.1007/s11103-014-0199-4.

Rastogi, K., Mankar, S.P. and Septiningsih, E.M. (2025) Genome-wide association study for traits related to cold tolerance and recovery during seedling stage in rice. *Crop Science* 65(1), 70003. https://doi.org/10.1002/csc2.70038.

Raza, A., Tabassum, J., Kudapa, H. and Varshney, R.K. (2021) Can omics deliver temperature resilient ready-to-grow crops? *Critical Reviews in Biotechnology* 41(8), 1209–1232. https://doi.org/10.1080/07388551.2021.1898332.

Ren, G., Yang, P., Cui, J., Gao, Y., Yin, C. *et al.* (2022) Multiomics analyses of two sorghum cultivars reveal the molecular mechanism of salt tolerance. *Frontiers in Plant Science* 13, 886805. https://doi.org/10.3389/fpls.2022.886805.

Saxena, K., Choudhary, A.K., Srivastava, R.K., Bohra, A., Saxena, R.K. *et al.* (2019) Origin of early maturing pigeonpea germplasm and its impact on adaptation and cropping systems. *Plant Breeding* 138(3), 243–251. https://doi.org/10.1111/pbr.12696.

Severin, A.J., Woody, J.L., Bolon, Y.T., Joseph, B., Diers, B.W. *et al.* (2010) RNA-Seq Atlas of *Glycine max*: A guide to the soybean transcriptome. *BMC Plant Biology* 10(1), 160. https://doi.org/10.1186/1471-2229-10-160.

Singh, V., Gupta, K., Singh, S., Jain, M. and Garg, R. (2023) Unravelling the molecular mechanism underlying drought stress response in chickpea via integrated multi-omics analysis. *Frontiers in Plant Science* 14, 1156606. https://doi.org/10.3389/fpls.2023.1156606.

Sivasakthi, K., Tharanya, M., Kholová, J., Wangari Muriuki, R., Thirunalasundari, T. *et al.* (2017) Chickpea genotypes contrasting for vigor and canopy conductance also differ in their dependence on different water transport pathways. *Frontiers in Plant Science* 8, 1663. https://doi.org/10.3389/fpls.2017.01663.

Sun, X., Xiong, H., Jiang, C., Zhang, D., Yang, Z. *et al.* (2022) Natural variation of DROT1 confers drought adaptation in upland rice. *Nature Communications* 13(1), 4265. https://doi.org/10.1038/s41467-022-31844-w.

Sunkar, R., Li, Y.F. and Jagadeeswaran, G. (2012) Functions of microRNAs in plant stress responses. *Trends in Plant Science* 17(4), 196–203. https://doi.org/10.1016/j.tplants.2012.01.010.

Susmitha, P., Kumar, P., Yadav, P., Sahoo, S., Kaur, G. *et al.* (2023) Genome-wide association study as a powerful tool for dissecting competitive traits in legumes. *Frontiers in Plant Science* 14, 1123631. https://doi.org/10.3389/fpls.2023.1123631.

Talwar, H.S., Elangovan, M. and Patil, J.V. (2013) Sorghum-A potential crop to adapt to future climate change scenario. In: Elangovan, M. and JV Patil (Eds.). Managing Intellectual Property under PVP and PGR. Directorate of Sorghum Research (DSR), Rajendranagar, Hyderabad 500030, India. ISBN: 81-89335-44-8. pp. 254–256.

Tettelin, H., Masignani, V., Cieslewicz, M.J., Donati, C., Medini, D. *et al.* (2005) Genome analysis of multiple pathogenic isolates of *Streptococcus agalactiae*: Implications for the microbial 'pan-genome'. *Proceedings of the National Academy of Sciences USA* 102, 13950–13955. https://doi.org/10.1073/pnas.0506758102.

Thorpe, P., Altmann, S., Lopez-Cobollo, R., Douglas, N., Iqbal, J. *et al.* (2024) Multi-omics approaches define novel aphid effector candidates associated with virulence and avirulence phenotypes. *BMC Genomics* 25(1), 1065. https://doi.org/10.1186/s12864-024-10984-x.

Timilsina, A.P., Baigorria, G.A., Wilhite, D., Shulski, M., Heeren, D. *et al.* (2023) Soybean response under climatic scenarios with changed mean and variability under rainfed and irrigated conditions in major soybean-growing states of the USA. The *Journal of Agricultural Science* 161(2), 157–174. https://doi.org/10.1017/S0021859623000011.

Varshney, R.K., Graner, A. and Sorrells, M.E. (2005) Genomics-assisted breeding for crop improvement. *Trends in Plant Science* 10, 621–630. https://doi.org/10.1016/j.tplants.2005.10.004.

Varshney, R.K., Kudapa, H., Pazhamala, L., Chitikineni, A., Thudi, M. *et al.* (2015) Translational genomics in agriculture: Some examples in grain legumes. *Critical Reviews in Plant Sciences* 34(1–3), 169–194. https://doi.org/10.1080/07352689.2014.897909.

Varshney, R.K., Roorkiwal, M., Sun, S., Bajaj, P., Chitikineni, A. *et al.* (2021) A chickpea genetic variation map based on the sequencing of 3,366 genomes. *Nature* 599(7886), 622–627. https://doi.org/10.1038/s41586-021-04066-1.

Varshney, R.K., Shi, C., Thudi, M., Mariac, C., Wallace, J. *et al.* (2017) Pearl millet genome sequence provides a resource to improve agronomic traits in arid environments. *Nature Biotechnology* 35(10), 969–976. https://doi.org/10.1038/nbt.3943.

Varshney, R.K., Song, C., Saxena, R.K., Azam, S., Yu, S. *et al.* (2013) Draft genome sequence of chickpea (*Cicer arietinum*) provides a resource for trait improvement. *Nature Biotechnology* 31(3), 240–246. https://doi.org/10.1038/nbt.2491.

Wang, J., Qi, M., Liu, J. and Zhang, Y. (2015) CARMO: A comprehensive annotation platform for functional exploration of rice multi-omics data. *The Plant Journal* 83(2), 359–374. https://doi.org/10.1111/tpj.12894.

Wang, J., Vanga, S.K., Saxena, R., Orsat, V. and Raghavan, V. (2018) Effect of climate change on the yield of cereal crops: A review. *Climate* 6(2), 41. https://doi.org/10.3390/cli6020041.

Wang, P., Yao, S., Kosami, K.I., Guo, T., Li, J. *et al.* (2020) Identification of endogenous small peptides involved in rice immunity through transcriptomics-and proteomics-based screening. *Plant Biotechnology Journal* 18(2), 415–428. https://doi.org/10.1111/pbi.13208.

Wang, Q., Xu, J., Pu, X., Lv, H. and Liu, Y. (2021) Maize DNA methylation in response to drought stress is involved in target gene expression and alternative splicing. *International Journal of Molecular Sciences* 22(15), 8285. https://doi.org/10.3390/ijms22158285.

West, P.T., Li, Q., Ji, L., Eichten, S.R., Song, J. *et al.* (2014) Genomic distribution of H3K9me2 and DNA methylation in a maize genome. *PLoS ONE* 9(8), e105267. https://doi.org/10.1371/journal.pone.0105267.

Wong, C.Y., Gilbert, M.E., Pierce, M.A., Parker, T.A., Palkovic, A. *et al.* (2023) Hyperspectral remote sensing for phenotyping the physiological drought response of common and tepary bean. *Plant Phenomics* 5, 0021. https://doi.org/10.34133/plantphenomics.0021.

Xu, Y., Song, D., Qi, X., Asad, M., Wang, S. *et al.* (2023) Physiological responses and transcriptome analysis of soybean under gradual water deficit. *Frontiers in Plant Science* 14, 1269884. https://doi.org/10.3389/fpls.2023.1269884.

Yadav, S., Yadava, Y.K., Meena, S., Kalwan, G. and Bharadwaj, C. (2024) Novel insights into drought-induced regulation of ribosomal genes through DNA methylation in chickpea. *International Journal of Biological Macromolecules* 266, 131380. https://doi.org/10.1016/j.ijbiomac.2024.131380.

Yan, H., Sun, M., Zhang, Z., Jin, Y., Zhang, A. *et al.* (2023) Pangenomic analysis identifies structural variation associated with heat tolerance in pearl millet. *Nature Genetics* 55(3), 507–518. https://doi.org/10.1038/s41588-023-01302-4.

Yang, Y., Saand, M.A., Huang, L., Abdelaal, W.B., Zhang, J. *et al.* (2021) Applications of multi-omics technologies for crop improvement. *Frontiers in Plant Science* 12, 563953. https://doi.org/10.3389/fpls.2021.563953.

Yizhak, K., Benyamini, T., Liebermeister, W., Ruppin, E. and Shlomi, T. (2010) Integrating quantitative proteomics and metabolomics with a genome-scale metabolic network model. *Bioinformatics* 26(12), i255–i260. https://doi.org/10.1093/bioinformatics/btq183.

Yu, T., Zhang, J., Ma, X., Cao, S., Li, W. *et al.* (2024) A multi-omics view of maize's (*Zea mays* L.) response to low temperatures during the seedling stage. *International Journal of Molecular Sciences* 25(22), 12273. https://doi.org/10.3390/ijms252212273.

Yung, W.S., Wang, Q., Chan, L.Y., Wang, Z. and Huang, M. (2024) DNA hypomethylation is one of the epigenetic mechanisms involved in salt-stress priming in soybean seedlings. *Plant, Cell & Environment,* 1–13. https://doi.org/10.1111/pce.15297.

Zenda, T., Liu, S., Dong, A., Li, J., Wang, Y. *et al.* (2021) Omics-facilitated crop improvement for climate resilience and superior nutritive value. *Frontiers in Plant Science* 12, 774994. https://doi.org/10.3389/fpls.2021.774994.

Zenda, T., Liu, S., Wang, X., Jin, H., Liu, G. *et al.* (2018) Comparative proteomic and physiological analyses of two divergent maize inbred lines provide more insights into drought-stress tolerance mechanisms. *International Journal of Molecular Sciences* 19(10), 3225. https://doi.org/10.3390/ijms19103225.

Zhang, R., Zhang, C., Yu, C., Dong, J. and Hu, J. (2022) Integration of multi-omics technologies for crop improvement: Status and prospects. *Frontiers in Bioinformatics* 2, 1027457. https://doi.org/10.3389/fbinf.2022.1027457.

Zhao, Q., Feng, Q., Lu, H., Li, Y., Wang, A. *et al.* (2018) Pan-genome analysis highlights the extent of genomic variation in cultivated and wild rice. *Nature Genetics*, *50*(2), 278–284. https://doi.org/10.1038/s41588-018-0041-z.

Zheng, X., Chen, L., Xia, H., Wei, H., Lou, Q. *et al.* (2017) Transgenerational epimutations induced by multi-generation drought imposition mediate rice plant's adaptation to drought condition. *Scientific Reports*, 7(1), 39843. https://doi.org/10.1038/srep39843.

12 Genome Editing for Crop Improvement and Sustainable Agriculture

Tanmoy Halder[1], Sonali Panda[1,2], Bijayalaxmi Sahoo[1], Chandana Ghosh[1], M.J. Baig[1] and Kutubuddin A. Molla[1,*]
[1]*ICAR-National Rice Research Institute, Cuttack, India, 753006;*
[2]*Ravenshaw University, Cuttack, India, 753006*

Abstract
World food security is increasingly threatened by the growing impacts of climate change. Traditional agricultural approaches often fall short in addressing the challenges posed by rapid climate change. Sustainable agriculture and crop improvement are critical to mitigating the effects of global warming on food production systems. Genome editing technologies, such as CRISPR-Cas9, TnpB, base editing and prime editing, hold immense promise for enhancing crop resilience and productivity. These technologies have rapidly advanced, enabling gene knockouts, targeted genetic variation, precise insertions/deletions, single-nucleotide changes and large-scale genetic alterations. This chapter reviews the remarkable progress in applying genome editing tools to crop improvement and sustainable agriculture. Additionally, it highlights potential research areas where these technologies can drive transformative advancements in modern crop breeding towards sustainability.

12.1 Introduction

Modern agriculture faces significant challenges due to factors such as climate shifts, temperature fluctuations, erratic rainfall, extreme weather events, floods, droughts, soil salinity and pest and disease outbreaks. Numerous recent studies have highlighted the impact of these issues in Asia, Sub-Saharan Africa and the European Union (EU). For instance, in 2018, the EU experienced severe heatwaves and droughts, reducing cereal production by approximately 8%, which exacerbated fodder shortages and led to increased commodity prices (Brás *et al.*, 2021). Tropical nations are disproportionately affected, as they grapple with resource disparities and imbalances between resilience and vulnerability.

Globally, an estimated 720–811 million people face hunger, and over 2 billion suffer from food deprivation. The COVID-19 pandemic worsened this crisis, adding approximately 130 million people to the food-insecure population (Ribeiro-Silva *et al.*, 2020). By 2050, the global population is projected to reach 9.7 billion, necessitating a staggering 70% increase in crop yields to meet the world's food demand (Expert Meeting on How to Feed the World in 2050. FAO Headquarters, 24–26 June 2009, n.d.).

Traditional agricultural methods often fall short in addressing these challenges. Plant breeding techniques, such as cross-breeding and mutation breeding, are labor-intensive, time-consuming and limited by the genetic variation available within species. Transgenic approaches,

*Corresponding author. Email: kutubuddin.molla@icar.org.in

DOI: 10.1079/9781800627185.0012

while efficient in introducing specific traits from other organisms, are excessively expensive, time-intensive and subject to stringent regulatory hurdles in many countries.

Genome editing technologies offer immense promise for enhancing crop resilience and productivity to sustainably feed the growing global population. Compared with conventional breeding techniques and transgenic approaches, genome editing overcomes several major bottlenecks and significantly reduces the time required to develop improved cultivars. Genome editing enables precise, site-specific modifications of the genome, creating novel and beneficial genetic variations. Unlike transgenic techniques, which randomly introduce DNA of interest from related or unrelated sources into the genome, genome editing employs site-directed nucleases (SDNs) to target and cleave specific DNA sequences. Error-prone repair of these cleaved sequences often generates insertions or deletions (indels), resulting in novel genetic variation (Molla and Yang, 2019).

In genome editing, SDNs and their components are initially introduced into plants as transgenes. However, once the desired edits are achieved, the transgenes encoding SDNs and their components can be segregated out, resulting in transgene-free mutants. Over time, five distinct families of SDNs have been utilized for genome editing: meganucleases, Zinc Finger Nucleases (ZFNs), Transcription Activator-Like Effector Nucleases (TALENs), CRISPR-associated nucleases (Cas) and transposon-associated TnpBs.

12.2 Available Genome Editing Tools

12.2.1 Meganucleases

Meganucleases were the first nucleases used for genome editing, discovered in the late 1980s (Rouet *et al.*, 1994). Highly specific, they recognize long DNA sequences ranging from 14–40 bp. An example of their application in plant genome editing is the use of the I-SceI meganuclease for genome modification in tomato (*Solanum lycopersicum*) and oilseed rape (*Brassica napus*) (Danilo *et al.*, 2022). However, meganucleases have limitations, including non-availability of recognition sequence in the desired locus.

12.2.2 Zinc Finger Nucleases (ZFNs)

Zinc finger nucleases (ZFNs) combine the DNA-binding domain of zinc finger (ZF) proteins with the cleavage domain of the FokI endonuclease. ZFs are designed to bind specific DNA sequences and, when fused with FokI, create targeted DNA cleavages. ZFNs have been tested in several crops (Wright *et al.*, 2005; Shukla *et al.*, 2009; Curtin *et al.*, 2011). However, their complex and time-consuming construction has limited their widespread adoption.

12.2.3 Transcription Activator-like Effector Nucleases (TALENs)

Transcription activator-like (TAL) effectors, derived from *Xanthomonas* spp., bind specific DNA sequences. Artificially designed TAL effectors fused to FokI dimers form TALENs (Christian *et al.*, 2010). TALENs offer flexible target design and have been tested in various plant species. For example, TALENs were successfully used to develop bacterial blight-resistant rice plants (T. Li *et al.*, 2012). However, TALEN-based genome engineering is hindered by their large size and difficulties in designing specific and efficient DNA-binding domains.

12.2.4 CRISPR-Cas

The CRISPR-Cas9 system, first identified in bacteria, serves as an adaptive immune defense mechanism. Its foundation lies in CRISPR arrays and Cas proteins that cut foreign DNA. In 2012, the system was reconstructed *in vitro* to demonstrate targeted DNA cleavage directed by a 20-nucleotide RNA. CRISPR-Cas9 utilizes a single guide RNA (sgRNA) to direct the Cas9 nuclease to specific genome locations via complementarity. Cas9 generates double-strand breaks (DSBs), repaired by the cell's DNA repair machinery. In higher eukaryotes, non-homologous end joining (NHEJ) often generates random insertions or deletions (indels), resulting in novel genetic variations or gene knockouts. Cas9 from *Streptococcus pyogenes* requires an 'NGG' protospacer adjacent motif (PAM) at the target genomic locus. However, Cas orthologs with alternative

PAM specificities have been discovered and engineered (Molla *et al.*, 2020).

12.2.5 TnpB and IscB

TnpB and IscB, transposon-encoded RNA-guided nucleases, are considered ancestors of Cas12 and Cas9, respectively (Altae-Tran *et al.*, 2023). Being significantly smaller (approximately two-fifths the size of Cas12a and Cas9), these hyper-compact nucleases are promising genome editing tools (Karvelis *et al.*, 2021; Xue *et al.*, 2024). Recently, a high-efficiency TnpB genome editor was developed for use in monocot and dicot systems, including protoplasts and regenerated plants (Karmakar *et al.*, 2024).

12.2.6 Base Editing

While NHEJ-mediated single nucleotide changes are predictable (Molla *et al.*, 2022a), canonical CRISPR-Cas9 cannot generate precise base substitutions. Base editing is a precise genome editing technique that does not require DSBs. A base editor consists of a nickase Cas9 fused with a cytidine deaminase or an adenosine deaminase. Different base editors have been developed to substitute specific bases within a small editing window (Molla and Yang, 2019; Molla *et al.*, 2021). For instance, a cytosine base editor can convert a C to a T (Komor *et al.*, 2016), while an adenine base editor can change an A to a G (Gaudelli *et al.*, 2017; Molla, Shih, *et al.*, 2020).

12.2.7 Prime Editing

Although base editors efficiently mediate specific base substitutions, they are unsuitable for introducing predefined indels. Predefined indels are typically introduced through the homology-directed repair (HDR) pathway, which requires DSBs and repair templates but suffers from extremely low efficiency in plants (Molla *et al.*, 2022b). Prime editing, an advanced genome editing technique, enables efficient insertion, deletion and all types of base substitutions without DSBs or donor DNA templates (Anzalone *et al.*, 2019). A prime editor consists of a nickase Cas9 fused to a reverse transcriptase enzyme and is guided by a prime editing guide RNA (pegRNA). The pegRNA specifies both the target genomic site and the desired genetic change.

The genome editing tools described above provide researchers with unparalleled capabilities to manipulate genomes, ranging from introducing random genetic variation to precise base substitutions and sequence alterations (Figure 12.1).

12.3 Editing for Abiotic Stress Tolerance

Abiotic stresses such as drought, salinity and temperature are major contributors to significant crop yield losses. In the modern era, genome editing technologies have emerged as a promising approach to mitigate these losses. Genome editing tools have been successfully implemented in diverse crops, including major staples. Various strategies have been employed to enhance abiotic stress tolerance in different plant species. Genes that negatively regulate abiotic stress responses are prime targets for genome editing, as their knockout can confer stress tolerance to plants (Table 12.1).

Among the different abiotic stresses, drought, salinity and temperature have the most severe impact on crop production in the context of climate change. Below, we discuss selected case studies of genome editing for enhanced drought stress tolerance in crops.

12.3.1 Drought Stress

Drought can cause growth, developmental and productive damage to plants, resulting in major yield loss. Drought stress adversely affects plants by decreasing germination, dwarfing growth and reducing leaf size. Drought causes closure of stomata, restricting CO_2 uptake and lowering chlorophyll content, thus affecting photosynthesis. Water deficit triggers oxidative stress, and reactive oxygen species (ROS) build up to harm cellular constituents. Long-term drought leads to leaf senescence, premature flowering and yield loss, ultimately impacting crop yield and survival. The frequency and severity of drought have increased due to climate change. With

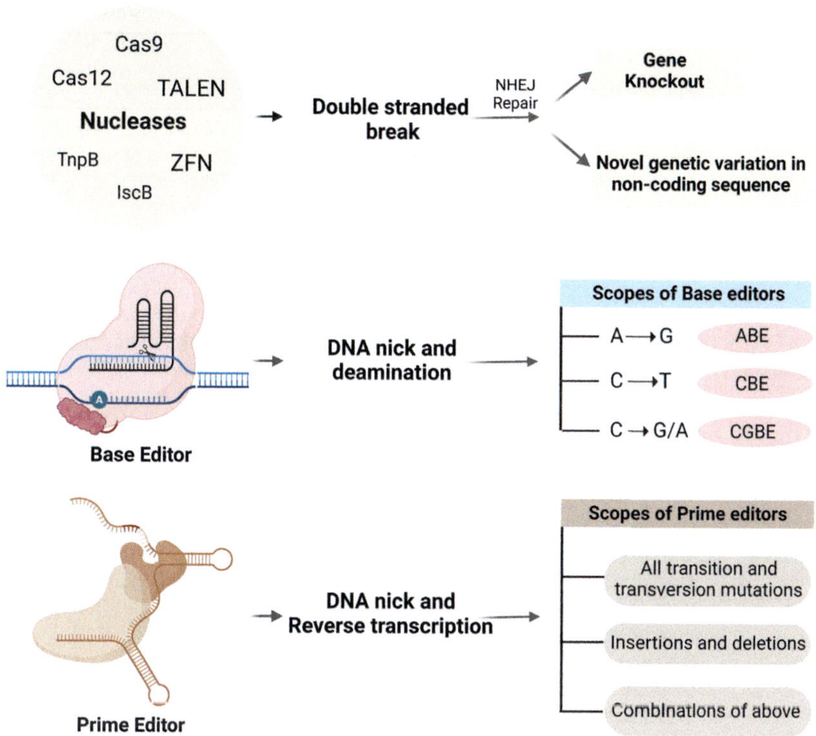

Figure 12.1. Image illustrating different genome editing tools and their specific scopes, including nucleases (ZFN, TALEN, Cas9, TnpB), base editors (ABE, CBE, CGBE) and prime editors. (Authors' own image.)

advancements in genome editing, researchers have focused on modifying genes involved in drought response pathways to enhance crop resilience. CRISPR-Cas9-mediated mutation of the *OST2* gene, which encodes a plasma membrane H+ ATPase crucial for stomatal regulation, significantly enhanced drought tolerance in *Arabidopsis*. The *ost2* mutant showed a higher degree of stomatal closure and reduced water loss under ABA-induced conditions compared with wild-type plants (Merlot *et al.*, 2007; Osakabe *et al.*, 2016).

Mutation of the *SlLBD40* gene, a lateral organ boundaries domain (LBD) transcription factor expressed in roots and fruits, significantly enhanced drought tolerance in tomato. The knockout mutants displayed a notably lower water loss rate compared with wild-type plants (L. Liu *et al.*, 2020). Similarly, editing the *SlMAPK3* and *SlARF4* genes improved drought tolerance (L. Wang *et al.*, 2017; M. Chen *et al.*, 2021). Silencing of the *SlNPR1* gene, primarily a plant defense gene, enhanced drought tolerance (R. Li *et al.*, 2019). Disruption of the gibberellin receptor gene *GID1* reduced water loss, maintaining higher leaf water content in tomato under drought conditions (Illouz Eliaz *et al.*, 2020).

Similarly, in rice, knockout of the *OsERA1* gene, a negative regulator of ABA signaling, enhanced drought tolerance by increasing stomatal closure under water-depleted conditions (Ogata *et al.*, 2020). Mutation of *Semi-rolled leaf 1* (*SRL1*) and *SRL2* resulted in a rolled-leaf phenotype, improving drought tolerance in rice (Liao *et al.*, 2019). Editing *OsDST*, a negative regulator of drought tolerance, enhanced stress resilience in the Indica mega rice cultivar MTU1010 (Kumar *et al.*, 2020).

Overexpression of *ARGOS8*, a negative regulator of ethylene responses, improved grain yield under drought conditions. Using CRISPR-HDR, the native *ARGOS8* promoter was swapped with a strong promoter (*PGOS2*), resulting in higher transcript levels and enhanced drought tolerance in maize (Shi *et al.*, 2017). In wheat, multiplex CRISPR-Cas9 editing of the *SAL1* gene, a

Table 12.1. Targeted genome editing for abiotic stress tolerance in plants. (Authors' own table.)

Name of gene	Organism	Type of abiotic stress	References
OST2	*Arabidopsis thaliana*	Drought	(Merlot *et al.*, 2007)
AITR	*Arabidopsis thaliana*	Drought	(S. Chen *et al.*, 2021)
AVP1	*Arabidopsis thaliana*	Drought	(Park *et al.*, 2017)
SlMAPK3	*Solanum lycopersicum* (Tomato)	Drought	(L. Wang *et al.*, 2017)
SlARF4	*Solanum lycopersicum*	Drought	(M. Chen *et al.*, 2021)
SlNPR1	*Solanum lycopersicum*	Drought	(R. Li *et al.*, 2019)
GID1	*Solanum lycopersicum*	Drought	(Illouz-Eliaz *et al.*, 2020)
OsERA1	*Oryza sativa* (Rice)	Drought	(Ogata *et al.*, 2020)
Semi-rolled leaf 1	*Oryza sativa*	Drought	(Liao *et al.*, 2019)
OsDST	*Oryza sativa*	Drought	(Santosh Kumar *et al.*, 2020a)
ARGOS8	*Zea mays* (Maize)	Drought	(Shi *et al.*, 2017)
SAL1	*Triticum aestivum* (Wheat)	Drought	(Abdallah *et al.*, 2022)
GhHB12	*Gossypium hirsutum* (Cotton)	Drought	(He *et al.*, 2020)
GmHdz4	*Glycine max* (Soybean)	Drought	(Zhong *et al.*, 2022)
TRE1	*Arabidopsis thaliana*	Drought	(Nuñez-Muñoz *et al.*, 2021)
AREB1	*Arabidopsis thaliana*	Drought	(Roca Paixão *et al.*, 2019)
SlLBD40	*Solanum lycopersicum*	Drought	(L. Liu *et al.*, 2020)
TaGW2	*Triticum aestivum*	Drought	(S. Li *et al.*, 2024)
OsPYL9	*Oryza sativa*	Drought	(Usman *et al.*, 2020)
SlHyPRP1	*Solanum lycopersicum*	Salinity	(Tran *et al.*, 2021)
OsRR22	*Oryza sativa*	Salinity	(A. Zhang *et al.*, 2019)
OsbHLH024	*Oryza sativa*	Salinity	(Alam *et al.*, 2022)
GmAITR	*Glycine max*	Salinity	(T. Wang *et al.*, 2021)
HKT1;5	*Oryza sativa*	Salinity	(Farhat *et al.*, 2019)
OsNAC45	*Oryza sativa*	Salinity	(X. Zhang *et al.*, 2020)
SlSOS1	*Solanum lycopersicum*	Salinity	(Z. Wang *et al.*, 2020)
DST	*Oryza sativa*	Salinity	(Santosh Kumar *et al.*, 2020b)
VDE	*Oryza sativa*	Salinity	(X. Wang *et al.*, 2022)
SlMAPK3	*Solanum lycopersicum*	High temperature	(L. Wang *et al.*, 2017)
SlAGL6	*Solanum lycopersicum*	High temperature	(Klap *et al.*, 2017)
SlCBF1	*Solanum lycopersicum*	Cold	(R. Li *et al.*, 2018)
OsMYB30	*Oryza sativa*	Cold	(Zeng *et al.*, 2020)
OsGS3	*Oryza sativa*	Cold	(Zeng *et al.*, 2020)
OsNRAMP1	*Oryza sativa*	Heavy metal	(Chang *et al.*, 2020)
OsNRAMP5	*Oryza sativa*	Heavy metal	(T. Wang *et al.*, 2019)
OsARM1	*Oryza sativa*	Heavy metal	(F. Z. Wang *et al.*, 2017)
OsHAK1	*Oryza sativa*	Heavy metal	(Nieves-Cordones *et al.*, 2017)

negative regulator of drought tolerance, induced a rolled-leaf phenotype, improving stress tolerance (Abdallah *et al.*, 2022). Likewise, silencing the *HB12* gene improved drought tolerance in cotton (He *et al.*, 2020). In soybean, editing the *GmHdz4* transcription factor enhanced drought tolerance (Zhong *et al.*, 2022).

These advancements highlight the potential of the CRISPR-Cas9 system to unravel and manipulate drought-responsive pathways in various plant species. This technology offers a promising solution to mitigate the adverse impacts of drought stress on agricultural productivity.

12.3.2 Salinity Stress Responses

Salinity is one of the most significant environmental stressors, degrading soil quality and reducing agricultural productivity, as most crops do not thrive under high-salinity. By 2050, half of the arable land is anticipated to be saline (Hassani *et al.*, 2021). Salinity stress, mediated through osmotic stress and nutrient and ion imbalances, impacts plant morphology, biochemical processes and biomass. This stress increases the production of ROS, which causes cellular damage through lipid peroxidation and the degradation of critical proteins and DNA.

The Cas9-mediated editing of the *OsRR22* gene showed better growth under 0.75% NaCl conditions, and the edited rice plants were more tolerant to salt stress (A. Zhang *et al.*, 2019). Similarly, the knockout of *OsbHLH024* also increased salt tolerance in rice (Alam *et al.*, 2022).

Recently, CRISPR-mediated editing of *DST* and *VDE* genes was shown to enhance salinity stress tolerance in rice (Santosh Kumar *et al.*, 2020a; X. Wang *et al.*, 2022). Along with rice, several other plants have been targeted for enhanced salt tolerance via genome editing approaches. Editing of the hybrid proline-rich protein (*SlHyPRP1*) gene in tomato enhanced salt tolerance (Tran *et al.*, 2021). Similarly, in soybean, seeds of the CRISPR-Cas9-generated *Gmaitr* mutants germinated at a higher rate than the wild type under salt treatment. The mutant *Gmaitr* plants had morphologically indistinguishable phenotypes from the wild type in normal soil but grew better in saline soil than the wild type (Wang *et al.*, 2021). These findings suggest that CRISPR-Cas9 targeting of *AITR* genes holds promise for improving salinity and drought tolerance across various crop species.

12.3.3 High-Temperature Stress

Heat stress is a serious threat to plant development, affecting various physiological and biochemical mechanisms. Heat stress impacts photosynthesis, chlorophyll content, turgor maintenance, nutrient uptake and gas exchange, which eventually reduces crop yields. Temperature differences can also impact plant development and productivity. Heat stress, defined as prolonged exposure to temperatures exceeding the plant's tolerance threshold, often involves a 10–15°C increase above the crop's optimal range. This "heat shock" response will cause irreversible damage if prolonged, resulting in increased mortality and deteriorating production quality. Notably, heat stress ranks third among the most damaging abiotic factors for agriculture, after drought and salinity. Research indicates that a rise in global temperature of 1°C can reduce the yield of wheat by 6% (Asseng *et al.*, 2014), rice by 10–20% (Y. Wang *et al.*, 2019) and corn by 21–31% (H. Yang *et al.*, 2017).

Advances in CRISPR-Cas9 technology now allow researchers to manipulate temperature-responsive genes precisely, thus enhancing heat tolerance. In tomato, CRISPR-Cas9-mediated *SlMAPK3* knockout lines exhibited better tolerance under elevated temperatures (W. Yu *et al.*, 2019). *SlMAPK3* is a negative regulator of heat stress tolerance. The mutant line also showed upregulation of heat stress transcription factors (HSFs) and heat shock proteins (HSPs) (W. Yu *et al.*, 2019).

The microsporogenesis process in tomato is highly temperature-sensitive. CRISPR-Cas9-mediated knockout of *SlAGAMOUS-LIKE 6* (*SlAGL6*) resulted in a parthenocarpic phenotype and exhibited better fruit development under heat stress conditions compared with the wild-type plant (Klap *et al.*, 2017).

In temperature tolerance, the application of CRISPR-Cas to generate stress-tolerant plants is still limited; however, massive efforts have focused on identifying several positive regulators of heat stress tolerance, such as *OsIAA7* in rice (R. Qiu *et al.*, 2025), *ZmCDPK7* and *CBK3* in maize (Ruan *et al.*, 2024) and ethylene response transcription factors (*ERFC1/F4/F5*) in tomato (Balyan *et al.*, 2020). These genes can be further manipulated to promote temperature stress tolerance in plants. Such advancements are crucial for developing heat-tolerant crops and maintaining food security during climate change when increasing temperatures threaten global crop productivity.

12.3.4 Cold Stress

Cold stress, comprising chilling and freezing temperatures (down to 0°C), has severe effects on the growth, productivity and ecological adaptability of plants. Exposure to cold impairs photosynthesis, biomass accumulation and grain yield by altering the metabolic responses of plants. It also induces osmotic and oxidative stress. Extreme cold triggers both mechanical damage and metabolic dysfunction, leading to arrested growth and a partial loss of resilience, especially in cold-prone areas.

A cold-responsive R2R3-type MYB gene, *OsMYB30*, was found to negatively affect cold stress tolerance in rice. *OsMYB30* mutants generated by CRISPR-Cas exhibited enhanced tolerance to cold (Y. Zeng *et al.*, 2020). In tomato plants, CRISPR-Cas9-generated mutants of the C-repeat binding factor 1 (*cbf1*) showed better tolerance to the adverse effects of cold stress. The edited plants demonstrated a remarkable reduction in electrolyte leakage (Li *et al.*, 2018). CRISPR-Cas-mediated editing holds great potential for enhancing cold stress tolerance in plants by targeting specific genes involved in cold-responsive pathways, thereby mitigating adverse situations.

12.3.5 Heavy Metals Stress

Heavy metals (HMs) such as Mn, Cu, Ni, Co, Cd, Fe, Zn and Hg pose a significant threat to sustainable agriculture. These metals damage plant growth and productivity. They accumulate in soil due to anthropogenic activities, including excessive use of fertilizers, improper disposal of industrial waste, unlawful sewage discharge and rapid dumping of vehicle waste. As a result, HMs either accumulate on the ground surface or seep into groundwater, causing environmental contamination.

Heavy metal accumulation in plants, particularly in roots, adversely affects root activity. Metals are excluded by the casparian strips or trapped within the cell walls of the root system, thereby lowering crop productivity. Consuming plants or plant parts contaminated with heavy metals leads to severe health effects. Recent studies have shown promise in improving plant tolerance to heavy metals using gene-editing techniques.

Rice grains with *OsNRAMP1* knocked out by CRISPR-Cas9 showed decreased levels of Cd (Chang *et al.*, 2020). CRISPR-mediated silencing of the R2R3 MYB transcription factor *OsARM1* in rice reduced the absorption and transportation of arsenic (F. Z. Wang *et al.*, 2017). Additionally, inactivating the cesium-permeable potassium transporter *OsHAK1* reduced cesium (Cs^+) absorption and translocation in rice (Nieves-Cordones *et al.*, 2017). CRISPR-Cas technology offers a sustainable approach for mitigating heavy metal toxicity in plants, thus promoting safer and more productive agriculture.

12.4 Genome Editing for Disease Resistance

Genome editing technologies offer several strategies for enhancing pathogen resistance in crops (Karmakar *et al.*, 2022). These include creating targeted mutations in genes critical for pathogen entry or proliferation, thereby reducing susceptibility. Additionally, genome editing can knock out genes that facilitate infection or interaction with pathogens, thus boosting resistance. By introducing or enhancing resistance genes, CRISPR-Cas9 improves immunity to specific pathogens. Furthermore, modifying regulatory pathways involved in disease resistance can lead to an overall enhancement of a crop's defense mechanisms. Below, we discuss some case studies in major crop species. Table 12.2 summarizes some notable studies of crop genome editing for biotic stress tolerance. A more comprehensive account of enhancing disease and pest resistance with crop genome editing can be found in an earlier report (Karmakar *et al.*, 2022).

12.4.1 Rice

CRISPR-Cas9 gene editing has proven useful in countering threats from diverse pathogens. Editing the promoters of *OsSWEET11*, *OsSWEET13* and *OsSWEET14* has conferred broad-spectrum resistance to *Xanthomonas oryzae* pv. *oryzae* (Blanvillain-Baufumé *et al.*, 2017; Kim *et al.*, 2019; Oliva *et al.*, 2019; Ancydiana *et al.*, 2022). Editing the OsSWEET14 gene itself also imposed similar resistance (Zeng *et al.*, 2020). Editing the ERF transcription factor gene *OsERF922* reduced the size of leaf blast lesions and enhanced resistance to rice blast (F. Wang *et al.*, 2016). Modifying eIF4G has made rice fully resistant to the rice tungro virus, thus increasing yields (Macovei *et al.*, 2018). CRISPR-mediated knockout mutants of the methyl esterase-like (*OsMESL*) gene have demonstrated enhanced resistance to *Rhizoctonia solani* infection (Hu *et al.*, 2021).

Table 12.2. Targeted genome editing for biotic stress tolerance in plants. (Authors' own table.)

Name of gene	Organism	Type of biotic stress	References
OsERF922	*Oryza sativa* (Rice)	Rice blast (*Magnaporthe oryzae*)	(F. Wang *et al.*, 2016)
Xa13	*Oryza sativa*	Bacterial blight (*Xanthomonas oryzae*)	(C. Li *et al.*, 2019)
OsSWEET14	*Oryza sativa*	Bacterial blight (*Xanthomonas oryzae*)	(Blanvillain-Baufumé *et al.*, 2017)
Pi21	*Oryza sativa*	Rice blast (*Magnaporthe oryzae*)	(Yang *et al.*, 2023)
OsMESL	*Oryza sativa*	*R. solani* infection	(Hu *et al.*, 2021)
Xa23	*Oryza sativa*	Bacterial blight (*Xanthomonas oryzae*)	(Zhu *et al.*, 2024)
OsCEBiP	*Oryza sativa*	Rice blast (*Magnaporthe oryzae*)	(Qiu *et al.*, 2022)
Pid2	*Oryza sativa*	Rice blast (*Magnaporthe oryzae*)	(Ren *et al.*, 2018)
OsBSR-K	*Oryza sativa*	Rice blast (*Magnaporthe oryzae*)	(Zhou *et al.*, 2018)
OseIF4G	*Oryza sativa*	Tungro virus	(Mushtaq *et al.*, 2020)
CYP71A1	*Oryza sativa*	Brown plant hoppers and stem borers	(Lu *et al.*, 2018)
TaMLO	*Triticum aestivum* (Wheat)	Powdery mildew	(Y. Wang *et al.*, 2014)
EDR1	*Triticum aestivum*	Powdery mildew	(Y. Zhang *et al.*, 2017)
LOX3	*Zea mays* (Maize)	Smut (*Ustilago maydis*)	(Pathi *et al.*, 2020)
ZmGDIα	*Zea mays*	Maize rough dwarf disease	(C. Liu *et al.*, 2022)
SlDMR6-1	*Solanum lycopersicum* (Tomato)	*Pseudomonas syringae* and *Xanthomonas gardneri*	(Paula De Toledo Thomazella *et al.*, 2021)
SlJAZ2	*Solanum lycopersicum*	Bacterial speck	(Ortigosa *et al.*, 2019)
SlMLO1	*Solanum lycopersicum*	Powdery mildew	(Nekrasov *et al.*, 2017)
SlWAT1	*Solanum lycopersicum*	*Verticillium dahliae* and *Fusarium oxysporum*	(Hanika *et al.*, 2021)
HvMORC1 and HvMORC6a	*Hordeum vulgare* (Barley)	*Bipolaris sorokiniana*, *Fusarium graminearum*	(Galli *et al.*, 2022)
DMR6	*Musa* sp. (Banana)	Musa downy mildew	(Tripathi *et al.*, 2021)
GmMLO	*Glycine max* (Soybean)	Powdery mildew	(Bui *et al.*, 2023)
GmF3H1, GmF3H2, GmFNSII-1	*Glycine max*	Soybean mosaic virus	(P. Zhang *et al.*, 2020)
CsLOB1	*Citrus sinensis* (sweet orange)	Citrus canker (*Xanthomonas citri* ssp. *citri*)	(Jia *et al.*, 2017)
CsWRKY22	*Citrus sinensis*	Citrus canker (*Xanthomonas citri* ssp. *citri*)	(L. Wang *et al.*, 2019)
Gh14–3–3d	*Gossypium* sp. (Cotton)	Verticillium wilt	(Z. Zhang *et al.*, 2018)
ClPSK1	*Citrullus lanatus* (Watermelon)	Fusarium wilt (*Fusarium oxysporum*)	(M. Zhang *et al.*, 2020)
TcNPR3	*Theobroma cacao* cocoa	Black Pod disease (*Phytophthora tropicalis*)	(Fister *et al.*, 2018)
VvMLO3	*Vitis vinifera* (Grapevine)	Powdery mildew	(Wan *et al.*, 2020)
MeSWEET10a	*Manihot esculenta* (Cassava)	Bacterial blight	(Veley *et al.*, 2023; Y. Wang *et al.*, 2024)

Continued

Table 12.2. Continued.

eIF4E	*Cucumis sativus* (Cucumber)	Wide range of viruses	(Chandrasekaran *et al.*, 2016)
CsaMLO1	*Cucumis sativus*	Powdery mildew (*Podosphaera xanthii*)	(Tek *et al.*, 2022)
nCBP-1nCBP-2	*Manihot esculenta*	Cassava brown streak disease	(Gomez *et al.*, 2019)
StDND1, StCHL1, and StDMR6–1	*Solanum tuberosum* (Potato)	Late blight (*Phytophthora infestans*)	(Kieu *et al.*, 2021)

12.4.2 Wheat

The first successful genetic modification of the wheat genome was conducted by Wang *et al.* (2014) by knocking out three *TaMLO* homoeologs using TALENs, thereby generating powdery mildew-resistant wheat. The MLO gene is a susceptibility factor, and its disruption leads to broad-spectrum resistance to powdery mildew in most plant species. However, earlier *MLO* knockouts resulted in growth and yield penalties. Researchers optimized the process by targeting the MLO-B1 locus, generating the tamlo-R32 mutant, which is resistant to powdery mildew with no negative effects on growth or yield (S. Li *et al.*, 2022). In 2017, Zhang *et al.* used CRISPR technology to knock out three homoeologs of the *EDR1* gene in wheat, resulting in plants resistant to powdery mildew (Y. Zhang *et al.*, 2017).

12.4.3 Sorghum

Striga is a parasitic weed that infects most of the sorghum produced in Africa and one of the severe challenges for *Sorghum* cultivation. Genome editing technologies open the possibility to provide effective solutions by targeting genes involved in *Striga*-host interactions. Strigolactone promotes the germination of *Striga*, therefore targeting the genes on the strigolactone biosynthetic pathway could be a potential step. Disruption of *SbCCD8* gene in sorghum reduced orobanchol synthesis and *Striga* germination (Hao *et al.*, 2023).

12.4.4 Maize

In maize, knockout lines of the *LIPOXYGENASE 3* (*LOX3*) gene showed reduced susceptibility to the smut pathogen *Ustilago maydis* (Pathi *et al.*, 2020). More recently, CRISPR-mediated targeted mutation of the *ZmGDIα* gene enhanced resistance to maize rough dwarf disease (MRDD) without compromising agronomic performance (C. Liu *et al.*, 2022).

12.4.5 Banana

Banana varieties resistant to several diseases have been generated using CRISPR-Cas9 technology. Knocking out the Musa downy mildew resistance 6 (*DMR6*) gene, which acts as a negative regulator of plant defense, resulted in enhanced resistance to wilt caused by *Xanthomonas campestris* (Tripathi *et al.*, 2021).

12.4.6 Soybean

In soybean, three genes, namely GmF3H1, GmF3H2 and GmFNSII-1, were targeted using CRISPR multiplexing. The results showed that T3 mutant plant lines exhibited increased isoflavone content and significantly reduced soybean mosaic virus (SMV) coat protein content, revealing enhanced resistance against SMV (P. Zhang *et al.*, 2020). Similarly, enhanced powdery mildew resistance was obtained in soybean by knocking out *GmMLO* genes (Bui *et al.*, 2023).

12.4.7 Tomato

Powdery mildew-resistant tomato lines were generated by CRISPR-Cas-mediated knockout of the SlMlo1 gene (Nekrasov *et al.*, 2017). Tomatoes were edited using the CRISPR-Cas9 system to target the *SlDMR6-1* gene, resulting in broad-spectrum resistance to pathogens, including *Pseudomonas syringae* and *Xanthomonas gard-*

neri (Paula De Toledo Thomazella *et al.*, 2021). The susceptibility gene *PMR4* was also targeted to develop resistance to the powdery mildew fungus (Santillán Martínez *et al.*, 2020). Additionally, increased resistance to bacterial speck was observed in tomato plants with edits in the *SlJAZ2* gene (Ortigosa *et al.*, 2019).

12.4.8 Cassava

Editing *eIF4E*, along with new cap-binding proteins, nCBPs, in cassava has been shown to improve resistance against cassava brown streak disease (CBSD) (Gomez *et al.*, 2019). Furthermore, the CRISPR-Cas9 editing of the promoter for the *MeSWEET10a* gene has showed resistance to cassava bacterial blight without negatively affecting other plant characteristics (Veley *et al.*, 2023; Y. Wang *et al.*, 2024).

12.5 Improvement in Crop Yield and Quality Enhancements using CRISPR

Global climate change is anticipated to decrease crop yields for several reasons. To feed the growing population, it is essential to produce more food from less available land. Genome editing technologies hold great promise in addressing this challenge by increasing yield and improving crop quality (Figure 12.2). All major crops have been edited to enhance yield.

12.5.1 Rice

In rice, mutations in genes such as *DEP1* and *Gn1a*, which are related to yield, have significantly increased productivity. New variant alleles of these genes, generated using CRISPR-Cas9, exhibit considerably higher yields compared with wild-type varieties (Li *et al.*, 2016). Similarly, knockouts of the *GW2* gene, which controls grain size, width and number, have improved yields and quality (enhanced protein and mineral content) (Achary and Reddy, 2021). Furthermore, the use of CRISPR-Cas9 to mutate class-I *PYL* genes, which are part of drought-signaling pathways, has allowed for better yields under well-watered conditions due to increased panicle size and a higher number of tillers (Miao *et al.*, 2018).

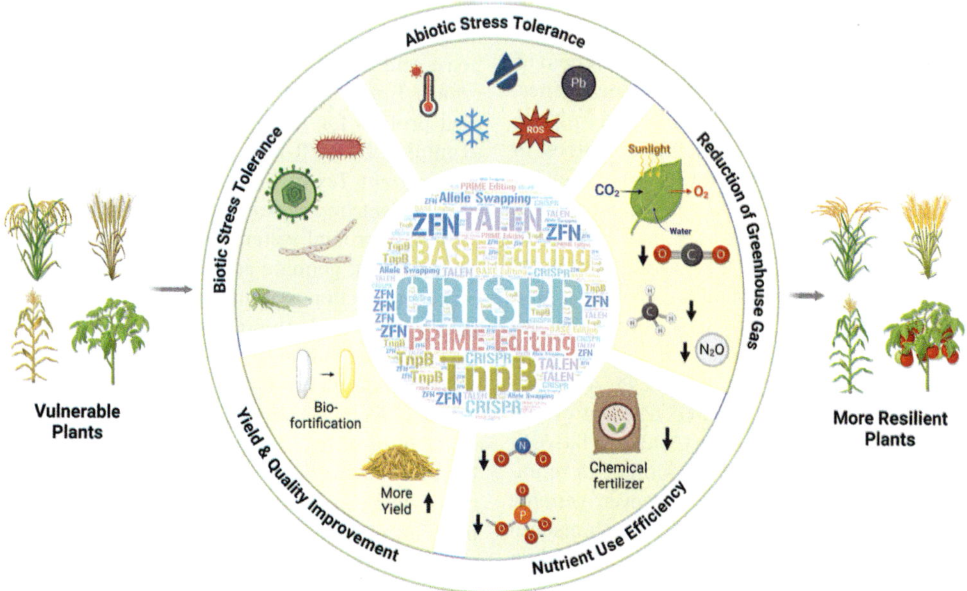

Figure 12.2. Crop genome editing for resilient agriculture and meeting the modern world's needs. (Authors' own image.)

12.5.2 Wheat

In wheat, genome editing has been used successfully to improve both yield and quality. For instance, low-gluten wheat suitable for individuals with celiac disease was developed by editing the α-gliadin gene (Sánchez-León *et al.*, 2018). High-amylose wheat, which enhances resistant starch content, was created by targeting the *TaSBEIIa* gene (Li *et al.*, 2021). The application of CRISPR-Cas9 has also improved wheat yields by knocking out negative regulatory genes such as *TaGASR7*, resulting in a significant increase in kernel weight (Zhang *et al.*, 2016).

12.5.3 Maize

In maize, high-amylopectin varieties have been developed by knocking out the *Waxy* gene. Field trials of these waxy maize varieties showed an increase in yield of 5.5 bushels per acre compared with crops developed through traditional breeding techniques (Gao *et al.*, 2020).

12.5.4 Tomato

CRISPR-Cas9 has been used to enhance both yield and quality in tomatoes. By editing the *SlCLV3* gene, which controls the development of floral stem cells, and other architectural genes, significant improvements were made in fruit size and plant morphology, enabling tomatoes to thrive better in urban conditions (Rodríguez-Leal *et al.*, 2017; Lemmon *et al.*, 2018). Sugar content, a key factor in consumer preference, was improved by editing *SlCDPK27* and *SlCDPK26*. Knockouts of these genes increased glucose and fructose contents by up to 30%, enhancing perceived sweetness without penalties to fruit weight or yield (Zhang *et al.*, 2024). Additionally, the CRISPR-Cas-mediated deletion of the autoinhibitory domain of SlGAD2 and SlGAD3 significantly increased GABA content in tomatoes (Nonaka *et al.*, 2017). Moreover, genome editing restructured vine-like tomato plants into compact, early-yielding plants suitable for urban agriculture (Kwon *et al.*, 2019).

12.6 Editing for Enhancing Carbon Sequestration

Reducing atmospheric carbon by capturing and sequestering it is a promising strategy to mitigate the worst impacts of climate change. CRISPR technology opens new prospects for addressing climate change by enhancing the natural capacity of plants to capture and store carbon. This can be achieved through improved biomass accumulation and carbon sequestration in soil. Genome editing tools can support various strategies, including: 1. Increasing root biomass and depth to store more carbon underground. 2. Enhancing photosynthesis to capture more CO_2, enabling greater carbon storage in leaves, stems and roots. 3. Influencing root secretions to increase populations of soil microbes that sequester carbon.

12.7 Reducing Dependency on Fossil Fuels: Editing for Biodiesel

Genome editing technologies, such as CRISPR-Cas9, offer transformative potential to reduce reliance on fossil fuels by enhancing biodiesel production from renewable sources. In oil-rich crops such as *Jatropha curcas*, soybean and rapeseed, genome editing can increase lipid biosynthesis and optimize fatty acid profiles for higher-quality biodiesel. For example, CRISPR-mediated disruption of *Transparent Testa 8* enhanced oil accumulation in *Camelina sativa* (Cai *et al.*, 2024). Similarly, in microbial systems such as yeasts, algae and bacteria, genome editing can boost triglyceride production, enhance photosynthetic efficiency and enable the use of waste biomass as feedstocks. Additionally, CRISPR-Cas-mediated reductions in lignin content are highly beneficial for biodiesel extraction. A CRISPR-Cas *CCR1* mutant line demonstrated tissue-specific lignin reduction without compromising biomass yield (Yu *et al.*, 2021). Several species of microalgae have been engineered with genome editing tools to increase oil content (Dhokane *et al.*, 2023).

12.8 Addressing Methane Emissions

Rice, a staple food for half of the world's population, accounts for approximately 12% of global methane emissions – a greenhouse gas 25 times more potent than carbon dioxide. This is largely due to the flooded conditions in which rice is commonly grown. These conditions cut roots off from oxygen, creating an ideal environment for anaerobic microbes to decompose organic matter and release methane. To promote sustainable agriculture, it is critical to address methane emissions from rice paddies. Scientists are leveraging CRISPR and advanced breeding techniques to develop rice varieties and farming practices that minimize methane emissions without reducing yields (Kwon *et al.*, 2023).

12.9 Reducing Chemical Fertilizer Dependence

Using CRISPR to improve nitrogen use efficiency (NUE) and phosphorus use efficiency (PUE) in crops is a promising strategy to reduce dependency on chemical fertilizers, lower agricultural costs and minimize environmental impacts. Nitrogen and phosphorus are essential nutrients for plant growth, but their excessive use in agriculture often leads to soil degradation, water pollution and greenhouse gas emissions. CRISPR can be utilized to modify genes regulating nutrient transporters in plants, creating crops that absorb nutrients more efficiently from the soil. Enhancing the expression of these transporters enables crops to uptake and utilize nutrients effectively, even with minimal fertilizer application.

CRISPR-mediated mutation of the *abnormal cytokinin response1 repressor1* (*ARE1*) gene resulted in increased NUE in wheat (J. Zhang *et al.*, 2021). A similar result was observed with CRISPR-Cas-mediated editing of the *HvARE1* gene, which led to higher nitrogen content in barley shoots under low nitrogen conditions (Karunarathne *et al.*, 2022). Base editing mediated allelic replacement of the NRT1.1B allele has been achieved in rice, which could improve NUE significantly (Li *et al.*, 2018). CRISPR-Cas9-targeted knockout of *FvPHO2* increased phosphorus content and enhanced fruit quality in woodland strawberry (Zhang *et al.*, 2023). Improving nutrient use efficiency through CRISPR provides a sustainable approach to feeding a growing population while ensuring environmental safety.

12.10 Feeding the Future: Strategies for Yield Maximization on Scarce Land

CRISPR can be used in vertical farming and urban agriculture to enhance crop yields. CRISPR–Cas9-mediated mutations in tomato genes controlling stem length (SlER), rapid flowering (SP5G) and precocious growth termination (SP) have transformed vine-like tomato plants into compact, early-yielding varieties suitable for urban farming (C.T. Kwon *et al.*, 2019). Field data confirmed that yields were maintained, and these varieties can be successfully cultivated in indoor farming systems.

A new Canadian start-up, Agrisea, is utilizing CRISPR gene-editing tools to develop salt-tolerant rice, enabling sea farming on floating platforms. This new rice variety requires no soil, fertilizers or fresh water, making it a promising solution for the future of food production. Agrisea plans to begin with small pilot farms before scaling up to large-scale production. The company envisions growing gene-edited rice in the ocean to feed billions worldwide while addressing environmental challenges.

12.11 Removing Breeding Bottlenecks

Genome-editing technologies enables precise and targeted genetic modifications with minimal unwanted variation, aiding in creating new traits, *de novo* domestication and overcoming breeding challenges. Specific applications of genome editing in improving breeding techniques include developing stable male-sterile lines, fixation of heterosis and developing apomictic hybrids, creating doubled haploids for faster trait fixation, generating haploid inducer lines, generating novel cis-alleles in the promoter to generate a gradient of phenotypes and inducing cross-over in recombination cold regions (Goyal *et al.*, 2024).

12.12 Rapid *de novo* Domestication

Genome editing has the potential to assist in crop domestication by significantly reducing the time required. Genome editing can rapidly introduce domestication traits into wild crop relatives, optimizing yield-related traits such as plant architecture, while preserving abiotic and biotic stress tolerance (Pattnaik *et al.*, 2023). Several studies have reported the use of CRISPR-Cas tools for the domestication of wild crops such as tomato (*Solanum pimpinellifolium*), groundcherry (*Physalis pruinosa*), tetraploid wild rice (*Oryza alta*), wild tetraploid *Solanum peruvianum* and wild sea barley grass, by targeting key domestication genes (Pattnaik *et al.*, 2023). Hence, the thoughtful use of editing tools can greatly aid in bringing wild and semi-domesticated orphan crops into mainstream agriculture.

12.13 Genome Editing in Forestry: Enhancing Tree Growth, Adaptability and Wood Quality

Genome editing technology provides a promising platform for forestry, enabling precise genetic modifications that can improve tree growth, adaptability and wood quality. The application of this technology to forest species is especially promising, as it addresses a number of challenges in forestry and industrial applications. CRISPR-Cas allows for the editing of growth rate-regulating genes to develop fast-growing tree species. For example, Yao *et al.* used a deactivated Cas9 (dCas9)-based system for CRISPR activation (CRISPRa) to precisely control gene expression in poplars (Yao *et al.*, 2023). Their target genes, *TPX2* and *LecRLK-G*, are crucial for plant growth and defense responses. A 4.5-fold increase in gene expression was achieved in the hybrid poplar clone "717-1B4" and poplar clone "WV94" (Yao *et al.*, 2023). Targeting hormone pathways, such as those in gibberellin and auxin synthesis, could be promising strategies for optimizing tree height and root structure.

Editing tools also enable precise alterations of genes involved in lignin biosynthesis, a crucial factor in wood quality and tree growth. In *Eucalyptus grandis*, an important species for forestry and industrial applications, genetic modification has been hindered by its recalcitrance to transformation. Dai *et al.* (2020) overcame this challenge by using CRISPR-Cas9 in a hairy root system, which enhanced the efficiency of editing wood-related traits. The study targeted the *CCR1* gene, important in lignin biosynthesis, and *IAA9A*, which plays a role in xylem development. The study by Sulis *et al.* (2023) demonstrates that strategic CRISPR genome editing in trees can significantly improve wood composition, boosting pulp mill production potential by 40%, creating high bioeconomic value and reducing global warming potential by 20%. By editing 3–6 members of the lignin gene family, the researchers generated over 170 novel edited lines, resulting in up to a 49.1% reduction in lignin content, a 228% increase in the carbohydrate-to-lignin (C/L) ratio, and an increase in the syringyl-to-guaiacyl (S/G) ratio from 2.7 in the wild type to 4.0 in the edited lines, offering high potential of environmental benefits. In a recent study, cytosine base editor-mediated modification of the *4CL1* gene in poplar wood resulted in a 12.8% reduction in lignin content, which could significantly reduce the need for chemical delignification (Liu *et al.*, 2024).

12.14 Conclusion

To move forward in agricultural sustainability, it is important to adopt and improve best practices and technologies that promote diverse production systems and minimize environmental footprints. There must be adaptable and responsive policies for the rapid changes occurring in agriculture and its environment. The integration of biotechnology, incorporating genome editing, provide considerable opportunities for improving crop productivity and sustainability. Taken together, sustainable agriculture can hold the way to meet current and future food demands in times of climate change and growing population numbers.

References

Abdallah, N.A., Elsharawy, H., Abulela, H.A., Thilmony, R., Abdelhadi, A.A. and Elarabi, N.I. (2022) Multiplex CRISPR/Cas9-mediated genome editing to address drought tolerance in wheat. *GM Crops and Food.* 16(1), 1–17. DOI: 10.1080/21645698.2022.2120313.

Achary, V.M.M. and Reddy, M.K. (2021) CRISPR-Cas9 mediated mutation in GRAIN WIDTH and WEIGHT2 (GW2) locus improves aleurone layer and grain nutritional quality in rice. *Scientific Reports* 11(1), 1–13. DOI: 10.1038/s41598-021-00828-z.

Alam, M.S., Kong, J., Tao, R., Ahmed, T., Alamin, M. *et al.* (2022) CRISPR/Cas9 mediated knockout of the OsbHLH024 transcription factor improves salt stress resistance in rice (*Oryza sativa* L.). *Plants* 11(9), 1184. DOI: 10.3390/plants11091184.

Alfatih, A., Wu, J., Jan, S.U., Zhang, Z.S., Xia, J.Q. and Xiang, C.B. (2020) Loss of rice PARAQUAT TOLERANCE 3 confers enhanced resistance to abiotic stresses and increases grain yield in field. *Plant Cell and Environment* 43(11), 2743–2754. DOI: 10.1111/pce.13856.

Altae-Tran, H., Shmakov, S.A., Makarova, K.S., Wolf, Y.I., Kannan, S., Zhang, F. and Koonin, E.V. (2023) Diversity, evolution, and classification of the RNA-guided nucleases TnpB and Cas12. *Proceedings of the National Academy of Sciences of the United States of America* 120(48). DOI: 10.1073/PNAS.2308224120/FORMAT/EPUB.

Ancydiana, P.G., Shanthinie, A., Arulganesh, T., Kumam, Y., Kumar, K.K. *et al.* (2022) Targeted editing of OsSWEET13, a bacterial leaf blight susceptible gene in rice using CRISPR tool. *Electronic Journal of Plant Breeding* 13(3), 772–779. DOI: 10.37992/2022.1303.139.

Anzalone, A.V., Randolph, P.B., Davis, J.R., Sousa, A.A., Koblan, L.W. *et al.* (2019) Search-and-replace genome editing without double-strand breaks or donor DNA. *Nature* 576(7785). DOI: 10.1038/s41586-019-1711-4.

Asseng, S., Ewert, F., Martre, P., Rötter, R.P., Lobell, D.B. *et al.* (2014) Rising temperatures reduce global wheat production. *Nature Climate Change* 5(2), 143–147. DOI: 10.1038/nclimate2470.

Balyan, S., Rao, S., Jha, S., Bansal, C., Das, J.R. and Mathur, S. (2020) Characterization of novel regulators for heat stress tolerance in tomato from Indian sub-continent. *Plant Biotechnology Journal* 18(10), 2118. DOI: 10.1111/PBI.13371.

Bassett, C.L., Baldo, A.M., Moore, J.T., Jenkins, R.M., Soffe, D.S. *et al.* (2014) Genes responding to water deficit in apple (Malus × domestica Borkh.) roots. *BMC Plant Biology* 14(1), 1–12. DOI: 10.1186/1471-2229-14-182/FIGURES/4.

Blanvillain-Baufumé, S., Reschke, M., Solé, M., Auguy, F., Doucoure, H. *et al.* (2017) Targeted promoter editing for rice resistance to *Xanthomonas oryzae* pv. oryzae reveals differential activities for SWEET14-inducing TAL effectors. *Plant Biotechnology Journal* 15(3), 306–317. DOI: 10.1111/pbi.12613.

Brás, T.A., Seixas, J., Carvalhais, N. and Jagermeyr, J. (2021) Severity of drought and heatwave crop losses tripled over the last five decades in Europe. *Environmental Research Letters* 16(6), 065012. DOI: 10.1088/1748-9326/ABF004.

Bui, T.P., Le, H., Ta, D.T., Nguyen, C.X., Le, N.T. *et al.* (2023) Enhancing powdery mildew resistance in soybean by targeted mutation of MLO genes using the CRISPR/Cas9 system. *BMC Plant Biology* 23(1), 1–13. DOI: 10.1186/S12870-023-04549-5/FIGURES/6.

Cai, Y., Liang, Y., Shi, H., Cui, J., Prakash, S. *et al.* (2024) Creating yellow seed Camelina sativa with enhanced oil accumulation by CRISPR-mediated disruption of transparent Testa 8. *Plant Biotechnology Journal* 22(10), 2773–2784. DOI: 10.1111/PBI.14403.

Chandrasekaran, J., Brumin, M., Wolf, D., Leibman, D., Klap, C. *et al.* (2016) Development of broad virus resistance in non-transgenic cucumber using CRISPR/Cas9 technology. *Molecular Plant Pathology* 17(7), 1140–1153. DOI: 10.1111/MPP.12375.

Chang, J.D., Huang, S., Yamaji, N., Zhang, W., Ma, J.F. and Zhao, F.J. (2020) OsNRAMP1 transporter contributes to cadmium and manganese uptake in rice. *Plant Cell and Environment* 43(10), 2476–2491. DOI: 10.1111/pce.13843.

Chen, M., Zhu, X., Liu, X., Wu, C., Yu, C. *et al.* (2021) Knockout of auxin response factor SlARF4 improves tomato resistance to water deficit. *International Journal of Molecular Sciences* 22(7), 3347. DOI: 10.3390/IJMS22073347.

Chen, S., Zhang, N., Zhou, G., Hussain, S., Ahmed, S., Tian, H. and Wang, S. (2021) Knockout of the entire family of AITR genes in Arabidopsis leads to enhanced drought and salinity tolerance without fitness costs. *BMC Plant Biology* 21(1), 137. DOI: 10.1186/s12870-021-02907-9.

Christian, M., Cermak, T., Doyle, E.L., Schmidt, C., Zhang, F. *et al.* (2010) Targeting DNA double-strand breaks with TAL effector nucleases. *Genetics* 186(2), 756–761. DOI: 10.1534/genetics.110.120717.

Curtin, S.J., Zhang, F., Sander, J.D., Haun, W.J., Starker, C. *et al.* (2011) Targeted mutagenesis of duplicated genes in soybean with zinc-finger nucleases. *Plant Physiology* 156(2), 466–473. DOI: 10.1104/PP.111.172981.

Dai, Y., Hu, G., Dupas, A., Medina, L., Blandels, N. *et al.* (2020) Implementing the CRISPR/Cas9 technology in eucalyptus hairy roots using wood-related genes. *International Journal of Molecular Sciences* 21(10), 3408. DOI: 10.3390/IJMS21103408.

Danilo, B., Montes, É., Archambeau, H., Lodé, M., Rousseau-Gueutin, M., Chèvre, A.M. and Mazier, M. (2022) I-SceI and customized meganucleases-mediated genome editing in tomato and oilseed rape. *Transgenic Research* 31(1), 87–105. DOI: 10.1007/S11248-021-00287-2.

Expert meeting on How to Feed the World in 2050. FAO headquarters, 24-26 June 2009. (n.d.). Available at: https://www.fao.org/4/ak542e/ak542e00.htm (accessed 21 December 2024).

Farhat, S., Jain, N., Singh, N., Sreevathsa, R., Dash, P.K. *et al.* (2019) CRISPR-Cas9 directed genome engineering for enhancing salt stress tolerance in rice. *Seminars in Cell & Developmental Biology* 96, 91–99. DOI: 10.1016/J.SEMCDB.2019.05.003.

Fister, A.S., Landherr, L., Maximova, S.N. and Guiltinan, M.J. (2018) Transient expression of CRISPR/Cas9 machinery targeting TcNPR3 enhances defense response in Theobroma cacao. *Frontiers in Plant Science* 9, 329023. DOI: 10.3389/FPLS.2018.00268/BIBTEX.

Galli, M., Martiny, E., Imani, J., Kumar, N., Koch, A., Steinbrenner, J. and Kogel, K.H. (2022) CRISPR/SpCas9-mediated double knockout of barley Microrchidia MORC1 and MORC6a reveals their strong involvement in plant immunity, transcriptional gene silencing and plant growth. *Plant Biotechnology Journal* 20(1), 89–102. DOI: 10.1111/PBI.13697.

Gao, H., Gadlage, M.J., Lafitte, H.R., Lenderts, B., Yang, M. *et al.* (2020) Superior field performance of waxy corn engineered using CRISPR–Cas9. *Nature Biotechnology* 38(5), 579–581. DOI: 10.1038/s41587-020-0444-0.

Gaudelli, N.M., Komor, A.C., Rees, H.A., Packer, M.S., Badran, A.H., Bryson, D.I. and Liu, D.R. (2017) Programmable base editing of T to G C in genomic DNA without DNA cleavage. *Nature* 551(7681), 464–471. DOI: 10.1038/nature24644.

Gomez, M.A., Lin, Z.D., Moll, T., Chauhan, R.D., Hayden, L. *et al.* (2019) Simultaneous CRISPR/Cas9-mediated editing of cassava eIF4E isoforms nCBP-1 and nCBP-2 reduces cassava brown streak disease symptom severity and incidence. *Plant Biotechnology Journal* 17(2), 421–434. DOI: 10.1111/pbi.12987.

Goyal, L., Mandal, M., Bhatia, D. and Molla, K.A. (2024) *Plant Breeding Becomes Smarter with Genome Editing*, pp. 113–147. DOI: 10.1007/978-981-99-9338-3_5.

Hanika, K., Schipper, D., Chinnappa, S., Oortwijn, M., Schouten, H.J., Thomma, B.P.H.J. and Bai, Y. (2021) Impairment of tomato WAT1 enhances resistance to vascular wilt fungi despite severe growth defects. *Frontiers in Plant Science* 12, 721674. DOI: 10.3389/FPLS.2021.721674/BIBTEX.

Hao, J., Yang, Y., Futrell, S., Kelly, E.A., Lorts, C.M. *et al.* (2023) CRISPR/Cas9-mediated mutagenesis of carotenoid cleavage dioxygenase (CCD) genes in sorghum alters Strigolactone biosynthesis and plant biotic interactions. *Phytobiomes Journal* 10(10), 339–351. DOI: 10.1094/PBIOMES-08-22-0053-R.

Hassani, A., Azapagic, A. and Shokri, N. (2021) Global predictions of primary soil salinization under changing climate in the 21st century. *Nature Communications* 12(1), 1–17. DOI: 10.1038/s41467-021-26907-3.

He, X., Luo, X., Wang, T., Liu, S., Zhang, X. and Zhu, L. (2020) GhHB12 negatively regulates abiotic stress tolerance in Arabidopsis and cotton. *Environmental and Experimental Botany* 176, 104087. DOI: 10.1016/J.ENVEXPBOT.2020.104087.

Hu, B., Zhou, Y., Zhou, Z., Sun, B., Zhou, F. *et al.* (2021) Repressed OsMESL expression triggers reactive oxygen species-mediated broad-spectrum disease resistance in rice. *Plant Biotechnology Journal* 19(8), 1511–1522. DOI: 10.1111/PBI.13566.

Illouz-Eliaz, N., Nissan, I., Nir, I., Ramon, U., Shohat, H. and Weiss, D. (2020) Mutations in the tomato gibberellin receptors suppress xylem proliferation and reduce water loss under water-deficit conditions. *Journal of Experimental Botany* 71(12), 3603–3612. DOI: 10.1093/JXB/ERAA137.

Jia, H., Zhang, Y., Orbović, V., Xu, J., White, F.F., Jones, J.B. and Wang, N. (2017) Genome editing of the disease susceptibility gene CsLOB1 in citrus confers resistance to citrus canker. *Plant Biotechnology Journal* 15(7), 817–823. DOI: 10.1111/PBI.12677.

Karmakar, S., Das, P., Panda, D., Xie, K., Baig, M. and Molla, K.A. A detailed landscape of CRISPR-Cas-mediated plant disease and pest management. *Plant Science* 323, 111376. Available at: https://www.sciencedirect.com/science/article/pii/S016894522200200X (accessed 18 November 2024).

Karmakar, S., Panda, D., Panda, S., Dash, M., Saha, R. *et al.* (2024) A miniature alternative to Cas9 and Cas12: Transposon-associated TnpB mediates targeted genome editing in plants. DOI: 10.21203/rs.3.rs-3878499/v1.

Karunarathne, S.D., Han, Y., Zhang, X.Q. and Li, C. (2022) CRISPR/Cas9 gene editing and natural variation analysis demonstrate the potential for HvARE1 in improvement of nitrogen use efficiency in barley. *Journal of Integrative Plant Biology* 64(3), 756–770. DOI: 10.1111/JIPB.13214/SUPPINFO.

Karvelis, T., Druteika, G., Bigelyte, G., Budre, K., Zedaveinyte, R. *et al.* (2021) Transposon-associated TnpB is a programmable RNA-guided DNA endonuclease. *Nature* 599(7886), 692–696. DOI: 10.1038/s41586-021-04058-1.

Kieu, N.P., Lenman, M., Wang, E.S., Petersen, B.L. and Andreasson, E. (2021) Mutations introduced in susceptibility genes through CRISPR/Cas9 genome editing confer increased late blight resistance in potatoes. *Scientific Reports* 11(1), 1–12. DOI: 10.1038/s41598-021-83972-w.

Klap, C., Yeshayahou, E., Bolger, A.M., Arazi, T., Gupta, S.K. *et al.* (2017) Tomato facultative parthenocarpy results from SlAGAMOUS-LIKE 6 loss of function. *Plant Biotechnology Journal* 15(5), 634–647. DOI: 10.1111/pbi.12662.

Komor, A.C., Kim, Y.B., Packer, M.S., Zuris, J.A. and Liu, D.R. (2016) Programmable editing of a target base in genomic DNA without double-stranded DNA cleavage. *Nature* 533(7603), 420–424. DOI: 10.1038/nature17946.

Kwon, C.T., Heo, J., Lemmon, Z.H., Capua, Y., Hutton, S.F. *et al.* (2019) Rapid customization of Solanaceae fruit crops for urban agriculture. *Nature Biotechnology* 38(2), 182–188. DOI: 10.1038/s41587-019-0361-2.

Kwon, Y., Lee, J.Y., Choi, J., Lee, S.M., Kim, D. *et al.* (2023) Loss-of-function gs3 allele decreases methane emissions and increases grain yield in rice. *Nature Climate Change* 13(12), 1329–1333. DOI: 10.1038/s41558-023-01872-5.

Lemmon, Z.H., Reem, N.T., Dalrymple, J., Soyk, S., Swartwood, K.E. *et al.* (2018) Rapid improvement of domestication traits in an orphan crop by genome editing. *Nature Plants* 4(10), 766–770. DOI: 10.1038/s41477-018-0259-x.

Li, C., Li, W., Zhou, Z., Chen, H., Xie, C. and Lin, Y. (2019) A new rice breeding method: CRISPR/Cas9 system editing of the Xa13 promoter to cultivate transgene-free bacterial blight-resistant rice. *Plant Biotechnology Journal* 18(2), 313. DOI: 10.1111/PBI.13217.

Li, J., Jiao, G., Sun, Y., Chen, J., Zhong, Y. *et al.* (2021) Modification of starch composition, structure and properties through editing of TaSBEIIa in both winter and spring wheat varieties by CRISPR/Cas9. *Plant Biotechnology Journal* 19(5), 937–951. DOI: 10.1111/PBI.13519.

Li, J., Zhang, X., Sun, Y., Zhang, J., Du, W. *et al.* (2018) Efficient allelic replacement in rice by gene editing: A case study of the NRT1.1B gene. *Journal of Integrative Plant Biology* 60(7), 536–540. DOI: 10.1111/JIPB.12650/SUPPINFO.

Li, M., Li, X., Zhou, Z., Wu, P., Fang, M. *et al.* (2016) Reassessment of the four yield-related genes Gn1a, DEP1, GS3, and IPA1 in rice using a CRISPR/Cas9 system. *Frontiers in Plant Science* 7(MAR2016), 186370. DOI: 10.3389/FPLS.2016.00377/BIBTEX.

Li, R., Liu, C., Zhao, R., Wang, L., Chen, L. *et al.* (2019) CRISPR/Cas9-mediated SlNPR1 mutagenesis reduces tomato plant drought tolerance. *BMC Plant Biology* 19(1), 1–13. DOI: 10.1186/S12870-018-1627-4/FIGURES/7.

Li, R., Zhang, L., Wang, L., Chen, L., Zhao, R., Sheng, J. and Shen, L. (2018) Reduction of tomato-plant chilling tolerance by CRISPR-Cas9-mediated SlCBF1 mutagenesis. *Journal of Agricultural and Food Chemistry* 66(34), 9042–9051. DOI: 10.1021/ACS.JAFC.8B02177.

Li, S., Lin, D., Zhang, Y., Deng, M., Chen, Y. *et al.* (2022) Genome-edited powdery mildew resistance in wheat without growth penalties. *Nature* 602(7897), 455–460. DOI: 10.1038/s41586-022-04395-9.

Li, S., Zhang, Y., Liu, Y., Zhang, P., Wang, X. *et al.* (2024) The E3 ligase TaGW2 mediates transcription factor TaARR12 degradation to promote drought resistance in wheat. *The Plant Cell* 36(3), 605–625. DOI: 10.1093/PLCELL/KOAD307.

Li, T., Liu, B., Spalding, M.H., Weeks, D.P. and Yang, B. (2012) High-efficiency TALEN-based gene editing produces disease-resistant rice. *Nature Biotechnology* 30(5), 390–392. DOI: 10.1038/nbt.2199.

Liao, S., Qin, X., Luo, L., Han, Y., Wang, X. *et al.* (2019) CRISPR/Cas9-induced mutagenesis of semi-rolled Leaf1,2 confers curled leaf phenotype and drought tolerance by influencing protein expression patterns and ROS scavenging in rice (*Oryza sativa* L.). *Agronomy* 9(11), 728. DOI: 10.3390/AGRONOMY9110728.

Liu, C., Kong, M., Yang, F., Zhu, J., Qi, X. *et al.* (2022) Targeted generation of null mutants in ZmGDIα confers resistance against maize rough dwarf disease without agronomic penalty. *Plant Biotechnology Journal* 20(5), 803. DOI: 10.1111/PBI.13793.

Liu, L., Zhang, J., Xu, J., Li, Y., Guo, L. *et al.* (2020) CRISPR/Cas9 targeted mutagenesis of SlLBD40, a lateral organ boundaries domain transcription factor, enhances drought tolerance in tomato. *Plant Science* 301, 110683. DOI: 10.1016/J.PLANTSCI.2020.110683.

Liu, X., Wu, D., Shan, T., Xu, S., Qin, R. *et al.* (2020) The trihelix transcription factor OsGTγ-2 is involved adaption to salt stress in rice. *Plant Molecular Biology* 103(4–5), 545–560. DOI: 10.1007/s11103-020-01010-1.

Liu, Y., Li, G., Mao, Y., Gao, Y., Zhao, M. *et al.* (2024) Genome-edited trees for high-performance engineered wood. *Matter* 7(10), 3658–3671. DOI: 10.1016/j.matt.2024.07.003.

Lu, H.P., Luo, T., Fu, H.W., Wang, L., Tan, Y.Y. *et al.* (2018) Resistance of rice to insect pests mediated by suppression of serotonin biosynthesis. *Nature Plants* 4(6), 338–344. DOI: 10.1038/s41477-018-0152-7.

Macovei, A., Sevilla, N.R., Cantos, C., Jonson, G.B., Slamet-Loedin, I. *et al.* (2018) Novel alleles of rice eIF4G generated by CRISPR/Cas9-targeted mutagenesis confer resistance to Rice tungro spherical virus. *Plant Biotechnology Journal* 16(11), 1918–1927. DOI: 10.1111/pbi.12927.

Merlot, S., Leonhardt, N., Fenzi, F., Valon, C., Costa, M. *et al.* (2007) Constitutive activation of a plasma membrane H+-ATPase prevents abscisic acid-mediated stomatal closure. *EMBO Journal* 26(13), 3216–3226. DOI: 10.1038/sj.emboj.7601750.

Miao, C., Xiao, L., Hua, K., Zou, C., Zhao, Y., Bressan, R.A. and Zhu, J.K. (2018) Mutations in a subfamily of abscisic acid recepto genes promote rice growth and productivity. *Proceedings of the National Academy of Sciences of the United States of America* 115(23), 6058–6063. DOI: 10.1073/pnas.1804774115.

Molla, K.A., Karmakar, S., Chanda, P.K., Sarkar, S.N., Datta, S.K. and Datta, K. (2016) Tissue-specific expression of Arabidopsis NPR1 gene in rice for sheath blight resistance without compromising phenotypic cost. *Plant Science* 250, 105–114. DOI: 10.1016/J.PLANTSCI.2016.06.005.

Molla, K.A., Karmakar, S. and Islam, M.T. (2020) Wide horizons of CRISPR-Cas-derived technologies for basic biology. *Agriculture, and Medicine.* 1–23. DOI: 10.1007/978-1-0716-0616-2_1.

Molla, K.A., Shih, J., Wheatley, M.S. and Yang, Y. (2022) Predictable NHEJ insertion and assessment of HDR editing strategies in plants. *Frontiers in Genome Editing* 4, 825236. DOI: 10.3389/FGEED.2022.825236/BIBTEX.

Molla, K.A., Shih, J. and Yang, Y. (2020) Single-nucleotide editing for zebra3 and wsl5 phenotypes in rice using CRISPR/Cas9-mediated adenine base editors. *aBIOTECH* 1(2), 106. DOI: 10.1007/S42994-020-00018-X.

Molla, K.A., Sretenovic, S., Bansal, K.C. and Qi, Y. (2021) Precise plant genome editing using base editors and prime editors. *Nature Plants* 7(9), 1166–1187. DOI: 10.1038/s41477-021-00991-1.

Molla, K.A. and Yang, Y. (2019) CRISPR/Cas-Mediated Base editing: Technical considerations and practical applications. *Trends in Biotechnology* 37(10), 1121–1142. DOI: 10.1016/J.TIBTECH.2019.03.008/ASSET/3B225794-5A6E-4C2F-96D8-31AA340ECAA3/MAIN.ASSETS/GR1B2.SML.

Mushtaq, M., Mukhtar, S., Sakina, A., Dar, A.A., Bhat, R. *et al.* (2020) Tweaking genome-editing approaches for virus interference in crop plants. *Plant Physiology and Biochemistry* 147, 242–250. DOI: 10.1016/J.PLAPHY.2019.12.022.

Nekrasov, V., Wang, C., Win, J., Lanz, C., Weigel, D. and Kamoun, S. (2017) Rapid generation of a transgene-free powdery mildew resistant tomato by genome deletion. *Scientific Reports* 7(1), 1–6. DOI: 10.1038/s41598-017-00578-x.

Nieves-Cordones, M., Mohamed, S., Tanoi, K., Kobayashi, N.I., Takagi, K. *et al.* (2017) Production of low-Cs+ rice plants by inactivation of the K+ transporter OsHAK1 with the CRISPR-Cas system. *Plant Journal* 92(1), 43–56. DOI: 10.1111/TPJ.13632.

Nonaka, S., Arai, C., Takayama, M., Matsukura, C. and Ezura, H. (2017) Efficient increase of γ-aminobutyric acid (GABA) content in tomato fruits by targeted mutagenesis. *Scientific Reports* 7(1), 1–14. DOI: 10.1038/s41598-017-06400-y.

Nuñez-Muñoz, L., Vargas-Hernández, B., Hinojosa-Moya, J., Ruiz-Medrano, R. and Xoconostle-Cázares, B. (2021) Plant drought tolerance provided through genome editing of the trehalase gene. *Plant Signaling and Behavior* 16(4), 1877005. DOI: 10.1080/15592324.2021.1877005.

Ogata, T., Ishizaki, T., Fujita, M. and Fujita, Y. (2020) CRISPR/Cas9-targeted mutagenesis of OsERA1 confers enhanced responses to abscisic acid and drought stress and increased primary root growth under nonstressed conditions in rice. *PLoS One* 15(12), e0243376. DOI: 10.1371/JOURNAL.PONE.0243376.

Ortigosa, A., Gimenez-Ibanez, S., Leonhardt, N. and Solano, R. (2019) Design of a bacterial speck resistant tomato by CRISPR/Cas9-mediated editing of SlJAZ2. *Plant Biotechnology Journal* 17(3), 665–673. DOI: 10.1111/pbi.13006.

Osakabe, Y., Watanabe, T., Sugano, S.S., Ueta, R., Ishihara, R., Shinozaki, K. and Osakabe, K. (2016) Optimization of CRISPR/Cas9 genome editing to modify abiotic stress responses in plants. *Scientific Reports* 6. DOI: 10.1038/srep26685.

Park, J.J., Dempewolf, E., Zhang, W. and Wang, Z.Y. (2017) RNA-guided transcriptional activation via CRISPR/dCas9 mimics overexpression phenotypes in Arabidopsis. *PLoS One* 12(6), e0179410. DOI: 10.1371/JOURNAL.PONE.0179410.

Pathi, K.M., Rink, P., Budhagatapalli, N., Betz, R., Saado, I. *et al.* (2020) Engineering smut resistance in maize by site-directed mutagenesis of LIPOXYGENASE 3. *Frontiers in Plant Science* 11, 543895. DOI: 10.3389/FPLS.2020.543895/BIBTEX.

Pattnaik, D., Avinash, S.P., Panda, S., Bansal, K.C., Chakraborti, M. *et al.* (2023) Accelerating crop domestication through genome editing for sustainable agriculture. *Journal of Plant Biochemistry and Biotechnology.* DOI: 10.1007/s13562-023-00837-1.

Paula De Toledo Thomazella, D., Seong, K., Mackelprang, R., Dahlbeck, D. *et al.* (2021) Loss of function of a DMR6 ortholog in tomato confers broad-spectrum disease resistance. *Proceedings of the National Academy of Sciences* 118(27), e2026152118. DOI: 10.1073/pnas.2026152118/-/DCSupplemental.

Qiu, J., Xie, J., Chen, Y., Shen, Z., Shi, H. *et al.* (2022) Warm temperature compromises JA-regulated basal resistance to enhance *Magnaporthe oryzae* infection in rice. *Molecular Plant* 15(4), 723–739. DOI: 10.1016/j.molp.2022.02.014.

Qiu, R., Yao, P., Yang, J., Hou, J., Xiao, H. *et al.* (2025) OsIAA7 enhances heat stress tolerance by inhibiting the activity of OsARF6 in rice. *International Journal of Biological Macromolecules* 288, 138746. DOI: 10.1016/J.IJBIOMAC.2024.138746.

Ren, B., Yan, F., Kuang, Y., Li, N., Zhang, D. *et al.* (2018) Improved base editor for efficiently inducing genetic variations in rice with CRISPR/Cas9-guided hyperactive hAID mutant. *Molecular Plant* 11(4), 623–626. DOI: 10.1016/j.molp.2018.01.005.

Ribeiro-Silva, R.d.C., Pereira, M., Campello, T., Aragão, É., Guimarães, J.M.d.M. *et al.* (2020) Covid-19 pandemic implications for food and nutrition security in Brazil. *Ciência & Saúde Coletiva* 25(9), 3421–3430. DOI: 10.1590/1413 81232020259.22152020.

Roca Paixão, J.F., Gillet, F.X., Ribeiro, T.P., Bournaud, C., Lourenço-Tessutti, I.T. *et al.* (2019) Improved drought stress tolerance in Arabidopsis by CRISPR/dCas9 fusion with a histone AcetylTransferase. *Scientific Reports* 9(1), 1–9. DOI: 10.1038/s41598-019-44571-y.

Rodríguez-Leal, D., Lemmon, Z.H., Man, J., Bartlett, M.E. and Lippman, Z.B. (2017) Engineering quantitative trait variation for crop improvement by genome editing. *Cell* 171(2), 470–480.e8. DOI: 10.1016/J.CELL.2017.08.030/ATTACHMENT/6A63EBB2-2975-454D-8CE3-34AEA81354FE/MMC1.PDF.

Rouet, P., Smih, F. and Jasin, M. (1994) Expression of a site-specific endonuclease stimulates homologous recombination in mammalian cells. *Proceedings of the National Academy of Sciences* 91(13), 6064–6068. DOI: 10.1073/PNAS.91.13.6064.

Ruan, M., Zhao, H., Wen, Y., Chen, H., He, F. *et al.* (2024) The complex transcriptional regulation of heat stress response in maize. *Stress Biology* 4(1), 1–14. DOI: 10.1007/S44154-024-00165-X.

Sánchez-León, S., Gil-Humanes, J., Ozuna, C.V., Giménez, M.J., Sousa, C., Voytas, D.F. and Barro, F. (2018) Low-gluten, nontransgenic wheat engineered with CRISPR/Cas9. *Plant Biotechnology Journal* 16(4), 902–910. DOI: 10.1111/pbi.12837.

Santillán Martínez, M.I., Bracuto, V., Koseoglou, E., Appiano, M., Jacobsen, E. *et al.* (2020) CRISPR/Cas9-targeted mutagenesis of the tomato susceptibility gene PMR4 for resistance against powdery mildew. *BMC Plant Biology* 20(1). DOI: 10.1186/s12870-020-02497-y.

Santosh Kumar, V.V., Verma, R.K., Yadav, S.K., Yadav, P., Watts, A., Rao, M.V. and Chinnusamy, V. (2020) CRISPR-Cas9 mediated genome editing of drought and salt tolerance (OsDST) gene in indica mega rice cultivar MTU1010. *Physiology and Molecular Biology of Plants* 26(6), 1099–1110. DOI: 10.1007/s12298-020-00819-w.

Shi, J., Gao, H., Wang, H., Lafitte, H.R., Archibald, R.L. *et al.* (2017) ARGOS8 variants generated by CRISPR-Cas9 improve maize grain yield under field drought stress conditions. *Plant Biotechnology Journal* 15(2), 207–216. DOI: 10.1111/pbi.12603.

Shukla, V.K., Doyon, Y., Miller, J.C., Dekelver, R.C., Moehle, E.A. *et al.* (2009) Precise genome modification in the crop species Zea mays using zinc-finger nucleases. *Nature* 459(7245), 437–441. DOI: 10.1038/nature07992.

Tek, M.I., Calis, O., Fidan, H., Shah, M.D., Celik, S. and Wani, S.H. (2022) CRISPR/Cas9 based mlo-mediated resistance against Podosphaera xanthii in cucumber (*Cucumis sativus* L.). *Frontiers Plant Science* 13, 1081506. DOI: 10.3389/FPLS.2022.1081506/BIBTEX.

Tran, M.T., Doan, D.T.H., Kim, J., Song, Y.J., Sung, Y.W. *et al.* (2021) CRISPR/Cas9-based precise excision of SlHyPRP1 domain(s) to obtain salt stress-tolerant tomato. *Plant Cell Reports* 40(6), 999–1011. DOI: 10.1007/s00299-020-02622-z.

Tripathi, J.N., Ntui, V.O., Shah, T. and Tripathi, L. (2021) CRISPR/Cas9-mediated editing of DMR6 orthologue in banana (Musa spp.) confers enhanced resistance to bacterial disease. *Plant Biotechnology Journal* 19(7), 1291. DOI: 10.1111/PBI.13614.

Usman, B., Nawaz, G., Zhao, N., Liao, S., Liu, Y. and Li, R. (2020) Precise editing of the OsPYL9 gene by RNA-guided Cas9 nuclease confers enhanced drought tolerance and grain yield in rice (*Oryza sativa* L.) by regulating circadian rhythm and abiotic stress responsive proteins. *International Journal of Molecular Sciences* 21(21), 7854. DOI: 10.3390/IJMS21217854.

Veley, K.M., Elliott, K., Jensen, G., Zhong, Z., Feng, S. *et al.* (2023) Improving cassava bacterial blight resistance by editing the epigenome. *Nature Communications* 14(1), 1–9. DOI: 10.1038/s41467-022-35675-7.

Wan, D.Y., Guo, Y., Cheng, Y., Hu, Y., Xiao, S., Wang, Y. and Wen, Y.Q. (2020) CRISPR/Cas9-mediated mutagenesis of VvMLO3 results in enhanced resistance to powdery mildew in grapevine (*Vitis vinifera*). *Horticulture Research* 7(1), 116. DOI: 10.1038/S41438-020-0339-8/41990461/41438_2020_ARTICLE_339.PDF.

Wang, F., Wang, C., Liu, P., Lei, C., Hao, W. *et al.* (2016) Enhanced rice blast resistance by CRISPR/Cas9-targeted mutagenesis of the ERF transcription factor gene OsERF922. *PLoS One* 11(4), e0154027. DOI: 10.1371/JOURNAL.PONE.0154027.

Wang, F.Z., Chen, M.X., Yu, L.J., Xie, L.J., Yuan, L.B. *et al.* (2017) OsARM1, an R2R3 MYB transcription factor, is involved in regulation of the response to arsenic stress in rice. *Frontiers in Plant Science* 8, 299307. DOI: 10.3389/FPLS.2017.01868/BIBTEX.

Wang, L., Chen, L., Li, R., Zhao, R., Yang, M., Sheng, J. and Shen, L. (2017) Reduced drought tolerance by CRISPR/Cas9-mediated SlMAPK3 mutagenesis in tomato plants. *Journal of Agricultural and Food Chemistry* 65(39), 8674–8682. DOI: 10.1021/ACS.JAFC.7B02745/SUPPL_FILE/JF7B02745_SI_001.PDF.

Wang, L., Chen, S., Peng, A., Xie, Z., He, Y. and Zou, X. (2019) CRISPR/Cas9-mediated editing of CsWRKY22 reduces susceptibility to Xanthomonas citri subsp. citri in Wanjincheng orange (*Citrus sinensis* (L.) Osbeck). *Plant Biotechnology Reports* 13(5), 501–510. DOI: 10.1007/S11816-019-00556-X/FIGURES/4.

Wang, T., Li, Y., Fu, Y., Xie, H., Song, S. *et al.* (2019) Mutation at different sites of metal transporter gene OsNramp5 affects cd accumulation and related agronomic traits in rice (*Oryza sativa* L.). *Frontiers Plant Science* 10, 472897. DOI: 10.3389/FPLS.2019.01081/BIBTEX.

Wang, T., Xun, H., Wang, W., Ding, X., Tian, H. *et al.* (2021) Mutation of GmAITR genes by CRISPR/Cas9 genome editing results in enhanced salinity stress tolerance in soybean. *Frontiers in Plant Science* 12, 779598. DOI: 10.3389/fpls.2021.779598.

Wang, X., Ren, P., Ji, L., Zhu, B. and Xie, G. (2022) OsVDE, a xanthophyll cycle key enzyme, mediates abscisic acid biosynthesis and negatively regulates salinity tolerance in rice. *Planta* 255(1). DOI: 10.1007/s00425-021-03802-1.

Wang, Y., Cheng, X., Shan, Q., Zhang, Y., Liu, J., Gao, C. and Qiu, J.L. (2014) Simultaneous editing of three homoeoalleles in hexaploid bread wheat confers heritable resistance to powdery mildew. *Nature Biotechnology* 32(9), 947–951. DOI: 10.1038/nbt.2969.

Wang, Y., Geng, M., Pan, R., Zhang, T., Lu, X. *et al.* (2024) Editing of the MeSWEET10a promoter yields bacterial blight resistance in cassava cultivar SC8. *Molecular Plant Pathology* 25(10), e70010. DOI: 10.1111/mpp.70010.

Wang, Y., Wang, L., Zhou, J., Hu, S., Chen, H. *et al.* (2019) Research progress on heat stress of rice at flowering stage. *Rice Science* 26(1), 1–10. DOI: 10.1016/J.RSCI.2018.06.009.

Wang, Z., Hong, Y., Li, Y., Shi, H., Yao, J. *et al.* (2020) Natural variations in SlSOS1 contribute to the loss of salt tolerance during tomato domestication. *Plant Biotechnology Journal* 19(1), 20. DOI: 10.1111/PBI.13443.

Wright, D.A., Townsend, J.A., Winfrey, R.J., Irwin, P.A., Rajagopal, J. *et al.* (2005) High-frequency homologous recombination in plants mediated by zinc-finger nucleases. *The Plant Journal* 44(4), 693–705. DOI: 10.1111/J.1365-313X.2005.02551.X.

Xue, N., Hong, D., Zhang, D., Wang, Q., Zhang, S. *et al.* (2024) Engineering IscB to develop highly efficient miniature editing tools in mammalian cells and embryos. *Molecular Cell* 84(16), 3128–3140.e4. DOI: 10.1016/j.molcel.2024.07.007.

Yang, H., Huang, T., Ding, M., Lu, D. and Lu, W. (2017) High temperature during grain filling impacts on leaf senescence in waxy maize. *Agronomy Journal* 109(3), 906–916. DOI: 10.2134/AGRONJ2016.08.0452.

Yang, J., Fang, Y., Wu, H., Zhao, N., Guo, X. *et al.* (2023) Improvement of resistance to rice blast and bacterial leaf streak by CRISPR/Cas9-mediated mutagenesis of Pi21 and OsSULTR3;6 in rice (*Oryza sativa* L.). *Frontiers Plant Science* 14, 1209384. DOI: 10.3389/FPLS.2023.1209384/BIBTEX.

Yao, T., Yuan, G., Lu, H., Liu, Y., Zhang, J. *et al.* (2023) CRISPR/Cas9-based gene activation and base editing in Populus. *Horticulture Research* 10(6). DOI: 10.1093/HR/UHAD085.

Yu, H., Liu, C. and Dixon, R.A. (2021) A gene-editing/complementation strategy for tissue-specific lignin reduction while preserving biomass yield. *Biotechnology for Biofuels* 14(1), 1–12. DOI: 10.1186/S13068-021-02026-5/FIGURES/6.

Yu, W., Wang, L., Zhao, R., Sheng, J., Zhang, S., Li, R. and Shen, L. (2019) Knockout of SlMAPK3 enhances tolerance to heat stress involving ROS homeostasis in tomato plants. *BMC Plant Biology* 19(1). DOI: 10.1186/s12870-019-1939-z.

Zeng, X., Luo, Y., Vu, N.T.Q., Shen, S., Xia, K. and Zhang, M. (2020) CRISPR/Cas9-mediated mutation of OsSWEET14 in rice cv. Zhonghua11 confers resistance to Xanthomonas oryzae pv. oryzae without yield penalty. *BMC Plant Biology* 20(1), 1–11. DOI: 10.1186/S12870-020-02524-Y/FIGURES/5.

Zeng, Y., Wen, J., Zhao, W., Wang, Q. and Huang, W. (2020) Rational improvement of rice yield and cold tolerance by editing the three genes OsPIN5b, GS3, and OsMYB30 with the CRISPR–Cas9 system. *Frontiers in Plant Science* 10, 1663. DOI: 10.3389/fpls.2019.01663.

Zhang, A., Liu, Y., Wang, F., Li, T., Chen, Z. *et al.* (2019) Enhanced rice salinity tolerance via CRISPR/Cas9-targeted mutagenesis of the OsRR22 gene. *Molecular Breeding* 39(3), 47. DOI: 10.1007/s11032-019-0954-y.

Zhang, C., Liu, Y., Wang, B., Li, H., Zhang, J. *et al.* (2023) CRISPR/Cas9 targeted knockout FvPHO2 can increase phosphorus content and improve fruit quality of woodland strawberry. *Scientia Horticulturae* 317, 112078. DOI: 10.1016/J.SCIENTA.2023.112078.

Zhang, J., Lyu, H., Chen, J., Cao, X., Du, R. *et al.* (2024) Releasing a sugar brake generates sweeter tomato without yield penalty. *Nature* 635(8039), 647–656. DOI: 10.1038/s41586-024-08186-2.

Zhang, J., Zhang, H., Li, S., Li, J., Yan, L. and Xia, L. (2021) Increasing yield potential through manipulating of an ARE1 ortholog related to nitrogen use efficiency in wheat by CRISPR/Cas9. *Journal of Integrative Plant Biology* 63(9), 1649–1663. DOI: 10.1111/JIPB.13151/SUPPINFO.

Zhang, M., Liu, Q., Yang, X., Xu, J., Liu, G. *et al.* (2020) CRISPR/Cas9-mediated mutagenesis of Clpsk1 in watermelon to confer resistance to *Fusarium oxysporum* f.sp. niveum. *Plant Cell Reports* 39(5), 589–595. DOI: 10.1007/S00299-020-02516-0/TABLES/1.

Zhang, P., Du, H., Wang, J., Pu, Y., Yang, C. *et al.* (2020) Multiplex CRISPR/Cas9-mediated metabolic engineering increases soya bean isoflavone content and resistance to soya bean mosaic virus. *Plant Biotechnology Journal* 18(6), 1384–1395. DOI: 10.1111/PBI.13302.

Zhang, X., Long, Y., Huang, J., & Xia, J. (2020). OsNAC45 is involved in ABA response and salt tolerance in rice. *Rice*, *13*(1), 1–13. DOI: 10.1186/S12284-020-00440-1/FIGURES/8

Zhang, Y., Bai, Y., Wu, G., Zou, S., Chen, Y., Gao, C. and Tang, D. (2017) Simultaneous modification of three homoeologs of TaEDR1 by genome editing enhances powdery mildew resistance in wheat. *Plant Journal* 91(4), 714–724. DOI: 10.1111/tpj.13599.

Zhang, Y., Liang, Z., Zong, Y., Wang, Y., Liu, J. *et al.* (2016) Efficient and transgene-free genome editing in wheat through transient expression of CRISPR/Cas9 DNA or RNA. *Nature Communications* 7(1), 1–8. DOI: 10.1038/ncomms12617.

Zhang, Z., Ge, X., Luo, X., Wang, P., Fan, Q. *et al.* (2018) Simultaneous editing of two copies of GH14-3-3D confers enhanced transgene-clean plant defense against Verticillium dahliae in allotetraploid upland cotton. *Frontiers in Plant Science* 9, 356214. DOI: 10.3389/FPLS.2018.00842/BIBTEX.

Zhong, X., Hong, W., Shu, Y., Li, J., Liu, L. *et al.* (2022) CRISPR/Cas9 mediated gene-editing of GmHdz4 transcription factor enhances drought tolerance in soybean (*Glycine max* [L.] Merr.). *Frontiers. Plant Science* 13, 988505. DOI: 10.3389/FPLS.2022.988505/BIBTEX.

Zhou, X., Liao, H., Chern, M., Yin, J., Chen, Y. *et al.* (2018) Loss of function of a rice TPR-domain RNA-binding protein confers broad-spectrum disease resistance. *Proceedings of the National Academy of Sciences of the United States of America* 115(12), 3174–3179. DOI: 10.1073/PNAS.1705927115/SUPPL_FILE/PNAS.1705927115.SD06.XLS.

Zhu, Y., Yang, X., Luo, P., Yan, J., Cao, X. *et al.* (2024) Creation of bacterial blight resistant rice by targeting homologous sequences of Xa13 and Xa25 genes. *Agronomy* 14(4), 800. DOI: 10.3390/AGRONOMY14040800/S1.

13 Microbiome Genomics and Climate Resilience

Ramesha H. Jayaramaiah[1,*], Pankaj K. Singh[2], Simranjit Kaur[3], Premchand Maisnam[4] and Brajesh K. Singh[2,5]

[1]SoilsWest, Centre for Sustainable Farming Systems, Food Futures Institute, Murdoch University, Perth, WA 6150, Australia; [2]The UWA Institute of Agriculture and School of Agriculture & Environment, The University of Western Australia, Perth WA 6001, Australia; [3]Austrian Institute of Technology, Tulln, 3430, Austria; [4]W. K. Kellogg Biological Station, Michigan State University, Michigan, USA; [5]Hawkesbury Institute for the Environment, Western Sydney University, Penrith, NSW 2753, Australia

Abstract

This chapter explores the potential of microbiome genomics as a transformative tool for enhancing crop resilience to climate change. Advances in metagenomics, transcriptomics and synthetic biology have significantly deepened our understanding of microbial communities and their role in improving plant stress tolerance, nutrient acquisition and disease suppression. By examining key microbial traits and metabolic pathways associated with drought tolerance, heat resistance and salinity adaptation, this chapter highlights strategies for optimizing microbial functions to support sustainable agriculture. Additionally, it addresses the challenges of translating genomic discoveries into field applications, proposing solutions to bridge the gap between laboratory research and large-scale implementation. An integrated framework for deploying microbiome-based solutions in climate-resilient agriculture is presented, emphasizing the need for collaboration among researchers, policymakers and agricultural practitioners to ensure effective adoption and scalability.

13.1 Introduction

The microbiome, a dynamic community of microorganisms including bacteria, fungi, archaea and viruses, is fundamental to both plant and soil health. In agricultural systems, these microbial communities influence plant productivity, stress resilience and ecological sustainability. The soil microbiome consists of microbial communities residing in the soil, particularly in the rhizosphere, playing a pivotal role in nutrient cycling, organic matter decomposition and plant-microbe interactions (Liu *et al.*, 2020; Kuang *et al.*, 2023). Meanwhile, plant microbiomes inhabit the plant tissues (endosphere), surfaces (phyllosphere) and root systems (rhizosphere), forming a functional continuum with the soil microbiome (Trivedi *et al.*, 2020). Together, they form a critical component of the plant-soil-microbe interface, shaping plant physiology and metabolism, and resilience to environmental stress. Recent advances in microbiome genomics have uncovered complex, mutually beneficial relationships between plants and their microbiota, offering new

*Corresponding author. Email: Ramesha.HJayaramaiah@murdoch.edu.au

 Genomic Solutions for Breeding Climate-resilient Crops (R.K. Varshney, R. Barmukh and R.J. Harper)
DOI: 10.1079/9781800627185.0013

opportunities for sustainable agriculture (Singh *et al.*, 2020b; Kaul *et al.*, 2021; Trivedi *et al.*, 2021; Hanif *et al.*, 2024). For example, root-associated microbes enhance nutrient acquisition through nitrogen fixation and phosphorus solubilization, while also acting as natural biocontrol agents (Ray *et al.*, 2020). Similarly, soil microbiomes contribute to soil structure by stabilizing aggregates which enhances water retention, regulates greenhouse gas emissions and mediates carbon sequestration, which are critical for combating the adverse effects of climate change (Verma, 2018; Santos and Olivares, 2021). Understanding the distinct yet interconnected roles of plant and soil microbiomes is essential for developing targeted, microbiome-based strategies aimed at improving crop productivity, enhancing resilience to climate-induced stress and supporting the transition to sustainable agricultural systems. This chapter explores how advances in microbiome genomics and associated tools can be harnessed to achieve these goals.

13.2 Plant Microbiomes and their Role in Plant Health and Performance

Plant microbiomes are essential for nutrient acquisition, stress tolerance and disease resistance (Liu *et al.*, 2020; Kuang *et al.*, 2023). These holobionts – the plant and all the microorganisms living in association with it, forming a single ecological unit, possess functional traits that contribute directly to plant resilience in fluctuating environments (Xiong *et al.*, 2021b). Within the plant microbiome, the "core microbiome" consists of microbial taxa that remain consistently associated with a host plant across environments, often contributing structural (e.g. *Bacillus*, *Pseudomonas*) or functional (e.g. pathogen suppression) benefits (Toju *et al.*, 2018; Trivedi *et al.*, 2020). Studies have shown that genetically related plants often harbor similar microbial communities (Ofek-Lalzar *et al.*, 2014; Bulgarelli *et al.*, 2015; Yeoh *et al.*, 2017), supporting the idea of functional redundancy, where multiple microbes can fulfill similar ecological role (Liu *et al.*, 2017; Trivedi *et al.*, 2020). The concept of the "Eco-holobiont" emphasizes that both host genetics and environmental factors such as soil properties and climate, govern microbiome composition (Singh *et al.*, 2020a). Some members of the core microbiome such as *Rhizobiales* and *Pseudomonales*, are found across a wide range of plants, including barley, rice, sugarcane, citrus, grapevine and soybean, reflecting overlap of the core microbiome across microbial kingdoms of both bacteria and fungi (Garrido-Oter *et al.*, 2018; Poole *et al.*, 2018; Lazcano *et al.*, 2021). While others are present only in particular plant species. The core microbiome also includes certain keystone microbial taxa that form the "hub microbiomes", which contribute to stability of core microbiome structure and in some cases, are essential for plant stress tolerance and microbiome assembly (Singh *et al.*, 2020a; Trivedi *et al.*, 2020). The removal of these taxa can disrupt microbiome functionality, leading to negative impacts on plant performance (Hol *et al.*, 2015).

Another important aspect in understanding the plant microbiome composition is the method of microbial transmission in plants, which occurs via vertical and horizontal pathways. Vertical transmission transfers microbes from parent to offspring through seed transmission, sometimes conferring traits such as drought or disease resistance (Lau and Lennon, 2012; Gehring *et al.*, 2017). In contrast, horizontal transmission primarily results from the bulk soil, which serves as the main reservoir from which plants recruit their microbes (Xiong *et al.*, 2021a; Singh *et al.*, 2023b). This transmission results in a significant overlap in microbial populations between soil and plant tissues, particularly in the rhizosphere and roots (Trivedi *et al.*, 2020). This is the reason plant-microbiome composition is influenced by both host and environmental factors, with epiphytes, in contrast to endophytes, being more strongly governed by external environmental conditions (Compant *et al.*, 2019).

13.3 Contributions of the Plant Microbiome in Mitigating Climate Change Impacts on Plants

Climate change significantly affects soil health and microbiomes, which are fundamental to sustainable agricultural productivity. Rising temperatures, altered precipitation patterns and increased atmospheric CO_2 reshape microbial community composition, function and resilience. For example, elevated temperatures accelerate organic matter decomposition and disrupt nutrient cycling and carbon sequestration, favoring heat-tolerant taxa while reducing overall diversity and functionality. This process also enhances microbial respiration, contributing to higher CO_2 emissions (Jansson and Hofmockel, 2020; Novielli *et al.*, 2024). Shifts in precipitation patterns, particularly increased drought intensity, can limit microbial activity and reduce functional redundancy, leading to lower resilience under subsequent environmental changes (Amante and Wedajo, 2024). Conversely, heavy rainfall causes soil erosion and nutrient leaching, further disrupting microbial habitats and community structure (Dubey *et al.*, 2019). Elevated CO_2 alters plant root exudation, thereby reshaping rhizosphere microbial assemblages and their interaction with host plants (Drigo *et al.*, 2009). While elevated CO_2 can stimulate microbial activity and nutrient cycling in some cases, long-term imbalances in soil carbon-nitrogen dynamics can compromise soil fertility and ecosystem stability (Shah *et al.*, 2022).

In response to these climatic pressures, plant-associated microbes adopt multiple mechanisms to enhance host plant tolerance to abiotic and biotic stress. These include the production of osmolytes and exopolysaccharides to enhance water retention under drought, modulation of stress-related phytohormones such as abscisic acid and jasmonic acid, and activation of antioxidant enzymes that reduce oxidative stress. Under saline conditions, microbes contribute to ion homeostasis and mitigate sodium toxicity through compatible solutes and Na^+/H^+ antiporter regulation. During heat stress, they promote expression of heat shock proteins and stabilize plant cellular structures. Additionally, many of these microbes facilitate nutrient acquisition, notably nitrogen and phosphorus through the secretion of siderophores, phosphatases and nitrogen-fixing enzymes. Climate change also intensifies biotic pressure, such as pathogen outbreaks and pest invasions, increasing the risk of crop failures (Boyle *et al.*, 2023). Plant microbiomes can help counter these threats by producing antimicrobial compounds, outcompeting pathogens for space and nutrients, and enhancing plant immunity through induced systemic resistance (ISR) (Dubey *et al.*, 2019; Zhu *et al.*, 2022). Although environmental extremes may influence microbial activity and diversity, growing evidence shows that targeted microbial interventions, such as inoculants, synthetic consortia and microbiome-informed management can restore or enhance beneficial functions under stress conditions (Peixoto *et al.*, 2022). By leveraging microbial traits related to hormonal modulation, antioxidant defense, osmotic adjustment, nutrient mobilization and immune priming, microbiomes present a powerful toolkit for strengthening crop resilience (Jansson and Hofmockel, 2020). The following sections examine how these mechanisms operate under drought, heat and salinity stress.

13.3.1 Microbiome-mediated Drought Resilience

Plant microbiomes play a vital role in enhancing drought resilience through multiple mechanisms, including modifying root architecture, boosting antioxidant production and regulating plant stress response genes, antioxidant enzyme production and secretion of extracellular polysaccharides that aid in soil moisture retention (Jochum *et al.*, 2019; Li *et al.*, 2019). However, most research has been conducted in model plants, limiting direct applications in crop systems (van der Heijden *et al.*, 2008; Jochum *et al.*, 2019; De Vries *et al.*, 2020). Beneficial microbes can induce drought-responsive genes such as aquaporins (TaTIP1;1) and helicases, improving water retention and stress tolerance (Chen *et al.*, 2022). Fungal species such as *Trichoderma hamatum* and *Piriformospora indica* enhance drought adaptation by modulating stomatal conductance, photosynthesis and ROS scavenging (Anith *et al.*, 2011; De Vries *et al.*, 2020; Mosa *et al.*, 2022). Similarly, drought-tolerant strains such as *Bacillus* sp. (12D6) and *Enterobacter* sp. (16i) promote root branching, hormone production and stress-responsive signaling in wheat and maize, significantly improving drought tolerance (Jochum *et al.*, 2019). In native arid ecosystems, plants such as *Alhagi sparsifolia* host distinct drought-adaptive microbial communities. Sequencing studies revealed enriched endosphere and rhizosphere bacteria promoting drought adaptation (Zhang *et al.*, 2020). Further genome sequencing of *Pseudomonas* LTGT-11-2Z uncovered key biosynthetic genes (ACC deaminase, siderophores and T6SS) facilitating biofilm formation and plant-microbe interactions (Zhang *et al.*, 2020). Additionally, the Earth Microbiome Project identified *Bacillus* and *Acidobacteria* species as widespread, stress-resilient microbes across diverse environments, offering potential for scalable drought and heat tolerance solutions (Gilbert *et al.*, 2018). A soil native species combination of *Pseudomonas* sp. MRBP4 and *Bacillus* sp. MRBP10 enhanced maize biomass, leaf area and water retention under 50% reduced irrigation, demonstrating their potential for mitigating drought stress (Ait-El-Mokhtar *et al.*, 2023). Likewise, *Bacillus cereus* (11MN1) and *Bacillus pseudomycoides* (21MN1B) demonstrated the stress response genes catalase (CAT2) and dehydrin (DHN2) modulation and regulation, which play roles in reducing oxidative stress in maize growing under drought conditions (Notununu *et al.*, 2024).

13.3.2 Salt-tolerant Plants with Optimized Microbiomes

Halophilic plant growth-promoting rhizobacteria (PGPR) derived from halophytes have shown great potential in enhancing salinity tolerance in non-halophytic crops. For instance, PGPR isolated from halophytes such as *Sesuvium portulacastrum* and *Arthrocnemum macrostachyum* improve germination, biomass and photosynthetic efficiency under high salinity by producing phytohormones and enhancing nutrient availability (Navarro-Torre *et al.*, 2017; John *et al.*, 2023). Root microbiomes from halophytes such as *Salicornia* are key tools for transferring salt tolerance traits to non-halophytic crops, helping maintain ion homeostasis and mitigate oxidative stress. These functions help crops to grow in saline soils and ensure better resilience to salt-induced stress. Synthetic microbial communities (SynComs), including key taxa from halophyte root microbiomes, have been identified as promising bioinoculants for enhancing crop productivity in saline environments (Akyol *et al.*, 2020). Halophytes such as *Prosopis strombulifera* harbor beneficial microbiomes that help plants adjust osmotically, reduce oxidative damage and enhance nutrient uptake, promoting crop resilience in saline soils (Ruppel *et al.*, 2013). Microbial genera such as *Halomonas* and *Pseudomonas*, often associated with halophytes, offer salt-tolerant properties. Incorporating these halotolerant bacteria into crop production has been shown to enhance growth and resilience under saline conditions (Etesami and Glick, 2020). For example, sugar beet inoculated with halotolerant strains exhibited improved germination, biomass and stress tolerance (Zhou *et al.*, 2017). Leveraging halophyte microbiomes presents a sustainable solution for improving crop resilience to salinity, offering a pathway for more productive agriculture in salinity-prone regions and supporting global food security.

13.3.3 Microbiome Strategies for Heat Stress Tolerance

Heat stress disrupts plant physiology by impairing photosynthesis, protein stability and cellular homeostasis, while also affecting the holobiont. Plant microbiomes play a significant role in alleviating heat stress through numerous mechanisms, including expression of heat shock proteins (HSPs), modulating phytohormone synthesis (e.g. indole acetic acid, cytokinin, salicylic acids), as well as relieving oxidative stress in plants by enhancing activity of antioxidant enzymes such as superoxide dismutase and catalase (Ahmad *et al.*, 2023; Parasar *et al.*, 2024; Anas *et al.*, 2025). Numerous studies have identified and utilized the antagonistic properties of plant growth promoting microorganisms (PGPMs). For example, *Bacillus* and *Pseudomonas* as part of a microbial consortium increased expression of HSPs in maize during early growth stages (Ahmad *et al.*, 2023). In wheat plants, multiple reports highlight the important roles of PGPMs such as *Azospirillum brasilense* NO40 and *Bacillus amyloliquefaciens* UCMB5113, which have been reported to assist wheat seedlings during heat tolerance (Abd El-Daim *et al.*, 2014). Apart from bacteria, inoculation with arbuscular mycorrhizal fungi (AMF) such as *Glomus intraradices* and *Glomus* sp. have been reported to improve performance of pepper plants under heat stress in comparison to non-mycorrhizal pepper plants

(Martin and Stutz, 2004). Similarly, improved resistance to heat stress in tomato plants was observed due to inoculation with two different species of *Septoglomus* (Duc *et al.*, 2018). These studies highlight the immense potential of beneficial microbiota in assisting plants with heat tolerance mechanisms. Microbes with synergistic traits can be further manipulated to design microbial consortia and SynComs to improve plant performance and improve adaptability. Future efforts should focus on optimizing microbe-host compatibility and integrating these strategies with breeding and agronomic practices for scalable field application.

13.3.4 Case Study: Cyanobacteria as Climate-resilient Microbial Group

Cyanobacteria, a diverse group of photosynthetic and nitrogen-fixing microorganisms, contribute to soil health, fertility and plant resilience under environmental stress (Singh *et al.*, 2016; Nawaz *et al.*, 2024). These microbes are well adapted to survive in extreme conditions such as salinity, drought, UV radiation and high temperatures, making them valuable allies in climate-resilient agriculture (Adessi *et al.*, 2018; Curatti *et al.*, 2024). Under salinity stress, cyanobacteria accumulate osmoprotectants such as glycine betaine, while UV radiation triggers the production of protective pigments such as scytonemin. Cyanobacteria also alter their morphology by forming colonies or filaments to survive harsh conditions. Additionally, they improve plant stress tolerance by releasing rhizosphere metabolites, which induce systemic plant responses (Kollmen and Strieth, 2022). At the molecular level, heat shock proteins and flavodoxins protect cyanobacteria from high temperatures, while superoxide dismutase (SOD) scavenges reactive oxygen species (ROS) under low-temperature stress. Their production of bioactive cyclic peptides also exhibits antifungal, pesticidal and pathogen-suppressing properties, making cyanobacteria valuable bioagents for sustainable agriculture (Singh *et al.*, 2016). Their ability to adapt to multiple abiotic stresses highlights their potential in climate-smart farming systems, offering solutions for enhancing soil fertility, reducing chemical inputs and improving resilience under shifting climatic conditions. Before exploring practical tools and applications, it is essential to assess the current state of microbiome research, highlighting existing knowledge gaps, and examine the challenges limiting field scale deployment.

13.4 Current Knowledge and Challenges in Microbiome Applications

13.4.1 Advances in Microbiome Genomics

The rapid evolution of high-throughput sequencing (HTS) technologies has significantly advanced our understanding of plant-microbe interactions, enabling the precise characterization of microbial communities and their roles in stress tolerance. Whole-genome sequencing (WGS), metagenomics and targeted sequencing have facilitated the identification of beneficial microbes, functional genes and pathways that enhance plant resilience to abiotic stresses such as drought, heat and salinity (Knief, 2014). Metagenomic binning has been particularly effective in reconstructing microbial genomes from complex environmental samples, shedding light on previously uncultivable microbes and their ecological functions (Srivastav and Suneja, 2019). Third-generation platforms, such as nanopore and PacBio sequencing, provide long reads and high accuracy, making them invaluable for studying intricate plant microbiomes, including rhizosphere and endosphere communities (Tedersoo *et al.*, 2021). Recent multi-omics approaches, integrating metagenomics, meta-transcriptomics and metabolomics, have further enhanced our ability to predict microbial functions and interactions at an unprecedented scale (Kumar *et al.*, 2023b; Jain *et al.*, 2024; Kimotho and Maina, 2024). These tools have helped identify microbial genes associated with ROS regulation, osmolyte biosynthesis and metabolite production, all of which contribute to plant stress tolerance. For example, metagenomic sequencing of drought-adapted *Bacillus pacificus* G124 revealed its capacity to enhance plant survival by boosting osmolyte accumulation (e.g. proline) and activating antioxidant enzyme pathways (Ma *et al.*, 2024). Similarly, genomic analysis of soybean PP2A-B″

genes demonstrated their critical role in ROS scavenging pathways, mitigating oxidative damage under drought and salinity stress (Xiong *et al.*, 2022).

Despite these advancements, challenges remain in applying microbiome genomics to agriculture at scale. The lack of comprehensive reference genomes and annotated metadata limits accurate taxonomic and functional characterization of microbial communities. Additionally, standardized bioinformatic pipelines are still lacking, creating barriers to data analysis and integration. Computational constraints further restrict access to sophisticated multi-omics tools, particularly in resource-limited settings. Addressing these issues requires Findable, Accessible, Interoperable and Reproducible (FAIR) solutions that improve data sharing and usability in microbiome research (Liao *et al.*, 2024). By overcoming these hurdles, microbiome genomics can be harnessed to develop climate-resilient crops, optimize plant-microbe interactions and support sustainable agricultural practices in the face of global climate change.

13.4.2 Unresolved Questions – Stability of Microbial Communities, Plant-specific Interactions and Scalability

The transition of microbiome-based solutions from controlled research settings to practical, large-scale agricultural applications presents both opportunities and challenges. A major obstacle is the stability and persistence of introduced microbial consortia, which must compete with native soil microbiota and adapt to variable environmental conditions, such as fluctuating moisture, temperature extremes and soil heterogeneity (Qiu *et al.*, 2019; Pino and Griffon, 2024). Many beneficial microbes exhibit strain-specific plant interactions, meaning their efficacy is often host-dependent, requiring tailored solutions for different crops and ecosystems. Scalability is another key issue. Most microbiome studies are conducted under controlled greenhouse or lab conditions, but translating these findings to field environments introduces unpredictable variables that can affect microbial establishment and function. Additionally, regulatory and market constraints slow adoption, as microbial-based solutions often fall outside conventional fertilizer and pesticide approval frameworks (Trivedi *et al.*, 2021). Advances in precision agriculture, such as soil microbiome diagnostics, remote sensing and AI-driven predictive models, are being developed to overcome these uncertainties by enabling real-time monitoring and site-specific interventions (French *et al.*, 2021). Engineered SynComs, tailored to specific crops and environmental conditions provide a promising path forward. However, collaborative research between microbiologists, agronomists and policymakers is essential to ensure these innovations transition effectively from research to practice (Batista and Singh, 2021; Callens *et al.*, 2022). Addressing these challenges requires adaptive microbiome management, where microbial formulations are fine-tuned to specific soil-climate-crop systems, supported by long-term field trials and policy incentives to drive commercial adoption. By integrating microbiome research with smart agricultural technologies, microbial solutions can become scalable, stable and tailored for diverse farming systems, paving the way for climate-resilient and sustainable agriculture.

13.4.3 Challenges in Field Application – Variability in Soil Conditions, Competition with Native Microbes and Reproducibility

While microbiome-based technologies hold great promise for enhancing crop resilience, their field application remains a major challenge due to context-dependent microbial functions and differences between lab and field environments (Bittleston *et al.*, 2020; Egidi *et al.*, 2024; Jayaramaiah *et al.*, 2025). Soil properties, climate variations and plant-microbe interactions strongly influence microbial survival and effectiveness, leading to inconsistent outcomes across different agroecosystems. Laboratory studies help identify microbial traits beneficial to plant growth and stress adaptation, but real-world conditions introduce additional complexities that complicate predictions (Goodrich *et al.*, 2014; Chen *et al.*, 2023). To improve reproducibility, researchers are combining laboratory screening with multi-location field trials. This hybrid approach allows for controlled microbial selection while validating their effectiveness in variable soil-climate conditions (Busby *et al.*, 2017).

Additionally, advanced modeling techniques and AI-driven microbiome predictions are being developed to anticipate microbial adaptation in different environments (McPherson *et al.*, 2018; Song *et al.*, 2020).

Another compounding issue is the complex interplay between introduced microbes and native soil microbial communities. Introduced microbial consortia often face competition from established microbial populations, which may inhibit their colonization and functionality. Further, inconsistencies in inoculum preparation and application methods hinder reproducibility (Albright *et al.*, 2022; Fachrul *et al.*, 2022). Unlike chemical inputs, which often deliver immediate and targeted results, microbiome-based solutions function as ecological interventions that require optimization through integrated crop management practices. Additionally, variability in environmental conditions, such as prolonged droughts or unexpected pathogen outbreaks, may diminish the short-term impacts of introduced microbial consortia (Pino and Griffon, 2024). Addressing these challenges requires long-term field trials, standardized methodologies and predictive tools such as genome-scale metabolic models, to optimize microbial applications for diverse environments (Ankrah *et al.*, 2021). By refining field application methods, improving microbial stability and integrating predictive models, microbiome-based interventions can transition from experimental concepts to scalable agricultural solutions, ensuring they deliver consistent benefits across diverse climates and soil types (Table 13.1).

13.4.4 Regulatory and Policy Considerations

The widespread adoption of microbiome-based solutions in agriculture requires robust regulatory frameworks to ensure biosafety, efficacy and environmental sustainability. While microbial inoculants and biofertilizers are inherently ecological in nature, they pose potential biological risks, including unintended ecosystem disruptions, horizontal gene transfer and microbial instability (Ke *et al.*, 2021; Vassileva *et al.*, 2022). Recognizing these challenges, global regulatory initiatives are emerging to integrate microbiome-based alternatives into mainstream farming. Recently, the European Commission put regulatory frameworks in place to legalize the integration of microbial solutions into farming to achieve a 50% reduction in chemicals and fertilizers in farming practices (Sundh *et al.*, 2021). Countries such as the USA, Brazil, Australia and India have introduced regulatory provisions for microbial applications such as biofertilizers, biopesticides and microbial inoculants. However, many of these policies were originally designed for chemical inputs, making regulatory approval complex and time-consuming for microbiome-based innovations. As climate change accelerates, adaptive regulatory frameworks must evolve to support microbiome-based technologies, ensuring they contribute to climate-resilient, low-input and sustainable agriculture while maintaining long-term biosafety and compliance.

13.5 Microbiome Informed Tools for Climate Resilience

Building on the current understanding and the gaps identified above, recent innovation has enabled the development of microbiome-informed tools to support climate-resilient agriculture. These tools aim to harness microbial functionality for enhancing crop performance. Key strategies include microbial inoculants, SynComs, metabolite-based bioinputs and breeding strategies, and are rapidly evolving to address the limitations identified in current practices. These innovations aim to translate microbial functionality into field-ready solutions for sustainable agriculture.

13.5.1 Microbial Inoculants and Biofertilizers

Advancements in genomics-driven microbiome research have underscored the potential of microbial inoculants and biofertilizers in enhancing crop resilience to climate stressors. Beneficial microbes, including plant growth-promoting rhizobacteria (PGPR), mycorrhizal fungi and rhizobia, contribute to nutrient acquisition, drought tolerance, salinity resistance and thermal adaptation (Backer *et al.*, 2018; Heydari

Table 13.1. Microbiome-based interventions for climate resilience. (Authors' own table.)

Intervention	Target crop	Description	Stress addressed	Outcome	References
Microbial consortia (PGPMs and SynComs)	Maize Wheat Rice Tomato	Engineered or naturally assembled microbial communities to enhance resilience	Drought, salinity, heat	Improved growth, water-use efficiency, and stress tolerance	(Durán *et al.*, 2018; Tabacchioni *et al.*, 2021; Wang *et al.*, 2022; Ahmad *et al.*, 2023)
Microbial inoculants and biofertilizers	Various	Beneficial microbes enhancing nutrient uptake and stress adaptation	Drought, salinity	Increased root biomass, nutrient availability, and yield stability	(Nephali *et al.*, 2021; Morcillo *et al.*, 2022; Noureen *et al.*, 2024)
PGPR and endophytes	Wheat Rice Legumes	Growth-promoting microbes improving drought and salinity tolerance	Drought, salinity	Enhanced root development, nutrient uptake, and osmotic adjustment	(Backer *et al.*, 2018; Fatima *et al.*, 2022; Al-Turki *et al.*, 2023; Ben Gaied *et al.*, 2024)
Biochar and microbial synergies	Wheat Maize	Biochar combined with PGPR for enhanced stress resilience	Drought	Increased soil moisture retention, root growth, and crop productivity	(Heydari *et al.*, 2023; Anbuganesan *et al.*, 2024; Noureen *et al.*, 2024)
Microbiome manipulation via soil practices	Rice Alfalfa Sorghum	Selective recruitment of beneficial microbes through conservation agriculture	Drought	Improved root colonization, water-use efficiency and stress adaptation	(Santos-Medellín *et al.*, 2021; Fan *et al.*, 2023; Pino and Griffon, 2024)
Phosphate-solubilizing microbes	Sorghum Legumes	Microbial solubilization of phosphorus under drought conditions	Drought, nutrient deficiency	Increased nutrient uptake, stress tolerance, and plant vigor	(Kour *et al.*, 2020; Wu *et al.*, 2024)
Microbial volatile organic compounds (MVOCs)	Various	Volatile compounds from microbes enhancing plant defense	Heat, drought	Improved plant signaling, antioxidant activity and stress response	(Kanchiswamy *et al.*, 2015; Montejano-Ramírez *et al.*, 2024)

et al., 2023). These microbes improve water-use efficiency, root development and plant immune responses, making them critical for sustainable agriculture, particularly in arid and degraded soils (Pan *et al.*, 2020). Genomic-assisted breeding programs now integrate plant-microbe interactions to optimize microbial recruitment, leading to holobiont-based breeding approaches that enhance stress tolerance and disease resistance (Wille *et al.*, 2019). Moreover, microbiome engineering, using synthetic biology and directed evolution, has led to the development of next-generation microbial consortia as biofertilizers and biostimulants, reducing reliance on chemical inputs (Pirttilä *et al.*, 2021). Integrating genomic tools, AI-driven predictive models and metagenomic screening allows for the identification of stress-resilient microbial strains, tailoring their application for enhanced soil health and crop productivity (Srivastava *et al.*, 2023). Additionally, research has shown that combining microbial inoculants with reduced chemical fertilizers improves crop yields and maintains soil microbiome stability (He *et al.*, 2024). This highlights the potential of microbial biofertilizers as a sustainable strategy for climate-resilient agriculture.

13.5.2 Synthetic Microbial Consortia (SynComs)

Synthetic microbial consortia are engineered microbial communities designed to enhance plant resilience against climate-induced stresses such as drought, salinity and extreme temperatures by optimizing nutrient uptake, root development and stress-responsive pathways (Martins *et al.*, 2023). Advances in metagenomics and functional genomics have identified microbial genes linked to ion homeostasis, osmoregulation and oxidative stress resistance, while horizontal gene transfer (HGT) has facilitated the exchange of beneficial stress-adaptive traits, including glutathione S-transferase (GST) and Na+/proline symporters, which enhance salinity and oxidative stress tolerance in plants (Yin *et al.*, 2022). The design of SynComs involves selecting core microbial taxa with complementary traits, including biofilm formation, secondary metabolite production and pathogen suppression, ensuring their stability and function in diverse agroecosystems (Martins *et al.*, 2023). These engineered microbial communities, often referred to as "DefenseBiomes," integrate multiple stress-adaptive microbes to deliver broad-spectrum resilience (de Souza *et al.*, 2020). Advances in machine learning and AI-driven metagenomic analysis have further refined the predictive design of SynComs, enabling the selection of optimal microbial combinations that improve crop performance under varying environmental conditions (Han, 2024). However, despite their potential, challenges such as field reproducibility, microbial competition and environmental variability hinder large-scale adoption (Shayanthan *et al.*, 2022). Integrating SynComs with other genomic tools, such as genome editing and predictive breeding, could provide next-generation microbiome-based solutions for climate-resilient agriculture, reducing reliance on chemical inputs while improving crop sustainability (de Souza *et al.*, 2020).

13.5.3 Microbial Metabolite-based Inputs

Root exudates and microbial secondary metabolites play a crucial role in shaping rhizosphere microbial interactions, which in turn influence plant stress responses (Figure 13.1). However, the composition, signaling mechanisms and specific roles of these exudates in stress adaptation remain underexplored (Lamichhane *et al.*, 2024; Mukherjee *et al.*, 2024). These exudates include primary metabolites (sugars, amino acids, organic acids) and secondary metabolites (flavonoids, phenolics and terpenoids) that act as chemical signals to recruit beneficial microbes and regulate soil pH, nutrient solubilization and microbial community composition (Baetz and Martinoia, 2014; Kumar *et al.*, 2023a; Lamichhane *et al.*, 2024). A few examples to understand this include the role of secreted flavonoids which induce Rhizobia-nodule formation (Mukherjee *et al.*, 2024). Plant microbiome manages pathogen invasion in root/soil via regulation of decreased salicylic acid concentrations, while jasmonic acid concentrations can protect against pathogens. Flavonoids play a key role in regulating plant growth under water-stressed conditions (Patil *et al.*, 2024). Additionally, citric acid and

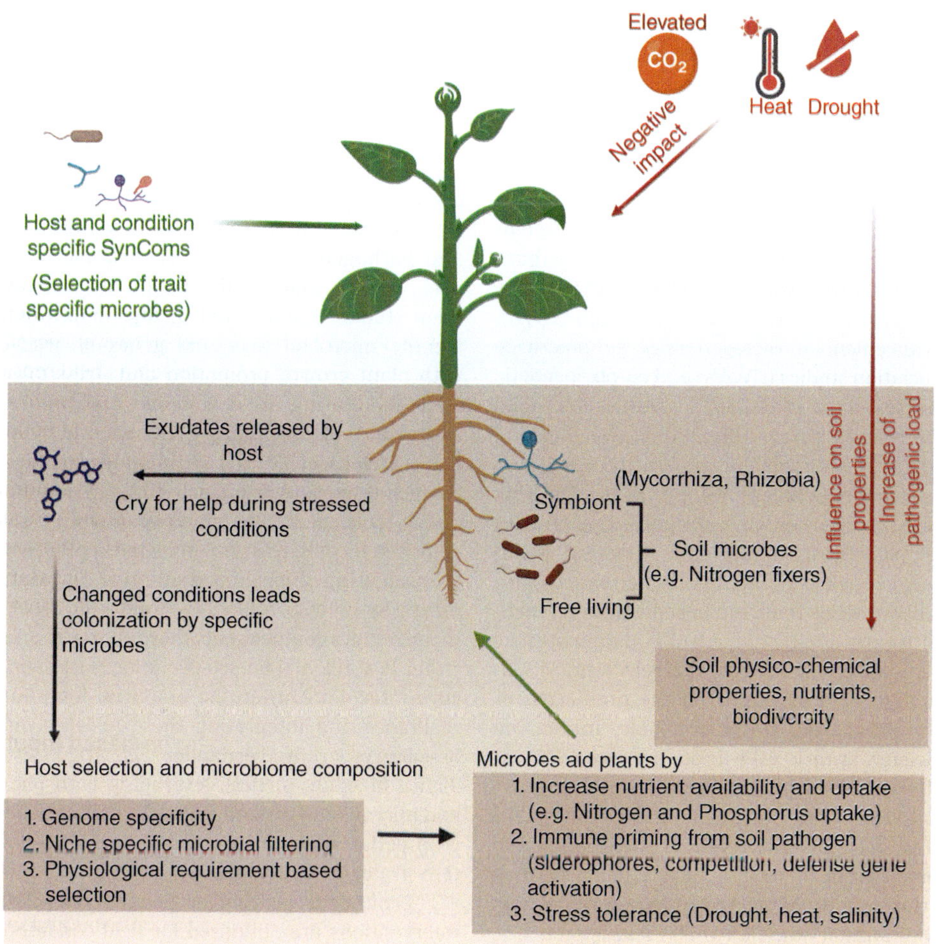

Figure 13.1. A conceptual framework illustrating how climate change stressors impact plants, triggering the release of exudates to recruit microbes. These exudates provide chemical signals and favorable conditions for beneficial microbes to colonize. Microbial recruitment is genome and niche specific, and is done based on the physiological requirement of the host plant. Recruited microbes aid the plant by enhancing nutrient availability, strengthening immune response and improving the stress tolerance. Based on their host and functional traits, these microbes can be further engineered into synthetic microbial communities to optimize plant resilience and agriculture sustainability. (Authors' own image.)

malic acid secretion mediates the recruitment of a higher number of Bacilli in chickpea crops (Mukherjee *et al.*, 2024). As discussed earlier, the role of *Bacillus* in drought resilience, combine with their spore-forming trait, enhances their survival under various abiotic stresses. Advances in metabolomics and root exudate profiling have allowed for the targeted manipulation of microbial recruitment, offering promising strategies for metabolite-driven bioinoculants (Salem *et al.*, 2022). Harnessing these microbial interactions through metabolite-based formulations provides an alternative to traditional fertilizers while enhancing plant resilience against climate stressors.

13.5.4 Microbiome-informed Plant Breeding

Microbiome-informed breeding represents an innovative approach to enhancing crop resilience and productivity by integrating plant genetics with the functional capabilities of their associated microbiomes. Holobiont breeding, which considers plants and their microbiomes as a single functional unit, offers a promising strategy for improving stress tolerance, nutrient use efficiency and disease resistance. Plants optimize the recruitment of beneficial microbes through traits such as root exudation patterns, enhancing nutrient cycling, pathogen suppression and stress resilience (Gopal and Gupta, 2016). Advancements in metagenomics, genome-wide association studies (GWAS) and co-phylogenetic analyses have identified keystone microbial species linked to plant functional traits, paving the way for microbiome-enabled breeding and SynCom development (Singh *et al.*, 2020b). These tailored SynComs, designed to align with host plant genetics and environmental conditions, act as bio-stimulants that enhance plant resilience while reducing dependence on chemical inputs (Arif *et al.*, 2020). For instance, microbiome-enabled genomic selection (MEGS) has significantly improved the prediction of nitrogen-related traits in maize, increasing efficiency by up to 40% under low-nitrogen conditions – demonstrating the critical role of beneficial rhizosphere microbes in nutrient uptake (Yang *et al.*, 2024). Despite its potential, microbiome-informed breeding faces challenges such as the stability of microbial partnerships and environmental variability. Multi-omics approaches, including genomics, metabolomics and machine learning-based predictive models, are crucial for refining microbial selection and ensuring consistency across agroecosystems (Ait-El-Mokhtar and Baslam, 2023). Integrating microbiome-based strategies with traditional crop improvement methods holds significant potential for plant resilience, reducing reliance on chemical fertilizers and aligning agriculture with global sustainability goals in the face of climate change and soil degradation.

13.6 Future Prospects: Toward Integrated Microbiome-based Strategies for Crop Resilience

The future of climate-resilient agriculture lies in the seamless integration of microbiome-based solutions with conventional crop improvement strategies encompassing agronomy, microbiology, ecology, genomics and plant sciences. Advances in genome editing, such as CRISPR-Cas, enable targeted modifications in plant traits to enhance microbial symbioses, optimizing root exudation patterns to recruit beneficial microbes for improved nutrient uptake, drought tolerance and pathogen suppression (Pourkheirandish *et al.*, 2020). Additionally, genome-wide association studies (GWAS) and metagenomics help identify microbial taxa and genes associated with plant growth promotion and stress tolerance, facilitating the development of tailored microbial tools, including SynComs and bioinoculants, for specific crop and climate conditions (Nwachukwu and Babalola, 2022). Computational tools, including machine learning and AI-driven models, refine microbiome applications by predicting microbial dynamics and optimizing SynCom formulations (Kaul *et al.*, 2021). Pangenomics complements these efforts by analyzing host-microbiome co-evolution, identifying core microbial functions essential for plant resilience and integrating these insights into breeding programs (Fernandez *et al.*, 2021). Predictive breeding models, leveraging both plant and microbiome genomic data, further enhance crop performance forecasts and stress adaptation strategies (Thriveni *et al.*, 2024).

Ecological principles ensure microbial interventions are optimized for field conditions (Table 13.1). Sustainable practices such as conservation agriculture, reduced chemical inputs and soil health management promote microbiome functionality and resilient agroecosystems (Trivedi *et al.*, 2021). Precision irrigation, biofertilizers and microbiome-informed breeding further enhance crop resilience while minimizing environmental impact (Pattnaik *et al.*, 2021). Aligning these strategies with global sustainability initiatives, such as One Health policies, can boost agricultural productivity and ecosystem restoration, and reduce reliance on chemical inputs (Singh *et al.*, 2023b). A structured roadmap is essential for translating scientific discoveries into real-world agricultural applications, integrating multi-omics, AI-driven modeling and microbiome engineering to enhance nutrient cycling, stress resilience and disease resistance (Figure 13.2). Key priorities include continuous monitoring through multi-omics technologies,

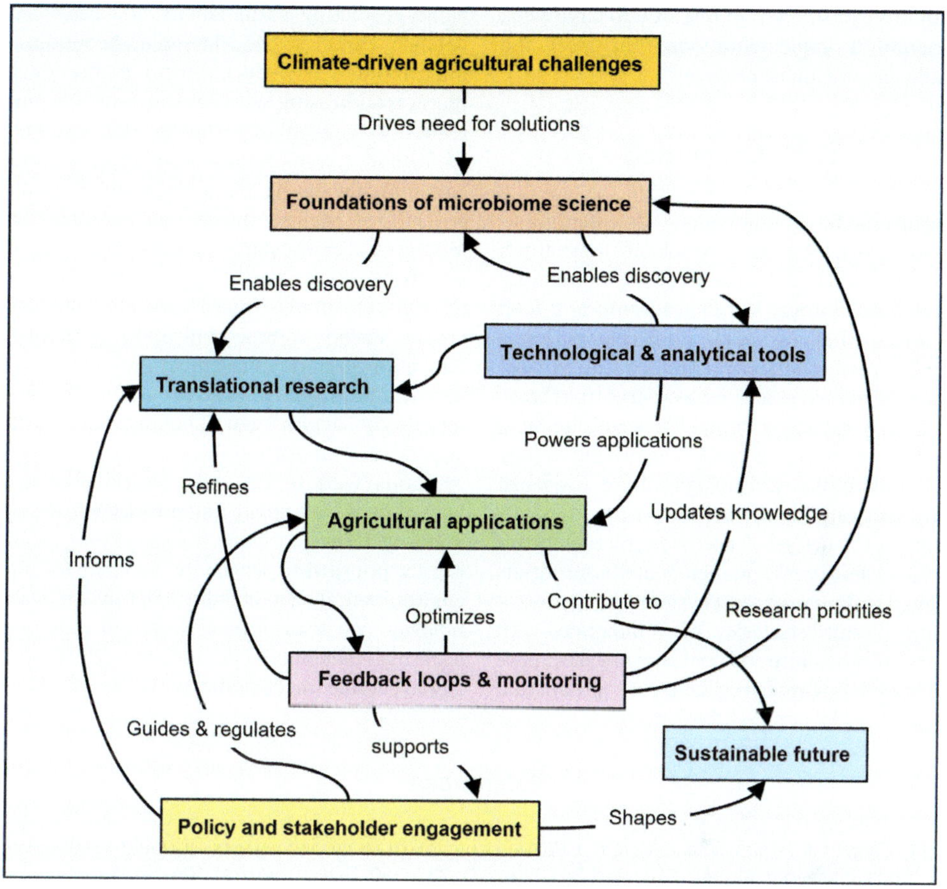

Figure 13.2. A novel framework for integrating microbiome science with climate-resilient agriculture. This roadmap presents a structured approach to enhance climate-resilient agriculture by bridging microbiome science with real-world farming practices. *Climate-driven agricultural challenges*, including increasing climate variability and rising food demand, necessitate innovative microbial solutions. *Foundations of microbiome science* encompasses genomic technologies (high-throughput sequencing, multi-omics) and microbial ecology (interaction networks, community dynamics), which inform translational applications. *Translational research* bridges the gap between basic science and practical applications through the development of microbial inoculants, identification of stress-tolerant strains and creation of predictive models. *Technological and analytical tools*, including real-time monitoring systems, bioinformatics pipelines, machine learning and AI-driven analytics, facilitate data-driven decision-making across research and implementation phases. *Agricultural applications* represent the practical field implementation of microbiome science, focusing on four key areas: climate stress resilience (drought, salinity, heat), nutrient cycling optimization, soil carbon sequestration and biological pest management. *Policy and stakeholder engagement* facilitates implementation through scientific collaboration networks and evidence-based policy guidelines. *Feedback loops and monitoring* enable adaptive management, creating an iterative improvement cycle across all system components. *Sustainable future* represents the goals to achieve enhanced crop resilience to climate stressors, improved ecosystem sustainability and strengthened food security through microbially optimized farming practices. Directional arrows indicate information flow and influence patterns, while feedback loops demonstrate the adaptive nature of the system. (Authors' own image.)

stakeholder collaboration to drive adoption and policy incentives to accelerate microbiome-based innovations. The successful adoption of these solutions requires interdisciplinary collaboration among breeders, microbiologists, ecologists and computational biologists, ensuring that microbial solutions transition from research to large-scale agricultural implementations that meet the challenges of climate change.

13.7 Conclusion

Microbiome-based innovations are emerging as a transformative frontier in sustainable agriculture, offering solutions to the growing challenges of climate change, soil degradation and food security. By harnessing plant-microbiome interactions, researchers can enhance nutrient acquisition, disease resistance and stress tolerance, ultimately reducing dependence on chemical inputs. Advances in genomics, synthetic biology and multi-omics technologies have deepened our understanding of microbiome functionality, enabling the design of precision microbial consortia, biofertilizers and metabolite-based inputs tailored to diverse agroecosystems. These microbiome-informed strategies, when integrated with precision agriculture and predictive breeding, pave the way for climate-resilient crop production while promoting sustainability and ecological balance. However, the widespread adoption of microbiome-based agricultural technologies requires addressing key technical, economic and regulatory challenges. Ensuring consistent field performance, establishing standardized protocols and navigating biosafety regulations are crucial for scaling microbiome applications. Moreover, interdisciplinary collaboration among scientists, policymakers and agricultural practitioners is essential to bridge knowledge gaps and create viable commercialization pathways. Education, farmer engagement and policy support will be critical in fostering acceptance and large-scale deployment. Looking ahead, the successful integration of microbiome-based solutions with holistic agricultural practices, backed by robust policy frameworks and innovations in delivery systems, holds immense promise. By prioritizing scalability, accessibility and environmental sustainability, microbiome technologies can transform agricultural practices, fostering resilient food production in the face of global climate uncertainties.

References

Abd El-Daim, I.A., Bejai, S. and Meijer, J. (2014) Improved heat stress tolerance of wheat seedlings by bacterial seed treatment. *Plant and Soil* 379(1–2), 337–350. DOI: 10.1007/s11104-014-2063-3.

Adessi, A., Cruz de Carvalho, R., De Philippis, R., Branquinho, C. and Marques da Silva, J. (2018) Microbial extracellular polymeric substances improve water retention in dryland biological soil crusts. *Soil Biology and Biochemistry* 116, 67–69. DOI: 10.1016/J.SOILBIO.2017.10.002.

Ahmad, M., Imtiaz, M., Nawaz, M.S., Mubeen, F., Sarwar, Y. *et al.* (2023) Thermotolerant PGPR consortium B3P modulates physio-biochemical and molecular machinery for enhanced heat tolerance in maize during early vegetative growth. *Annals of Microbiology* 73(1), 34. DOI: 10.1186/s13213-023-01736-5.

Ait-El-Mokhtar, M. and Baslam, M. (2023) Holo-omic applications to unveil microbiome shaping sustainable horticultural production. *Frontiers in Sustainable Food Systems* 7, 1151367. DOI: 10.3389/FSUFS.2023.1151367.

Ait-El-Mokhtar, M., Ben-Laouane, R., Anli, M., Subramanian, S., Babalola, O.O. and Ojuederie, O.B. (2023) Growth enhancement and extenuation of drought stress in maize inoculated with multifaceted ACC deaminase producing rhizobacteria. *Frontiers in Sustainable Food Systems* 6, 1076844. DOI: 10.3389/FSUFS.2022.1076844.

Akyol, T.Y., Sato, S. and Turkan, I. (2020) Deploying root microbiome of halophytes to improve salinity tolerance of crops. *Plant Biotechnology Reports* 14(2), 143–150. DOI: 10.1007/S11816-020-00594-W.

Albright, M.B.N., Louca, S., Winkler, D.E., Feeser, K.L., Haig, S.J. *et al.* (2022) Solutions in microbiome engineering: Prioritizing barriers to organism establishment. *The ISME Journal* 16(2), 331–338. DOI: 10.1038/S41396-021-01088-5.

Al-Turki, A., Murali, M., Omar, A.F., Rehan, M. and Sayyed, R.Z. (2023) Recent advances in PGPR-mediated resilience toward interactive effects of drought and salt stress in plants. *Frontiers in Microbiology* 14, 1214845. DOI: 10.3389/FMICB.2023.1214845.

Amante, G. and Wedajo, M. (2024) Impacts of climate change on soil microbial diversity, distribution and abundance. *International Journal on Food, Agriculture and Natural Resources* 5(2), 158–168. DOI: 10.46676/IJ-FANRES.V5I2.342.

Anas, M., Khalid, A., Saleem, M.H., Ali Khan, K., Ahmed Khattak, W. and Fahad, S. (2025) Symbiotic synergy: Unveiling plant-microbe interactions in stress adaptation. *Journal of Crop Health* 77(1), 18. DOI: 10.1007/s10343-024-01070-z.

Anbuganesan, V., Vishnupradeep, R., Bruno, L.B., Sharmila, K., Freitas, H. and Rajkumar, M. (2024) Combined application of biochar and plant growth-promoting rhizobacteria improves heavy metal and drought stress tolerance in *Zea mays*. *Plants* 13(8), 1143. DOI: 10.3390/PLANTS13081143.

Anith, K.N., Faseela, K.M., Archana, P.A. and Prathapan, K.D. (2011) Compatibility of *Piriformospora indica* and *Trichoderma harzianum* as dual inoculants in black pepper (*Piper nigrum* L.). *Symbiosis* 55(1), 11–17. DOI: 10.1007/S13199-011-0143-1.

Ankrah, N.Y.D., Bernstein, D.B., Biggs, M.B., Carey, M.A., Engevik, M. *et al.* (2021) Enhancing microbiome research through genome-scale metabolic modeling. *mSystems* 6, 00599–21. DOI: 10.1128/MSYSTEMS.00599-21.

Arif, I., Batool, M. and Schenk, P.M. (2020) Plant microbiome engineering: Expected benefits for improved crop growth and resilience. *Trends in Biotechnology* 38(12), 1385–1396. DOI: 10.1016/J.TIBTECH.2020.04.015.

Backer, R., Rokem, J.S., Ilangumaran, G., Lamont, J., Praslickova, D. *et al.* (2018) Plant growth-promoting rhizobacteria: Context, mechanisms of action, and roadmap to commercialization of biostimulants for sustainable agriculture. *Frontiers in Plant Science* 871, 402666. DOI: 10.3389/FPLS.2018.01473.

Baetz, U. and Martinoia, E. (2014) Root exudates: The hidden part of plant defense. *Trends in Plant Science* 19(2), 90–98. DOI: 10.1016/J.TPLANTS.2013.11.006.

Batista, B.D. and Singh, B.K. (2021) Realities and hopes in the application of microbial tools in agriculture. *Microbial Biotechnology* 14(4), 1258–1268. DOI: 10.1111/1751-7915.13866.

Ben Gaied, R., Sbissi, I., Tarhouni, M. and Brígido, C. (2024) Bacterial endophytes from legumes native to arid environments are promising tools to improve Mesorhizobium-chickpea symbiosis under salinity. *Biology* 13, 96. DOI: 10.3390/BIOLOGY13020096.

Bittleston, L.S., Gralka, M., Leventhal, G.E., Mizrahi, I. and Cordero, O.X. (2020) Context-dependent dynamics lead to the assembly of functionally distinct microbial communities. *Nature Communications* 11(1), 1–10. DOI: 10.1038/s41467-020-15169-0.

Boyle, J.A., Murphy, B.K., Ensminger, I., Stinchcombe, J.R. and Frederickson, M.E. (2023) Resistance and resilience of soil microbiomes under climate change. *Ecosphere* 15, e70077. DOI: 10.1002/ECS2.70077.

Bulgarelli, D., Garrido-Oter, R., Münch, P.C., Weiman, A., Dröge, J. *et al.* (2015) Structure and function of the bacterial root microbiota in wild and domesticated barley. *Cell Host and Microbe* 17(3), 392–403. DOI: 10.1016/j.chom.2015.01.011.

Busby, P.E., Soman, C., Wagner, M.R., Friesen, M.L., Kremer, J. *et al.* (2017) Research priorities for harnessing plant microbiomes in sustainable agriculture. *PLoS Biology* 15, e2001793. DOI: 10.1371/JOURNAL.PBIO.2001793.

Callens, K., Fontaine, F., Sanz, Y., Bogdanski, A., D'Hondt, K. *et al.* (2022) Microbiome-based solutions to address new and existing threats to food security, nutrition, health and agrifood systems' sustainability. *Frontiers in Sustainable Food Systems* 6, 1047765. DOI: 10.3389/FSUFS.2022.1047765.

Chen, Y., Yao, Z., Sun, Y., Wang, E., Tian, C. *et al.* (2022) Current studies of the effects of drought stress on root exudates and rhizosphere microbiomes of crop plant species. *International Journal of Molecular Sciences* 23, 2374. DOI: 10.3390/IJMS23042374.

Chen, J., Zhang, Y., Kuzyakov, Y., Wang, D. and Olesen, J.E. (2023) Challenges in upscaling laboratory studies to ecosystems in soil microbiology research. *Global Change Biology* 29(3), 569–574. DOI: 10.1111/GCB.16537.

Compant, S., Samad, A., Faist, H. and Sessitsch, A. (2019) A review on the plant microbiome: Ecology, functions, and emerging trends in microbial application. *Journal of Advanced Research* 19, 29–37. DOI: 10.1016/j.jare.2019.03.004.

Curatti, L., Do Nascimento, M., Pagnussat, L.A., Sanchez Rizza, L., Sanchez, A.O., Garcia Martinez, L. and Hernandez, J.A. (2024) Filamentous nitrogen-fixing cyanobacteria: Contributing to filling nitrogen and water gaps in a context of climate change. *Reviews in Environmental Science and Bio/Technology* 23(2), 291–320. DOI: 10.1007/S11157-024-09690-7.

de Souza, R.S.C., Armanhi, J.S.L. and Arruda, P. (2020) From microbiome to traits: Designing synthetic microbial communities for improved crop resiliency. *Frontiers in Plant Science* 11, 01179. DOI: 10.3389/FPLS.2020.01179.

De Vries, F.T., Griffiths, R.I., Knight, C.G., Nicolitch, O. and Williams, A. (2020) Harnessing rhizosphere microbiomes for drought-resilient crop production. *Science* 368(6488), 270–274. DOI: 10.1126/SCIENCE.AAZ5192.

Drigo, B., Van Veen, J.A. and Kowalchuk, G.A. (2009) Specific rhizosphere bacterial and fungal groups respond differently to elevated atmospheric CO_2. *The ISME Journal* 3(10), 1204–1217. DOI: 10.1038/ISMEJ.2009.65.

Dubey, A., Malla, M.A., Khan, F., Chowdhary, K., Yadav, S. *et al.* (2019) Soil microbiome: A key player for conservation of soil health under changing climate. *Biodiversity and Conservation* 28(8–9), 2405–2429. DOI: 10.1007/S10531-019-01760-5.

Duc, N.H., Csintalan, Z. and Posta, K. (2018) Arbuscular mycorrhizal fungi mitigate negative effects of combined drought and heat stress on tomato plants. *Plant Physiology and Biochemistry* 132, 297–307. DOI: 10.1016/j.plaphy.2018.09.011.

Durán, P., Thiergart, T., Garrido-Oter, R., Agler, M., Kemen, E., Schulze-Lefert, P. and Hacquard, S. (2018) Microbial interkingdom interactions in roots promote *Arabidopsis* survival. *Cell* 175(4), 973–983.e14. DOI: 10.1016/J.CELL.2018.10.020.

Egidi, E., Bristol, D., Hassan, K., Tissue, D., Wright, I.J. and Nielsen, U.N. (2024) The influence of plant traits on soil microbial communities varies between arid and mesic grasslands. *Plant and Soil* 509, 501–521. DOI: 10.1007/S11104-024-06876-4.

Etesami, H. and Glick, B.R. (2020) Halotolerant plant growth–promoting bacteria: Prospects for alleviating salinity stress in plants. *Environmental and Experimental Botany* 178, 104124. DOI: 10.1016/J.ENVEXPBOT.2020.104124.

Fachrul, M., Méric, G., Inouye, M., Pamp, S.J. and Salim, A. (2022) Assessing and removing the effect of unwanted technical variations in microbiome data. *Scientific Reports* 12(1), 1–15. DOI: 10.1038/s41598-022-26141-x.

Fan, W., Tang, F., Wang, J., Dong, J., Xing, J. and Shi, F. (2023) Drought-induced recruitment of specific root-associated bacteria enhances adaptation of alfalfa to drought stress. *Frontiers in Microbiology* 14, 1114400. DOI: 10.3389/FMICB.2023.1114400.

Fatima, T., Sharma, S., Bano, A., Srivastava, D., Verma, I. and Singh, P.C. (2022) Microbial endophytes: A hidden plant resident, application and their role in abiotic stress management in plants. *Journal of Ecophysiology and Occupational Health* 22(3), 127–140. DOI: 10.18311/JEOH/2022/30619.

Fernandez, C.T., Marsh, J., Danilevicz, M.F., Mercé, C. and Edwards, D. (2021) Application of pangenomics for wheat molecular breeding. In *Molecular Breeding in Wheat, Maize and Sorghum: Strategies for Improving Abiotic Stress Tolerance and Yield*, CABI, Wallingford UK, pp. 236–246. DOI: 10.1079/9781789245431.0013.

French, E., Kaplan, I., Iyer-Pascuzzi, A., Nakatsu, C.H. and Enders, L. (2021) Emerging strategies for precision microbiome management in diverse agroecosystems. *Nature Plants* 7(3), 256–267. DOI: 10.1038/S41477-020-00830-9.

Garrido-Oter, R., Nakano, R.T., Dombrowski, N., Ma, K.W., McHardy, A.C. and Schulze-Lefert, P. (2018) Modular traits of the Rhizobiales root microbiota and their evolutionary relationship with symbiotic rhizobia. *Cell Host and Microbe* 24(1), 155–167.e5. DOI: 10.1016/j.chom.2018.06.006.

Gehring, C.A., Sthultz, C.M., Flores-Rentería, L., Whipple, A.V. and Whitham, T.G. (2017) Tree genetics defines fungal partner communities that may confer drought tolerance. *Proceedings of the National Academy of Sciences USA* 114(42), 11169–11174. DOI: 10.1073/pnas.1704022114.

Gilbert, J.A., Jansson, J.K. and Knight, R. (2018) Earth microbiome project and global systems biology. *mSystems* 3, 10–1128. DOI: 10.1128/mSystems.00217-17.

Goodrich, J.K., Di Rienzi, S.C., Poole, A.C., Koren, O., Walters, W.A. *et al.* (2014) Conducting a microbiome study. *Cell* 158(2), 250–262. DOI: 10.1016/J.CELL.2014.06.037.

Gopal, M. and Gupta, A. (2016) Microbiome selection could spur next-generation plant breeding strategies. *Frontiers in Microbiology* 7(DEC), 209912. DOI: 10.3389/FMICB.2016.01971.

Han, L. (2024) Optimizing synthetic microbial communities for sustainable agriculture: Design, functionality, and field performance. *Molecular Microbiology Research*. DOI: 10.5376/MMR.2024.14.0004.

Hanif, M.S., Tayyab, M., Baillo, E.H., Islam, M.M., Islam, W. and Li, X. (2024) Plant microbiome technology for sustainable agriculture. *Frontiers in Microbiology* 15, 1500260. DOI: 10.3389/FMICB.2024.1500260.

He, S., Zhang, Y., Yang, X., Li, Q., Li, C. and Yao, T. (2024) Effects of microbial inoculants combined with chemical fertilizer on growth and soil nutrient dynamics of timothy (*Phleum pratense* L.). *Agronomy* 14, 1016. DOI: 10.3390/AGRONOMY14051016.

Heydari, M., Hajinia, S., Jafarian, N., Karamian, M., Mosa, Z. *et al.* (2023) Synergistic use of biochar and the plant growth-promoting rhizobacteria in mitigating drought stress on oak (*Quercus brantii* Lindl.) seedlings. *Forest Ecology and Management* 531, 120793. DOI: 10.1016/J.FORECO.2023.120793.

Hol, W.H.G., de Boer, W., de Hollander, M., Kuramae, E.E., Meisner, A. and van der Putten, W.H. (2015) Context dependency and saturating effects of loss of rare soil microbes on plant productivity. *Frontiers in Plant Science* 6, 485. DOI: 10.3389/fpls.2015.00485.

Jain, A., Sarsaiya, S., Singh, R., Gong, Q., Wu, Q. and Shi, J. (2024) Omics approaches in understanding the benefits of plant-microbe interactions. *Frontiers in Microbiology* 15, 1391059. DOI: 10.3389/FMICB.2024.1391059.

Jansson, J.K. and Hofmockel, K.S. (2020) Soil microbiomes and climate change. *Nature Reviews Microbiology* 18(1), 35–46. DOI: 10.1038/S41579-019-0265-7.

Jayaramaiah, R.H., Martins, C.S.C., Egidi, E., Macdonald, C.A., Wang, J.T. *et al.* (2025) Soil function-microbial diversity relationship is impacted by plant functional groups under climate change. *Soil Biology and Biochemistry* 200, 109623. DOI: 10.1016/J.SOILBIO.2024.109623.

Jochum, M.D., McWilliams, K.L., Pierson, E.A. and Jo, Y.K. (2019) Host-mediated microbiome engineering (HMME) of drought tolerance in the wheat rhizosphere. *PLoS ONE* 14, e0225933. DOI: 10.1371/JOURNAL.PONE.0225933.

John, J.E., Maheswari, M., Kalaiselvi, T., Prasanthrajan, M., Poornachandhra, C. *et al.* (2023) Biomining *Sesuvium portulacastrum* for halotolerant PGPR and endophytes for promotion of salt tolerance in *Vigna mungo* L. *Frontiers in Microbiology* 14, 1085787. DOI: 10.3389/FMICB.2023.1085787.

Kanchiswamy, C.N., Malnoy, M. and Maffei, M.E. (2015) Chemical diversity of microbial volatiles and their potential for plant growth and productivity. *Frontiers in Plant Science* 6(MAR), 133272. DOI: 10.3389/FPLS.2015.00151.

Kaul, S., Choudhary, M., Gupta, S. and Dhar, M.K. (2021) Engineering host microbiome for crop improvement and sustainable agriculture. *Frontiers in Microbiology* 12, 635917. DOI: 10.3389/FMICB.2021.635917.

Ke, J., Wang, B. and Yoshikuni, Y. (2021) Microbiome engineering: Synthetic biology of plant-associated microbiomes in sustainable agriculture. *Trends in Biotechnology* 39(3), 244–261. DOI: 10.1016/J.TIBTECH.2020.07.008.

Kimotho, R.N. and Maina, S. (2024) Unraveling plant–microbe interactions: Can integrated omics approaches offer concrete answers? *Journal of Experimental Botany* 75(5), 1289–1313. DOI: 10.1093/JXB/ERAD448.

Knief, C. (2014) Analysis of plant microbe interactions in the era of next generation sequencing technologies. *Frontiers in Plant Science* 5, 00216. DOI: 10.3389/FPLS.2014.00216.

Kollmen, J. and Strieth, D. (2022) The beneficial effects of cyanobacterial co-culture on plant growth. *Life* 12(2), 223. DOI: 10.3390/LIFE12020223.

Kour, D., Rana, K.L., Kaur, T., Sheikh, I., Yadav, A.N. *et al.* (2020) Microbe-mediated alleviation of drought stress and acquisition of phosphorus in great millet (*Sorghum bicolor* L.) by drought-adaptive and phosphorus-solubilizing microbes. *Biocatalysis and Agricultural Biotechnology* 23, 101501. DOI: 10.1016/J.BCAB.2020.101501.

Kuang, L., Li, T., Wang, B., Peng, J., Li, J., Li, P. and Jiang, J. (2023) Diseased-induced multifaceted variations in community assembly and functions of plant-associated microbiomes. *Frontiers in Microbiology* 14, 1141585. DOI: 10.3389/FMICB.2023.1141585.

Kumar, G.A., Kumar, S., Bhardwaj, R., Swapnil, P., Meena, M., Seth, C.S. and Yadav, A. (2023a) Recent advancements in multifaceted roles of flavonoids in plant–rhizomicrobiome interactions. *Frontiers in Plant Science* 14. DOI: 10.3389/FPLS.2023.1297706.

Kumar, U., Raj, S., Sreenikethanam, A., Maddheshiya, R., Kumari, S. *et al.* (2023b) Multi-omics approaches in plant–microbe interactions hold enormous promise for sustainable agriculture. *Agronomy* 13(7), 1804. DOI: 10.3390/AGRONOMY13071804.

Lamichhane, J.R., Barbetti, M.J., Chilvers, M.I., Pandey, A.K. and Steinberg, C. (2024) Exploiting root exudates to manage soil-borne disease complexes in a changing climate. *Trends in Microbiology* 32(1), 27–37. DOI: 10.1016/J.TIM.2023.07.011.

Lau, J.A. and Lennon, J.T. (2012) Rapid responses of soil microorganisms improve plant fitness in novel environments. *Proceedings of the National Academy of Sciences USA* 109(35), 14058–14062. DOI: 10.1073/pnas.1202319109.

Lazcano, C., Boyd, E., Holmes, G., Hewavitharana, S., Pasulka, A. and Ivors, K. (2021) The rhizosphere microbiome plays a role in the resistance to soil-borne pathogens and nutrient uptake of strawberry cultivars under field conditions. *Scientific Reports* 11(1), 1–17. DOI: 10.1038/s41598-021-82768-2.

Li, J., Meng, B., Chai, H., Yang, X., Song, W. *et al.* (2019) Arbuscular mycorrhizal fungi alleviate drought stress in C3 (*Leymus chinensis*) and C4 (*Hemarthria altissima*) grasses via altering antioxidant enzyme activities and photosynthesis. *Frontiers in Plant Science* 10, 499. DOI: 10.3389/fpls.2019.00499.

Liao, X., Ederveen, T.H.A., Niehues, A., de Visser, C., Huang, J. *et al.* (2024) FAIR data cube, a FAIR data infrastructure for integrated multi-omics data analysis. *Journal of Biomedical Semantics* 15(1), 1–13. DOI: 10.1186/S13326-024-00321-2.

Liu, H., Brettell, L.E., Qiu, Z. and Singh, B.K. (2020) Microbiome-mediated stress resistance in plants. *Trends in Plant Science* 25(8), 733–743. DOI: 10.1016/J.TPLANTS.2020.03.014.

Liu, H., Carvalhais, L.C., Crawford, M., Singh, E., Dennis, P.G., Pieterse, C.M.J. and Schenk, P.M. (2017) Inner plant values: Diversity, colonization and benefits from endophytic bacteria. *Frontiers in Microbiology* 8(DEC), 322792. DOI: 10.3389/FMICB.2017.02552.

Ma, X., Zhang, B., Xiang, X., Li, W., Li, J. *et al.* (2024) Characterization of *Bacillus pacificus* G124 and its promoting role in plant growth and drought tolerance. *Plants* 13, 2864. DOI: 10.3390/PLANTS13202864.

Martin, C.A. and Stutz, J.C. (2004) Interactive effects of temperature and arbuscular mycorrhizal fungi on growth, P uptake and root respiration of *Capsicum annuum* L. *Mycorrhiza* 14(4), 241–244. DOI: 10.1007/s00572-003-0261-6.

Martins, S.J., Pasche, J., Silva, H.A.O., Selten, G., Savastano, N. *et al.* (2023) The use of synthetic microbial communities (SynComs) to improve plant health. *Phytopathology* 113(8), 1369–1379. DOI: 10.1094/PHYTO-01-23-0016-IA.

McPherson, M.R., Wang, P., Marsh, E.L., Mitchell, R.B. and Schachtman, D.P. (2018) Isolation and analysis of microbial communities in soil, rhizosphere, and roots in perennial grass experiments. *Journal of Visualized Experiments: JoVE* 2018, 57932. DOI: 10.3791/57932.

Montejano-Ramírez, V., Ávila-Oviedo, J.L., Campos-Mendoza, F.J. and Valencia-Cantero, E. (2024) Microbial volatile organic compounds: Insights into plant defense. *Plants* 13(15), 2013. DOI: 10.3390/PLANTS13152013.

Morcillo, R.J.L., Baroja-Fernández, E., López-Serrano, L., Leal-López, J., Muñoz, F.J. *et al.* (2022) Cell-free microbial culture filtrates as candidate biostimulants to enhance plant growth and yield and activate soil- and plant-associated beneficial microbiota. *Frontiers in Plant Science* 13, 1040515. DOI: 10.3389/FPLS.2022.1040515.

Mosa, W.F.A., Behiry, S.I., Ali, H.M., Abdelkhalek, A., Sas-Paszt, L. *et al.* (2022) Pomegranate trees quality under drought conditions using potassium silicate, nanosilver, and selenium spray with valorization of peels as fungicide extracts. *Scientific Reports* 12(1), 1–16. DOI: 10.1038/s41598-022-10354-1.

Mukherjee, A., Singh, B.N., Kaur, S., Sharma, M., Ferreira de Araújo, A.S. *et al.* (2024) Unearthing the power of microbes as plant microbiome for sustainable agriculture. *Microbiological Research* 286, 127780. DOI: 10.1016/J.MICRES.2024.127780.

Navarro-Torre, S., Barcia-Piedras, J.M., Mateos-Naranjo, E., Redondo-Gómez, S., Camacho, M. *et al.* (2017) Assessing the role of endophytic bacteria in the halophyte *Arthrocnemum macrostachyum* salt tolerance. *Plant Biology* 19(2), 249–256. DOI: 10.1111/PLB.12521.

Nawaz, T., Saud, S., Gu, L., Khan, I., Fahad, S. and Zhou, R. (2024) Cyanobacteria: Harnessing the power of microorganisms for plant growth promotion, stress alleviation, and phytoremediation in the era of sustainable agriculture. *Plant Stress* 11, 100399. DOI: 10.1016/J.STRESS.2024.100399.

Nephali, L., Moodley, V., Piater, L., Steenkamp, P., Buthelezi, N. *et al.* (2021) A metabolomic landscape of maize plants treated with a microbial biostimulant under well-watered and drought conditions. *Frontiers in Plant Science* 12, 676632. DOI: 10.3389/FPLS.2021.676632.

Notununu, I., Moleleki, L., Roopnarain, A. and Adeleke, R. (2024) Enhancing maize drought and heat tolerance: Single vs combined plant growth promoting rhizobacterial inoculation. *Frontiers in Plant Science* 15, 1480718. DOI: 10.3389/FPLS.2024.1480718.

Noureen, S., Iqbal, A. and Muqeet, H.A. (2024) Potential of drought tolerant rhizobacteria amended with biochar on growth promotion in wheat. *Plants* 13, 1183. DOI: 10.3390/PLANTS13091183.

Novielli, P., Magarelli, M., Romano, D., de Trizio, L., Di Bitonto, P. *et al.* (2024) Climate change and soil health: Explainable artificial intelligence reveals microbiome response to warming. *Machine Learning and Knowledge Extraction* 6(3), 1564–1578. DOI: 10.3390/MAKE6030075.

Nwachukwu, B.C. and Babalola, O.O. (2022) Metagenomics: A tool for exploring key microbiome with the potentials for improving sustainable agriculture. *Frontiers in Sustainable Food Systems* 6, 886987. DOI: 10.3389/FSUFS.2022.886987.

Ofek-Lalzar, M., Sela, N., Goldman-Voronov, M., Green, S.J., Hadar, Y. and Minz, D. (2014) Niche and host-associated functional signatures of the root surface microbiome. *Nature Communications* 5(1), 1–9. DOI: 10.1038/ncomms5950.

Pan, J., Huang, C., Peng, F., Zhang, W., Luo, J., Ma, S. and Xue, X. (2020) Effect of arbuscular mycorrhizal fungi (AMF) and plant growth-promoting bacteria (PGPR) inoculations on *Elaeagnus angustifolia* L. in saline soil. *Applied Sciences* 10, 945. DOI: 10.3390/APP10030945.

Parasar, B.J., Kashyap, S., Sharma, I., Marme, S.D., Das, P. and Agarwala, N. (2024) Microbe mediated alleviation of drought and heat stress in plants – current understanding and future prospects. *Discover Plants* 1(1), 26. DOI: 10.1007/s44372-024-00022-1.

Patil, J.R., Kuldeep, M.J., Yadav, K., Lal, Y.S., Srivastava, S. and Nikalje, G.C. (2024) Flavonoids in plant-environment interactions and stress responses. *Discover Plants* 1(1), 1–19. DOI: 10.1007/S44372-024-00063-6.

Pattnaik, S., Mohapatra, B. and Gupta, A. (2021) Plant growth-promoting microbe mediated uptake of essential nutrients (Fe, P, K) for crop stress management: Microbe–soil–plant continuum. *Frontiers in Agronomy* 3, 689972. DOI: 10.3389/FAGRO.2021.689972.

Peixoto, R.S., Voolstra, C.R., Sweet, M., Duarte, C.M., Carvalho, S. *et al.* (2022) Harnessing the microbiome to prevent global biodiversity loss. *Nature Microbiology* 7(11), 1726–1735. DOI: 10.1038/s41564-022-01173-1.

Pino, C. and Griffon, D. (2024) Scaling up: Microbiome manipulation for climate change adaptation in large organic vineyards. *Frontiers in Sustainable Food Systems* 8, 1285981. DOI: 10.3389/FSUFS.2024.1285981.

Pirttilä, A.M., Tabas, H.M.P., Baruah, N. and Koskimäki, J.J. (2021) Biofertilizers and biocontrol agents for agriculture: How to identify and develop new potent microbial strains and traits. *Microorganisms* 9(4), 817. DOI: 10.3390/MICROORGANISMS9040817/S1.

Poole, P., Ramachandran, V. and Terpolilli, J. (2018) Rhizobia: From saprophytes to endosymbionts. *Nature Reviews Microbiology* 16(5), 291–303. DOI: 10.1038/nrmicro.2017.171.

Pourkheirandish, M., Golicz, A.A., Bhalla, P.L. and Singh, M.B. (2020) Global role of crop genomics in the face of climate change. *Frontiers in Plant Science* 11, 00922. DOI: 10.3389/FPLS.2020.00922.

Qiu, Z., Egidi, E., Liu, H., Kaur, S. and Singh, B.K. (2019) New frontiers in agriculture productivity: Optimised microbial inoculants and in situ microbiome engineering. *Biotechnology Advances* 37, 107371. DOI: 10.1016/J.BIOTECHADV.2019.03.010.

Ray, P., Lakshmanan, V., Labbé, J.L. and Craven, K.D. (2020) Microbe to microbiome: A paradigm shift in the application of microorganisms for sustainable agriculture. *Frontiers in Microbiology* 11, 622926. DOI: 10.3389/FMICB.2020.622926.

Ruppel, S., Franken, P. and Witzel, K. (2013) Properties of the halophyte microbiome and their implications for plant salt tolerance. *Functional Plant Biology* 40(9), 940–951. DOI: 10.1071/FP12355.

Salem, M.A., Wang, J.Y. and Al-Babili, S. (2022) Metabolomics of plant root exudates: From sample preparation to data analysis. *Frontiers in Plant Science* 13, 1062982. DOI: 10.3389/FPLS.2022.1062982.

Santos, L.F. and Olivares, F.L. (2021) Plant microbiome structure and benefits for sustainable agriculture. *Current Plant Biology* 26, 100198. DOI: 10.1016/J.CPB.2021.100198.

Santos-Medellín, C., Liechty, Z., Edwards, J., Nguyen, B., Huang, B., Weimer, B.C. and Sundaresan, V. (2021) Prolonged drought imparts lasting compositional changes to the rice root microbiome. *Nature Plants* 7(8), 1065–1077. DOI: 10.1038/S41477-021-00967-1.

Shayanthan, A., Ordoñez, P.A.C. and Oresnik, I.J. (2022) The Role of Synthetic Microbial Communities (SynCom) in Sustainable Agriculture. *Frontiers in Agronomy* 4. DOI: 10.3389/FAGRO.2022.896307.

Shah, A.M., Khan, I.M., Shah, T.I., Bangroo, S.A., Kirmani, N.A. *et al.* (2022) Soil microbiome: A treasure trove for soil health sustainability under changing climate. *Land* 11, 1887. DOI: 10.3390/LAND11111887.

Singh, P.K., Egidi, E., Macdonald, C.A. and Singh, B.K. (2023a) Host selection has a stronger impact on leaf microbiome assembly compared to land-management practices. *Journal of Sustainable Agriculture and Environment* 2(2), 99–111. DOI: 10.1002/sae2.12043.

Singh, J.S., Kumar, A., Rai, A.N. and Singh, D.P. (2016) Cyanobacteria: A precious bio-resource in agriculture, ecosystem, and environmental sustainability. *Frontiers in Microbiology* 7(APR), 529. DOI: 10.3389/FMICB.2016.00529.

Singh, B.K., Liu, H. and Trivedi, P. (2020a) Eco-holobiont: A new concept to identify drivers of host-associated microorganisms. *Environmental Microbiology* 22(2), 564–567. DOI: 10.1111/1462-2920.14900.

Singh, B.K., Trivedi, P., Egidi, E., Macdonald, C.A. and Delgado-Baquerizo, M. (2020b) Crop microbiome and sustainable agriculture. *Nature Reviews Microbiology* 18(11), 601–602. DOI: 10.1038/s41579-020-00446-y.

Singh, B.K., Yan, Z.Z., Whittaker, M., Vargas, R. and Abdelfattah, A. (2023b) Soil microbiomes must be explicitly included in one health policy. *Nature Microbiology* 8(8), 1367–1372. DOI: 10.1038/s41564-023-01386-y.

Song, C., Zhu, F., Carrión, V.J. and Cordovez, V. (2020) Beyond plant microbiome composition: Exploiting microbial functions and plant traits via integrated approaches. *Frontiers in Bioengineering and Biotechnology* 8, 00896. DOI: 10.3389/FBIOE.2020.00896.

Srivastav, R. and Suneja, G. (2019) Recent advances in microbial genome sequencing. In *Microbial Genomics in Sustainable Agroecosystems*, pp. 131–144. Singapore: Springer Singapore. DOI: 10.1007/978-981-32-9860-6_8.

Srivastava, P., Jamir, A., Jamir, S., Uikey, P., Singh, B.V., Sulochna, and Saikanth, D.R.K. (2023) Harnessing microorganisms for sustainable agriculture: Promoting environmental protection and soil health. *Bionature* 43(1), 26–29. DOI: 10.56557/BN/2023/V43I11851.

Sundh, I., Del Giudice, T. and Cembalo, L. (2021) Reaping the benefits of microorganisms in cropping systems: Is the regulatory policy adequate? *Microorganisms* 9, 1437. DOI: 10.3390/MICROORGANISMS9071437.

Tabacchioni, S., Passato, S., Ambrosino, P., Huang, L., Caldara, M. *et al.* (2021) Identification of beneficial microbial consortia and bioactive compounds with potential as plant biostimulants for a sustainable agriculture. *Microorganisms* 9(2), 426. DOI: 10.3390/MICROORGANISMS9020426.

Tedersoo, L., Albertsen, M., Anslan, S. and Callahan, B. (2021) Perspectives and benefits of high-throughput long-read sequencing in microbial ecology. *Applied and Environmental Microbiology* 87(17), 1–19. DOI: 10.1128/AEM.00626-21.

Thriveni, V., Teotia, J., Hazra, S., Bharti, T., Kumar, M., Lallawmkimi, M.C. and Panwar, D. (2024) A review on integrating bioinformatics tools in modern plant breeding. *Archives of Current Research International* 24(9), 293–308. DOI: 10.9734/ACRI/2024/V24I9894.

Toju, H., Peay, K.G., Yamamichi, M., Narisawa, K., Hiruma, K. *et al.* (2018) Core microbiomes for sustainable agroecosystems. *Nature Plants* 4(5), 247–257. DOI: 10.1038/s41477-018-0139-4.

Trivedi, P., Leach, J.E., Tringe, S.G., Sa, T. and Singh, B.K. (2020) Plant–microbiome interactions: From community assembly to plant health. *Nature Reviews Microbiology* 18(11), 607–621. DOI: 10.1038/s41579-020-0412-1.

Trivedi, P., Mattupalli, C., Eversole, K. and Leach, J.E. (2021) Enabling sustainable agriculture through understanding and enhancement of microbiomes. *New Phytologist* 230(6), 2129–2147. DOI: 10.1111/NPH.17319.

van der Heijden, M.G.A.A., Bardgett, R.D.R., Van Straalen, N.M., van der Heijden, M., Bardgett, R.D.R. and van Straalen, N. (2008) The unseen majority: Soil microbes as drivers of plant diversity and productivity in terrestrial ecosystems. *Ecology Letters* 11(3), 296–310. DOI: 10.1111/j.1461-0248.2007.01139.x.

Vassileva, M., Mocali, S., Canfora, L., Malusá, E., García del Moral, L.F. *et al.* (2022) Safety level of microorganism-bearing products applied in soil-plant systems. *Frontiers in Plant Science* 13, 862875. DOI: 10.3389/FPLS.2022.862875.

Verma, J.P. (2018) Functional importance of the plant microbiome: Implications for agriculture, forestry and bioenergy: A book review. *Journal of Cleaner Production* 178, 877–879. DOI: 10.1016/J.JCLEPRO.2018.01.043.

Wang, Y., Li, H., Liu, Y., Zhou, M., Ding, M. and Yuan, Y. (2022) Construction of synthetic microbial consortia for 2-keto-L-gulonic acid biosynthesis. *Synthetic and Systems Biotechnology* 7(1), 481–489. DOI: 10.1016/J.SYNBIO.2021.12.001.

Wille, L., Messmer, M.M., Studer, B. and Hohmann, P. (2019) Insights to plant-microbe interactions provide opportunities to improve resistance breeding against root diseases in grain legumes. *Plant, Cell & Environment* 42(1), 20–40. DOI: 10.1111/PCE.13214.

Wu, B., Luo, Z., Yang, P., Lei, N., Pei, X., Guo, S. and Du, J. (2024) Phosphorus solubilizing bacteria combined with sticky glutinous rice paste accelerated the vegetation recovery of spoil material under drought conditions. *Applied Soil Ecology* 203, 105647. DOI: 10.1016/J.APSOIL.2024.105647.

Xiong, Y., Fan, X.H., Wang, Q., Yin, Z.G., Sheng, X.W. *et al.* (2022) Genomic analysis of soybean PP2A-B" family and its effects on drought and salt tolerance. *Frontiers in Plant Science* 12. DOI: 10.3389/FPLS.2021.784038.

Xiong, J., Lu, J., Li, X., Qiu, Q., Chen, J. and Yan, C. (2021a) Effect of rice (*Oryza sativa* L.) genotype on yield: Evidence from recruiting spatially consistent rhizosphere microbiome. *Soil Biology and Biochemistry* 161, 108395. DOI: 10.1016/j.soilbio.2021.108395.

Xiong, C., Zhu, Y.-G., Wang, J.-T., Singh, B., Han, L.-L. *et al.* (2021b) Host selection shapes crop microbiome assembly and network complexity. *The New Phytologist* 229, 1091–1104. DOI: 10.1111/nph.16890.

Yang, Z., Zhao, T., Cheng, H. and Yang, J. (2024) Microbiome-enabled genomic selection improves prediction accuracy for nitrogen-related traits in maize. *G3 Genes|Genomes|Genetics* 14, jkad286. DOI: 10.1093/G3JOURNAL/JKAD286.

Yeoh, Y.K., Dennis, P.G., Paungfoo-Lonhienne, C., Weber, L., Brackin, R. *et al.* (2017) Evolutionary conservation of a core root microbiome across plant phyla along a tropical soil chronosequence. *Nature Communications* 8(1), 1–9. DOI: 10.1038/s41467-017-00262-8.

Yin, C., Hagerty, C.H. and Paulitz, T.C. (2022) Synthetic microbial consortia derived from rhizosphere soil protect wheat against a soilborne fungal pathogen. *Frontiers in Microbiology* 13, 908981. DOI: 10.3389/FMICB.2022.908981.

Zhang, L., Zhang, W., Li, Q., Cui, R., Wang, Z. *et al.* (2020) Deciphering the root endosphere microbiome of the desert plant *Alhagi sparsifolia* for drought resistance-promoting bacteria. *Applied and Environmental Microbiology* 86, e02863-19. DOI: 10.1128/AEM.02863-19.

Zhou, N., Zhao, S. and Tian, C.Y. (2017) Effect of halotolerant rhizobacteria isolated from halophytes on the growth of sugar beet (*Beta vulgaris* L.) under salt stress. *FEMS Microbiology Letters* 364, 91. DOI: 10.1093/FEMSLE/FNX091.

Zhu, L., Huang, J., Lu, X. and Zhou, C. (2022) Development of plant systemic resistance by beneficial rhizobacteria: Recognition, initiation, elicitation and regulation. *Frontiers in Plant Science* 13, 952397. DOI: 10.3389/FPLS.2022.952397.

14 Nanotechnology in Agriculture: Climate-responsive Applications

G.E.J. Poinern[1]*, S. Brundavanam[1], A.F.M. Halim[1], W. Chamsa-ard[2] and D. Fawcett[1]

[1]*Murdoch Applied Innovation Nanotechnology Research Group, School of Mathematics and Statistics, Chemistry and Physics, Murdoch University, 90 South Street, Perth WA 6150, Western Australia, Australia;* [2]*School of Renewable Energy and Smart Grid Technology (SGtech), Naresuan University, Phitsanulok, Thailand, 65000*

Abstract

Climate change poses significant threats to global agriculture through rising temperatures, erratic rainfall, soil degradation and increased biotic and abiotic stresses. This chapter explores the potential use of nanotechnology as a transformative tool to create climate-resilient, sustainable agricultural systems. It begins with an overview of the historical development of nanotechnology and its evolution into modern applications. Key innovations such as nanofertilizers, nanopesticides and nanosensors are discussed, highlighting their role in improving nutrient use efficiency, reducing chemical inputs and enabling precision farming. Furthermore, the chapter examines nanotechnology's role in enhancing plant tolerance to environmental stressors, improving soil health, managing water efficiently and supporting carbon sequestration. Furthermore, the convergence of nanotechnology with advanced biotechnological approaches and gene-editing platforms such as CRISPR-Cas9 presents promising avenues for the development of climate-resilient crop varieties. The chapter also emphasizes the importance of addressing several challenges with nanotechnologies, including environmental safety, public acceptance, regulatory frameworks and cost-effective scalability. Through this comprehensive analysis, the chapter highlights the transformative potential of frontier nanotechnologies in advancing agriculture toward greater resilience, efficiency and climate adaptability.

14.1 Nanotechnology

The pioneering conceptualization of nanotechnology is attributed to the renowned Nobel laureate, Professor Richard Feynman. A versatile scientist whose contributions span multiple domains – including the development of the atomic bomb and advancements in quantum electrodynamics, for which he was awarded the Nobel Prize in Physics in 1965. Feynman significantly influenced the foundation of this emerging field. During his address at the Annual Meeting of the American Physical Society in 1959, titled "There is Plenty of Room at the Bottom," he articulated visionary ideas about a nascent science focused on the manipulation of matter at extremely small scales. Feynman demonstrated that atomic-level manipulation was feasible, exemplified by his bold proposition of a $1000 prize to anyone who could, using existing technology, reduce a page of text to an area 1/25,000th smaller in linear dimensions – such

*Corresponding author. Email: g.poinern@murdoch.edu

 Genomic Solutions for Breeding Climate-resilient Crops (R.K. Varshney, R. Barmukh and R.J. Harper)
DOI: 10.1079/9781800627185.0014

that it could still be read under an electron microscope.

It was not until 1974 that the term "nanotechnology" was formally introduced by Japanese physicist Professor Norio Taniguchi, who, working within the semiconductor industry, used it to describe processes capable of fabricating semiconductor structures with nanometer-scale precision employing techniques such as focused ion beam milling and atomic layer deposition. Subsequently, the concept was expanded by American engineer Eric Drexler in his seminal work, *Engines of Creation: The Coming Era of Nanotechnology*, where he envisioned nanoscale assemblers capable of self-replication and capable of orchestrating the construction of complex molecular structures.

Bining and Rohrer invented the scanning tunneling microscope (STM) in 1981, revolutionizing surface science by enabling the visualization of individual atoms on conductive surfaces. Their innovative approach utilized a biased (charged) sharp metallic tip brought extremely close to a sample surface, allowing quantum tunneling currents to be measured and mapped with atomic and molecular resolution. This technology enables scientists across disciplines to image atoms, molecules, macro-molecular assemblies all the way down to DNA and other oligomers in vacuum and in their native state. Building on this foundation, they also contributed to the development of the atomic force microscope (AFM), which extends the capability to image a wide range of materials, including insulators, by measuring the mechanical forces between a sharp tip and the sample surface. Their innovations expanded the frontiers of nanotechnology and surface analysis. In recognition of this pioneering contribution, they were awarded the Nobel Prize in Physics in 1986, sharing the honor with Ernst Ruska for his development in electron optics that opened the way for the scanning electron microscope.

In contemporary usage, nanotechnology broadly encompasses the manipulation and fabrication of materials possessing at least one dimension within the nanometer range ($1–100 \times 10^{-9}$ meters). The field includes not only the development of novel nanomaterials but also the technological frameworks that enable their creation. Nanomaterials are defined as clusters or assemblies of atoms and molecules whose properties differ markedly from their bulk counterparts, owing to quantum effects and high surface-to-volume ratios. These unique properties such as increased chemical reactivity, enhanced electrical conductivity, superior mechanical strength and magnetic behavior offer promising avenues for scientific and technological advancement. Furthermore, nanomaterials exhibit novel opto-electronic properties that are unprecedented in natural materials, thus opening new frontiers across multiple disciplines (Shah *et al*., 2015). As Feynman envisioned, these principles and properties present opportunities for harnessing the peculiarities of the nanoscale to benefit society at large.

Nanotechnology's application in agriculture encompasses a broad and multifaceted scope, addressing critical issues such as food security, sustainable production, environmental protection and agricultural testing. Key innovations include nanofertilizers, nanopesticides and nanosensors, which aim to enhance resource use efficiency and crop productivity while reducing environmental impacts. Nanofertilizers enable controlled nutrient release, thereby decreasing chemical input requirements and supporting soil health (Reddy and Chhabra, 2022; Garg *et al*., 2023). Conversely, the unregulated application of agrochemicals, particularly nitrogen fertilizers, has contributed to soil acidification, groundwater contamination and atmospheric pollution, underscoring the need for more sustainable nanotechnology-based solutions (Sekhon, 2014; Kim *et al*., 2017; Prasad *et al*., 2017).

14.2 Scope of Nanotechnology in Agriculture

Plants are sessile organisms that thrive in soil and must manufacture their own sustenance through photosynthesis (CO_2, H_2O and sunlight) for their life processes. Each plant's ability to grow and propagate is dependent on several factors and the plant endures two types of stresses: abiotic and biotic. Biotic pressures are from pathogens such as bacteria, viruses, fungi and nematodes causing diseases. Insects and pests such as aphids, beetles, caterpillars and other herbivores feeding on plant tissues are threats to certain plants. Moreover, weeds and competing plants that reduce access to nutrients, water and light to these plants are stresses to contend with. Finally, plants can be affected by parasitic plants that extract nutrients from their host plants. In the category of abiotic stresses, temperature extremes, flooding, droughts, salinity, heavy metals toxicity, wind and radiation are threats to plant life and have major effects on the use of land for agriculture across the globe.

Plants' reliance on self-produced energy makes them vulnerable to environmental stresses such as salinity, drought, temperature changes and flooding, which hinder their growth and development. These stresses cause physiological and biochemical disruptions, reducing photosynthesis, nutrient uptake and crop yields. Understanding plant stress responses is essential for developing resilient crops and ensuring food security amid climate change. Continued research at both academic and industrial levels on stress tolerance and adaptation is vital for sustainable agriculture.

Agriculture is one of the biggest demands on Earth's resources, accounting for 29% of greenhouse gas emissions, 30% of energy usage, 33% of land utilization, 70% of groundwater withdrawals and 75% of deforestation as well as affecting soil salinity in negative ways (Lowry *et al.*, 2019).

This pressure is increasing as the current global population of 8.01 billion is expected to reach nearly 9.8 billion by 2050, primarily driven by growth in developing regions. To meet the nutritional needs of this expanding global population, agricultural production including food and other related products must increase drastically. Technology is poised to significantly enhance current agricultural practices. While discussions about the use of nanotechnology in agriculture have been ongoing since 1999, successful industry applications have been relatively limited compared with other fields such as medicine, material science applications, semiconductor and other optoelectronics applications.

Several barriers, including low public funding, consumers' acceptance of technology driven agriculture and the slow adoption of innovations by agricultural research organizations, are now diminishing. As a result, the field has seen a significant surge in research and development in recent years, evidenced by the rapidly increasing number of publications and patents (Kah *et al.*, 2019). Nanotechnology is a crucial component in addressing the current challenges facing crop production and agricultural practices, offering innovative solutions to enhance efficiency, sustainability and productivity (Figures 14.1 and 14.2).

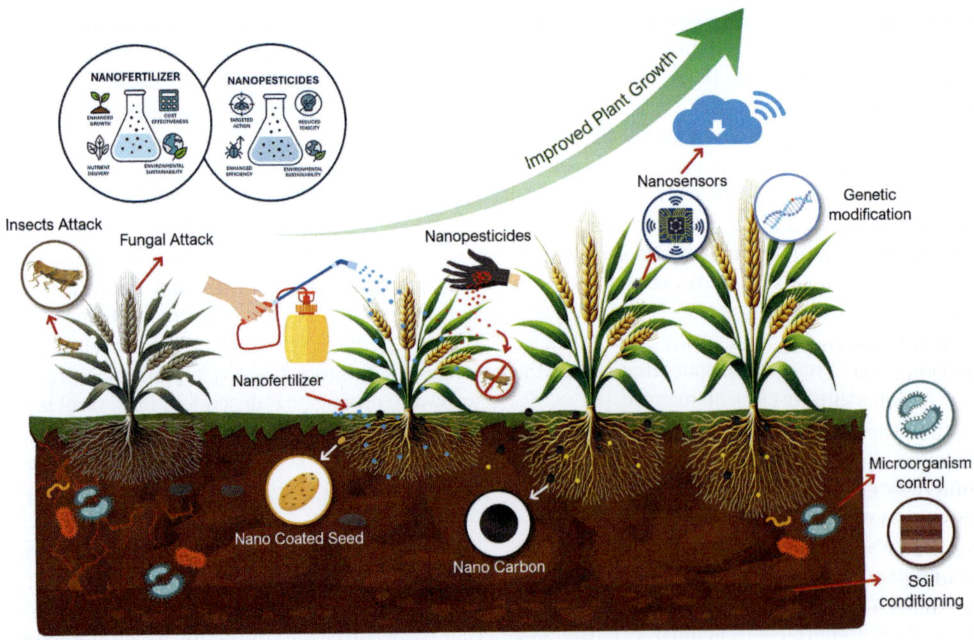

Figure 14.1. Schematic diagram of the different applications of nanotechnology in agriculture. (Authors' own image.)

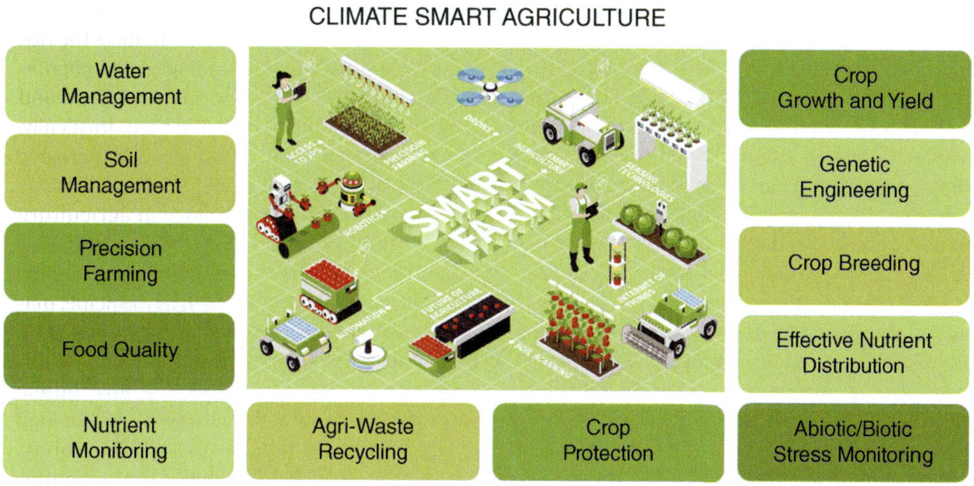

Figure 14.2. Different areas of climate-smart agriculture: a visual representation of diverse agricultural zones implementing sustainable practices to enhance productivity, resilience and carbon reduction. (Authors' own image.)

Nanomaterials can be deployed directly to plants via foliage, stem and the soil to access the root system of plants and crops. These nanoscale materials can be made via bottom up or top to bottom technologies. Nanoscale matter can be combined with other components to produce effective agricultural formulations. Moreover, nanotechnology facilitates the creation of nanosensors that can monitor soil conditions, detect plant diseases and optimize irrigation practices. These sensors provide real-time data, enabling farmers to make informed decisions that enhance resource utilization and crop management (Prasad *et al.*, 2017; Reddy and Chhabra, 2022). The integration of nanotechnology into agricultural systems also supports precision farming, where inputs are applied more accurately based on the specific needs of crops, leading to improved productivity and sustainability (Dutta *et al.*, 2022; El-Saadony *et al.*, 2022).

Another significant aspect of nanotechnology in agriculture is its role in biotic and abiotic stress management. Nanoparticles have been shown to enhance plant tolerance to various stresses, such as drought and salinity, by improving water retention and nutrient uptake (El-Saadony *et al.*, 2022; Giri *et al.*, 2023). This capability is crucial in the context of climate change, where agricultural systems are increasingly challenged by extreme weather conditions (Shang *et al.*, 2019).

Furthermore, the application of biopolymeric nanocarriers for nutrient delivery and crop biofortification represents a promising avenue for enhancing the nutritional quality of crops. These nanocarriers can encapsulate essential nutrients, ensuring their effective delivery to plants and improving overall crop health (Dutta *et al.*, 2022; Garg *et al.*, 2023). This innovation not only addresses nutrient deficiencies but also contributes to the development of fortified food products that can combat malnutrition.

In addition to crop production, nanotechnology also finds applications in food packaging and safety. Nanomaterials can be utilized to create active packaging solutions that extend the shelf life of agricultural products while ensuring food safety through antimicrobial properties (Zare *et al.*, 2022). This aspect of nanotechnology is particularly important in reducing food waste and enhancing the overall efficiency of the food supply chain.

14.3 Importance of Climate-responsive Applications

Climate change poses significant challenges to agriculture and the lives of many, impacting food security, crop yields and farming practices worldwide. Rising temperatures, shifting precipitation patterns, and an increase in extreme weather events such as droughts and floods threaten agricultural productivity. In lower-latitude regions, staple crops such as maize and wheat have already experienced yield declines due to climate-related stressors. Additionally, water scarcity is becoming more severe, particularly in regions reliant on irrigation. The nutritional quality of crops is also at risk, as elevated CO_2 levels can reduce essential nutrients such as protein, zinc and iron in staple foods. These challenges demand urgent adaptation strategies, including improved land management, sustainable farming techniques and climate-resilient crop varieties. Without proactive measures, climate change will continue to exacerbate food insecurity, disproportionately affecting vulnerable populations (Masson-Delmotte, 2022).

Global climate change has led to unprecedented droughts, extreme temperature fluctuations, excessive rainfall and unforeseen storms, triggering disasters unlike anything previously experienced. Considering these challenges, it is crucial to implement environmentally sustainable mechanisms to mitigate its impact and safeguard ecosystems for the future. Climate change is increasingly challenging sustainable global agriculture, requiring the adoption of climate-responsive strategies and technologies to improve resilience, optimize resource use and secure food supply.. Climate-smart technologies, such as satellite-guided advisories for drought and flood risks, enable early interventions that can mitigate productivity losses and enhance adaptive capacity (Janbandhu *et al.*, 2024). Furthermore, the development of genome-edited climate-resilient cultivars represents a promising avenue for long-term solutions to climate-induced agricultural challenges, although regulatory barriers may hinder their widespread adoption (Janbandhu *et al.*, 2024).

Understanding the relationship between climate and crop productivity is crucial for enhancing resilience in agricultural systems. Research indicates that adapting current crop varieties to withstand climate-induced stresses through improved management practices can significantly enhance productivity (Ullah *et al.*, 2017). Strategies such as intercropping, crop diversification and climate-resilient varieties can boost yields and resource use amid climate change. Combining indigenous knowledge with technology also helps local communities develop effective adaptation strategies, improving food security and sustainability (Negash and Fininsa, 2015). In addition, the role of gender in climate change adaptation is also critical, as disparities in resource access and responsibilities can influence the adoption of climate-responsive practices (Huyer *et al.*, 2021).

Nanotechnology has emerged as a transformative tool in the agricultural sector, offering innovative solutions to mitigate the impacts of climate change. The application of nanotechnology in developing nanofertilizers and nanopesticides can enhance nutrient utilization and improve pest control, thereby reducing the environmental footprint of agricultural practices (Tang *et al.*, 2023). Additionally, the integration of nanotechnology with artificial intelligence and precision agriculture allows for real-time monitoring and management of crops, optimizing resource use and minimizing waste (Zhang *et al.*, 2021). This convergence of technologies (Artificial Intelligence and Data Sciences) not only enhances productivity but also contributes to sustainable agricultural practices that are resilient to climate variability.

The urgency of addressing climate change through innovative agricultural practices cannot be overstated. Climate change adaptation plans must include a comprehensive set of measures, such as climate-smart agriculture, waste management and the promotion of innovative technologies (Durán-Sandoval *et al.*, 2023).

Prioritizing climate-responsive applications strengthens agricultural resilience, ensures food security and promotes sustainable development amid environmental challenges. They enhance resilience, optimize resources and address climate change impacts. Integrating advanced technologies including nanotechnology, promoting gender equity and inclusive policies are key to developing effective strategies.

14.4 Overview of Current Challenges in Agriculture due to Climate Change

In recent years, climate change has emerged as a critical challenge to global agriculture, significantly impacting food security, livelihoods and the sustainability of agricultural practices. The agricultural sector is inherently vulnerable to climate change, which manifests through increased temperatures, altered precipitation patterns and the frequency of extreme weather events. These changes threaten the stability of food production systems, particularly in regions that rely heavily on rainfed agriculture. The Intergovernmental Panel on Climate Change (IPCC) has projected that climate change could reduce agricultural productivity by as much as 21% since 1961, with developing regions such as sub-Saharan Africa facing even more severe impacts, potentially up to 40% (Brenton *et al.*, 2022).

One of the most immediate challenges posed by climate change is its detrimental effect on crop yields. Research indicates that rising temperatures and altered precipitation patterns can lead to reduced agricultural productivity, particularly for staple crops such as wheat, rice and maize (Phalkey *et al.*, 2015). For instance, smallholder farmers, who often rely on rainfed agriculture, are particularly vulnerable to climate variability, as their crop yields are directly tied to rainfall patterns. This review by Palkey *et al*, highlighted that climate change impacts on crop yields could lead to alarming risks for food availability, particularly for subsistence farmers and their families (Phalkey *et al.*, 2015). Moreover, studies have shown that the adverse effects of climate change on crop yields are not uniform across regions. Some areas may experience increased yields due to longer growing seasons while others, in arid and semi-arid regions, can face significant declines. This variability makes it harder to ensure food security, especially for regions already at risk of climate-induced shocks.

14.5 Nanotechnology in Crop Production

Numerous promising nanotechnologies have been proposed and evaluated at various scales. However, several barriers to their adoption must be addressed, including efficient field-scale delivery, regulatory and safety concerns, and consumer acceptance. (Hofmann *et al.*, 2020). Nanotechnology is still revolutionizing crop production by introducing innovative solutions as scientific discoveries are made and these enhance agricultural productivity and sustainability. The agricultural sector faces significant challenges due to increasing global populations, climate change and the need for sustainable practices. Traditional agricultural methods often lead to inefficiencies, such as nutrient runoff, excessive pesticide use and soil degradation. In this context, nanotechnology offers innovative solutions that can enhance crop production while minimizing environmental impacts. Nanotechnology involves manipulating materials at the nanoscale (1–100 nm), which can lead to unique properties and functionalities that are not present in bulk materials (Abd-Elrahman and Mostafa, 2015; Liu *et al.*, 2021). The various aspects of where nanotechnology is applied are further detailed below.

14.5.1 Nanofertilizers

Fertilizers play a crucial role in supporting healthy plant growth and maximizing agricultural productivity. Plants require key nutrients such as nitrogen (N), phosphorus (P) and potassium (K), along with other micronutrients, for various physiological functions. Fertilizers provide these nutrients in adequate amounts, ensuring optimal growth and yield. With time however, nutrients in the soil can become depleted due to continuous cropping, leaching or erosion. Fertilizers replenish these nutrients, maintaining soil fertility and enabling plants to thrive and proper fertilization results in higher crop yields by promoting vigorous growth, better flowering and fruiting. Proper input of fertilizers can improve the size, color, flavor and nutritional quality of fruits, vegetables and other crops, making them more appealing and nutritious for consumers. This is especially important to meet the food demands of the growing global population. At present there are issues with over-fertilizing plants and crops that can lead to several issues: including nutrient runoff, environmental pollution and plant health problems. Excess fertilizers can leach into waterways, causing water pollution and phenomena such as algal blooms. Additionally, over-fertilization can result in nutrient imbalances, making plants more susceptible to diseases and reducing their overall health. Thus, providing the right amount at the right time of the cropping season is a crucial part of sustainable farming.

Nanofertilizers represent a significant advancement in efficient nutrient delivery systems to plants (Table 14.1). Unlike conventional fertilizers, which can lead to nutrient leaching and environmental pollution, nanofertilizers are designed for controlled release, ensuring that nutrients are available to plants when needed (Tang *et al.*, 2023; Gowda *et al.*, 2024). Research has shown that nanofertilizers can enhance nutrient use efficiency, resulting in improved crop yields and reduced application rates (Sekhon, 2014; Singh and Kumar, 2024). For instance, studies indicate that the application of nano-sized nutrients can lead to better absorption by plant roots, thereby increasing the overall effectiveness of fertilization (Duhan *et al.*, 2017; Kumar, 2021).

14.5.2 Nanopesticides

Pest are an ever-present threat to food production and the use of nanopesticides is another promising application of nanotechnology in crop protection. These formulations allow for targeted delivery of active ingredients, reducing the overall quantity of pesticides required, minimizing their impact on non-target organisms and reducing costs for the farmer (Ghidan and Antary, 2020; Tang *et al.*, 2023) (Table 14.2). Nanopesticides can also enhance the efficacy of pest control by improving the solubility and stability of active compounds, leading to better pest management outcomes (Saikanth *et al.*, 2023; Singh and Kumar, 2024). Furthermore, the development of biodegradable nanopesticides can contribute to more sustainable pest management practices by reducing chemical residues in the environment (Sekhon, 2014; Kah, 2015).

14.5.3 Nanosensors and Plant Diagnostics

Historically, farmers have monitored key environmental parameters such as temperature, humidity, soil moisture, illumination, wind speed and rainfall using macroscale technologies. However, the emergence of micro sensors in the mid-20th century, which gained significant traction from the 1980s onward, has revolutionized agricultural food production by enabling more precise data collection to enhance crop yields. As technological advancements continue and our understanding of plant physiological processes deepens, it has become increasingly important to track additional parameters. These include soil pH levels, nutrient concentrations (such as NPK and micronutrients), air quality, evaporation rates, pollination activity, disease spread, CO_2 concentrations, dissolved oxygen and potential contaminants. Furthermore, crop-specific indicators such as phenological stages, pest presence and disease symptoms provide a comprehensive view of crop health.

Table 14.1. Classification of nanofertilizer materials, associated crops, application pathways and principal findings. (Authors' own table.)

Type of nanofertilizer	Crop	Application method	Key findings	Reference
Nano-zinc	Triticale	Soil application	Increased grain yield by 39% under salinity stress	Hussein and Abou-Baker (2018)
Nano-urea	Rice, maize	Foliar spray	Enhanced nitrogen use efficiency and crop yield; reduced conventional fertilizer use	Sadhukhan *et al.* (2022) and Kumar *et al.* (2023)
Nano-chitosan NPK	Wheat	Foliar application	Improved grain yield and protein content compared to conventional fertilizers	Abdel-Aziz *et al.* (2018)
Nano-potassium	Various crops	Soil application	Improved plant growth and nutrient uptake efficiency	Sebastian *et al.* (2023)
Nano-urea and Nano-DAP	Cauliflower, tomato	Foliar spray	Significant crop biomass and improved soil fertility	Sharma *et al.* (2024)

Table 14.2. Classification of nanopesticides, associated crops, application pathways and principal findings. (Authors' own table.)

Type of nanopesticide	Crop	Application method	Key findings	Reference
Nano-formulated atrazine	Various crops	Soil application	Decreased the required concentration of atrazine while affecting rhizosphere bacterial communities	Zhai *et al.* (2020)
Acetamiprid-loaded silica	Tea plants	Foliar spray	Enhanced absorption and distribution of acetamiprid, improving pest control efficacy	Wang *et al.* (2021)
Copper hydroxide nanoparticles	Various crops	Soil application	Improved pest control while assessing impacts on soil enzyme activity and bacterial communities	Zhang *et al.* (2020)
Nano-encapsulated Chlorantraniliprole	Vegetables	Foliar spray	Increased effectiveness in pest control with reduced environmental impact	Feng *et al.* (2021)
Zein nanoparticles	Various crops	Foliar application	Eco-friendly carrier systems for botanical repellents, enhancing pest management efficacy	de Oliveira *et al.* (2018)
Bacillus thuringiensis nano-bioformulations	Various crops	Soil application	Effective against major agricultural pests, providing a biological pest control alternative	Kumar *et al.* (2021)
Nano-encapsulated lambda cyhalothrin	Various crops	Foliar spray	Synergistic pest control effects with reduced pesticide load	Feng *et al.* (2021)
Nano-agrochemicals	Various crops	Various methods	Enhanced efficacy and reduced environmental contamination compared with conventional pesticides	Khan *et al.* (2022)
NanopesticidesI	Organic crops	Foliar and soil application	Improved pest management while minimizing adverse environmental impacts	Tailor *et al.* (2022)

Collectively, these measurements facilitate a holistic approach to precision agriculture, optimizing management strategies and improving overall productivity. With the continual developments in nanotechnology, nanosensors for these various parameters are now available on the market for farmers to apply to crop protection and production.

Nanosensors play a crucial role in precision agriculture by enabling real-time monitoring of soil and crop health. These sensors can detect nutrient levels, moisture content and the presence of pathogens, allowing farmers to make informed decisions about irrigation, fertilization and pest control (Ghidan and Antary, 2020; Liu *et al.*, 2021). Agricultural-based nanosensors can lead to more efficient resource use and improved crop management, ultimately enhancing productivity and sustainability (Kim *et al.*, 2017; Gowda *et al.*, 2024). For example, nanosensors can provide early warnings of plant stress, enabling timely interventions that can prevent yield losses (Saikanth *et al.*, 2023; Mohanty *et al.*, 2024). Moreover, with the integration with real-time monitoring via satellite tracking, it affords the farm manager to fully utilize all the resources to increase yield.

Nanotechnology-based plant disease diagnostics provide a highly sensitive, accurate and rapid approach to detecting pathogens, addressing the limitations of traditional methods such as microbial culturing and visual inspections. Nanotechnology integrates nanobiosensors, nanoparticle-based assays and nanomaterial-enhanced imaging to enable early disease detection. Nanobiosensors, combining nanomaterials with biological recognition elements, detect pathogens through electrical, optical or colorimetric signals in real-time (Table 14.3). Electrochemical nanobiosensors use gold nanoparticles or carbon nanotubes for enhanced signal detection, while optical nanobiosensors utilize quantum dots for fluorescence-based pathogen identification (Zhang *et al.*, 2024). Additionally, surface-enhanced Raman spectroscopy sensors amplify signals using metallic nanoparticles, allowing ultra-sensitive detection (Kashyap *et al.*, 2019).

Nanoparticle-based assays improve diagnostics by using gold nanoparticles, which change color upon binding to pathogen-specific molecules (Najafabad *et al.*, 2022), and magnetic nanomaterials to isolate pathogens from plant samples, increasing accuracy (Yadav and Yadav, 2018). Quantum dot-based assays provide fluorescence labeling for precise identification, while imaging techniques like nanoparticle-enhanced fluorescence imaging (Lv *et al.*, 2023) and nanostructured scanning electron microscopy offer high-resolution visualization of plant-pathogen interactions, aiding early-stage detection. These methods reduce detection time and allow farmers to take quick preventive measures, limiting disease spread.

Nano-based diagnostics offer key advantages such as high specificity, rapid detection, portability and reduced chemical use, making them valuable for sustainable agriculture. However, challenges such as high nanomaterial costs, regulatory concerns and scalability issues persist. By enabling early and precise pathogen detection, nano-based diagnostics help protect crop yields, reduce pesticide dependency and enhance global food security.

14.6 Nanobiotechnology

Over the past 50 years, biotechnology has profoundly transformed agriculture, leading to increased productivity, sustainability and crop resilience. Some of the key impacts include the

Table 14.3. Classification of nanosensors, applications, targeted parameters and principal findings. (Authors' own table.)

Type of nanosensor	Application	Target parameter	Key findings	Reference
Soil moisture nanosensor	Soil health monitoring	Soil moisture content	Enabled real-time monitoring of soil moisture, improving irrigation efficiency	Zhang *et al.* (2025)
pH nanosensor	Soil quality assessment	Soil pH	Provided accurate pH measurements, facilitating better nutrient management	Tiwari *et al.* (2020)
Pathogen detection nanosensor	Disease management	Pathogen presence	Successfully identified bacterial pathogens in soil, aiding in early disease detection	Kumari *et al.* (2023)
Temperature nanosensor	Climate monitoring	Soil and air temperature	Improved understanding of temperature effects on crop growth and development	Singh *et al.* (2022).
Volatile organic compound nanosensor (VOC)	Pest monitoring	VOC emissions from plants	Detected VOCs emitted during pest attacks, enabling timely pest management	Xu *et al.* (2022).
Humidity nanosensor	Environmental monitoring	Humidity levels	Provided critical data on humidity, aiding in optimal growth conditions for crops	Haseeb *et al.* (2020).
Nanoscale biosensors	Crop health assessment	Biochemical markers	Monitored plant stress responses, allowing for adaptive management strategies	Naseem *et al.* (2024)
Nanosensor networks	Precision agriculture	Multiple parameters	Integrated data collection from various sensors, enhancing decision-making processes	Upadhyay *et al.* (2023)

development of genetically modified (GM) crops, with the introduction of crops such as soybean, maize, cotton and canola with enhanced traits such as pest resistance (e.g. Bt crops), herbicide tolerance (e.g. glyphosate-resistant varieties) and crops with improved yield potential to farms. These modifications have reduced the need for chemical pesticides and herbicides, lowering environmental impact and production costs. Biotechnology has enabled the development of crops resistant to pests and diseases, drought-tolerant corn and rice varieties, allowing cultivation in marginal areas and reducing crop losses. Another example is the impact to improved nutritional content of plants such as the biofortification efforts which resulted in Golden Rice enriched with vitamin A, aimed at combatting nutritional deficiencies in developing countries. Pest-resistant crops have decreased pesticide use, leading to benefits for farmers' health and ecosystems and herbicide-tolerant crops have streamlined weed management. Biotechnology has greatly supported the accelerated breeding and genetic improvement in crops. Biotechnology tools such as marker-assisted selection and gene editing (e.g. CRISPR-Cas9) have shortened breeding cycles and introduced precise genetic changes. This accelerates the development of new varieties tailored to specific needs.

Nanobiotechnology encompasses the use of nanomaterials in genetic engineering and plant breeding. This approach allows for precise manipulation of plant genomes, facilitating the development of crops that are more resilient to environmental stresses, such as drought and disease (Abd-Elrahman and Mostafa, 2015; Ghidan and Antary, 2020). By employing nanoparticles as carriers for genetic material, researchers can enhance the efficiency of gene transfer techniques, leading to the creation of improved crop varieties (Table 14.4). This application of nanotechnology holds great promise for addressing the challenges posed by climate change and ensuring food security.

14.7 Nanotechnology Climate Resilience Responses

14.7.1 Nanomaterials

Nanotechnology has emerged as a transformative approach in agriculture, particularly in enhancing crop resilience to climate stress. The integration of nanotechnology into agricultural practices offers innovative solutions to address the challenges posed by climate change, including abiotic stresses such as drought, salinity and

Table 14.4. Nanoparticles as carriers of genetic material across plant growth in the agricultural production cycle. (Authors' own table.)

Application	Description	Key findings	Reference
Gene delivery systems	Use of nanoparticles for targeted delivery of genetic material	Enhanced efficiency of gene editing and transformation in crops	Allafchian *et al.* (2018); Abdalla *et al.* (2019).
Stress tolerance enhancement	Application of nanomaterials to improve abiotic stress tolerance	Improved resilience to drought and salinity in various crops	Halake and Haro (2022); Raza *et al.* (2023).
Bioremediation	Use of nanoparticles for soil and water remediation	Effective removal of pollutants and restoration of soil health	Shahcheraghi *et al.* (2022); Singh and Kumar (2024).
Plant growth promotion	Nanoparticles used to stimulate plant growth and development	Enhanced seed germination and root development	Sharma and Gupta (2021); Salimi *et al.* (2024).
Disease resistance	Nanobiotechnology applied to enhance plant disease resistance	Development of disease-resistant crop varieties through genetic modification	Jenne *et al.* (2018); Fazil *et al.* (2024).
Smart packaging	Use of nanotechnology in food packaging	Improved shelf life and safety of agricultural products	Alvarado *et al.*, 2019; Kansotia *et al.*, 2024).

extreme temperatures. Some nanomaterials improve nutrient delivery and uptake efficiency, which is crucial for enhancing plant growth and resilience under stress conditions. For instance, studies have shown that nanoparticles can facilitate faster seed germination and improve resistance to both biotic and abiotic stresses, ultimately leading to increased crop yields and quality (Liu *et al.*, 2021). The use of zinc oxide nanoparticles, for example, has been documented to alleviate heat stress in plants, enhancing their growth and productivity. Furthermore, nanotechnology enables the targeted application of agrochemicals, reducing environmental contamination and improving soil health (Ahmad *et al.*, 2022).

Additionally, nanotechnology plays a significant role in the development of stress-smart crops through advanced breeding techniques and genome editing. Nanoparticles have been shown to induce metabolic and transcriptional changes in plants, enhancing their tolerance to salinity and other stressors (Yan *et al.*, 2022; Raza *et al.*, 2023). The CRISPR-Cas9 system, when combined with nanobiotechnology, offers a promising avenue for developing crops that can withstand the multifaceted challenges posed by climate change (Raza *et al.*, 2023). These approaches not only focus on enhancing individual stress tolerance but also address the complex interactions between different stressors, which is critical for maintaining crop productivity in fluctuating environmental conditions (Rivero *et al.*, 2021).

14.7.2 Nanocoatings

Nanocoatings for climate-resilient crops constitute a burgeoning domain within agricultural nanotechnology, aiming to augment plant resilience and productivity amidst increasingly adverse environmental conditions. These coatings, predominantly formulated from nanoparticles, are applied either directly onto plant surfaces or seeds, or integrated into the soil matrix to confer a multitude of protective functionalities.

Current investigative efforts are centered on developing innovative nanocoatings to mitigate the detrimental impacts of climate change on agricultural systems. Notable nanomaterials employed include silica nanoparticles (Mushtaq *et al.*, 2018), titanium dioxide nanoparticles (Tan *et al.*, 2017) and biopolymer-based nanocomposites (Manjunatha *et al.*, 2016). These materials are engineered to bolster crop survival under extreme abiotic stresses such as drought, elevated temperatures and heightened ultraviolet (UV) radiation levels. An innovative research stream is the fabrication of "smart" seed coatings – nanopolymer layers designed to facilitate germination only under favorable environmental conditions, thereby optimizing seedling establishment (Abd-Elrahman and Mostafa, 2015).

Nanocoatings also serve to enhance water retention capabilities and diminish evaporation rates, which is particularly advantageous in arid and water-scarce regions, thus aiding in maintaining adequate soil moisture levels (Hatti *et al.*, 2020). Moreover, these coatings can form protective barriers against UV-induced damage, shielding plants from intensified solar radiation due to ozone layer depletion (Zambrano-Zaragoza *et al.*, 2021).

Empirical studies have demonstrated that nanocoatings improve plant resistance to various abiotic stresses by establishing physical and chemical shields that preserve cellular integrity under adverse conditions (Hayat *et al.*, 2023). Additionally, nanocoatings can function as carriers for nutrients and micronutrients, facilitating more efficient nutrient delivery and uptake. This targeted nutrient release from nano-encapsulated fertilizers enhances soil fertility and supports plant growth in nutrient-deficient environments, while also reducing nutrient runoff and environmental pollution (Guo *et al.*, 2018).

Furthermore, the development of biodegradable nanocoatings offers an environmentally sustainable alternative to traditional chemical treatments, aligning with principles of eco-friendly agriculture (Ramos *et al.*, 2014). Ongoing research aims to optimize the composition, structure and functional properties of these nanocoatings to improve their efficacy, durability and ecological compatibility, thereby contributing to the advancement of resilient agricultural practices in the face of climate change.

14.7.3 Water Management and Soil Health

Nanotechnology is revolutionizing water management and soil remediation by providing innovative solutions to address global challenges related to water scarcity, soil contamination and environmental degradation. Recent research has shown that nanomaterials, due to their unique properties such as high surface area, reactivity and size, offer significant potential for improving both water quality and soil health. In water management, nanotechnology plays a critical role in enhancing the efficiency of water purification processes, improving water storage and enabling more precise irrigation techniques (Dasgupta *et al.*, 2017). Nanomaterials such as carbon nanotubes, Fullerene, graphene and nanoscale zero-valent iron (nZVI) are being explored for their ability to remove pollutants such as heavy metals, pesticides and organic compounds from contaminated water (Singh *et al.*, 2022; Krishna *et al.*, 2023). These materials work by adsorbing or chemically reacting with contaminants, making them more efficient than conventional methods. For instance, nZVI nanoparticles have demonstrated excellent performance in removing toxic substances such as arsenic and chromium from groundwater (Li *et al.*, 2020; Yin *et al.*, 2022), while novel graphene oxide-based filters offer a promising approach to desalination and wastewater treatment by selectively allowing water molecules to pass while blocking contaminants ((Li *et al.*, 2019).

In addition to purification, nanotechnology is being integrated into precision irrigation systems, which use nanosensors to monitor soil moisture levels and weather conditions, thereby optimizing water usage. These sensors can provide real-time data to farmers, allowing for the efficient application of water only where and when it is needed, thus reducing water waste and improving crop yield. Furthermore, nano-enabled hydrogels are being developed to improve water retention in soil, making them particularly useful in arid regions where water scarcity is a major concern. These hydrogels can absorb large amounts of water and release it slowly, ensuring that plants have access to water during dry spells (Albalasmeh *et al.*, 2022). Research shows that nano-biochar increases the water-holding capacity of soils, making it particularly useful in degraded lands (Abed Hussein *et al.*, 2022).

In soil remediation, nanotechnology offers innovative approaches for cleaning up contaminated soils, especially in areas affected by heavy metals, organic pollutants and hazardous waste (Rabbani *et al.*, 2016). Nanoparticles such as nZVI are used for soil decontamination through a process known as chemical reduction, where the nanoparticles interact with organic contaminants and transform them into less harmful forms (Li *et al.*, 2021). Additionally, nanomaterials can enhance phytoremediation, a process where plants are used to absorb and remove toxins from the soil. Nanoparticles can be applied to plant roots to increase their uptake of pollutants, such as heavy metals, and improve the overall efficiency of the phytoremediation process.

Recent research on nanoclay materials highlights their potential in improving soil structure and fertility in degraded lands. Nanoclays, due to their high surface area and unique physicochemical properties, have been shown to enhance soil porosity and water retention and to reduce compaction. Nanoclay and its nanocomposites have good adsorption capacity and have been modified to separate pollutants, notably, heavy metals (Cd, Pd, Co, Cu, As, Hg, Ni, Zn and Cr), ammonia, nitrate, fluoride, dyes, bacteria and other emerging pollutants (Manna *et al.*, 2021; Abd-Elsalam *et al.*, 2024). The application of nanoclay polymer composites as water management materials for the renewal of arid and desert environments has recently received a lot of attention (Abd-Elsalam *et al.*, 2024). Promising results have been observed, as they can reduce irrigation water consumption, improve nutrient retention in soil, reduce plant death and increase crop growth (Mukhopadhyay and De, 2014; Nayak *et al.*, 2025). Moreover, these studies indicate that nanoclays can adsorb essential nutrients such as nitrogen, phosphorus and potassium, improving their availability to plants and reducing nutrient leaching. Additionally, nanoclays promote beneficial microbial activity, which is vital for soil health and nutrient cycling (Merino *et al.*, 2020).

14.7.4 Enhancing Carbon Sequestration in Soils

As soils are among the largest terrestrial carbon sinks, improving their ability to store carbon is a key focus of current research in soil science and nanotechnology. Carbon sequestration, the process by which carbon dioxide (CO_2) is captured and stored in soil, is a critical strategy for reducing atmospheric CO_2 levels, which contribute to global warming. Recent advances in nanotechnology have shown great promise in enhancing carbon sequestration by improving soil structure, promoting organic matter stabilization and facilitating more efficient carbon storage (Pramanik *et al.*, 2020).

One of the primary ways nanotechnology aids in carbon sequestration is through the development of nanomaterials that can enhance the bioavailability and stability of soil organic matter, which is crucial for long-term carbon storage. Nanomaterials, such as nanoclays and bio-based nanocomposites, are being incorporated into soils to improve their physical and chemical properties (Mondal *et al.*, 2021; Abd-Elsalam *et al.*, 2024). Nanoparticles such as nano-silica, nano-zeolite, nano-ZnO and nano-Fe particles can promote soil aggregation and carbon build-up in agricultural soil (Pramanik *et al.*, 2020). Additionally, nano-biochar has been shown to improve soil's ability to store carbon by increasing its surface area and porosity. Nano-biochar not only improves the stability of carbon but also enhances nutrient retention and microbial activity in soils, promoting long-term carbon sequestration (Sultan *et al.*, 2024). However, while nanotechnology offers promising solutions for soil carbon sequestration, further research is required to assess its long-term ecological impacts and potential risks to soil microbial communities.

14.7.5 Mitigation of Greenhouse Gas (GHG) Emissions

Climate change is an important threat to agriculture. Increases in extreme weather events such as floods, droughts and heatwaves due to climate change effects adversely affects agriculture. Therefore, accelerated increase in the concentration of greenhouse gases (GHGs) in the atmosphere is a cause of concern. Nanotechnology offers innovative solutions for mitigating GHG emissions, particularly by enhancing soil management practices that reduce nitrous oxide (N_2O) and methane (CH_4) emissions. Nanofertilizers such as nano-chitosan have demonstrated significant potential in improving nitrogen use efficiency, thereby reducing excess nitrogen in the soil and minimizing N_2O emissions (Abuelsoud *et al.*, 2023; Omoregie *et al.*, 2025). Furthermore, direct application of nanomaterials such as nZVI has been shown to enhance microbial denitrification, converting nitrogen compounds into less harmful forms and reducing N_2O emissions from pesticide-contaminated paddy feild soil (Su *et al.*, 2023). Research indicates that nZVI can effectively control methane emissions in wastewater treatment settings (Cen *et al.*, 2025).

Nanomaterials due to their unique properties are reported to enhance carbon sequestration in soil and aid in mitigation of GHG emissions. However, contradictory reports on the potential impact of nanomaterials on soil microorganisms are one of the major reasons to limit the adoption of this technology at large scale (Pramanik *et al.*, 2020). Concerns regarding nanoparticle toxicity, bioaccumulation and ecological risks must be addressed before its large-scale implementation.

14.7.6 Nanotechnology-based Genetic Modifications

Nanotechnology-based genetic modifications in agriculture offer a ground-breaking approach to addressing global food security amid climate change. As climate patterns become increasingly unpredictable: with extreme weather events such as droughts, floods and temperature fluctuations, traditional farming methods are struggling to meet rising food demands. Nanotechnology, involving the manipulation of materials at the molecular or atomic scale, can be harnessed to enhance crop genetics, improving their resilience to these changing conditions. One of the key advantages of nanotechnology in genetic modification is its ability to enable precise delivery systems for genes, allowing researchers to insert or modify specific genes in crops with greater accuracy and efficiency (Manjunatha *et al.*, 2016). This targeted approach minimizes the risk of unintended genetic effects, which are often a challenge in conventional genetic modification techniques.

CRISPR-Cas9 is a ground-breaking gene-editing technology that allows scientists to make precise modifications to the DNA of living organisms, including plants. CRISPR-Cas9 was discovered through the collective efforts of multiple research teams. The foundational work identifying the CRISPR (Clustered Regularly Interspaced Short Palindromic Repeats) sequences were carried out by Francisco Mojica and colleagues in the late 1990s and early 2000s. They observed these sequences in the genomes of bacteria and proposed their role in bacterial immune defense. The breakthrough in harnessing CRISPR-Cas9 as a gene-editing tool is credited primarily to Jennifer Doudna and Emmanuelle Charpentier. In 2012, they published a landmark paper demonstrating that the Cas9 enzyme could be programmed with RNA to target and cut specific DNA sequences, effectively developing a versatile gene-editing system. Their work laid the foundation for the subsequent widespread use of CRISPR-Cas9 in biotechnology, medicine and agriculture.

In agriculture, this tool has revolutionized crop breeding by enabling the development of crops that are more resilient to pests, diseases and environmental stresses such as drought and salinity. Unlike traditional breeding methods, which can take many years to achieve desired traits, CRISPR-Cas9 accelerates this process by directly editing specific genes, thus saving time and resources. This precision also reduces the likelihood of unintended genetic changes, leading to safer and more targeted improvements in crop varieties.

The application of CRISPR-Cas9 in crop farming holds significant potential for enhancing food security and sustainability. It allows for the creation of high-yield, nutrient-rich and climate-adapted crops that can thrive in challenging growing conditions. Additionally, gene editing can reduce the reliance on chemical pesticides and herbicides by developing pest-resistant and herbicide-tolerant crops, which benefits the environment and reduces farming costs.

Overall, CRISPR-Cas9 offers a promising tool for sustainable agriculture by enabling the rapid development of innovative crop solutions to meet the growing global demand for food production. Researchers employed CRISPR-Cas9 to edit genes associated with water use efficiency and stress response, resulting in rice plants that can better withstand drought conditions. For instance, a study demonstrated how editing the OsSAPK2 gene in rice improved drought tolerance without sacrificing yield, helping farmers cope with water scarcity linked to climate change (Lou *et al.*, 2017; Li *et al.*, 2022). Another example involves wheat, where CRISPR-Cas9 has been used to modify genes related to heat tolerance. Scientists edited specific genes to produce wheat varieties that maintain productivity under higher temperature stresses, which are becoming more frequent due to global warming.

These advancements illustrate how CRISPR-Cas9 can be strategically used to develop crop varieties better adapted to the changing climate, thereby supporting agricultural resilience and food security.

14.8 Challenges and Future Perspectives in Nanotechnology-enhanced Agriculture

Nanotechnology holds immense promise for revolutionizing crop production and addressing pressing environmental challenges such as climate change. By manipulating materials at the molecular or atomic scale, nanotechnology offers innovative solutions to improve resource efficiency, enhance crop resilience and enable precision agriculture. Moreover, nanotools such as the atomic force and other scanning probe microscopies will enable agricultural scientists to further understand the physiology of plants, insects and pests and other pathogens and deepen our knowledge base. Applications including nanofertilizers, nanopesticides, nanosensors, soil remediation, water management and carbon sequestration can significantly mitigate the impacts on agriculture from climate change – such as reduced crop yields, soil degradation and water scarcity, while promoting sustainable farming practices.

However, despite these potential benefits, several challenges must be overcome to realize nanotechnology's full impact in agriculture. Concerns regarding the environmental and human health impacts of nanomaterials, particularly their toxicity and long-term effects, necessitate rigorous research and robust regulatory frameworks. Additionally, the economic feasibility and scalability of nanotechnology based solutions still face critical hurdles, especially for widespread adoption in developing countries where agricultural challenges are most acute.

The successful incorporation of nanotechnology into agriculture depends on positive collaboration among researchers, policymakers and farmers to ensure that innovations are safe, accessible and beneficial for all stakeholders. Future research should prioritize understanding the long-term impacts of nanomaterials on soil health, plants, ecosystem dynamics and crop yields. Developing affordable, scalable nanotech solutions will be crucial for promoting sustainable and climate-resilient agriculture worldwide. As advancements continue – not only in applying nanotechnology to agriculture but also in expanding our understanding of nanoscience related to plant physiology, weeds, insects, pests and other pathogens – agriculturists and farm managers will gain a more comprehensive perspective. Addressing safety concerns and establishing regulatory frameworks will be essential for enabling nanotechnology to effectively contribute to environmental management, food security and the mitigation of global environmental and climatic challenges.

References

Abdalla, N., Ragab, M.I., Fári, M., El-Ramady, H., Alshaal, T. *et al.* (2019) Nanobiotechnology for plants. *Environment, Biodiversity and Soil Security* 2, 155–174. DOI: 10.21608/jenvbs.2018.6711.1041.

Abd-Elrahman, S. and Mostafa, M. (2015) Applications of nanotechnology in agriculture: An overview. *Egyptian Journal of Soil Science* 55(2), 197–214. DOI: 10.21608/ejss.2015.324.

Abd-Elsalam, K.A., Mehmood, M.A., Ashfaq, M., Abdelkhalek, T.E., Hassan, R.K. and Ravichandran, M. (2024) Liquid nanoclay: Synthesis and applications to transform an arid desert into fertile land. *Soil Systems* 8(3), 73–73. DOI: 10.3390/soilsystems8030073.

Abdel-Aziz, H., Hasaneen, M.N. and Omar, A. (2018) Effect of foliar application of nano chitosan NPK fertilizer on the chemical composition of wheat grains. *Egyptian Journal of Botany* 58(1), 87–95. DOI: 10.21608/ejbo.2018.1907.1137.

Abed Hussein, B., Mahdi, A.B., Emad Izzat, S., Acwin Dwijendra, N.K., Romero Parra, R.M. *et al.* (2022) Production, structural properties of nano biochar, and effects of nano biochar in soil: A review. *Egyptian Journal of Chemistry* 65(12), 607–618. DOI: 10.21608/ejchem.2022.131162.5772.

Abuelsoud, W., Saleh, A.M., Mohammed, A.E., Alotaibi, M.O. and Abd Elgawad, H. (2023) Chitosan nanoparticles upregulate C and N metabolism in soybean plants grown under elevated levels of atmospheric carbon dioxide. *International Journal of Biological Macromolecules* 252, 126434. DOI: 10.1016/j.ijbiomac.2023.126434.

Ahmad, Z., Tahseen, S., Wasi, A., Ganie, I.B., Shahzad, A. *et al.* (2022) Nanotechnological interventions in agriculture. *Nanomaterials* 12(15), 2667. DOI: 10.3390/nano12152667.

Albalasmeh, A., Mohawesh, O., Gharaibeh, M.A., Alghamdi, A.G., Alajlouni, M.A. and Alqudah, A.M. (2022) Effect of hydrogel on corn growth, water use efficiency, and soil properties in a semi-arid region. *Journal of the Saudi Society of Agricultural Sciences* 21(8), 518–524. DOI: 10.1016/j.jssas.2022.03.001.

Allafchian, A., Jalali, S.A.H., Aghaei, F. and Farhang, H.R. (2018) Green synthesis of silver nanoparticles using *Glaucium corniculatum* (L.) Curtis extract and evaluation of its antibacterial activity. *IET Nanobiotechnology* 12(5), 574–578. DOI: 10.1049/iet-nbt.2017.0265.

Alvarado, K., Bolaños, M., Camacho, C., Quesada, E.M. and Vega-Baudrit, J.R. (2019) Nanobiotechnology in agricultural sector: Overview and novel applications. *Journal of Biomaterials and Nanobiotechnology* 10(02), 120–141. DOI: 10.4236/jbnb.2019.102007.

Brenton, P., Chemutai, V. and Pangestu, M. (2022) Trade and food security in a climate change-impacted world. *Agricultural Economics* 53(4), 580–591. DOI: 10.1111/agec.12727.

Cen, X., Hu, Z., Yu, Z., Huang, X., Zuo, Z. *et al.* (2025) Integrated application of nanoscale zero-valent iron for sulfide and methane control in sewers and improved wastewater treatment. *Water Research* 276, 123248–123248. DOI: 10.1016/j.watres.2025.123248.

Dasgupta, N., Ranjan, S. and Ramalingam, C. (2017) Applications of nanotechnology in agriculture and water quality management. *Environmental Chemistry Letters* 15(4), 591–605. DOI: 10.1007/s10311-017-0648-9.

de Oliveira, J.L., Campos, E.V.R., Pereira, A.E.S., Pasquoto, T., Lima, R. *et al.* (2018) Zein nanoparticles as eco-friendly carrier systems for botanical repellents aiming sustainable agriculture. *Journal of Agricultural and Food Chemistry* 66(6), 1330–1340. DOI: 10.1021/acs.jafc.7b05552.

Duhan, J., Kumar, R., Kumar, N., Kaur, P., Nehra, K. and Duhan, S. (2017) Nanotechnology: The new perspective in precision agriculture. *Biotechnology Reports* 15, 11–23. DOI: 10.1016/j.btre.2017.03.002.

Durán-Sandoval, D., Uleri, F., Romero, G. and López, A. (2023) Food, climate change, and the challenge of innovation. *Encyclopedia* 3(3), 839–852. DOI: 10.3390/encyclopedia3030060.

Dutta, S., Pal, S., Panwar, P., Sharma, R.K. and Bhutia, P.L. (2022) Biopolymeric nanocarriers for nutrient delivery and crop biofortification. *ACS Omega* 7(30), 25909–25920. DOI: 10.1021/acsomega.2c02494.

El-Saadony, M.T., Saad, A.M., Soliman, S.M., Salem, H.M., Desoky, E.-S.M. *et al.* (2022) Role of nanoparticles in enhancing crop tolerance to abiotic stress: A comprehensive review. *Frontiers in Plant Science* 13, 946717. DOI: 10.3389/fpls.2022.946717.

Fazil, M.M., Gul, A. and Jawed, H. (2024) Optimization of silver nanoparticles synthesis via Plackett–Burman experimental design: In vitro assessment of their efficacy against oxidative stress-induced disorders. *RSC Advances* 14(29), 20809–20823. DOI: 10.1039/d4ra02774d.

Feng, B., Zhi, H., Chen, H., Cui, B., Zhao, X. *et al.* (2021) Development of chlorantraniliprole and lambda cyhalothrin double-loaded nano-microcapsules for synergistical pest control. *Nanomaterials* 11(10), 2730–2730. DOI: 10.3390/nano11102730.

Garg, D., Sridhar, K., Inbaraj, B.S., Chawla, P., Tripathi, M. and Sharma, M. (2023) Nano-biofertilizer formulations for agriculture: A systematic review on recent advances and prospective applications. *Bioengineering* 10(9), 1010–1010. DOI: 10.3390/bioengineering10091010.

Ghidan, A. and Antary, T. (2020) Applications of nanotechnology in agriculture. *IntechOpen* 1–14. DOI: 10.5772/intechopen.88390.

Giri, V.P., Shukla, P., Tripathi, A., Verma, P., Kumar, N. *et al.* (2023) A review of sustainable use of biogenic nanoscale agro-materials to enhance stress tolerance and nutritional value of plants. *Plants* 12(4), 815. DOI: 10.3390/plants12040815.

Gowda, R., Roopashree, B., Umarani, K. and Sowmya, K. (2024) Redgram resurgence – A perspective of nanotechnological approach. *Environment and Ecology* 42(1A), 248–254. DOI: 10.60151/envec/ LUIT8549.

Guo, H., White, J.C., Wang, Z. and Xing, B. (2018) Nano-enabled fertilizers to control the release and use efficiency of nutrients. *Current Opinion in Environmental Science & Health* 6, 77–83. DOI: 10.1016/j.coesh.2018.07.009.

Halake, N.H. and Haro, J.M. (2022) Role of nanobiotechnology towards agri-food system. *Journal of Nanotechnology* 2022, 1–7. DOI: 10.1155/2022/6108610.

Haseeb, K., Din, I.U., Almogren, A. and Islam, N. (2020) An energy efficient and secure IoT-based WSN framework: An application to smart agriculture. *Sensors* 20(7), 2081. DOI: 10.3390/s20072081.

Hatti, V., Raghavendra, M., Singh, Y.V., Sharma, S.K., Halli, H.M. and Goud, B.R. (2020) Application of nano technology in crop production to enhance resource use efficiency with special reference to soil and water conservation: A review. *Journal of Soil and Water Conservation* 19(4), 375–381. DOI: 10.5958/2455-7145.2020.00050.8.

Hayat, F., Khanum, F., Li, J., Iqbal, S., Khan, U. *et al.* (2023) Nanoparticles and their potential role in plant adaptation to abiotic stress in horticultural crops: A review. *Scientia Horticulturae* 321, 112285. DOI: 10.1016/j.scienta.2023.112285.

Hofmann, T., Lowry, G.V., Ghoshal, S., Tufenkji, N., Brambilla, D. *et al.* (2020) Technology readiness and overcoming barriers to sustainably implement nanotechnology-enabled plant agriculture. *Nature Food* 1(7), 416–425. DOI: 10.1038/s43016-020-0110-1.

Hussein, M.M. and Abou-Baker, N.H. (2018) The contribution of nano-zinc to alleviate salinity stress on cotton plants. *Royal Society Open Science* 5(8), 171809. DOI: 10.1098/rsos.171809.

Huyer, S., Simelton, E., Chanana, N., Mulema, A.A. and Marty, E. (2021) Expanding opportunities: A framework for gender and socially-inclusive climate resilient agriculture. *Frontiers in Climate* 3, 718240. DOI: 10.3389/fclim.2021.718240.

Janbandhu, M.S., Mehta, A., Beese, S., Pandey, S.K., Singh, B., Patel, A. and Singh, B.P. (2024) Advances and emerging trends in horticultural production and management. *Journal of Experimental Agriculture International* 46(3), 47–69. DOI: 10.9734/jeai/2024/v46i32325.

Jenne, M., Kambham, M., Tollamadugu, N.V.K.V.P., Karanam, H.P., Tirupati, M.K. *et al.* (2018) The use of slow releasing nanoparticle encapsulated Azadirachtin formulations for the management of *Caryedon serratus* O. (groundnut bruchid). *IET Nanobiotechnology* 12(7), 963–967. DOI: 10.1049/iet-nbt.2017.0276.

Kah, M. (2015) Nanopesticides and nanofertilizers: Emerging contaminants or opportunities for risk mitigation? *Frontiers in Chemistry* 3, 00064. DOI: 10.3389/fchem.2015.00064.

Kah, M., Tufenkji, N. and White, J.C. (2019) Nano-enabled strategies to enhance crop nutrition and protection. *Nature Nanotechnology* 14(6), 532–540. DOI: 10.1038/s41565-019-0439-5.

Kansotia, K., Naresh, R., Sharma, Y., Sekhar, M., Sachan, P., Baral, K. and Pandey, S.K. (2024) Nanotechnology-driven solutions: Transforming agriculture for a sustainable and productive future. *Journal of Scientific Research and Reports* 30(3), 32–51. DOI: 10.9734/jsrr/2024/v30i31856.

Kashyap, P.L., Kumar, S., Jasrotia, P., Singh, D.P. and Singh, G.P. (2019) *Nanosensors for Plant Disease Diagnosis: Current Understanding and Future Perspectives*. Springer eBooks, pp. 189–205. Cham, Switzerland. DOI: 10.1007/978-3-319-97852-9_9.

Khan, F., Pandey, P. and Upadhyay, T.K. (2022) Applications of nanotechnology-based agrochemicals in food security and sustainable agriculture: An overview. *Agriculture* 12(10), 1672. DOI: 10.3390/agriculture12101672.

Kim, D., Kadam, A., Shinde, S., Saratale, R., Patra, J. and Ghodake, G. (2017) Recent developments in nanotechnology transforming the agricultural sector: A transition replete with opportunities. *Journal of the Science of Food and Agriculture* 98(3), 849–864. DOI: 10.1002/jsfa.8749.

Krishna, R.H., Chandraprabha, M.N., Samrat, K., Murthy, T.K., Manjunatha, C. and Kumar, S.G. (2023) Carbon nanotubes and graphene-based materials for adsorptive removal of metal ions – A review on surface functionalization and related adsorption mechanism. *Applied Surface Science Advances* 16, 100431. DOI: 10.1016/j.apsadv.2023.10043.

Kumar, Y. (2021) Nanofertilizers and their role in sustainable agriculture. *Annals of Plant and Soil Research* 23(3), 238–255. DOI: 10.47815/apsr.2021.10067.

Kumar, P., Pandhi, S., Mahato, D.K., Kamle, M. and Mishra, A. (2021) Bacillus-based nano-bioformulations for phytopathogens and insect–pest management. *Egyptian Journal of Biological Pest Control* 31(1), 128. DOI: 10.1186/s41938-021-00475-6.

Kumar, M., Singh, R., Maurya, S.K., Maurya, S.K., Maurya, D.K. *et al.* (2023) Efficient use of nano-fertilizer for increasing productivity and profitability along with maintain sustainability in rice crop: A review. *International Journal of Enviornment and Climate Change* 13(10), 1358–1368. DOI: 10.9734/ijecc/2023/v13i102788.

Kumari, R., Suman, K., Karmakar, S., Mishra, V., Lakra, S.G., Saurav, G.K. and Mahto, B.K. (2023) Regulation and safety measures for nanotechnology-based agri-products. *Frontiers in Genome Editing* 5, 1200987. DOI: 10.3389/fgeed.2023.1200987.

Li, J., Fan, M., Li, M. and Liu, X. (2020) Cr (VI) removal from groundwater using double surfactant-modified nanoscale zero-valent iron (nZVI): Effects of materials in different status. *Science of the Total Environment* 717, 137112. DOI: 10.1016/j.scitotenv.2020.137112.

Li, X., Xu, S., Fuhrmann-Aoyagi, M.B., Yuan, S., Iwama, T., Kobayashi, M. and Miura, K. (2022) CRISPR/Cas9 technique for temperature, drought, and salinity stress responses. *Current Issues in Molecular Biology* 44(6), 2664–2682. DOI: 10.3390/cimb44060182.

Li, Y., Zhao, H.P. and Zhu, L. (2021) Remediation of soil contaminated with organic compounds by nanoscale zero-valent iron: A review. *Science of the Total Environment* 760, 143413. DOI: 10.1016/j.scitotenv.2020.143413.

Li, X., Zhu, B. and Zhu, J. (2019) Graphene oxide based materials for desalination. *Carbon* 146, 320–328. DOI: 10.1016/j.carbon.2019.02.007.

Liu, C., Zhou, H. and Zhou, J. (2021) The applications of nanotechnology in crop production. *Molecules* 26(23), 7070. DOI: 10.3390/molecules26237070.

Lou, D., Wang, H., Liang, G. and Yu, D. (2017) OsSAPK2 confers abscisic acid sensitivity and tolerance to drought stress in rice. *Frontiers in Plant Science* 8, 993. DOI: 10.3389/fpls.2017.00993.

Lowry, G.V., Avellan, A. and Gilbertson, L.M. (2019) Opportunities and challenges for nanotechnology in the agri-tech revolution. *Nature Nanotechnology* 14(6), 517–522. DOI: 10.1038/s41565-019-0461-7.

Lv, Y., Fan, J., Zhao, M., Wu, R. and Li, L.S. (2023) Recent advances in quantum dot-based fluorescence-linked immunosorbent assays. *Nanoscale* 15(12), 5560–5578. DOI: 10.1039/D2NR07247E.

Manjunatha, S.B., Biradar, D.P. and Aladakatti, Y.R. (2016) Nanotechnology and its applications in agriculture: A review. *Journal of Farm Sciences* 29(1), 1–13.

Manna, S., Das, P., Basak, P., Sharma, A.K., Singh, V.K. *et al.* (2021) Separation of pollutants from aqueous solution using nanoclay and its nanocomposites: A review. *Chemosphere* 280, 130961. DOI: 10.1016/j.chemosphere.2021.130961.

Masson-Delmotte, V. and Intergovernmental Panel on Climate Change, contributor (2022) *Climate Change and Land : An IPCC Special Report on Climate Change, Desertification, Land Degradation, Sustainable Land Management, Food Security, and Greenhouse Gas Fluxes in Terrestrial Ecosystems/Edited by Valérie Masson-Delmotte [and Twenty-One Others]*. 1st edn, Cambridge University Press, Cambridge, United Kingdom.

Merino, D., Tomadoni, B., Salcedo, M.F., Mansilla, A.Y., Casalongué, C.A. and Alvarez, V.A. (2020) Nanoclay as carriers of bioactive molecules applied to agriculture. In *Handbook of Nanomaterials and Nanocomposites for Energy and Environmental Applications*, pp. 1–22. Springer International Publishing, Cham, Switzerland. DOI: 10.1007/978-3-030-11155-7_62-1.

Mohanty, L., Singh, A., Pandey, A., Kumar, R., Ramesh, G. *et al.* (2024) Harnessing nanotechnology for eco-friendly crop enhancement and sustainable agriculture. *Journal of Experimental Agriculture International* 46(7), 154–167. DOI: 10.9734/jeai/2024/v46i72568.

Mondal, M., Biswas, B., Garai, S., Sarkar, S., Banerjee, H. *et al.* (2021) Zeolites enhance soil health, crop productivity and environmental safety. *Agronomy* 11(3), 448. DOI: 10.3390/agronomy11030448.

Mukhopadhyay, R. and De, N. (2014) Nano clay polymer composite: Synthesis, characterization, properties and application in rainfed agriculture. *Global Journal of Bio-Science and BioTechnology* 3(2), 133–138.

Mushtaq, A., Jamil, N., Rizwan, S., Mandokhel, F., Riaz, M. *et al.* (2018) Engineered silica nanoparticles and silica nanoparticles containing controlled release fertilizer for drought and saline areas. In *IOP Conference Series: Materials Science and Engineering*. Vol. 414, IOP Publishing, 012029. Bristol, England.

Najafabad, M.B., Rastin, S.J., Taghvaei, F. and Khiyavi, A.A. (2022) A review on applications of gold nanoparticles-based biosensor for pathogen detection. *Advances in Natural Sciences: Nanoscience and Nanotechnology* 13(3), 033002. DOI: 10.1088/2043-6262/ac79ae.

Naseem, J., Shah, A.A., Usman, S., Gatasheh, M.K., Shafique, S. and Javad, S. (2024) Green synthesized Fe-Nps ameliorate drought stress in *Spinacia oleracea* L. through improved photosynthetic capacity, redox balance, and antioxidant defense. *Scientific Reports* 15, 1782. DOI: 10.21203/rs.3.rs-4798044/v1.

Nayak, B.K., Chatterjee, D., Paul, R., Das, S.R., Adak, T. *et al.* (2025) Review of nano-clay polymer composites for controlled nitrogen release: Prospects and limitations. *Agricultural Research* 14(4), 675–696. DOI: 10.1007/s40003-025-00848-5.

Negash, H. and Fininsa, C. (2015) Effect of climate change resilience strategies on productivity of common bean (*Phaseolus vulgaris* L.) in semi-arid areas of Eastern Hararghe, Ethiopia. *African Journal of Agricultural Research* 10(15), 1852–1862. DOI: 10.5897/ajar2015.9634.

Omoregie, I.P., Oyegbade, S.A., Oyewole, O.A. and Adetunji, C.O. (2025). In Adetunji, C.O., Shah, M.P., Bello, Y.M., Hefft, D., Singh, J., Pandey, S.S. and Singh, R.P. (eds) *Reduction of Agricultural Greenhouse Gas Emissions by Nanochitosan*, pp. 69–87. Wiley, New Jersey, USA. DOI: 10.1002/9781394214945.ch3.

Phalkey, R.K., Aranda-Jan, C., Marx, S., Höfle, B. and Sauerborn, R. (2015) Systematic review of current efforts to quantify the impacts of climate change on undernutrition. *Proceedings of the National Academy of Sciences USA* 112(33), E4522–E4529. DOI: 10.1073/pnas.1409769112.

Pramanik, P., Ray, P., Maity, A., Das, S., Ramakrishnan, S. and Dixit, P. (2020) Nanotechnology for improved carbon management in soil. In *Carbon Management in Tropical and Sub-Tropical Terrestrial Systems*, pp. 403–415. Springer Nature Singapore, Singapore. DOI: 10.1007/978-981-13-9628-1_24.

Prasad, R., Bhattacharyya, A. and Nguyen, Q.D. (2017) Nanotechnology in sustainable agriculture: Recent developments, challenges, and perspectives. *Frontiers in Microbiology* 8, 1014. DOI: 10.3389/fmicb.2017.01014.

Rabbani, M.M., Ahmed, I. and Park, S.J. (2016) Application of nanotechnology to remediate contaminated soils. In *Environmental Remediation Technologies for Metal-Contaminated Soils*, pp. 219–229. Springer, Tokyo, Japan. DOI: 10.1007/978-4-431-55759-3_10.

Ramos, O.L., Pereira, R.N., Rodrigues, R., Teixeira, J.A., Vicente, A.A. and Malcata, F.X. (2014) Physical effects upon whey protein aggregation for nano-coating production. *Food Research International* 66, 344–355. DOI: 10.1016/j.foodres.2014.09.036.

Raza, A., Charagh, S., Salehi, H., Abbas, S., Saeed, F. *et al.* (2023) Nano-enabled stress-smart agriculture: Can nanotechnology deliver drought and salinity-smart crops? *Journal of Sustainable Agriculture and Environment* 2(3), 189–214. DOI: 10.1002/sae2.12061.

Reddy, S. and Chhabra, V. (2022) Nanotechnology: Its scope in agriculture. *Journal of Physics: Conference Series* 2267(1), 012112. DOI: 10.1088/1742-6596/2267/1/012112.

Rivero, R., Mittler, R., Blumwald, E. and Zandalinas, S. (2021) Developing climate-resilient crops: Improving plant tolerance to stress combination. *The Plant Journal* 109(2), 373–389. DOI: 10.1111/tpj.15483.

Sadhukhan, R., Sharma, L., Sen, S., Karmakar, S., Banerjee, K. and Baral, K. (2022) *Enhancing the Productivity of Field Crops through Nano-Fertilizer*. IntechOpen eBooks: London, United Kingdom. DOI: 10.5772/intechopen.101146.

Saikanth, D., Gawande, V., Chaithanya, G., Mishra, S., Swami, S., Singh, B.V. and Hemanth Kumar, M.P. (2023) Development and current state of nanotechnology in the agricultural and food sectors: A review. *International Journal of Environment and Climate Change* 13(9), 1602–1612. DOI: 10.9734/ijecc/2023/v13i92392.

Salimi, F., Ahari, H., Anvar, S.A.A. and Moghanjoghi, A.M. (2024) A comparative study on the anti-bacterial properties of three types of silver nano packaging based on (PLA, TiO2, PE) on *Husinae acipenseridae* (*Huso huso*). *SPE Polymers* 5(4), 590–609. DOI: 10.1002/pls2.10148.

Sebastian, A., Devika, P.S., Nair, P.P. and Devadas, V.S. (2023) Green synthesis of potassium-infused nitrogenous nano-fertilizer for enhanced plant growth. *Chemical Papers* 78(3), 1481–1492. DOI: 10.1007/s11696-023-03175-7.

Sekhon, B. (2014) Nanotechnology in agri-food production: An overview. *Nanotechnology, Science and Applications* 7, 31. DOI: 10.2147/nsa.s39406.

Shah, M., Fawcett, D., Sharma, S., Tripathy, S.K. and Poinern, G.E.J. (2015) Green Synthesis of Metallic Nanoparticles via Biological Entities. *Materials* 8, 7278–7308. DOI: 10.3390/ma8115377.

Shahcheraghi, N., Golchin, H., Sadri, Z., Tabari, Y., Borhanifar, F. and Makani, S. (2022) Nano-biotechnology, an applicable approach for sustainable future. *3 Biotech* 12(3), 65. DOI: 10.1007/s13205-021-03108-9.

Shang, Y., Hasan, M.K., Ahammed, G.J., Li, M., Yin, H. and Zhou, J. (2019) Applications of nanotechnology in plant growth and crop protection: A review. *Molecules* 24(14), 2558. DOI: 10.3390/molecules24142558.

Sharma, R. and Gupta, H. (2021) Green synthesis of silver, copper and gold nanoparticles using *Terminalia arjuna* bark and their effect on seed germination. *Nanoscience & Nanotechnology-Asia* 11(2), 243–247. DOI: 10.2174/2210681210999200521131404.

Sharma, A., Mittal, V., Grover, R., Sharma, D., Gupta, V. and Kumar, K. (2024) Applications of nanotechnology in phytoremediation. In *Phytoremediation: Biological Treatment of Environmental Pollution*, pp. 291–313. Springer, Cham, Switzerland.

Singh, N. and Kumar, A. (2024) Enhancing crop productivity through nanotechnology: A comprehensive review of strategies and results. *Journal of Experimental Agriculture International* 46(5), 435–458. DOI: 10.9734/jeai/2024/v46i52395.

Singh, K.K., Singh, A. and Rai, S. (2022) A study on nanomaterials for water purification. *Materials Today: Proceedings* 51, 1157–1163. DOI: 10.1016/j.matpr.2021.07.116.

Su, G., Chen, B., Wu, X., Xu, J., Yang, K. and Lin, D. (2023) nZVI decreases N_2O emission from pesticide-contaminated paddy soil. *Science of the Total Environment* 892, 164613. DOI: 10.1016/j.scitotenv.2023.164613.

Sultan, H., Li, Y., Ahmed, W., Shah, A., Faizan, M. *et al.* (2024) Biochar and nano biochar: Enhancing salt resilience in plants and soil while mitigating greenhouse gas emissions: A comprehensive review. *Journal of Environmental Management* 355, 120448. DOI: 10.1016/j.jenvman.2024.120448.

Tailor, S., Jain, K., Marwal, A., Meena, M., Kumarasamy, A. and Gaur, R.K. (2022) Outlooks of nanotechnology in organic farming management. *Defence Life Science Journal* 7(1), 52–60. DOI: 10.14429/dlsj.7.16763.

Tan, W., Du, W., Barrios, A.C., Armendariz, R.Jr, Zuverza-Mena, N., Ji, Z., Chang, C.H., Zink, J.I., Hernandez-Viezcas, J.A., Peralta-Videa, J.R. and Gardea-Torresdey, J.L., (2017). Surface coating changes the physiological and biochemical impacts of nano-TiO_2 in basil (*Ocimum basilicum*) plants. *Environmental Pollution*, 222, 64–72. DOI:10.1016/j.envpol.2017.01.002.

Tang, Y., Zhao, W., Zhu, G., Tan, Z., Huang, L. *et al.* (2023) Nano-pesticides and fertilizers: Solutions for global food security. *Nanomaterials* 14(1), 90. DOI: 10.3390/nano14010090.

Tiwari, E., Mondal, M., Singh, N., Khandelwal, N., Monikh, F.A. and Darbha, G.K. (2020) Effect of the irrigation water type and other environmental parameters on CeO_2 nanopesticide–clay colloid interactions. *Environmental Science: Processes & Impacts* 22(1), 84–94. DOI: 10.1039/c9em00428a.

Ullah, A., Ahmad, A., Khaliq, T. and Akhtar, J. (2017) Recognizing production options for pearl millet in Pakistan under changing climate scenarios. *Journal of Integrative Agriculture* 16(4), 762–773. DOI: 10.1016/s2095-3119(16)61450-8.

Upadhyay, P.K., Dey, A., Singh, V.K., Dwivedi, B.S., Singh, T. *et al.* (2023) Conjoint application of nano-urea with conventional fertilizers: An energy efficient and environmentally robust approach for sustainable crop production. *PLoS ONE* 18(7), e0284009–e0284009. DOI: 10.1371/journal.pone.0284009.

Wang, X., Yan, M., Zhou, J., Song, W., Xiao, Y. *et al.* (2021) Delivery of acetamiprid to tea leaves enabled by porous silica nanoparticles: Efficiency, distribution and metabolism of acetamiprid in tea plants. *BMC Plant Biology* 21(1), 337. DOI: 10.1186/s12870-021-03120-4.

Xu, L., Wang, X., Shi, H., Hua, B., Burken, J.G. *et al.* (2022) Uptake of engineered metallic nanoparticles in soil by lettuce in single and binary nanoparticle systems. *ACS Sustainable Chemistry & Engineering* 10(50), 16692–16700. DOI: 10.1021/acssuschemeng.2c04748.

Yadav, A. and Yadav, K. (2018) Nanoparticle-based plant disease management: Tools for sustainable agriculture. *Nanotechnology in the Life Sciences*, 29–61. DOI: 10.1007/978-3-319-91161-8_2.

Yan, X., Chen, S., Pan, Z., Zhao, W., Rui, Y. and Zhao, L. (2022) AgNPs-triggered seed metabolic and transcriptional reprogramming enhanced rice salt tolerance and blast resistance. *ACS Nano* 17(1), 492–504. DOI: 10.1021/acsnano.2c09181.

Yin, L., Liu, L., Lin, S., Owens, G. and Chen, Z. (2022) Synthesis and characterization of nanoscale zero-valent iron (nZVI) as an adsorbent for the simultaneous removal of As (III) and As (V) from groundwater. *Journal of Water Process Engineering* 47, 102677. DOI: 10.1016/j.jwpe.2022.102677.

Zambrano-Zaragoza, M.L., Quintanar-Guerrero, D., González-Reza, R.M., Cornejo-Villegas, M.A., Leyva-Gómez, G. and Urbán-Morlán, Z. (2021) Effects of UV-C and edible nano-coating as a combined strategy to preserve fresh-cut cucumber. *Polymers* 13(21), 3705. DOI: 10.3390/polym13213705.

Zare, M., Namratha, K., Ilyas, S., Sultana, A., Hezam, A. *et al.* (2022) Emerging trends for ZnO nanoparticles and their applications in food packaging. *ACS Food Science and Technology* 2(5), 763–781. DOI: 10.1021/acsfoodscitech.2c00043.

Zhai, Y., Monikh, F.A., Wu, J., Grillo, R., Arenas-Lago, D. *et al.* (2020) Interaction between a nano-formulation of atrazine and rhizosphere bacterial communities: Atrazine degradation and bacterial community alterations. *Environmental Science: Nano* 7(11), 3372–3384. DOI: 10.1039/d0en00638f.

Zhang, X., Xu, Z., Qian, X., Lin, D., Zeng, T., Filser, J., Li, L. and Kah, M. (2020) Assessing the Impacts of $Cu(OH)^2$ Nanopesticide and Ionic Copper on the Soil Enzyme Activity and Bacterial Community. *Journal of Agricultural and Food Chemistry* 68(11), 3372–3381. DOI:10.1021/acs.jafc.9b06325.

Zhang, P., Guo, Z., Ullah, S., Melagraki, G., Afantitis, A. and Lynch, I. (2021) Nanotechnology and artificial intelligence to enable sustainable and precision agriculture. *Nature Plants* 7(7), 864–876. DOI: 10.1038/s41477-021-00946-6.

Zhang, F., Li, D., Li, G. and Xu, S. (2025) New horizons in smart plant sensors: Key technologies, applications, and prospects. *Frontiers in Plant Science* 15, 1490801. DOI: 10.3389/fpls.2024.1490801.

Zhang, M., Ma, W., Tao, R., Fan, Q., Zhang, M. *et al.* (2024) Nanomaterials: Recent advances in plant disease diagnosis and treatment. *Nano Today* 57, 102326. DOI: 10.1016/j.nantod.2024.102326.

15 Policy, Governance and Future Pathways for Climate-Resilient Crop Production

Oluwaseun Deborah Adedeji[1,*], Akinbode Okunola[2], Omotola Dorcas Olaoye[3] and Richardson Uchenna Okechukwu[4]

[1]Department of Economics; Agricultural and Food Economics, Swedish University of Agricultural Sciences, Uppsala, Sweden; [2]Institute of Agriculture and Natural Resources, University of Nebraska-Lincoln, NE, USA; [3]West Africa Centre for Crop Improvement (WACCI), College of Basic and Applied Sciences, University of Ghana, Legon, Ghana and International Institute of Tropical Agriculture, Ibadan, Nigeria; [4]International Institute of Tropical Agriculture, Onne, Nigeria

Abstract

Climate change poses a significant threat to global food security, with far-reaching consequences for agricultural productivity, rural livelihoods and economic stability. Staple crops such as wheat, maize and rice, the cornerstones of the global food supply, are increasingly becoming vulnerable to climatic fluctuations including rising temperatures, erratic rainfall and extreme weather events. This growing susceptibility highlights the urgent need for adaptive strategies that enhance the resilience of food systems. Addressing these challenges requires a multidimensional approach where policy and governance play pivotal roles in shaping sustainable agricultural responses. By integrating climate considerations into agricultural policy frameworks and governance structures, stakeholders can mitigate risks and build long-term resilience. This chapter examines critical policy directions, institutional mechanisms and innovative governance pathways aimed at supporting climate-resilient crop production. Through this lens, it offers actionable insights for policymakers, researchers and practitioners seeking to safeguard food systems in an era of climate uncertainty.

15.1 Introduction

Climate change represents one of the most pressing threats to global food security, with profound implications for agricultural productivity, rural livelihoods and economic stability (Nguyen *et al.*, 2023; Bouteska *et al.*, 2024). Agricultural production is inherently sensitive to climatic variables, and recent decades have seen increasing disruptions in crop yields due to extreme weather events, including prolonged droughts, floods, heatwaves and erratic rainfall patterns (Lobell and Gourdji, 2012; Alotaibi, 2023). These climatic stressors are further exacerbated by shifts in temperature regimes, alterations in growing seasons and the intensification of extreme weather events. The cumulative effects of rising temperatures, unpredictable precipitation patterns, prolonged

*Corresponding author. Email: oluwaseun.adedeji@slu.se

 Genomic Solutions for Breeding Climate-resilient Crops (R.K. Varshney, R. Barmukh and R.J. Harper)
DOI: 10.1079/9781800627185.0015

droughts and heightened pest and disease outbreaks place considerable strain on conventional farming systems, reducing their ability to maintain stable yields (Kumar *et al.*, 2022; Saleem *et al.*, 2025).

Many staple crops, including wheat, maize and rice, which form the backbone of global food systems, are particularly vulnerable to these climatic fluctuations (Rosenzweig *et al.*, 2013; Zhao *et al.*, 2017; Kidane *et al.*, 2024). Temperature increases and water scarcity pose significant threats to food security, as these factors directly impact plant growth, flowering and grain development. Research indicates that for every degree Celsius rise in global temperatures, wheat yields could decline by approximately 6%, maize by 7% and rice by 3% (Zhao *et al.*, 2017). These reductions in yield could have far-reaching consequences, particularly in regions already grappling with food insecurity. Moreover, extreme weather events, such as floods and hurricanes, can cause sudden crop failures, leading to supply chain disruptions and price volatility in global food markets (Tchonkouang, *et al.*, 2024). The increasing vulnerability of staple crops to climate change underscores the urgent need for adaptive agricultural strategies that strengthen the resilience of food systems.

Agricultural systems exhibit resilience when they can endure environmental challenges, adjust to changing climatic conditions and recover efficiently from disruptions such as extreme weather events, pest infestations and market volatility (FAO, 2018; Chavez-Miguel *et al.*, 2024; Mekonnen *et al.*, 2024). Resilient agriculture anticipates, absorbs, adapts to and recovers from climate-related stresses while sustaining productivity, ensuring food security and maintaining ecosystem health (International Institute for Sustainable Development (IISD), 2008; Douxchamps *et al.*, 2017; Safdar *et al.*, 2024). This resilience is achieved through the implementation of strategies that enhance the ability of crops to withstand environmental stressors while sustaining or improving yield potential.

The resilience of crop production is a cornerstone of global food security, particularly in the face of climate change, resource constraints and economic instability. Smallholder farmers, who form the backbone of agricultural production in many regions, often lack the financial capacity to recover from climate- or market-induced shocks, underscoring the urgent need to enhance economic and ecological resilience across farming systems (OECD, 2020). Agroecological practices such as crop diversification, intercropping and conservation agriculture have shown promise in enhancing soil health, suppressing pests and stabilizing yields under variable climatic conditions (Altieri and Nicholls, 2017). However, while these sustainable approaches – along with improved irrigation and climate-resilient crop varieties – have helped mitigate some impacts of climate stress, they are not enough on their own.

The transformational potential of these practices can only be fully realized when supported by robust, coherent policy frameworks. Integrating climate-smart agricultural innovations – including genomic solutions – into national strategies requires targeted investments, institutional capacity-building and coordinated action across public, private and community stakeholders. In developing countries especially, the absence of enabling policies continues to constrain the adoption and scaling of resilience-enhancing technologies. Proactive and inclusive policy interventions are therefore essential to drive the systemic change needed to ensure that agriculture remains productive, sustainable and adaptive in a rapidly changing climate.

A comprehensive examination of policy and governance strategies offers critical insights into mitigating climate risks through sustainable agricultural practices (Altieri and Nicholls, 2017; FAO, 2018). This chapter therefore explores key policy considerations, governance mechanisms and future pathways essential for fostering climate-resilient crop production. It examines global and national policies that support climate adaptation in agriculture and innovative solutions for enhancing resilience. Furthermore, the chapter serves as a resource for policymakers seeking to strengthen the resilience of global food systems. Aligning with global sustainability efforts, particularly Sustainable Development Goal (SDG) 2 (Zero Hunger) and SDG 13 (Climate Action), this chapter discusses emerging strategies such as AI-driven precision farming and digital solutions in climate adaptation. These approaches provide a strategic framework for developing a more adaptive and sustainable agricultural sector (Vermeulen *et al.*, 2012; Rosenzweig *et al.*, 2013).

The remainder of the chapter is structured as follows: section 2 outlines policy frameworks for climate-resilient agriculture, section 3 explores technological innovations towards climate resilience, section 4 presents challenges in policy implementation and section 5 concludes and presents future pathways.

15.2 Policy Frameworks for Climate-Resilient Agriculture

15.2.1 Global and National Policies Supporting Climate Adaptation in Agriculture

Achieving climate-resilient agriculture requires policy-driven interventions at local, national and global levels to address food security challenges (Adger *et al.*, 2005; Climato and Mullan, 2010; Edwards *et al.*, 2024). Policies must evolve to meet changing climatic conditions (Urwin and Jordan, 2007). According to Adger *et al.* (2005), climate-resilient agricultural policies focus on raising awareness and building adaptive capacity to mitigate climate-induced vulnerabilities. For farming households and communities, adapting to climate variability is critical to their livelihoods, as they largely depend on agricultural income (Ngigi, 2009; Okonya *et al.*, 2013). A structured policy framework, reinforced by national and international initiatives, is crucial for strengthening farmers' adaptive capacity, facilitating investments and incentivizing the adoption of climate-resilient agricultural practices (Burton *et al.*, 2002; Hassan and Nhemachena, 2008).

Key global climate frameworks include the Paris Agreement, the United Nations (UN) Sustainable Development Goals (SDGs) and the Food and Agriculture Organization's (FAO) Climate-Smart Agriculture (CSA) initiative. The Paris Agreement seeks to significantly reduce greenhouse gas emissions and enhance climate adaptation by fostering cooperation between developed and developing nations in advancing environmental sustainability. It also establishes a framework for developed countries to provide financial and technical support to the Global South in their climate adaptation efforts (Dong, 2022; Heyl *et al.*, 2022; Hu and Denchak, 2025). The UN's SDGs emphasize the interdependence of food production and climate change, integrating objectives related to food security, water management, energy sustainability and climate resilience (Newell *et al.*, 2019). Complementing these efforts, the FAO's CSA initiative, guided by its Strategic Framework, promotes sustainable agricultural practices that align with the SDGs, aiming to enhance productivity, resilience and environmental sustainability in global food systems.

The implementation of national climate-resilient strategies has contributed to improved agricultural productivity, enhanced food security and greater dietary diversity (Akter *et al.*, 2022; Santalucia and Sibhatu, 2023; Vatsa *et al.*, 2023; Omotoso and Omotayo, 2025). Across the globe, national governments have introduced a variety of policies aimed at strengthening climate resilience in agriculture. These policies commonly support the development and dissemination of climate-smart technologies, promote adaptive farming practices and enhance institutional capacities to manage climate-related risks (Clar *et al.*, 2013). In India, for example, the government has implemented climate-smart village programs through its national agricultural research and extension systems, integrating weather-based advisories, improved seed varieties and water-saving technologies into localized policy frameworks (Hariharan *et al.*, 2018; Alam and Sikka, 2019).

In the European Union, the Common Agricultural Policy (CAP) was introduced to boost agricultural productivity through innovation and efficient resource utilization. In 2000, the CAP was restructured to incorporate environmental sustainability objectives. This led to the establishment of two policy pillars: Pillar I, which provides incentives for farmers to adopt good agricultural and environmental practices, and Pillar II, which supports sustainable resource use. In 2014, Green Direct Payments were introduced under Pillar I to promote sustainable agricultural practices such as crop diversification (Heyl *et al.*, 2020; Heyl *et al.*, 2022; Bernini and Galli, 2024). Although the CAP operates under supranational governance, it is executed through national strategic plans tailored to each member state's environmental and climate priorities (Heyl *et al.*, 2022).

In Africa, the Comprehensive African Agricultural Development Programme (CAADP), a key initiative under the 2063 Agenda, aims to eliminate hunger and poverty through agriculture-led economic growth. CAADP supports climate-resilient farming by promoting disaster preparedness, early warning systems, social safety nets and enhanced agricultural research (Esche and Jabril, 2017). In the United States, the Environmental Protection Agency's Climate Adaptation Plan (EPA's CAP) 2024–2027 outlines a four-year strategy to enhance climate resilience in agriculture. The plan focuses on building a climate-ready workforce, strengthening resilient infrastructure and securing sustainable supply chains.

Many other national governments have embedded climate-smart agricultural approaches within broader development strategies, often in partnership with civil society organizations and international agencies (Waters-Bayer *et al.*, 2015; Brown, 2016). These nationally anchored efforts reflect growing recognition of the role policy plays in enabling agricultural systems to adapt to climate change while supporting food and nutritional security.

15.2.2 Climate-Smart Agriculture (CSA) Policies and Incentives

Policy instruments are the practical tools through which governments operationalize climate-smart agriculture strategies. These instruments – ranging from financial incentives and regulatory frameworks to collaborative mechanisms – play a central role in shaping farmer behavior, guiding market dynamics and aligning agricultural development with environmental goals. Among the most influential instruments are subsidies, agricultural insurance and trade-related sanitary and phytosanitary (SPS) measures, all of which directly impact food security, environmental sustainability and the uptake of climate-resilient crops. Production subsidies reduce input costs or offer direct support for crop output. They are commonly used to encourage the adoption of improved agricultural practices (Jansson *et al.*, 2020; Heyl *et al.*, 2022). Agricultural insurance further enhances resilience by mitigating financial risks associated with climate variability, thereby enabling farmers to invest in more sustainable farming systems. At the international level, SPS measures establish quality and safety standards for traded agricultural goods, influencing both environmental outcomes and food security by regulating access to climate-resilient crop varieties.

15.2.2.1 Production Subsidies

Production subsidies, whether input- or output-focused, influence farmers' decision-making, productivity and resource allocation (Heyl *et al.*, 2022; Bernini and Galli, 2024). However, they can also lead to inefficiencies by encouraging investment in less productive activities, overreliance on subsidized inputs, or reduced incentives to adopt cost-effective climate-smart practices (Alston and James, 2002; Rizov *et al.*, 2013; Minviel and Latruffe, 2016; Skevas *et al.*, 2018). Additionally, subsidies can contribute to negative environmental externalities, such as land degradation, biodiversity loss and ecological imbalances (Saman, 2021; Heyl *et al.*, 2022). Despite these challenges, well-structured subsidies can enhance crop productivity and address market failures related to financial inequality and imperfect crop insurance markets (Roche, 2004; Ciaian and Swinnen, 2009; Skevas and Lansink, 2020). In the EU, for instance, CAP subsidies have supported the adoption of environmentally sustainable farming practices (Varela-Candamio *et al.*, 2018; Aubert and Enjolras, 2021; Saman, 2021; Balogh, 2022).

15.2.2.2 Trade Policies

The global food market has increasingly integrated biotechnology in agricultural production, with genetically modified (GM) crops emerging as a pivotal innovation for climate resilience (Hedoui and Beghin, 2025). GM crops, engineered for improved yield stability, pest resistance and environmental sustainability, have demonstrated significant potential in reducing crop losses, mitigating the impacts of climate variability, and enhancing food security, particularly among smallholder farmers (Finger *et al.*, 2011; Smyth *et al.*, 2015). In 2019, biotech crops were cultivated on 190 million hectares by 17 million farmers across 29 countries, underscoring their expanding role in global agriculture (ISAAA,

2019). Despite these benefits, the adoption and international trade of GM crops remain contentious due to divergent regulatory frameworks and public perceptions regarding their environmental and human health implications (Smyth *et al.*, 2004; Adams, 2014). While some countries actively promote GM crop cultivation and trade, others impose stringent regulatory barriers, including bans, labeling mandates and traceability requirements, citing precautionary principles and consumer safety concerns (Davison, 2009; Viju *et al.*, 2011; Vigani and Olper, 2013). These regulatory disparities contribute to additional non-tariff barriers in international trade, exacerbating transaction costs and market segmentation (Wohlers, 2010; Hedoui and Beghin, 2025). The harmonization of trade policies concerning GM crops is essential to facilitate their broader adoption and ensure global food system resilience. Multilateral agreements, such as those governed by the World Trade Organization (WTO), have attempted to address trade-related sanitary and phytosanitary (SPS) measures to promote transparency and scientific risk assessment in regulatory decision-making (Smyth *et al.*, 2015). However, challenges persist in aligning national policies with international standards, necessitating continued dialogue among policymakers, researchers and industry stakeholders to balance innovation with precautionary regulatory approaches.

15.2.2.3 Public-private Partnerships

Public-private partnerships (PPPs) play an increasingly critical role in advancing climate-resilient food production by facilitating climate adaptation and mitigation efforts globally (Aseete *et al.*, 2022). These partnerships leverage both public and private resources to address challenges in agricultural sustainability and climate finance (Casady *et al.*, 2024). Achieving comprehensive climate mitigation and adaptation measures requires substantial financial investment, estimated at approximately US$5 trillion annually (IRENA, 2023). For instance, in some regions, meeting the target of a 55% reduction in greenhouse gas emissions by 2030 necessitates additional annual capital investments of around €336 billion (Casady *et al.*, 2024). In the context of genetically modified (GM) crops, PPPs foster innovation through joint research initiatives, the transfer of genetic traits and licensing agreements for biotechnologies (Byerlee and Fischer, 2002; Spielman and Grebmer, 2006). These collaborations facilitate the development of climate-smart crop varieties designed to withstand extreme weather conditions, reduce dependence on chemical inputs and improve overall agricultural resilience. PPPs are also essential for optimizing limited public resources and mobilizing private capital, expertise and innovation to support climate-resilient food systems (Gasparro *et al.*, 2022). These partnerships can enhance financial sustainability by distributing investment risks between public and private entities while providing contractual stability to mitigate uncertainties associated with climate change (Aseete *et al.*, 2022; Casady *et al.*, 2024).

15.3 Technological and Agroecological Innovations Towards Climate Resilience

Genomic solutions for improved productivity are primarily adaptation efforts against climate change. Genomic-assisted breeding is an approach that enhances crop production by developing resilience varieties that cope with stressors such as drought, heat, salt, pests, floods, submersion, cold and more through marker-assisted selection, genomic selection and genetic modification (Pereira, 2016). Genomic analysis provides an effective framework for investigating stress response mechanisms in cultivated crops, their adaptive traits and wild relatives. By leveraging this approach, researchers can identify key genes and quantitative trait loci (QTLs) associated with desirable characteristics, positioning genomics as a transformative tool for deciphering stress adaptation pathways and advancing crop resilience (Prabhu *et al.*, 2023). Genomic solutions can also contribute indirectly to mitigation. For instance, increased productivity on existing farmland reduces the need to clear additional land, helping to preserve forests that act as carbon sinks. Improved nutrient use efficiency in crops can lower greenhouse gas emissions from fertilizers. Thus genomics, which is the study of deoxyribonucleic acid (DNA) involving structural, functional, mapping and characterization of genes and genetic variants, stands as a promising tool for enhancing crop stress tolerance and resilience.

15.3.1 Advances in Crop Breeding and Genetic Modification for Resilience

Some aspects of crop improvement that employ the latest advancements in genomics include pangenomes, genome editing, genomic selection and multi-omics machine learning which combines phenomics and genomics.

15.3.1.1 Pangenomics

Crop pangenomics identifies key genetic factors, such as single nucleotide polymorphisms (SNPs), mutations and structural variants (SVs), that are associated with important traits of interest within a population (Saad *et al.*, 2022). The identification of resistance genes within the germplasm and their strategic integration into breeding programs are essential for developing stress tolerant varieties (Budhlakoti *et al.*, 2022). Recently, genus-wide pangenomes known as super-pangenomes have been proposed to incorporate the genetic diversity of wild germplasm (Khan *et al.*, 2020). In peach, structural variants and single nucleotide polymorphisms (SNP) have revealed genetic associations linked to early fruit maturity (Guo *et al.*, 2020), an important resilience trait for escaping and adapting to unfavorable environmental conditions. Thus, pangenomes are important genomic resources because effective deployment of genomic-assisted breeding and genome editing for developing resilient crops requires the prior identification of markers, loci or genes linked to key traits of interest.

15.3.1.2 Genome Editing

Some progress has been made in developing stress-tolerant plants, leading to increased crop yields (Mickelbart *et al.*, 2015). However, novel strategies are desirable to overcome limitations such as lack of precision and requirement of substantial time to increase crop production in the current climate change and ever-increasing population scenario. The emergence of genome editing tools presents an opportunity to introduce precise, targeted modifications in plant genomes. These advancements not only enhance the understanding of gene function across diverse crop species but also make it possible to develop crop varieties with improved tolerance to abiotic stresses (Jain, 2015). One of such is CRISPR (clustered regularly interspaced short palindromic repeats), which uses protein-coding genes to modify genomes (Rai *et al.*, 2023).

CRISPR was employed to develop drought tolerance in rice (Lou *et al.*, 2017) and to improve drought and salt tolerance in soybean (Curtin *et al.*, 2017; Du *et al.*, 2018). Promising genes for targeted modification in wheat for drought and herbicide tolerance have been proposed (Zhang *et al.*, 2019; Zheng *et al.*, 2021). In vegetables, especially tomatoes, CRISPR technology has been used for a range of abiotic stresses such as herbicides, drought and heat (Yin *et al.*, 2018; Chen *et al.*, 2021; Yang *et al.*, 2022). These studies highlight the transformative potential of genome editing technology in enhancing crop traits and developing climate-resilient crops.

15.3.1.3 Genomic Selection

Genomic selection is a tool proposed to facilitate rapid breeding for quantitative traits (polygenic traits that are controlled by many minor genes) (Meuwissen *et al.*, 2001). Most of the stress tolerance traits are quantitative traits such as drought in maize (Vivek *et al.*, 2017; Das *et al.*, 2020), wheat (Rutkoski *et al.*, 2016; Sukumaran *et al.*, 2018) and rice (Bhandari *et al.*, 2019); this makes genomic selection a relevant approach for breeding climate-resilient crops. The primary advantage of genomic selection (GS) is its ability to shorten the breeding cycle compared with traditional breeding methods, reducing reliance and associated cost with extensive phenotyping (Heffner *et al.*, 2010). This efficiency enhances crop improvement efforts and ability to increase food production especially with the changing climate. The genomic selection process begins with the development of a training population consisting of individuals with both genotypic and phenotypic data. These data are used to develop a predictive model, where phenotypic traits serve as the response variable and genotypic markers act as predictors. Once the model is developed, it is applied to a breeding population consisting of individuals with only genotypic data to estimate their genomic estimated breeding values (GEBVs). This enables efficient selection without the need for extensive phenotyping (Budhlakoti *et al.*, 2022).

15.3.2 Precision Agriculture, AI and Digital Solutions in Climate Adaptation

Advancements in technology have further accelerated the expansion of omics applications. Omics is a broad and rapidly evolving field that integrates specialized high-throughput biotechnological methods, instruments and computational algorithms. These techniques are instrumental in analyzing and understanding intricate biological systems and their interactions (Jung *et al.*, 2020). Omics encompasses multiple disciplines, including genomics, transcriptomics, proteomics, metabolomics and phenomics among others. Each of these domains contains numerous subfields, each requiring specialized analytical and computational approaches to unlock their full potential (Grapov *et al.*, 2018). The increasing growth of "omics" data has made it too large and complex to be effectively analyzed using conventional statistical methods. This has driven the adoption of Artificial Intelligence (AI), which can manage vast datasets, extracting meaningful insights, and, most importantly, continuously improving its performance through machine learning based on training data (Grapov *et al.*, 2018; Khan *et al.*, 2022).

AI has been applied across various research in phenomics and genomics, demonstrating its versatility and impact. Key applications include genome assembly analysis and the development of genome-specific algorithms (Esposito *et al.*, 2019); broad-scale data analysis to address complex biological challenges in metabolomics, proteomics, genomics, transcriptomics and systems biology (Libbrecht and Noble, 2015); interpretation of gene expression cascades (Ni *et al.*, 2016; Kwon *et al.*, 2020); identification of significant SNPs in polyploid plants (Korani *et al.*, 2019); and high-throughput crop stress phenotyping (Zhao *et al.*, 2016; Selvaraj *et al.*, 2020).

Studies have shown that AI-driven crop phenotyping enhances crop improvement and prediction accuracy of crop performance (Brichet *et al.*, 2017; Sadeghi-Tehran *et al.*, 2017; Mochida *et al.*, 2018; Selvaraj *et al.*, 2020; Jung *et al.*, 2020; Khan *et al.*, 2022). For instance, Selvaraj *et al.* (2020) reported that machine learning algorithms such as k-Nearest Neighbors (kNN), Random Forest (RF) and Support Vector Machines (SVM) exhibited superior performance in predicting root yield in cassava (*Manihot esculenta* Crantz). Similarly, AI-assisted high-throughput phenotyping systems have been successfully applied to various crops, including wheat and maize, for tasks such as plant growth stage identification (Sadeghi-Tehran *et al.*, 2017), plant image segmentation (Brichet *et al.*, 2017) and improvement of plant productivity (Mochida *et al.*, 2018; Jung *et al.*, 2020).

15.3.3 Agroecology, Permaculture and Traditional Knowledge Integration

Agroecology integrates agronomic and ecological principles, emphasizing biodiversity and ecosystem services to enhance agricultural productivity (Jeanneret *et al.*, 2021). One of the key movements within agroecology is permaculture, which is rooted in diversified farming systems, ethical stewardship and practical design strategies (Fiebrig *et al.*, 2020). Permaculture represents a paradigm shift, offering partial solutions to pressing agricultural challenges such as soil degradation and habitat loss, which threaten the long-term sustainability of food production (McLennon *et al.*, 2021).

However, debates persist regarding the scalability and applicability of permaculture to large-scale agricultural systems (Fiebrig *et al.*, 2020). Relying solely on conventional agricultural practices will not be sufficient to ensure global food security in the future. Instead, a holistic and interdisciplinary approach that integrates agroecological principles with technological advancements and sustainable management practices is essential for building a resilient and self-sustaining food system.

15.4 Challenges in Policy Implementation

The implementation of policies supporting climate-resilient agriculture faces significant challenges due to the interplay between agricultural systems, market dynamics and environmental sustainability objectives. A key issue is the heterogeneity of agricultural landscapes and the unequal distribution of technological advancements, which create disparities in access to climate-adaptive innovations. The adoption of genetically modified (GM) crops, for instance, varies widely across regions due to differences in regulatory frameworks, trade policies and public acceptance (Viju *et al.*, 2011; Vigani and Olper, 2013).

Public perceptions of GM crops present additional hurdles to policy implementation. While scientific evidence highlights their potential to enhance food security and reduce reliance on chemical inputs (Finger *et al.*, 2011; Smyth *et al.*, 2015), consumer skepticism and resistance from advocacy groups have contributed to restrictive policies in certain regions (Wohlers, 2010; Adams, 2014). This has resulted in fragmented regulatory environments, where some countries actively promote GM crop adoption while others impose strict bans, labeling requirements or additional compliance barriers (Davison, 2009; Viju *et al.*, 2011). These disparities complicate global trade and limit the ability of farmers in highly regulated markets to benefit from agricultural innovations.

Trade policies governing climate-resilient crops also present both opportunities and constraints. While sanitary and phytosanitary (SPS) measures aim to protect biodiversity and food safety, they can inadvertently restrict access to improved crop varieties in regions where they are needed most (Hedoui and Beghin, 2025). For instance, stringent SPS regulations in some markets create additional non-tariff barriers that limit the international flow of climate-smart agricultural products, increasing costs and reducing efficiency in global food supply chains.

Public-private partnerships are also often constrained by financial, political and structural challenges, despite their potential. Public funding for agricultural biotechnology research is frequently limited due to high opportunity costs, competing budgetary demands and the political complexities associated with increasing tax burdens to sustain research initiatives (CPI, 2019). In contrast, private sector investments tend to prioritize commercially viable, lower-risk projects, often favoring climate mitigation over adaptation strategies (Bisaro and Hinkel, 2018; Global Infrastructure Hub, 2022). This preference can result in an underinvestment in agricultural adaptation technologies, particularly in developing regions where the need for climate resilience is most pressing.

Another major concern for policymakers is the integration of environmental sustainability with productivity growth (Lankoski and Thiem, 2020; Bernini and Galli, 2024). While incentives such as subsidies can encourage farmers to adopt climate-smart practices, socially optimal outcomes are often hindered by market failures, weak institutional frameworks and disparities in adaptive capacity (Climato and Mullan, 2010). Addressing these challenges requires policies that promote resilient food production, develop climate-resistant technologies and enhance productivity while reducing environmental degradation (World Bank, 2024).

15.5 Conclusion and Future Pathways

Climate change poses an existential threat to agriculture globally, with disproportionately severe impacts in the global south, for instance, in sub-Saharan Africa, due to high exposure, low adaptive capacity and weak institutional support structures (IPCC, 2022). Developing and implementing climate-resilient agricultural policies is thus critical to safeguarding food security in the region. A central pillar of this strategy must be a comprehensive national seed system policy that encompasses key stakeholders – breeders, foundation seed producers, commercial seed entrepreneurs, extension agents and farmers. Such a policy must cater to both true seed crops and vegetatively propagated crops, which are often neglected despite their importance. Tools like the International Institute of Tropical Agriculture's (IITA) SeedTracker have shown promise in tracking and promoting adoption of resilient clones, particularly for vegetatively propagated crops (IITA, 2021).

Climate change-induced erosion of crop genetic diversity highlights the urgency of national genetic resources conservation initiatives. Conservation of plant genetic resources provides essential raw materials for developing stress-resilient crop varieties (FAO, 2010). Establishing national gene banks and community seed banks, and ensuring their integration into seed policy frameworks, will be key to long-term resilience.

In sub-Saharan Africa, the limited reach and effectiveness of extension systems hinder adoption of improved climate-resilient varieties. As a result, many farmers continue to recycle obsolete, low-yielding seed varieties. Strategic investments in demonstration farms, farmer field schools and the use of mass media and mobile technologies are recommended to bridge this knowledge gap. An example can be seen with rice farmers in Nigeria's Benue and Nasarawa states who remain heavily dependent on FARO 44, a popular but flood-susceptible variety. Increased awareness and dissemination of flood-tolerant varieties such as FARO 66 and FARO 67 could significantly reduce losses and improve resilience. In the case of drought, an enabling irrigation and water management policy is also essential. Increased reliance on rain-fed agriculture exposes farmers to unpredictable rainfall patterns. National strategies must prioritize the development and equitable distribution of affordable irrigation technologies, water harvesting systems and sustainable water use practices to reduce climate vulnerability. Agroecological transformation is another critical area. Regions traditionally reliant on vulnerable grain crops must transition towards more resilient staples such as cassava and potato, while others should expand cultivation of climate-sensitive crops using precision and climate-smart agriculture techniques.

Lastly, climate-resilient agricultural transformation requires policy coherence, institutional alignment and sustained political commitment. While many developing countries have formulated climate and agriculture strategies, these often lack the coordination and enforcement mechanisms needed for effective implementation. Without unified efforts across the seed engineering system, irrigation, pest control and land use domains, progress will remain fragmented and unsustainable. To this end, governments must prioritize climate-resilient agriculture in national development plans, ensure that policies are inclusive and participatory, and strengthen implementation capacity through adequate funding, monitoring systems and local ownership. Policymakers must balance innovation with public concerns by fostering transparent risk assessment frameworks and ensuring that climate-resilient agricultural policies align with both trade and sustainability goals. Climate resilience is not a single intervention but a systemic shift – one that requires long-term vision, multi-stakeholder engagement and adaptive governance.

References

Adams, M. (2014) Biotech genocide, Monsanto collaborators and the Nazi legacy of 'science' as justification for murder. Natural News, July 21 [https://skepdic.com/adamsmonsanto.html] (accessed April 10, 2025).

Adger, W.N., Arnell, N.W. and Tompkins, E.L. (2005) Successful adaptation to climate change across scales. *Global Environmental Change* 15(2), 77–86. DOI: 10.1016/j.gloenvcha.2004.12.005.

Akter, A. *et al.* (2022) Income and yield effects of climate-smart agriculture (CSA) adoption in flood prone areas of Bangladesh: Farm level evidence. *Climate Risk Management* 37, 100455. DOI: 10.1016/j.crm.2022.100455.

Alam, M.F. and Sikka, A.K. (2019) Prioritizing land and water interventions for climate-smart villages. *Irrigation and Drainage* 68(4), 714–728. DOI: 10.1002/ird.2366.

Alotaibi, M. (2023) Climate change, its impact on crop production, challenges, and possible solutions. *Notulae Botanicae Horti Agrobotanici Cluj-Napoca* 51(1), 13020. DOI: 10.15835/nbha51113020.

Alston, J.M. and James, J.S. (2002). Chapter 33 The incidence of agricultural policy. In *Handbook of Agricultural Economics* (pp. 1689–1749). Elsevier. https://doi.org/10.1016/s1574-0072(02)10020-x.

Altieri, M.A., Nicholls, C.I. and Montalba, R. (2017) Technological Approaches to Sustainable Agriculture at a Crossroads: An Agroecological Perspective. *Sustainability* 9(3), 1–13. https://doi.org/10.3390/su9030349.

Aseete, P. *et al.* (2022) Public–private partnership generates economic benefits to smallholder bean growers in Uganda. *Food Security* 15(1), 201–218. DOI: 10.1007/s12571-022-01309-5.

Aubert, M. and Enjolras, G. (2021) Intensive and extensive impacts of EU subsidies on pesticide expenditures at the farm level. *Journal of Environmental Economics and Policy* 11(2), 218–234. DOI: 10.1080/21606544.2021.1955749.

Balogh, J. (2022) The impact of agricultural subsidies on environmental pollution in the European Union. *AgEcon* 69(4), 140–150. https://doi.org/10.22004/ag.econ.321223.

Bernini, C. and Galli, F. (2024) Economic and environmental efficiency, subsidies and spatio-temporal effects in agriculture. *Ecological Economics* 218, 108120. DOI: 10.1016/j.ecolecon.2024.108120.

Bhandari, A. *et al.* (2019) Selection of trait-specific markers and multi-environment models improve genomic predictive ability in rice. *PLoS ONE* 14(5), e0208871. DOI: 10.1371/journal.pone.0208871.

Bisaro, A. and Hinkel, J. (2018). Mobilizing private finance for coastal adaptation: A literature review. *Wiley Interdisciplinary Reviews Climate Change* 9(3), 1757–7780. https://doi.org/10.1002/wcc.514.

Bouteska, A. *et al.* (2024) Impacts of the changing climate on agricultural productivity and food security: Evidence from Ethiopia. *Journal of Cleaner Production* 449, 141793. DOI: 10.1016/j.jclepro.2024.141793.

Brichet, N. *et al.* (2017). A robot-assisted imaging pipeline for tracking the growths of maize ear and silks in a high-throughput phenotyping platform. *Plant Methods* 13(1), 1–12. https://doi.org/10.1186/s13007-017-0246-7.

Brown, T. (2016) Civil society organizations for sustainable agriculture: Negotiating power relations for pro-poor development in India. *Agroecology and Sustainable Food Systems* 40(4), 381–404. DOI: 10.1080/21683565.2016.1139648.

Budhlakoti, N. *et al.* (2022) Genomic Selection: a tool for accelerating the efficiency of molecular breeding for development of Climate-Resilient crops. *Frontiers in Genetics* 13, 1–17. https://doi.org/10.3389/fgene.2022.832153.

Burton, I. *et al.* (2002) From impacts assessment to adaptation priorities: The shaping of adaptation policy. *Climate Policy* 2(2–3), 145–159. DOI: 10.3763/cpol.2002.0217.

Byerlee, D. and Fischer, K. (2002). Accessing modern science: policy and institutional options for agricultural biotechnology in developing countries. *World Development* 30(6), 931–948. https://doi.org/10.1016/S0305-750X(02)00013-X.

Casady, C.B., Cepparulo, A. and Giuriato, L. (2024) Public-private partnerships for low-carbon, climate-resilient infrastructure: Insights from the literature. *Journal of Cleaner Production* 470, 143338. DOI: 10.1016/j.jclepro.2024.143338.

Chavez-Miguel, G. *et al.* (2024) Local food system resilience in the context of shocks and crises: Vulnerabilities and responses of agroecology-based farmers in Peru, Germany, and the United States. *Agroecology and Sustainable Food Systems* 48(6), 876–897. DOI: 10.1080/21683565.2024.2341987.

Chen, M. *et al.* (2021) Knockout of AUXIN response factor SLARF4 improves tomato resistance to water deficit. *International Journal of Molecular Sciences* 22(7), 3347. DOI: 10.3390/ijms22073347.

Ciaian, P. and Swinnen, J.F.M. (2009) Credit market imperfections and the distribution of policy rents. *American Journal of Agricultural Economics* 91(4), 1124–1139. DOI: 10.1111/j.1467-8276.2009.01311.x.

Cimato, F. and Mullan, M. (2010) Adapting to climate change: analysing the role of government. *Defra Evidence and Analysis Series, Paper*, 1. https://assets.publishing.service.gov.uk/media/5a79c1b4ed915d07d35b7dff/pb13341-analysing-role-government-100122.pdf.

Clar, C., Prutsch, A. and Steurer, R. (2013) Barriers and guidelines for public policies on climate change adaptation: A missed opportunity of scientific knowledge-brokerage. *Natural Resources Forum* 37(1), 1–18. DOI: 10.1111/1477-8947.12013.

CPI (2019) Global landscape of climate finance (B. Buchner, A. Clark, A. Falconer, R. Macquarie, C. Meattle, R. Tolentino, & C. Wetherbee). Climate Policy Initiative, London

Curtin, S.J. *et al.* (2017) CRISPR/Cas 9 and TALENs generate heritable mutations for genes involved in small RNA processing of *Glycine max* and *Medicago truncatula*. *Plant Biotechnology Journal* 16(6), 1125–1137. DOI: 10.1111/pbi.12857.

Das, R.R. *et al.* (2020). Genetic gains with rapid-cycle genomic selection for combined drought and waterlogging tolerance in tropical maize (*Zea mays* L.). The Plant *Genome* 13(3), 1–15. https://doi.org/10.1002/tpg2.20035.

Davison, J. (2009) GM plants: Science, politics and EC regulations. *Plant Science* 178(2), 94–98. DOI: 10.1016/j.plantsci.2009.12.005.

Dong, J. (2022) Impact of the Paris agreement on agriculture, energy, and economy. *BCP Business & Management* 34, 1370–1379. DOI: 10.54691/bcpbm.v34i.3188.

Douxchamps, S. *et al.* (2017) Monitoring and evaluation of climate resilience for agricultural development – A review of currently available tools. *World Development Perspectives* 5, 10–23. DOI: 10.1016/j.wdp.2017.02.001.

Du, Y.-T. *et al.* (2018) Identification and characterization of GmMYB118 responses to drought and salt stress. *BMC Plant Biology* 18(1), 1–18. https://doi.org/10.1186/s12870-018-1551-7.

Edwards, F., Sonnino, R. and Cifuentes, M.L. (2024) Connecting the dots: Integrating food policies towards food system transformation. *Environmental Science & Policy* 156, 103735. DOI: 10.1016/j.envsci.2024.103735.

Esche, S. and Jabril, S. (2017) African Union: Climate Change Adaptation in Agriculture. Deutsche Gesellschaft FüR, Internationale Zusammenarbeit (GIZ) GmbH, Germany. https://www.giz.de/en/downloads/giz2017-en-climate.pdf (accessed April 10, 2025).

Esposito, S. *et al.* (2019) Applications and trends of machine learning in genomics and phenomics for next-generation breeding. *Plants* 9(1), 34. DOI: 10.3390/plants9010034.

FAO (2010) *The Second Report on the State of the World's Plant Genetic Resources for Food and Agriculture*. Food and Agriculture Organization, Rome.

FAO (2018) *Climate-Smart Agriculture Case Studies 2018*. Successful approaches from different regions. 44 pp.

Fiebrig, I. *et al.* (2020) Perspectives on permaculture for commercial farming: Aspirations and realities. *Organic Agriculture* 10(3), 379–394. DOI: 10.1007/s13165-020-00281-8.

Finger, R. *et al.* (2011) A meta analysis on farm-level costs and benefits of GM crops. *Sustainability* 3(5), 743–762. DOI: 10.3390/su3050743.

Gasparro, K. *et al.* (2022) Vanguard projects as intermediation spaces in sustainability transitions. *Project Management Journal* 53(2), 196–210. DOI: 10.1177/87569728221077011.

Global Infrastructure Hub (2022) Infrastructure Monitor 2022: Global trends in private investment in infrastructure. https://www.gihub.org/infrastructure-monitor/ (accessed April 12, 2025).

Grapov, D. *et al.* (2018) Rise of deep learning for genomic, proteomic, and metabolomic data integration in precision medicine. *OMICS a Journal of Integrative Biology* 22(10), 630–636. DOI: 10.1089/omi.2018.0097.

Guo, J. *et al.* (2020) An integrated peach genome structural variation map uncovers genes associated with fruit traits. *Genome Biology* 21(1), 1–19. https://doi.org/10.1186/s13059-020-02169-y.

Hariharan, V.K. *et al.* (2018) Does climate-smart village approach influence gender equality in farming households? A case of two contrasting ecologies in India. *Climatic Change* 158(1), 77–90. DOI: 10.1007/s10584-018-2321-0.

Hassan, R.M. and Nhemachena, C. (2008) Determinants of African farmers' strategies for adapting to climate change: Multinomial choice analysis. *African Journal of Agricultural and Resource Economics* 2(1), 83–104. DOI: 10.22004/ag.econ.56969.

Hedoui, M.A. and Beghin, J.C. (2025) Decomposing the impact of genetically modified organism regulation on bilateral trade: An application to corn trade. *Australian Journal of Agricultural and Resource Economics* 69, 157–177. https://doi.org/10.1111/1467-8489.12591.

Heffner, E.L. *et al.* (2010) Plant breeding with genomic selection: Gain per unit time and cost. *Crop Science* 50(5), 1681–1690. DOI: 10.2135/cropsci2009.11.0662.

Heyl, K. *et al.* (2020) The common agricultural policy beyond 2020: A critical review in light of global environmental goals. *Review of European Comparative & International Environmental Law* 30(1), 95–106. DOI: 10.1111/reel.12351.

Heyl, K. *et al.* (2022) Potentials and limitations of subsidies in sustainability governance: The example of agriculture. *Sustainability* 14(23), 15859. DOI: 10.3390/su142315859.

Hu, S. and Denchak, M. (2025) Paris climate agreement: Everything you need to know. Available at: https://www.nrdc.org/stories/paris-climate-agreement-everything-you-need-know (accessed April 15, 2025).

IITA (2021) *Seed Tracker: Digital Tool for Monitoring the Diffusion of Vegetative Propagated Crops*. International Institute of Tropical Agriculture, Ibadan, Nigeria.

International Institute for Sustainable Development (IISD) (2008) in Philander, S., *Encyclopedia of Global Warming and Climate Change*. Thousand Oaks, CA: Sage. DOI: 10.4135/9781412963893.n343.

IPCC (2022) *Climate Change 2022: Impacts, Adaptation and Vulnerability*. Intergovernmental Panel on Climate Change Sixth Assessment Report. Cambridge University Press. Cambridge, UK and New York, USA. DOI 10.1017/9781009325844.

IRENA (2023), World Energy Transitions Outlook 2023: 1.5°C Pathway, International Renewable Energy Agency, Abu Dhabi. https://www.irena.org/Digital-Report/World-Energy-Transitions-Outlook-2023 (accessed April 8, 2025).

ISAAA. International Service for the Acquisition of Agri-biotech Applications (2019). *ISAAA.org*. Available from: https://www.isaaa.org/ (accessed April 8, 2025).

Jain, M. (2015) Function genomics of abiotic stress tolerance in plants: a CRISPR approach. *Frontiers in plant science* 6(375), 1–4. https://doi.org/10.3389/fpls.2015.00375.

Jansson, T. *et al.* (2020) Coupled agricultural subsidies in the EU undermine climate efforts. *Applied Economic Perspectives and Policy* 43(4), 1503–1519. DOI: 10.1002/aepp.13092.

Jeanneret, P. *et al.* (2021) Agroecology landscapes. *Landscape Ecology* 36(8), 2235–2257. DOI: 10.1007/s10980-021-01248-0.

Jung, J. *et al.* (2020) The potential of remote sensing and artificial intelligence as tools to improve the resilience of agriculture production systems. *Current Opinion in Biotechnology* 70, 15–22. DOI: 10.1016/j.copbio.2020.09.003.

Khan, A.W. *et al.* (2020) Super-pangenome by integrating the wild side of a species for accelerated crop improvement. *Trends in Plant Science* 25(2), 148–158. DOI: 10.1016/j.tplants.2019.10.012.

Khan, M.H.U. *et al.* (2022) Applications of artificial intelligence in climate-resilient smart-crop breeding. *International Journal of Molecular Sciences* 23(19), 11156. DOI: 10.3390/ijms231911156.

Kidane, B. *et al.* (2024) Nutritional challenges of staple crops due to increasing atmospheric carbon dioxide levels: Case of sub-Saharan Africa. *Journal of Agriculture and Food Research*, 101592. DOI: 10.1016/j.jafr.2024.101592.

Korani, W. *et al.* (2019) Machine learning as an effective method for identifying true single nucleotide polymorphisms in polyploid plants. *The Plant Genome* 12(1), 1–10. https://doi.org/10.3835/plantgenome2018.05.0023.

Kumar, L. *et al.* (2022) Climate change and future of agri-food production. *Future Foods, Academic Press* 49–79. https://doi.org/10.1016/b978-0-323-91001-9.00009-8.

Kwon, M.S. *et al.* (2020) Modeling regulatory networks using machine learning for systems metabolic engineering. *Current Opinion in Biotechnology* 65, 163–170. DOI: 10.1016/j.copbio.2020.02.014.

Lankoski, J. and Thiem, A. (2020) Linkages between agricultural policies, productivity and environmental sustainability. *Ecological Economics* 178, 106809. DOI: 10.1016/j.ecolecon.2020.106809.

Libbrecht, M.W. and Noble, W.S. (2015) Machine learning applications in genetics and genomics. *Nature Reviews Genetics* 16(6), 321–332. DOI: 10.1038/nrg3920.

Lobell, D.B. and Gourdji, S.M. (2012) The influence of climate change on global crop productivity. *Plant Physiology* 160(4), 1686–1697. DOI: 10.1104/pp.112.208298.

Lou, D. *et al.* (2017). OSSAPK2 confers abscisic acid sensitivity and tolerance to drought stress in rice. *Frontiers in Plant Science* 8, 1–15. https://doi.org/10.3389/fpls.2017.00993.

McLennon, E. *et al.* (2021) Regenerative agriculture and integrative permaculture for sustainable and technology driven global food production and security. *Agronomy Journal* 113(6), 4541–4559. DOI: 10.1002/agj2.20814.

Mekonnen, S.A., Jalata, D.D. and Onyeaka, H. (2024). Building resilience in Sub-Saharan Africa's food systems: Diversification, traceability, capacity building and technology for overcoming challenges. *Food and Energy Security* 13(4), 1–17. https://doi.org/10.1002/fes3.563.

Meuwissen, T.H.E., Hayes, B.J. and Goddard, M.E. (2001) Prediction of total genetic value using genome-wide dense marker maps. *Genetics* 157(4), 1819–1829. DOI: 10.1093/genetics/157.4.1819.

Mickelbart, M.V., Hasegawa, P.M. and Bailey-Serres, J. (2015) Genetic mechanisms of abiotic stress tolerance that translate to crop yield stability. *Nature Reviews Genetics* 16(4), 237–251. DOI: 10.1038/nrg3901.

Minviel, J.J. and Latruffe, L. (2016) Effect of public subsidies on farm technical efficiency: A meta-analysis of empirical results. *Applied Economics* 49(2), 213–226. DOI: 10.1080/00036846.2016.1194963.

Mochida, K. *et al.* (2018) Computer vision-based phenotyping for improvement of plant productivity: a machine learning perspective. *GigaScience* 8(1), 1–12. https://doi.org/10.1093/gigascience/giy153.

Newell, P. *et al.* (2019) Climate Smart Agriculture? Governing the Sustainable Development Goals in Sub-Saharan Africa. *Frontiers in Sustainable Food Systems* 3, 1–15. https://doi.org/10.3389/fsufs.2019.00055.

Ngigi, S.N. (2009) *Climate Change Adaptation Strategies: Water Resources Management Options for Smallholder Farming Systems in sub-Saharan Africa.* Available at: http://hdl.handle.net/10919/68902.

Nguyen, T.T. *et al.* (2023) Security risks from climate change and environmental degradation: Implications for sustainable land use transformation in the global south. *Current Opinion in Environmental Sustainability* 63, 101322. DOI: 10.1016/j.cosust.2023.101322.

Ni, Y. *et al.* (2016) A machine learning approach to predict gene regulatory networks in seed development in Arabidopsis. *Frontiers in Plant Science* 7, 1–13. https://doi.org/10.3389/fpls.2016.01936.

OECD (2020) Strengthening Agricultural Resilience in the Face of Multiple Risks. OECD Publishing, Paris. https://doi.org/10.1787/2250453e-en.

Okonya, J.S., Syndikus, K. and Kroschel, J. (2013) Farmers' perception of and coping strategies to climate change: evidence from six Agro-Ecological Zones of Uganda. *Journal of Agricultural Science* 5(8), 252–263. https://doi.org/10.5539/jas.v5n8p252.

Omotoso, A.B. and Omotayo, A.O. (2025) Enhancing dietary diversity and food security through the adoption of climate-smart agricultural practices in Nigeria: a micro level evidence. *Environ Dev Sustain* 27, 17077–17094. https://doi.org/10.1007/s10668-024-04681-8.

Pereira, A. (2016) Plant Abiotic Stress Challenges from the Changing Environment. *Frontiers in Plant Science* 7, 1–3. https://doi.org/10.3389/fpls.2016.01123.

Prabhu, K.R. *et al.* (2023) A review on conventional and modern breeding approaches for developing climate resilient crop varieties. *Journal of Applied and Natural Science* 15(3), 987–997. DOI: 10.31018/jans.v15i3.4653.

Rai, G.K. *et al.* (2023) Enhancing crop resilience to drought stress through CRISPR-Cas9 genome editing. *Plants* 12(12), 2306. DOI: 10.3390/plants12122306.

Rizov, M., Pokrivcak, J. and Ciaian, P. (2013) CAP subsidies and productivity of the EU farms. *Journal of Agricultural Economics* 64(3), 537–557. DOI: 10.1111/1477-9552.12030.

Roche, M.J. (2004) Riskier product portfolio under decoupled payments. *European Review of Agricultural Economics* 31(2), 111–123. DOI: 10.1093/erae/31.2.111.

Rosenzweig, C. *et al.* (2013) Assessing agricultural risks of climate change in the 21st century in a global gridded crop model intercomparison. *Proceedings of the National Academy of Sciences USA* 111(9), 3268–3273. DOI: 10.1073/pnas.1222463110.

Rutkoski, J. *et al.* (2016) Canopy temperature and vegetation indices from high-throughput phenotyping improve accuracy of pedigree and genomic selection for grain yield in wheat. *G3 Genes Genomes Genetics* 6(9), 2799–2808. DOI: 10.1534/g3.116.032888.

Saad, N.S.M. *et al.* (2022) Advancing designer crops for climate resilience through an integrated genomics approach. *Current Opinion in Plant Biology* 67, 102220. DOI: 10.1016/j.pbi.2022.102220.

Sadeghi-Tehran, P. *et al.* (2017) Automated method to determine two critical growth stages of wheat: heading and flowering. *Frontiers in Plant Science* 8, 1–14. https://doi.org/10.3389/fpls.2017.00252.

Safdar, M., Shahid, M.A., Yang, C., Rasul, F., Tahir, M., Raza, A. and Sabir, R. M. (2024) Climate Smart Agriculture and Resilience. In J. Garwi, R. Masengu, & O. Chiwaridzo (Eds.), *Emerging Technologies and Marketing Strategies for Sustainable Agriculture* (pp. 28–52). IGI Global Scientific Publishing. https://doi.org/10.4018/979-8-3693-4864-2.ch002.

Saleem, A., Anwar, S., Nawaz, T. *et al.* (2025). Securing a sustainable future: the climate change threat to agriculture, food security, and sustainable development goals. *J.Umm Al-Qura Univ. Appll. Sci.* 11, 595–611. https://doi.org/10.1007/s43994-024-00177-3.

Saman, C. (2021) Does agricultural subsidies in the EU improved environmental outcomes? *Transformations in Business & Economics* 20(3C (54C)), 642–652.

Santalucia, S. and Sibhatu, K.T. (2023) Nourishing the farms, nourishing the plates: Association of climate-smart agricultural practices with household dietary diversity and food security in smallholders. *Agribusiness* 40(2), 513–533. DOI: 10.1002/agr.21892.

Selvaraj, M.G. *et al.* (2020) Machine learning for high-throughput field phenotyping and image processing provides insight into the association of above and below-ground traits in cassava (*Manihot esculenta* Crantz). *Plant Methods* 16(1), 1–19. https://doi.org/10.1186/s13007-020-00625-1.

Skevas, I. and Lansink, A.O. (2020) Dynamic inefficiency and spatial spillovers in Dutch dairy farming. *Journal of Agricultural Economics* 71(3), 742–759. DOI: 10.1111/1477-9552.12369.

Skevas, I., Zhu, X., Shestalova, V. and Emvalomatis, G. (2018). The Impact of Agri-Environmental Policies and Production Intensification on the Environmental Performance of Dutch Dairy Farms. *Journal of Agricultural and Resource Economics* 43(3), 423–440. https://www.jstor.org/stable/26797529.

Smyth, S.J., Kerr, W.A. and Phillips, P.W.B. (2015) Global economic, environmental and health benefits from GM crop adoption. *Global Food Security* 7, 24–29. DOI: 10.1016/j.gfs.2015.10.002.

Smyth, S. *et al.* (2004) Regulating the liabilities of agricultural biotechnology. DOI: 10.1079/9780851998152.0000.

Spielman, D.J. and von Grebmer, K. (2006). Public–private partnerships in international agricultural research: an analysis of constraints. *The Journal of Technology Transfer* 31(2), 291–300. https://doi.org/10.1007/s10961-005-6112-1.

Sukumaran, S. *et al.* (2018). Genomic-enabled prediction accuracies increased by modeling genotype × environment interaction in durum wheat. *The Plant Genome* 11(2), 1–11. https://doi.org/10.3835/plantgenome2017.12.0112.

Tchonkouang, R.D., Onyeaka, H. and Nkoutchou, H. (2024) Assessing the vulnerability of food supply chains to climate change-induced disruptions. *The Science of the Total Environment* 920, 171047. DOI: 10.1016/j.scitotenv.2024.171047.

Urwin, K. and Jordan, A. (2007) Does public policy support or undermine climate change adaptation? Exploring policy interplay across different scales of governance. *Global Environmental Change* 18(1), 180–191. DOI: 10.1016/j.gloenvcha.2007.08.002.

Varela-Candamio, L., Calvo, N. and Novo-Corti, I. (2018) The role of public subsidies for efficiency and environmental adaptation of farming: A multi-layered business model based on functional foods and rural women. *Journal of Cleaner Production* 183, 555–565. DOI: 10.1016/j.jclepro.2018.02.109.

Vatsa, P. *et al.* (2023) Climate-smart agricultural practices for promoting sustainable agrifood production: Yield impacts and implications for food security. *Food Policy* 121, 102551. DOI: 10.1016/j.foodpol.2023.102551.

Vermeulen, S.J., Aggarwal, P.K., Ainslie, A., Angelone, C., Campbell, B.M., Challinor, A.J. and Wollenberg, E. (2012) Options for support to agriculture and food security under climate change. *Environmental Science & Policy* 15(1), 136–144. https://doi.org/10.1016/j.envsci.2011.09.003.

Vigani, M. and Olper, A. (2013) GMO standards, endogenous policy and the market for information. *Food Policy* 43, 32–43. DOI: 10.1016/j.foodpol.2013.08.001.

Viju, C., Yeung, M.T. and Kerr, W.A. (2011). Post-Moratorium EU Regulation of Genetically Modified Products: Trade Concerns. Commissioned Paper-Canadian Agricultural Trade Policy Research Network (CATPRN) 2011-02 [Preprint]. https://ideas.repec.org/p/ags/catptp/116853.html (accessed April 10, 2025).

Vivek, B.S. *et al.* (2017) Use of genomic estimated breeding values results in rapid genetic gains for drought tolerance in maize. *The Plant Genome* 10(1), 1–8. https://doi.org/10.3835/plantgenome2016.07.0070.

Waters-Bayer, A. *et al.* (2015) Exploring the impact of farmer-led research supported by civil society organisations. *Agriculture & Food Security* 4(1), 1–7. https://doi.org/10.1186/s40066-015-0023-7.

Wohlers, A.E. (2010) Regulating genetically modified food: Policy trajectories, political culture, and risk perceptions in the U.S., Canada, and EU. *Politics and the Life Sciences* 29(2), 17–39. DOI: 10.2990/29_2_17.

World Bank (2024) *Climate-smart agriculture*. World Bank Group. https://www.worldbank.org/en/topic/climate-smart-agriculture (accessed April 10, 2025).

Yang, S.H. *et al.* (2022) Selection of the high efficient sgRNA for CRISPR-Cas9 to edit herbicide related genes, PDS, ALS, and EPSPS in tomato. *Applied Biological Chemistry* 65(1), 1–10. https://doi.org/10.1186/s13765-022-00679-w.

Yin, Y. *et al.* (2018) BZR1 Transcription Factor Regulates Heat Stress Tolerance Through FERONIA Receptor-Like Kinase-Mediated Reactive Oxygen Species Signaling in Tomato. *Plant and Cell Physiology* 59(11), 2239–2254, https://doi.org/10.1093/pcp/pcy146.

Zhang, A. *et al.* (2019) Enhanced rice salinity tolerance via CRISPR/Cas9-targeted mutagenesis of the OsRR22 gene. *Molecular Breeding* 39(3), 1–10. https://doi.org/10.1007/s11032-019-0954-y.

Zhao, J., Bodner, G. and Rewald, B. (2016) Phenotyping: Using machine learning for improved pairwise genotype classification based on root traits. *Frontiers in Plant Science* 7, 1–17. https://doi.org/10.3389/fpls.2016.01864.

Zhao, C. *et al.* (2017) Temperature increase reduces global yields of major crops in four independent estimates. *Proceedings of the National Academy of Sciences USA* 114(35), 9326–9331. DOI: 10.1073/pnas.1701762114.

Zheng, M. *et al.* (2021) Histone acetyltransferase TaHAG1 acts as a crucial regulator to strengthen salt tolerance of hexaploid wheat. *Plant Physiology* 186(4), 1951–1969. DOI: 10.1093/plphys/kiab187.

Index

Note: Page numbers in **bold** and *italics* refer to tables and figures, respectively.